LONDON MATHEMATICAL SOCIETY L~~ECTURE~~

Managing Editor:
Professor N.J. Hitchin,
Mathematical Institute, 24–29 St. Giles, Oxford OX1 3DP, UK

All the titles listed below can be obtained from good booksellers or from
Cambridge University Press. For a complete series listing visit
`http://publishing.cambridge.org/stm/mathematics/lmsn/`

London Mathematical Society Lecture Note Series: 331

Foundations of Computational Mathematics, Santander 2005

Edited by

Luis M. Pardo
University of Cantabria

Allan Pinkus
Technion

Endre Süli
University of Oxford

Michael J. Todd
Cornell University

CAMBRIDGE
UNIVERSITY PRESS

CAMBRIDGE UNIVERSITY PRESS

Cambridge, New York, Melbourne, Madrid, Cape Town, Singapore, São Paulo

Cambridge University Press

The Edinburgh Building, Cambridge CB2 2RU, UK

Published in the United States of America by Cambridge University Press, New York

www.cambridge.org

Information on this title: www.cambridge.org/9780521681612

© Cambridge University Press 2006

First published 2006

Printed in the United Kingdom at the University Press, Cambridge

A catalogue record for this book is available from the British Library

ISBN-13: 978-0-521-68161-2 paperback

ISBN-10: 0-521-68161-8 paperback

Contents

Preface

The Society for the Foundations of Computational Mathematics supports and promotes fundamental research in computational mathematics and its applications, interpreted in the broadest sense. It fosters interaction among mathematics, computer science and other areas of computational science through its conferences, workshops and publications. As part of this endeavour to promote research across a wide spectrum of subjects concerned with computation, the Society brings together leading researchers working in diverse fields. Major conferences of the Society have been held in Park City (1995), Rio de Janeiro (1997), Oxford (1999), Minneapolis (2002), and Santander (2005). The next conference is expected to be held in 2008. More information about FoCM is available at its website http://www.focm.net.

The conference in Santander on June 30 – July 9, 2005, was attended by several hundred scientists. FoCM conferences follow a set pattern: mornings are devoted to plenary talks, while in the afternoon the conference divides into a number of workshops, each devoted to a different theme within the broad theme of foundations of computational mathematics. This structure allows for a very high standard of presentation, while affording endless opportunities for cross-fertilization and communication across subject boundaries. Workshops at the Santander conference were held in the following twenty-one fields:

- information-based complexity
- special functions and orthogonal polynomials
- computational algebraic geometry
- multiresolution and adaptivity in numerical PDEs
- geometric modelling and animation
- computational geometry and topology
- mathematical control theory and applications
- geometric integration and computational mechanics
- learning theory
- optimization
- relations with computer science: algorithmic game theory and metric embeddings
- image and signal processing
- symbolic analysis

- random matrices
- foundations of numerical PDEs
- approximation theory
- computational number theory
- numerical linear algebra
- real-number complexity
- computational dynamics
- stochastic computation.

In addition to the workshops, eighteen plenary lectures, covering a broad spectrum of topics connected to computational mathematics, were delivered by some of the world's foremost researchers. This volume is a collection of articles, based on the plenary talks presented at FoCM 2005. The topics covered in the lectures and in this volume reflect the breadth of research within computational mathematics as well as the richness and fertility of interactions between seemingly unrelated branches of pure and applied mathematics.

We hope that this volume will be of interest to researchers in the field of computational mathematics and also to non-experts who wish to gain some insight into the state of the art in this active and significant field.

Like previous FoCM conferences, the Santander gathering proved itself as a unique meeting point of researchers in computational mathematics and of theoreticians in mathematics and computer science. While presenting plenary talks by foremost world authorities and maintaining the highest technical level in the workshops, the conference, like previous meetings, laid emphasis on multidisciplinary interaction across subjects and disciplines in an informal and friendly atmosphere.

We wish to express our gratitude to the local organizers and administrative staff of our hosts the Universidad Cantabria, and wish to thank the Departamento de Matemáticas, Estadística y Computación of Universidad Cantabria, the Vicerrectorado de Investigación de la Universidad de Cantabria, and Proyecto Nacional MTM2004-20180-E for their financial assistance and for making FoCM 2005 such an outstanding success. We would like to thank the authors of the articles in this volume for producing in short order such excellent contributions. Above all, however, we wish to express our gratitude to all the participants of FoCM 2005 for attending the meeting and making it such an exciting, productive and scientifically stimulating event.

On the Complexity of Non Universal Polynomial Equation Solving: Old and New Results

C. Beltrán

Departamento de Matemáticas
Universidad de Cantabria
Santander, Spain
e-mail: beltranc@unican.es

L. M. Pardo

Departamento de Matemáticas
Universidad de Cantabria
Santander, Spain
e-mail: luis.pardo@unican.es

Abstract

These pages summarize some results on the efficiency of polynomial equation solving. We focus on semantic algorithms, i.e., algorithms whose running time depends on some intrinsic/semantic invariant associated with the input data. Both computer algebra and numerical analysis algorithms are discussed. We show a probabilistic and positive answer to Smale's 17th problem. Estimates of the probability distribution of the condition number of singular complex matrices are also exhibited.

1.1 Introduction

These pages summarize some results on upper and lower complexity bounds in Elimination Theory. They are a revision of the program stated in Pardo (1995).

We focus on *Efficient Polynomial Equation Solving*. This is one of the challenges in the recent history of Computational Mathematics. Two main frameworks in scientific computing deal with this problem. Following different approaches, symbolic/algebraic computing and numerical analysis developed their own techniques for solving polynomial equations. We survey statements of both approaches. New results are contained in Sections 1.4 and 1.5.

Multivariate Polynomial Equation Solving is a central topic both in Computational Mathematics and Computational Algebraic Geometry

(Elimination Theory in nineteenth century terminology). Its origin goes back to work of Sturm, Hermite, Cayley, and Sylvester, among others. Elimination Theory consists of the preparation of input data (polynomial equations and inequalities) to answer questions involving quantifiers. This approach also underlies Kronecker (1882), Hilbert (1890) and further developments in Algebraic Geometry. A central problem in Elimination Theory is the following:

Problem 1 (Hilbert's Nullstellensatz) *Design an efficient algorithm that performs the following task:*
Given a system of multivariate polynomial equations

$$f_1, \ldots, f_s \in \mathbb{C}[X_1, \ldots, X_n],$$

decide whether the following algebraic variety is empty or not:

$$V(f_1, \ldots, f_s) := \{x \in \mathbb{C}^n \ : \ f_i(x) = 0, 1 \leq i \leq s\}.$$

Here the term efficient refers to computational complexity. In the words of Traub & Werschultz (1998): *"computational complexity is a measure of the intrinsic computational resources required to solve a mathematical problem"*. Computational resources are measured in terms of a computational model or computational device that performs the corresponding algorithm that solves the problem. Intrinsic here means that we measure resources required by the problem and not the concrete algorithm. Hence, computational complexity is the design and analysis of an optimal algorithm (in terms of computational resources) that solves a mathematical problem.

The notion of computational resource requirements has been present in the mathematical literature for many years, although not always in an explicit form. For instance, we cite Galois who explicitly described computational requirements in his *Mémoire sur la Résolubilité des Équations par Radicaux*. Galois wrote: *"En un mot, les calculs sont impracticables"*. Galois had developed an algorithm that decided whether a univariate polynomial equation was solvable by radicals, but he realized that the computational complexity required by his procedure was excessive. The phrase thus means that he declined to perform calculations. In fact, he had discovered a central subject in computational complexity: *Intractability*.

In Galois' time, neither the notion of algorithm nor that of a complexity measure had been established. This relevant step in the history of mathematics was made around 1933. The works of Gödel, Church and Turing established the notion of algorithm which in later years lead to the existence of computers. We note that Turing's work and his machine concept of algorithm also became the standard pattern for computational complexity. In these pages, we shall measure computational resources in terms of Turing machines as much as is possible.

Computational resources are measured as functions of the input length. The input length is the time we need to write down the data. The (running) time function is the function that relates input length and running time under a concrete computational model.

Intractability is one of the frustrating aspects of computational complexity studies. A mathematical problem is intractable if the computational resources required to solve it are so excessive that there is no hope of solving the problem in practice. Observe that intractability is independent of the algorithm we design. For example, mathematical problems whose running time are at least exponential in the input length are naturally intractable. These are called *exponential problems* and there is no hope of solving them in any real or future computer. The reason is that this exponential time requirement is intrinsic to the problem and not to the concrete algorithm or computer.

Tractable problems are those mathematical problems whose time function is bounded by a polynomial of the input length. Between tractable and intractable problems there lies a large number of problems for which it is not known as yet whether they are tractable. We call them the Boundary of Intractability (cf. Garey & Johnson (1979)). Hilbert's Nullstellensatz lies in this boundary. This simply means that no-one has yet designed a tractable algorithm that solves Hilbert's Nullstellensatz, and it also means that no-one has yet proved that this problem is intractable. That is, it is not known whether there is an algorithm that solves **HN** in running time which depends polynomially on the number of variables.

There are several strategies for studying the computational complexity of Hilbert's Nullstellensatz. We classify them in two main groups: *syntactical* and *semantical*. Although these pages are mainly concerned with semantical strategies, we shall sketch some of the syntactical achievements in the study of **HN**.

Syntactical strategies are characterized by the fact that polynomials are considered as lists of coefficients (dense encoding) in certain vector spaces. They are then treated as vectors and linear algebra methods are applied to answer questions (mainly those involving quantifiers).

Historically, the first syntactical algorithm for **HN** goes back to Hilbert and his student Hermann (cf. Hermann (1926)). Hilbert and Hermann reduced **HN** to the consistency of a system of linear equations.

Hilbert's Nullstellensatz (Hilbert (1890)) states that *given a list of polynomials* $f_1, \ldots, f_s \in \mathbb{C}[X_1, \ldots, X_n]$ *of degree at most d, the complex algebraic variety they define* $V(f_1, \ldots, f_s) \subseteq \mathbb{C}^n$ *is empty if an only if there are polynomials* $g_1, \ldots, g_s \in \mathbb{C}[X_1, \ldots, X_n]$ *such that the following equality holds:*

$$1 = g_1 f_1 + \cdots + g_s f_s. \tag{1.1}$$

Identities such as (1.1) are called *Bézout Identities.*

From Hermann's work, we know that there is a function $D(d, n)$ which depends only on the number of variables and the maximum of the degrees, such that the following equivalence holds:

The variety $V(f_1, \ldots, f_s) \subseteq \mathbb{C}^n$ *is empty if and only if there exist polynomials* $g_1, \ldots, g_s$ *in* $\mathbb{C}[X_1, \ldots, X_n]$ *of degree at most* $D(d, n)$ *satisfying identity (1.1).*

Let us observe that Hermann's bound $D(d, n)$ reduces **HN** to the consistency question of a system of linear equations. The unknowns are the coefficients of the (possibly existing) polynomials $g_1, \ldots, g_s$ occurring in (1.1). The linear equations are determined by linear functions in the coefficients of the input polynomials $f_1, \ldots, f_s$. This approach reduces **HN** to the problem of deciding consistency of the linear system given by (1.1) involving

$$s \binom{D(d, n) + n}{n}$$

variables and equations. Its running time is obviously polynomial in this quantity. Hence, sharp upper bounds for the function $D(d, n)$ also imply sharp upper complexity bounds for this approach to solving **HN**. Studies on sharp upper bounds for $D(d, n)$ are called *Effective Nullstellensätze.* We cite Brownawell (1987), Caniglia, Galligo & J. Heintz (1988), Kollár (1988), Berenstein & Yger (1991, 1991a), Krick & Pardo (1996), Hägele, Morais, Pardo & M. Sombra (2000), Krick, Pardo & Sombra (2001) and their references. The known bounds for $D(d, n)$ can be summarized by the following inequalities:

$$d^{n-1} \leq D(d, n) \leq d^n.$$

Thus, this approach is neither efficient nor applicable since the time complexity is of order

$$\binom{d^n + n}{n} \approx d^{n^2}.$$

For example, deciding consistency of a system of cubic polynomial equations in 20 variables by this method requires deciding consistency of a system of more than 3^{400} linear equations in a similar number of variables. This is intractable in any actual or future computer.

In Hägele, Morais, Pardo & Sombra (2000) a simply exponential time algorithm (time of order d^n) to compute Bézout identities was shown, although the technique used in that paper is not syntactical but semantical.

A second syntactical strategy to deal with **HN** is due to rewriting techniques. The most frequently used rewriting method is that of the standard/Gröbner basis algorithms. Since the works Hironaka (1964) and Buchberger (1965), a huge list of references has been produced (cf. for instance Becker & Weispfenning (1993), Cox, Little & O'Shea (1997), Mora (2003), Vasconcelos (1998) and references therein). Most of these references discuss algorithms that compute Gröbner bases of an ideal. This strategy has also been fruitful in terms of implementations. Gröbner basis algorithmics is a standard primitive implemented in most computer algebra packages (Maple, Magma or Mathematica, for example). Most efficient implementations are due to Faugère (the *FGb* series). This approach has a serious drawback in terms of computational complexity. Since Mayr & Meyer (1982), we know that computing with Gröbner bases is exponential space complete and this is even worse than the running time of methods based on Effective Nullstellensätze. Computing with Gröbner bases involving more than 15 variables is not yet available. Thus, purely syntactical Gröbner bases techniques do not seem to be the best methods of dealing with **HN**.

A third syntactical strategy uses the underlying concepts of Structural Complexity. Namely, problems are classified into complexity classes and the study of the complexity of a problem consists in locating the appropriate class where this problem is complete. In Blum, Shub & Smale (1989), the authors proved that **HN** is complete in the class **NP**$_\mathbb{C}$ of nondeterministic polynomial time under the abstract model of complex Turing machines (cf. also Blum, Cucker, Shub & Smale (1998)). Other authors studied the complexity of **HN** within the more realistic Turing machine framework. In Koiran (1996) (see also Rojas (2001, 2003)) the author proved that **HN** belongs to the complexity class **PH** (polynomial hierarchy).

Nevertheless, all these syntactical strategies seem to forget that we are dealing with geometric objects (algebraic varieties) and regular mappings (polynomials viewed as functions and not as mere lists of coefficients). Algebraic varieties and regular mappings are mathematical objects rich in terms of semantic invariants. They have been studied for years in an attempt to describe their topological, geometrical and arithmetical properties. These studies have generated a large number of semantic invariants that must be related to computational complexity. This idea of relating semantical invariants to complexity is not completely new. In fact, semantical invariants of geometrical objects have been used to show lower complexity bounds for computational problems (see, for example, Montaña, Morais & Pardo (1996) and references therein). The converse problem was to design an algorithm that solves **HN** in time which depends polynomially on some semantical invariants of the input list of multivariate polynomials. This was achieved by the TERA experience. In fact, the TERA experience was more a current of thought than a research project, that was active during the nineties. Some of its achievements will be described in Section 1.2.

Somewhere between syntactical and semantical strategies, we may find "sparse" elimination techniques as in Sturmfels (1996) and references therein. However, we do not discuss sparse elimination here.

The rest of the chapter is structured as follows. In Section 1.2 we present an overview of some of the achievements of the TERA experience. In Section 1.3 we discuss an exponential lower time bound for universal algorithms in Elimination Theory. In Section 1.4 we show a positive answer to Smale's 17th Problem. Finally, in Section 1.5 we show sharp upper bounds for the probability distribution of the condition number of singular matrices.

1.2 Semantic Algorithms

In the middle nineties, the notion of *semantic algorithms in Elimination Theory* was introduced. Two works initiated this new generation of algorithms : Pardo (1995) and Giusti, Heintz, Morais & Pardo (1995). The first paper established a program, whereas the second one exhibited the first example of a semantical algorithm for Elimination Theory. This program was achieved in the series of papers Giusti, Heintz, Morais, Morgenstern & Pardo (1998), Giusti, Hägele, Heintz, Montaña, Morais & Pardo (1997), Giusti, Heintz, Morais & Pardo (1997). These works became the basis of the research experience called TERA. This section is devoted to a brief sketch of some of these achievements.

First of all, we reformulate Hilbert's Nullstellensatz in the following form.

Problem 2 *Design an efficient algorithm that performs the following task: Given a list of polynomials* $f_1, \ldots, f_s, g \in \mathbb{C}[X_1, \ldots, X_n]$ *of degree at most* d, *decide whether the polynomial* g *vanishes at some point of the algebraic variety* $V(f_1, \ldots, f_s) \subseteq \mathbb{C}^n$.

This is the usual formulation of elimination polynomials (like resultants and discriminants) in classical Elimination Theory. This is also the usual formulation of **NP**–complete problems (cf. Heintz & Morgenstern (1993) or Pardo (1995) and references therein). Note that all **NP**–complete problems are particular instances of Problem 2 above.

In this formulation, the role played by g and the list of $f_1, \ldots, f_s$ seems to be different.

From a list of polynomials like $f_1, \ldots, f_s$ we want to compute some *information* concerning the variety $V := V(f_1, \ldots, f_s)$ of its common zeros. The information we compute is expected to be used to answer further questions involving the variety V. This information is commonly called a *solution* of the input list of polynomials $f_1, \ldots, f_s$. For instance, in Problem 2, a solution of $f_1, \ldots, f_s$ should be used to decide whether a new polynomial g vanishes at some point in V. The way we choose to represent this information in a computer may be called the *encoding* of the solution variety V.

Obviously, different questions will condition the way we represent the information on the variety V in a computer. Hence, different notions of solution lead to different kinds of algorithms and different encodings of algebraic varieties. In Section 1.4 we recall the Shub–Smale notion of solution (approximate zeros) whose potentiality is still unexplored.

The proposal of TERA consisted in the design and analysis of a semantic algorithm that performs the following task:

From an input list $f_1, \ldots, f_s$, the algorithm outputs a description of the solution variety $V(f_1, \ldots, f_s)$.

This algorithm must satisfy two main properties:

- Its running time should be bounded by some intrinsic/semantic quantity that depends on the input list.
- Its output must contain sufficient information to answer any kind of elimination question like the one described in Problem 2.

These two properties lead to a notion of solution that we briefly sketch here. It is called *Kronecker's encoding of an affine algebraic variety* (cf. Kronecker (1882)).

Let $f_1, \ldots, f_i \in \mathbb{C}[X_1, \ldots, X_n]$ be a sequence of polynomials defining a radical ideal $(f_1, \ldots, f_i)$ of codimension i. Let $V := V(f_1, \ldots, f_i) \subseteq \mathbb{C}^n$ be the complex algebraic variety of dimension $n - i$ given by its common zeros. A *Kronecker encoding* of V is a birational isomorphism of V with some complex algebraic hypersurface in some affine complex space of dimension $n - i + 1$.

Technically, this is expressed as follows. Firstly, let us assume that the variables $X_1, \ldots, X_n$ are in Noether position with respect to the variety V. Namely, we assume that the following is an integral ring extension:

$$\mathbb{C}[X_1, \ldots, X_{n-i}] \hookrightarrow \mathbb{C}[X_1, \ldots, X_n]/(f_1, \ldots, f_i).$$

Let $u := \lambda_{n-i+1} X_{n-i+1} + \cdots + \lambda_n X_n \in \mathbb{Q}[X_1, \ldots, X_n]$ be a linear form in the dependent variables $\{X_{n-i+1}, \ldots, X_n\}$. A Noether's normalization and the linear mapping u define a linear projection:

$$\mathcal{U} : \mathbb{C}^n \longrightarrow \mathbb{C}^{n-i+1} \ : \ (x_1, \ldots, x_n) \longmapsto (x_1, \ldots, x_{n-i}, u(x_1, \ldots, x_n)) \,.$$

Let $\mathcal{U} \mid_V : V \longrightarrow \mathbb{C}^{n-i+1}$ be the restriction of the projection $\mathcal{U}$ to the variety V. The image set of the projection $\mathcal{U} \mid_V$ is a complex hypersurface H_u in $\mathbb{C}^{n-i+1}$. Let us denote by $\chi_u \in \mathbb{C}[X_1, \ldots, X_{n-i}, T]$ the minimal equation of H_u. The polynomial χ_u is called the elimination polynomial of u with respect to V.

The linear form u is called a *primitive element* if and only if the projection $\mathcal{U} \mid_V$ defines a birational isomorphism of V with H_u.

A Kronecker solution of the system of polynomial equations $f_1 = 0, \ldots, f_i = 0$ consists of a description of the Noether normalization, the primitive element u, the hypersurface H_u and a description of the inverse of the birational isomorphism, i.e., a description of $(\mathcal{U} \mid_V)^{-1}$. Formally, this list of items can be given as follows:

- The list of variables in Noether position $X_1, \ldots, X_n$ (which includes a description of the dimension of V). It is just a regular matrix that defines a linear change of coordinates that puts the variables in Noether position.
- The primitive element $u := \lambda_{n-i+1} X_{n-i+1} + \cdots + \lambda_n X_n$ given by its coefficients in $\mathbb{Z}$ (or any other computable subfield of $\mathbb{C}$).
- The minimal equation χ_u of the hypersurface H_u.
- A description of $(\mathcal{U} \mid_V)^{-1}$. This description can be given by the following list of polynomials:
  - A nonzero polynomial $\rho \in \mathbb{C}[X_1, \ldots, X_{n-i}]$.
  - A list of polynomials $v_j \in \mathbb{C}[X_1, \ldots, X_{n-i}, T]$, $n - i + 1 \leq j \leq n$.

These polynomials must satisfy the equality:

$$(\mathcal{U} \mid_V)^{-1}(x, t) = \left(x_1, \ldots, x_{n-i}, \rho^{-1}(x) v_{n-i+1}(x, t), \ldots, \rho^{-1}(x) v_n(x, t) \right),$$

for all $x := (x_1, \ldots, x_{n-i}) \in \mathbb{C}^{n-i}$, $t \in \mathbb{C}$, such that $(x, t) \in H_u$, $\rho(x) \neq 0$.

In 1882, Kronecker conceived an iterative procedure for solving multivariate systems of equations $F := [f_1, \ldots, f_n]$ defining zero–dimensional complex varieties. Kronecker's idea can be sketched in the following terms:

First, the procedure starts with system $[f_1]$ and it "solves" the equidimensional variety of codimension one $V(f_1) \subseteq \mathbb{C}^n$. Then the procedure runs iteratively: From a Kronecker encoding of $V(f_1, \ldots, f_i)$, the procedure "eliminates" the polynomial f_{i+1} to obtain a Kronecker encoding of the "next" variety $V(f_1, \ldots, f_{i+1})$. Proceed until $i = n$ is reached.

This iterative procedure has two main drawbacks, which can be explained in the following terms:

- First of all, the storage problem arising with the encoding of the intermediate polynomials. The polynomials χ_u, ρ and v_j are polynomials of high degree (eventually of degree d^i) involving $n - i + 1$ variables. Thus, to compute with them, the procedure has to handle all their coefficients, which amounts to

$$\binom{d^i + n - i + 1}{n - i + 1}$$

in number, For example, for $i := n/2$ the procedure must save more than $d^{n^2/4}$ coefficients. Handling such polynomials also requires a time complexity of similar order. This does not seem to be more efficient than the original treatment based on the Effective Nullstellensätze (cf. Section 1.1).

- Secondly, Kronecker's iterative procedure introduces a nesting of interpolation procedures. This nesting is demanded by the iterative process. Every time the procedure computes a new set of variables in Noether position, the procedure makes a recursive call of previously computed objects. This increases the time complexity function to $d^{O(n^2)}$.

The procedure was therefore forgotten by contemporary mathematicians and is hardly mentioned in the literature of Algebraic Geometry. Macaulay quotes Kronecker's procedure in Macaulay (1916) and so does König (1903). But both of them thought that this procedure would require excessive running time to be efficient, and so references to it have progressively vanished from the literature. Traces of this procedure can be found spread over the Algebraic Geometry literature without giving the required reference to it. For example, Kronecker's notion of solution was used in Zariski (1995) to define a notion of dimension for algebraic varieties, claiming that it was also used in the same form by Severi and others.

In Giusti, Heintz, Morais & Pardo (1995) and Pardo (1995), Kronecker's approach for solving was rediscovered without previous knowledge of this ancestor. These two works were able to overcome the first drawback (space problem of representation) of the previous methods. The technical trick was the use of a data structure coming from semi–numerical modeling: straight–line programs. This idea of representing polynomials by programs evaluating them goes back to previous work of the same research group (such as Giusti, Heintz (1991, 1993) or Krick & Pardo (1996), see also the references given in Pardo (1995)).

To overcome the second drawback (nesting), the authors introduced a method based on nonarchimedean Newton's operator. The approximate zeros in the corresponding nonarchimedean basin of attraction were called *Lifting Fibers* in Giusti, Hägele, Heintz, Morais, Montaña & Pardo (1997) solving the problem of nesting of interpolation procedures by Hensel's Lemma (also called the Implicit Mapping Theorem).

Unfortunately, Giusti, Hägele, Heintz, Morais, Montaña & Pardo (1997) introduced (for the Lifting Fibers) running time requirements which depend on the heights of the intermediate varieties in the sense of Bost, Gillet & Soulé (1994) or Philippon (1991, 1994, 1995). This drawback was finally overcome in Giusti, Heintz, Morais & Pardo (1997), where integer numbers were represented by straight–line programs and the following result was finally established:

Theorem 1.1 *(Giusti, Heintz, Morais & Pardo (1997)) There exists a bounded error probability Turing machine M which performs the following task: Given a system of multivariate polynomial equations $F := (f_1, \ldots, f_n)$, satisfying the following properties*

- $\deg(f_i) \leq d$ *and* $ht(f_i) \leq h$ *for* $1 \leq i \leq n$ *(h is the bit length of the coefficients),*
- *the ideals* $(f_1, \ldots, f_i)$ *are radical ideals of codimension i in the ring* $\mathbb{Q}[X_1, \ldots, X_n]$ *for* $1 \leq i \leq n - 1$,
- *the variety* $V(f_1, \ldots, f_n) \subseteq \mathbb{C}^n$ *is a zero–dimensional complex algebraic variety,*

then the machine M outputs a Kronecker solution of the variety

$$V(f_1, \ldots, f_n).$$

The running time of the machine M is polynomial in the quantities

$$\delta(F), \ n, \ h, \ d, \ L,$$

where $\delta(F)$ is the maximum of the degrees of the intermediate varieties (in the sense of Heintz (1983)), namely

$$\delta(F) := \max\{\deg(V(f_1, \ldots, f_i)) : 1 \leq i \leq n - 1\},$$

and L is the input length in any natural encoding of multivariate polynomials.

It must be said that the coefficients of the polynomials involved in a Kronecker solution of the variety $V(f_1, \ldots, f_n)$ are given by straight–line programs. However, the complexity estimates for the Turing machine M are independent of the height.

The quantity $\delta(F)$ becomes a kind of condition number for symbolic methods to solve systems of multivariate polynomial equations by Kronecker's deformation technique.

After Giusti, Heintz, Morais, & Pardo (1997), several new authors got involved in the TERA experience, with several technical improvements, mainly on the exponents occurring in the polynomial upper time bound quoted in Theorem 1.1. Among them we can cite Giusti & Schost (1999), Lecerf (2001), Heintz, Matera & Weissbein (2001), for instance. This dependence on a semantic invariant was also translated to the problem of computing real solutions of real polynomial equations in the series of papers Bank, Giusti, Heintz & Mbakop (1997,2001), Bank, Giusti, Heintz & Pardo (2004, 2005). The algorithm was successfully implemented by Lecerf and Salvy. This implementation, involving some technical variations, was presented in Giusti, Lecerf & Salvy (2001).

Despite the expected good behavior in practical applications, the package Kronecker was not sufficiently efficient to deal with a reasonable number of variables. Hence, a deeper revision of the original goals was needed. Firstly, the reader should observe that the geometric degree of any input system $\delta(F)$ is generically equal to its worst case value (the *Bézout number* $\mathcal{D} := \prod_{i=1}^{n} deg(f_i)$). Secondly, the Bézout number is exponential in the number of variables and so is its running time on average. Thus, Kronecker's solving can only be efficient for a few particular instances (when $\delta(F)$ is "small"). Up to now, we have not found a good class of natural problems with "small" geometric degree $\delta(F)$.

1.3 Universal Solving

The TERA experience and the results of the package Kronecker lead to two central questions:

- *Is the Bézout number $\mathcal{D}$ a barrier for the complexity of polynomial equation solvers?*
- *In case of a positive answer, explain the meaning of this barrier.*

An attempt to answer these two questions was the notion of *Universal Algorithms* and the results shown in Heintz, Matera, Pardo & Wachenchauzer (1998), Pardo (2000), and Castro, Giusti, Heintz, Matera & Pardo (2003). Roughly speaking, a polynomial equation solver is universal if its output contains sufficient information to answer all elimination questions concerning the solution variety. All known algorithms (either syntactical or semantical) in Elimination Theory are universal. Formalizing this idea requires some additional terminology.

Another feature of semantic algorithms is that they can be adapted to any particular data structure used to represent input polynomials. Data structures of input polynomials are typically defined by a regular morphism from some space of parameters to some space of input data. This can be formalized as follows.

Let $P_d \subseteq \mathbb{C}[X_1, \ldots, X_n]$ be the vector space of all complex polynomials of degree at most d. For a list of degrees $(d) := (d_1, \ldots, d_n)$, let $\mathcal{P}_{(d)}$ be the Cartesian product

$$\mathcal{P}_{(d)} := \prod_{i=1}^{n} P_{d_i}.$$

The vector space of *dense input encoding* $\mathcal{P}_{(d)}$ represents the class of systems of multivariate polynomials $F := [f_1, \ldots, f_n] \in \mathcal{P}_{(d)}$. We denote by $V(F) \subseteq \mathbb{C}^n$ the set of its common zeros, if any. Namely,

$$V(F) := \{x \in \mathbb{C}^n \ : \ f_i(x) = 0, 1 \leq i \leq n\}.$$

For every constructible subset $W \subseteq \mathbb{C}^m$, a *data structure for input systems and parameters in W* is a regular mapping

$$\Phi : W \subseteq \mathbb{C}^m \longrightarrow \mathcal{P}_{(d)}.$$

The mapping Φ associates to every parameter $\alpha \in W$ a system of multivariate polynomial equations $F_\alpha \in \mathcal{P}_{(d)}$. The image set $Im(\Phi)$ is the particular class of systems we want to solve and the varieties $V(F_\alpha) \subseteq \mathbb{C}^n$ are the solution varieties. The constructible set W is called the source space and its dimension $dim(W) \leq m$ is called the *source dimension*. In standard applications, the dimension m of the source space is much smaller than the dimension N of $\mathcal{P}_{(d)}$.

A polynomial equation solver adapted to the unirational family Φ takes as input a system F_α in $Im(\Phi)$ and outputs some encoding of the solution variety $V(F_\alpha) \subseteq \mathbb{C}^n$. The encoding of $V(F_\alpha)$ is written as a point in some affine space $\mathbb{C}^M$. Once again, the dimension M is usually much greater than the source dimension.

For example, the semantic algorithm described in Section 1.2 associates to every $\alpha \in W$ a Kronecker description of $V(F_\alpha)$. Namely, we represent $V(F_\alpha)$ by the list of coefficients of all polynomials occurring in a Kronecker description of $V(F_\alpha)$. This can be done in some affine space $\mathbb{C}^M$, where M is a quantity polynomial in the number of variables and linear in some quantity $\delta(\Phi)$ given as the maximum of the geometric degrees of input systems in $Im(\Phi)$. Namely,

$$\delta(\Phi) := \{deg(V(F_\alpha)) \ : \ \alpha \in W\}.$$

Generically, $\delta(\Phi)$ equals the Bézout number $\mathcal{D} := \prod_{i=1}^{n} d_i$ and, hence, it is exponential in the number of variables n. A similar phenomenon can be observed using either Cayley–Chow encoding or Macaulay's encoding of equidimensional algebraic varieties.

This yields the model described in Castro, Giusti, Heintz, Matera & Pardo (2003) that we briefly sketch. In the sequel, a *unirational family of elimination problems* is a regular morphism

$$\varepsilon : W \subseteq \mathbb{C}^m \longrightarrow \mathbb{C}^M.$$

The space $\mathbb{C}^M$ is called the *target space*, a point $y \in Im\ \varepsilon \subseteq \mathbb{C}^M$ is called a *target point* (also a semantical object), and the dimension M of the target space is called the *target dimension*. In the previous notation, for every $\alpha \in W$, $\varepsilon(\alpha)$ is the encoding of the solution variety $V(F_\alpha)$, where $F_\alpha \in Im(\Phi) \subseteq \mathcal{P}_{(d)}$.

Given a unirational family of elimination problems ε, a *mathematical question* concerning target points $y \in Im\ \varepsilon \subseteq \mathbb{C}^M$ is simply a *transformation* of the target space in a neighborhood of $(Im\ \varepsilon, y)$. Namely, a transformation is the germ of a mapping

$$\theta : (\mathbb{C}^M, y) \longrightarrow (\mathbb{C}^\ell, q).$$

The space $\mathbb{C}^\ell$ is called the *space of answers*, and its dimension ℓ is called the dimension of the space of answers. Usual mathematical questions concern spaces of answers of small dimension (with respect to the target dimension). For example, *decisional questions* are transformations of the semantical object into some unidimensional space of answers, i.e., transformations of the form

$$\theta : (\mathbb{C}^M, y) \longrightarrow (\mathbb{C}, q).$$

We claim that the goal of Elimination Theory is the design of algorithms that answer questions concerning target points of unirational families of polynomials.

As the target dimension is usually too big, efficient elimination procedures evaluate an alternative mapping:

$$\mu : W \subseteq \mathbb{C}^m \longrightarrow \mathbb{C}^s.$$

We call μ a *black-box*. It is usually evaluated by an algorithm whose particular form will not be discussed here.

A *versal black-box* associated with a unirational family of elimination problems $\varepsilon : W \longrightarrow \mathbb{C}^M$ is a mapping $\mu : W \longrightarrow \mathbb{C}^s$ such that the following property holds:

For every source point $\alpha \in W$ and every question $\theta : (\mathbb{C}^M, \varepsilon(\alpha)) \longrightarrow (\mathbb{C}^\ell, z)$ there is a germ of a mapping $\rho : (\mathbb{C}^s, \mu(\alpha)) \longrightarrow (\mathbb{C}^\ell, z)$ such that the following diagram commutes:

$$(W, \alpha) \xrightarrow{\varepsilon} (Im\ \varepsilon, \varepsilon(\alpha)) \xrightarrow{\theta} (\mathbb{C}^\ell, z)$$

$$\mu \searrow \qquad \qquad \nearrow \rho$$

$$(Im\ \mu, \mu(\alpha))$$

For every source point α, the point $\mu(\alpha) \in \mathbb{C}^s$ is called the *output encoding* of the target $\varepsilon(\alpha)$. The number s of coordinates of $\mu(\alpha)$ is called the *output length*.

Proposition 1.1 *For every unirational family of elimination problems ε : $W \longrightarrow \mathbb{C}^M$ and a black–box $\mu : W \longrightarrow \mathbb{C}^s$, the following properties are equivalent:*

1. *μ is a versal black–box associated with ε.*
2. *For every source point $\alpha \in W$ there is a mapping germ $\rho_\alpha : (\mathbb{C}^s, \mu(\alpha)) \longrightarrow (\mathbb{C}^M, \varepsilon(\alpha))$ such that the following diagram commutes.*

$$(W, \alpha) \xrightarrow{\varepsilon} (Im\ \varepsilon, \varepsilon(\alpha))$$

$$\mu \searrow \qquad \uparrow \rho_\alpha$$

$$(Im\ \mu, \mu(\alpha))$$

The germ ρ_α is called the interpolation procedure of the versal black–box μ at α.

Let $\varepsilon : W \longrightarrow \mathbb{C}^M$ be a unirational family of elimination problems and let $\mu : W \longrightarrow \mathbb{C}^s$ be some versal black–box associated with ε. We say that μ is *certified* if there is a mapping

$$\varphi : Im\ \varepsilon \longrightarrow Im\ \mu,$$

such that $\mu = \varphi \circ \varepsilon$.

Definition 1.2 *Let $\varepsilon : W \longrightarrow \mathbb{C}^M$ be a unirational family of elimination problems. A universal black–box associated with ε is a versal and certified black–box $\mu : W \longrightarrow \mathbb{C}^s$ such that the following properties hold :*

1. *The black–box μ is holomorphic.*
2. *For every source point $\alpha \in W$ the interpolation procedure ρ_α of μ is the germ of a holomorphic mapping.*

Definition 1.3 *A polynomial equation solver is called universal if for every unirational family of elimination problems $\varepsilon : W \longrightarrow \mathbb{C}^M$ the procedure generates a universal black–box $\mu : W \longrightarrow \mathbb{C}^s$ associated with ε.*

Theorem 1.2 *There is a sequence $\left(\varepsilon_n : W_n \subseteq \mathbb{C}^{m(n)} \longrightarrow \mathbb{C}^{M(n)}\right)_{n \in \mathbb{N}}$ of unirational families of polynomials such that for every $n \in \mathbb{N}$ the following holds:*

1. *The input length is linear in n. Namely, $m(n) = O(n)$.*
2. *The degree of the input space W_n and that of the regular mapping ε_n are also linear in n.*
3. *There is an explicit description of the input space W_n of length linear in n.*
4. *The target dimension $M(n)$ is exponential in n.*
5. *For every $n \in \mathbb{N}$ and every universal black–box $\mu_n : W_n \longrightarrow \mathbb{C}^{s_n}$ associated with ε_n the output length s_n is exponential in the source dimension, i.e.,*

$$s_n \geq 2^n.$$

This technical statement can also be stated in the following terms.

Corollary 1.1 *(Castro, Giusti, Heintz, Matera & Pardo (2003)) Every universal polynomial equation solver requires exponential running time. In particular, the procedure described in Section 1.2 above is essentially optimal as a universal polynomial equation solver.*

As shown in Castro, Giusti, Heintz, Matera & Pardo (2003), we can always associate to every unirational family of elimination problems, a certified black–box whose output length is linear in the source dimension. Namely, the smoothness condition on ρ_α is a necessary condition for Theorem 1.2 to hold. However, nonsmooth exact interpolation procedures are difficult to comprehend.

It is also possible to define the notion of a universal solver in a numerical analysis context. For example, the algorithms implemented by Verschelde and collaborators (cf. Verschelde (2000) and references therein) are universal polynomial solvers. A universal numerical analysis solver takes as input a list of multivariate polynomial equations $F \in \mathcal{P}_{(d)}$ and outputs approximations for all zeros $\zeta \in V(F)$. Since the average number of zeros equals the Bézout number $\mathcal{D}$, it is immediate that universal numerical solvers also require exponential running time.

Corollary 1.1 must not be understood as a negative result. It is asking for a new generation of algorithms: *Nonuniversal polynomial equation solvers.*

The output of a nonuniversal solver will contain only partial information about the variety of solutions. This simple idea also leads to a long series of new problems and questions. The obvious and first is nevertheless the most difficult to answer: *Which questions can be answered with the information provided by a nonuniversal algorithm?* Much more experience with nonuniversal solvers is still required before dealing with this question.

An example of a symbolic, semantic and nonuniversal polynomial equation solver was given in San Martín & Pardo (2004). However, the worst case complexity of this algorithm is also exponential in the number of variables and, hence, intractable.

The search for nonuniversal solvers naturally leads to numerical analysis polynomial system solvers.

1.4 Shub & Smale Approximate Zero Theory: Bézout $5\frac{1}{2}$

In the first half of the nineties, Shub and Smale introduced a seminal conception of the foundations of numerical analysis. They focused on a theory of numerical polynomial equation solvers in the series of papers Shub & Smale (1993a, 1993b, 1993c, 1994, 1996). Other authors that also took this approach are Blum, Cucker, Shub & Smale (1998), Dedieu (2001b, 2003), Kim (1988, 1989), Malajovich (1994), Malajovich & Rojas (2002), Yakoubsohn (1995) and references therein.

Shub and Smale's theory of *approximate zeros* provides an answer to the barrier question stated in Section 1.3. The new results of this section are taken from the still unpublished manuscript Beltrán & Pardo (2005).

As in Shub & Smale (1994), the input space is the space of systems of multivariate homogeneous polynomials with dense encoding and fixed degree list. Namely, for every positive integer $d \in \mathbb{N}$, let $H_d \subseteq \mathbb{C}[X_0, \ldots, X_n]$ be the vector space of all homogeneous polynomials of degree d. For a list of degrees $(d) := (d_1, \ldots, d_n) \in \mathbb{N}^n$, let $\mathcal{H}_{(d)}$ be the set of all systems $F := [f_1, \ldots, f_n]$ of homogeneous polynomials of respective degrees $\deg(f_i) = d_i$, $1 \leq i \leq n$. In other words, $\mathcal{H}_{(d)} := \prod_{i=1}^{n} H_{d_i}$.

We denote by $N + 1$ the complex dimension of the vector space $\mathcal{H}_{(d)}$. Note that $N + 1$ is the input length for dense encoding of multivariate polynomials. For every system $F \in \mathcal{H}_{(d)}$, we also denote by $V(F)$ the projective algebraic variety of its common zeros. Namely,

$$V(F) := \{x \in \mathbb{P}_n(\mathbb{C}) \ : \ f_i(x) = 0, 1 \leq i \leq n\}.$$

Note that with this notation $V(F)$ is always a nonempty projective algebraic variety.

In Beltrán & Pardo (2005), the following statement is proven. It represents a positive answer to Problem 17 of Smale (2000).

Theorem 1.3 *There is a bounded error probabilistic numerical analysis procedure that solves most systems of multivariate polynomial equations with running time that is polynomial in*

$$n, N, d.$$

The probability that a system $F \in \mathcal{H}_{(d)}$ is solved by this procedure is greater than

$$1 - \frac{1}{N}.$$

In this statement the term "solves" means the "algorithm outputs nonuniversal information" about the variety of solutions, whereas the term "most" means "with high probability of success". The precise meaning of this theorem requires some additional technical notions.

A nonuniversal numerical analysis solver takes as input a system $F \in \mathcal{H}_{(d)}$ and outputs local information on some (mostly just one) of the zeros $\zeta \in V(F)$. The local information (close to a zero) we compute is the information provided by an *approximate zero* $z \in \mathbb{P}_n(\mathbb{C})$ of F associated with some zero $\zeta \in V(F)$ (in the sense of Shub & Smale (1993b) or Shub (1993)).

For every input system $F \in \mathcal{H}_{(d)}$, let N_F be the projective Newton operator as introduced in Shub (1993). According to Shub & Smale (1993a), an approximate zero $z \in \mathbb{P}_n(\mathbb{C})$ of a system $F \in \mathcal{H}_{(d)}$ with associated zero $\zeta \in V(F) \subseteq \mathbb{P}_n(\mathbb{C})$ is a projective point such that the sequence of iterates $(N_F^k(z))_{k \in \mathbb{N}}$ is well–defined and converges to the actual zero $\zeta \in V(F)$ at a speed which is doubly exponential in the number of iterations. In this sense, the approximate zero z is rich in local information about the zero $\zeta \in V(F)$. In Castro, Hägele, Morais & Pardo (2001), the authors also observed that approximate zeros with rational coordinates contain not only local information about the associated zero, but also algebraic information. However, we will not discuss these aspects here.

These basic notions stated, a nonuniversal numerical analysis solver is an algorithm that has the following input/output structure:

Input: A system of homogeneous polynomial equations $F \in \mathcal{H}_{(d)}$.

Output: An approximate zero $z \in \mathbb{P}_n(\mathbb{C})$ of F associated with some zero $\zeta \in V(F)$.

Such kinds of algorithms are not conceived for solving all input systems but a large subclass of them. In principle, singular systems are not intended to be solved by our procedure: this requires further (and more delicate) analysis.

Let $\Sigma \subseteq \mathcal{H}_{(d)}$ be the class of systems F such that $V(F)$ contains a singular zero. We call Σ the discriminant variety. These pages are mainly concerned with procedures that solve systems without singular zeros (i.e., systems $F \in \mathcal{H}_{(d)} \setminus \Sigma$).

Our main algorithmic scheme is Newton's Homotopic Deformation in the projective space (as described in Shub & Smale (1996)): Given $F, G \in \mathcal{H}_{(d)} \setminus \Sigma$, we consider the "segment" of systems "between" F and G,

$$\Gamma := \{F_t := (1-t)G + tF, t \in [0,1]\}.$$

If $\Gamma \cap \Sigma = \emptyset$, there are nonintersecting and smooth curves of equations–solutions associated with this segment:

$$C_i(\Gamma) := \{(F_t, \zeta_t) : \zeta_t \in V(F_t), t \in [0,1]\}, \quad 1 \le i \le \mathcal{D} := \prod_{i=1}^{n} d_i.$$

Then, Newton's operator may be used to follow closely one of these curves $C_i(\Gamma)$ in the incidence variety. This procedure computes some approximate zero z_1 associated with some zero of F (i.e., $t = 1$) starting at some approximate zero z_0 associated with G (i.e., from $t = 0$). The following definition formalizes this strategy based on a Newton Homotopic Deformation Technique.

Definition 1.4 *A Newton's Homotopic Deformation scheme (NHD for short) with initial data $(G, z_0) \in \mathcal{H}_{(d)} \times \mathbb{P}_n(\mathbb{C})$ and resource function $\varphi : \mathcal{H}_{(d)} \times \mathbb{R}^+ \longrightarrow \mathbb{R}^+$ is an algorithmic scheme based on the following strategy:*

Input: $F \in \mathcal{H}_{(d)}$, $\varepsilon \in \mathbb{R}^+$.

Perform $\varphi(F, \varepsilon)$ "homotopic steps" following the segment

$$(1-t)G + tF, \qquad t \in [0,1],$$

starting at (G, z_0), where z_0 is an approximate zero of G associated with some zero $\zeta_0 \in V(G)$.

Output:

   either failure, or
   an approximate zero $z_1 \in \mathbb{P}_n(\mathbb{C})$ of F.

An algorithm following the NHD scheme is an algorithm that constructs a polygonal P with $\varphi(F, \varepsilon)$ vertices. The initial vertex of P is the point (G, z_0) and its final vertex is the point (F, z_1) for some $z_1 \in \mathbb{P}_n(\mathbb{C})$. The output of the algorithm is the value $z_1 \in \mathbb{P}_n(\mathbb{C})$. The polygonal is constructed by "homotopic steps" (path-following methods) that go from one vertex to the next. Hence, $\varphi(F, \varepsilon)$ is the number of homotopic steps performed by the algorithm. Different subroutines have been designed to perform each one of these "homotopic steps". One of them is the projective Newton operator as described in Shub & Smale (1993b), Shub (1993), Malajovich (1994).

The positive real number ε is currently used both to control the number of steps (through the function $\varphi(F, \varepsilon)$) and the probability of failure (i.e., the probability that a given input $F \in \mathcal{H}_{(d)}$ is not solved in $\varphi(F, \varepsilon)$ steps with initial pair (G, z_0)).

Initial pairs with optimal tradeoff between the number of steps and the probability of failure are required. The following notion is an attempt to fix what this means.

Definition 1.5 *Let $\varepsilon > 0$ be a positive real number. We say that an initial pair $(G, z_0) \in \mathcal{H}_{(d)} \times \mathbb{P}_n(\mathbb{C})$ is ε-efficient for the NHD scheme if there is an algorithm based on the NHD scheme with initial pair (G, z_0) such that the following properties hold:*

1. *The resource function (i.e., the number of steps) $\varphi(F, \varepsilon)$ is bounded by a polynomial in the quantities $\varepsilon^{-1}, n, N, d$, where $d := max\{d_i : 1 \leq i \leq n\}$.*

2. *The probability of "failure" (i.e., the probability that a system is not solved) is at most ε.*

Observe that a pair $(G, z_0) \in \mathcal{H}_{(d)} \times \mathbb{P}_n(\mathbb{C})$ may be ε-efficient for some positive real number $\varepsilon > 0$ and not ε'-efficient for $\varepsilon' < \varepsilon$.

Moreover, the main outcome in Shub & Smale (1994) proves that *for every positive real number $\varepsilon > 0$, there is at least one ε-efficient initial pair $(G_\varepsilon, \zeta_\varepsilon) \in \mathcal{H}_{(d)} \times \mathbb{P}_n(\mathbb{C})$.* This statement is an absolute breakthrough regarding the efficiency of numerical analysis polynomial equation solving. It leads to the following procedure based on the NHD scheme:

Input: $F \in \mathcal{H}_{(d)}$, $\varepsilon \in \mathbb{R}^+$.

- *Compute $(G_\varepsilon, \zeta_\varepsilon)$ (the ε-efficient initial pair whose existence is guaranteed by Shub & Smale (1994)).*

- *Perform a polynomial (in $\varepsilon^{-1}, n, N, d$) number of homotopic steps following the segment $(1-t)G + tF$, $t \in [0, 1]$, starting at $(G_\varepsilon, \zeta_\varepsilon)$.*

Output:

> *either failure, or*
> *an approximate zero $z \in \mathbb{P}_n(\mathbb{C})$ of F.*

The procedure seems to give an answer since it may compute approximate zeros for most systems of homogeneous polynomial equations. Here, most means with probability greater than $1 - \varepsilon$.

However, the procedure has three main drawbacks. First of all, Shub & Smale (1994) prove the existence of some ε–efficient initial pair, but they give no hint as to how to compute such a pair $(G_\varepsilon, \zeta_\varepsilon)$. Note that if there is no method of computing $(G_\varepsilon, \zeta_\varepsilon)$, then the previous scheme is not a proper algorithm (one cannot "write" $(G_\varepsilon, \zeta_\varepsilon)$ and then one cannot start to compute). Shub & Smale (1994) used the term "quasi–algorithm" to explain the result they obtained, whereas Problem 17 in Smale (2000) asks for a "uniform algorithm". In a broad sense, this scheme is close to an "oracle machine" where the initial pair $(G_\varepsilon, \zeta_\varepsilon)$ is given by some undefinable oracle. Moreover, the lack of hints on ε–efficient initial pairs leads both to "Shub & Smale's Conjecture" (Shub & Smale (1994)) and to Smale's 17th problem.

A second drawback is the dependence of $(G_\varepsilon, \zeta_\varepsilon)$ on the value ε.

Thirdly, the reader should observe that the initial pair $(G_\varepsilon, \zeta_\varepsilon)$ must be solved before we can perform any computation. Namely, ζ_ε must be an approximate zero of G_ε. In fact, Shub & Smale (1994) proved the existence of such $(G_\varepsilon, \zeta_\varepsilon)$ assuming that ζ_ε is a true zero of G_ε (i.e., $\zeta_\varepsilon \in V(G_\varepsilon)$). This means that we not only need to start at some approximate zero of G_ε but it seems that we need to start at a true and exact zero of this initial system.

Thus, any algorithm based on this version of NHD requires some "a priori" tasks not all of them simple:

First, one has to detect some system of equations G_ε such that some of its zeros ζ_ε yields an ε–efficient initial pair $(G_\varepsilon, \zeta_\varepsilon)$. Second, one needs to "solve" the system G_ε in order to compute the "exact" solution ζ_ε.

As "exact" solutions do not seem a good choice, we must proceed in the opposite manner. We must start at some complex point $\zeta_\varepsilon \in \mathbb{P}_n(\mathbb{C})$, given a priori. Then we must prove that there is a system G_ε vanishing at ζ_ε such that $(G_\varepsilon, \zeta_\varepsilon)$ is an ε–efficient initial pair. The existence of such a kind of system G_ε for any given $\zeta_\varepsilon \in \mathbb{P}_n(\mathbb{C})$ easily follows from the arguments in Shub & Smale (1994). But, once again, no hint on how to find G_ε from ζ_ε seems to be known.

In Beltrán & Pardo (2005) we exhibit a solution to these drawbacks. We found a probabilistic approach and, hence, we can give an efficient uniform (i.e., true) algorithm that solves most systems of multivariate polynomial equations. This is achieved using the following notation.

Definition 1.6 *A class* $\mathcal{G} \subseteq \mathcal{H}_{(d)} \times \mathbb{P}_n(\mathbb{C})$ *is called a correct test class (also questor set) for efficient initial pairs if for every* $\varepsilon > 0$ *the probability that a randomly chosen pair* $(G, \zeta) \in \mathcal{G}$ *is* ε*-efficient is greater than*

$$1 - (nNd)^{O(1)}\varepsilon,$$

where $O(1)$ *denotes some fixed constant independent of* ε, d *and* n.

Note the analogy between these classes of efficient initial systems and the classes of "correct test sequences" (also "questor sets") for polynomial zero tests (as in Heintz & Schnorr (1982), Krick & Pardo (1996) or Castro, Giusti, Heintz, Matera & Pardo (2003)). The following is shown in Beltrán & Pardo (2005).

Theorem 1.4 *For every degree list* $(d) = (d_1, \ldots, d_n)$ *there is a questor set* $\mathcal{G}_{(d)}$ *for efficient initial pairs that solves most of the systems in* $\mathcal{H}_{(d)}$ *in time which depends polynomially on the input length* N *of the dense encoding of multivariate polynomials.*

The existence of a questor set for initial pairs $\mathcal{G}_{(d)} \subseteq \mathcal{H}_{(d)} \times \mathbb{P}_n(\mathbb{C})$ yields another variation (of a probabilistic nature) on the algorithms based on NHD schemes. First of all, note that the class $\mathcal{G}_{(d)}$ does not depend on the positive real number $\varepsilon > 0$ under consideration. Thus, we can define the following NHD scheme based on some fixed questor set $\mathcal{G}_{(d)}$.

Input: $F \in \mathcal{H}_{(d)}$, $\varepsilon \in \mathbb{R}^+$.

- *Guess at random* $(G, \zeta) \in \mathcal{G}_{(d)}$.
- *Perform a polynomial (in* $\varepsilon^{-1}, n, N, d$*) number of homotopic steps following the segment* $(1 - t)G + tF$, $t \in [0, 1]$, *starting at* (G, ζ).

Output:

        either failure, or
        an approximate zero $z \in \mathbb{P}_n(\mathbb{C})$ *of* F.

Observe that the questor set $\mathcal{G}_{(d)}$ is independent of the value ε under consideration. However, the existence of such a questor set does not imply the existence of an algorithm. In fact, a simple existence statement as Theorem 1.4 will not be better than the main outcome in Shub & Smale (1994).

In Beltrán & Pardo (2005), we exhibit an algorithmically tractable subset $\mathcal{G}_{(d)}$ which is proven to be a questor set for efficient initial pairs. This rather technical class can be defined as follows:

Let Δ be the Kostlan matrix as defined in Shub & Smale (1993a). Using this matrix, Shub & Smale (1993a) define a Hermitian product $\langle \cdot, \cdot \rangle_\Delta$ on $\mathcal{H}_{(d)}$ which is invariant under certain natural actions of the unitary group $\mathcal{U}_{n+1}$ on $\mathcal{H}_{(d)}$. We denote by $||\cdot||_\Delta$ the norm on $\mathcal{H}_{(d)}$ defined by $\langle \cdot, \cdot \rangle_\Delta$. This Hermitian product $\langle \cdot, \cdot \rangle_\Delta$ also defines a complex Riemannian structure on the complex projective space $\mathbb{P}(\mathcal{H}_{(d)})$. This complex Riemannian structure on $\mathbb{P}(\mathcal{H}_{(d)})$ induces a volume form $d\nu_\Delta$ on $\mathbb{P}(\mathcal{H}_{(d)})$ and hence a measure on this manifold. The measure on $\mathbb{P}(\mathcal{H}_{(d)})$ also induces a probability on this complex Riemannian manifold. Moreover, for every subset $A \subseteq \mathbb{P}_n(\mathbb{C})$ the probability $\nu_\Delta[A]$ induced by $d\nu_\Delta$ agrees with the Gaussian measure of its projecting cone $\hat{A} \subseteq \mathcal{H}_{(d)}$. In the sequel, volumes and probabilities in $\mathcal{H}_{(d)}$ and $\mathbb{P}(\mathcal{H}_{(d)})$ always refers to these probabilities and measures defined by $\langle \cdot, \cdot \rangle_\Delta$.

Let us now fix a projective point $e_0 := (1 : 0 : \cdots : 0) \in \mathbb{P}_n(\mathbb{C})$. Let $L_0 \subseteq \mathcal{H}_{(d)}$ be the class of systems of homogeneous polynomial equations given by the property:
A system $F := [\ell_1, \ldots, \ell_n] \in \mathcal{H}_{(d)}$ belongs to L_0 if and only if for every i, $1 \leq i \leq n$, there is a linear mapping $\lambda_i : \mathbb{C}^n \longrightarrow \mathbb{C}$ such that the following equality holds:

$$\ell_i := X_0^{d_i - 1} \lambda_i(X_1, \ldots, X_n).$$

Let $V_0 \subseteq \mathcal{H}_{(d)}$ be the class of all homogeneous systems $F \in \mathcal{H}_{(d)}$ that vanish at e_0. Namely,

$$V_0 := \{F \in \mathcal{H}_{(d)} \ : \ e_0 \in V(F)\}.$$

Note that L_0 is a vector subspace of V_0.

Next, let $L_0^\perp$ be the orthogonal complement of L_0 in V_0 with respect to Kostlan's metric $\langle \cdot, \cdot \rangle_\Delta$. Note that $L_0^\perp$ is the class of all systems $F \in \mathcal{H}_{(d)}$ that vanishes at e_0 and such that its derivative $DF(e_0)$ also vanishes at e_0. Namely, it is the class of all systems of polynomial equations of order at least 2 at e_0.

Let Y be the following convex affine set, obtained as the product of closed balls:

$$Y := [0, 1] \times B^1(L_0^\perp) \times B^1(\mathcal{M}_{n \times (n+1)}(\mathbb{C})) \subseteq \mathbb{R} \times \mathbb{C}^{N+1},$$

where $B^1(L_0^\perp)$ is the closed ball of radius one in $L_0^\perp$ with respect to the canonical Hermitian metric and $B^1(\mathcal{M}_{n \times (n+1)}(\mathbb{C}))$ is the closed ball of radius one in the space of $n \times (n+1)$ complex matrices with respect to the standard Frobenius norm. We assume Y is endowed with the product of the respective Riemannian structures and the corresponding measures and probabilities.

Let $\tau \in \mathbb{R}$ be the real number given by

$$\tau := \sqrt{\left(\frac{n^2 + n}{N} \right)}.$$

Now, let us fix any mapping $\phi : \mathcal{M}_{n \times (n+1)}(\mathbb{C}) \longrightarrow \mathcal{U}_{n+1}$ such that for every matrix $M \in \mathcal{M}_{n \times (n+1)}(\mathbb{C})$ of maximal rank, ϕ associates a unitary matrix $\phi(M) \in \mathcal{U}_{n+1}$ verifying $M\phi(M)e_0 = 0$. Namely, $\phi(M)$ transforms e_0 into a vector in the kernel $Ker(M)$ of M. Our statements below are independent of the chosen mapping ϕ that satisfies this property.

Next, let us denote by $e_0^\perp$ the orthogonal complement of the affine point $(1, 0, \ldots, 0) \in \mathbb{C}^{n+1}$ with respect to the standard Hermitian metric in $\mathbb{C}^{n+1}$. Note that we may identify $e_0^\perp$ with the tangent space $T_{e_0}\mathbb{P}_n(\mathbb{C})$ to the complex manifold $\mathbb{P}_n(\mathbb{C})$ at $e_0 \in \mathbb{P}_n(\mathbb{C})$.

For every matrix $M \in \mathcal{M}_{n \times (n+1)}(\mathbb{C})$ of maximal rank, we may define a linear isomorphism $\ell_M := M\phi(M) : e_0^\perp \longrightarrow \mathbb{C}^n$.

Let us define a mapping $\psi_0 : \mathcal{M}_{n \times (n+1)}(\mathbb{C}) \longrightarrow L_0$ in the following terms. For every matrix $M \in \mathcal{M}_{n \times (n+1)}(\mathbb{C})$, we associate the system of homogeneous polynomial equations $\psi_0(M) \in L_0$ given by the equality:

$$\psi_0(M) := [X_0^{d_1-1} d_1^{1/2} \lambda_1(X_1, \ldots, X_n), \ldots, X_0^{d_n-1} d_n^{1/2} \lambda_n(X_1, \ldots, X_n)],$$

where $\ell_M := [\lambda_1, \ldots, \lambda_n] : e_0^\perp \longrightarrow \mathbb{C}^n$ is the linear mapping defined by the matrix $M\phi(M)$.

Define a mapping $G_{(d)} : Y \longrightarrow V_0$ in the following terms. For every $(t, h, M) \in Y$, let $G_{(d)}(t, h, M) \in V_0$ be the system of homogeneous polynomial equations given by the identity:

$$G_{(d)}(t, h, M) := \left(1 - \tau^2 t^{\frac{1}{n^2+n}} \right)^{1/2} \frac{\Delta^{-1} h}{||h||_2} + \tau t^{\frac{1}{2n^2+2n}} \psi_0 \left(\frac{M}{||M||_F} \right) \in V_0.$$

Finally, let $\mathcal{G}_{(d)}$ be the class defined by the identity:

$$\mathcal{G}_{(d)} = Im(G_{(d)}) \times \{e_0\} \subseteq \mathcal{H}_{(d)} \times \mathbb{P}_n(\mathbb{C}). \tag{1.2}$$

Note that $\mathcal{G}_{(d)}$ is included in the incidence variety and that all systems in $Im(G_{(d)})$ share a common zero e_0. Hence initial pairs in $(G, z) \in \mathcal{G}_{(d)}$ always use the same exact zero $z = e_0$. In particular, they are all solved by construction.

We assume that the set $\mathcal{G}_{(d)}$ is endowed with the pull–back probability distribution obtained from Y via $G_{(d)}$. Namely, in order to choose a random point in $\mathcal{G}_{(d)}$, we choose a random point $y \in Y$, and we compute $(G_{(d)}(y), e_0) \in \mathcal{G}_{(d)}$.

The following statement has been shown in Beltrán & Pardo (2005).

Theorem 1.5 (Main) *With the above notation, the class $\mathcal{G}_{(d)}$ defined by identity (1.2) is a questor set for efficient initial pairs in $\mathcal{H}_{(d)}$.*

More precisely, for every positive real number $\varepsilon > 0$, the probability that a randomly chosen data $(G, e_0) \in \mathcal{G}_{(d)}$ is ε–efficient is greater than

$$1 - \varepsilon.$$

Additionally, for these ε–efficient pairs $(G, e_0) \in \mathcal{G}_{(d)}$, the probability that a randomly chosen input $F \in \mathcal{H}_{(d)}$ is solved by NHD with initial data (G, e_0) performing $O(n^5 N^2 d^4 \varepsilon^{-2})$ steps is at least

$$1 - \varepsilon.$$

As usual, the existence of questor sets immediately yields a probabilistic algorithm. This is Theorem 1.3 above, which is an immediate consequence of Theorem 1.5. The following corollary shows how this statement applies.

Corollary 1.2 *There is a bounded error probability algorithm that solves most homogeneous systems of cubic equations (namely inputs are in $\mathcal{H}_{(3)}$) in time of order*

$$O(n^{13} \varepsilon^{-2}),$$

with probability greater than $1 - \varepsilon$.

Taking $\varepsilon = \frac{1}{n^4}$, for instance, this probabilistic algorithm solves a cubic homogeneous system in running time at most $O(n^{21})$ with probability greater than $1 - \frac{1}{n^4}$.

However, randomly choosing a pair $(G, e_0) \in \mathcal{G}_{(d)}$ is not exactly what a computer can perform. We need a discrete class of ε–efficient initial systems. This is achieved by the following argument (that follows those in Castro, San Martín & Pardo (2002,2003)).

Observe that $Y \subseteq \mathbb{R} \times \mathbb{C}^{N+1}$ may be seen to be a real semi–algebraic set under the identification $\mathbb{R} \times \mathbb{C}^{N+1} \equiv \mathbb{R}^{2N+3}$. Let $H \geq 0$ be a positive integer number. Let $\mathbb{Z}^{2N+3} \subseteq \mathbb{R}^{2N+3}$ be the lattice consisting of the integer points in $\mathbb{R}^{2N+3}$. Let Y^H be the set of points defined as follows:

$$Y^H := Y \cap \mathbb{Z}^{2N+3} \left[\frac{1}{H} \right],$$

where $\mathbb{Z}^{2N+3}[\frac{1}{H}]$ is the lattice given by the equality:

$$\mathbb{Z}^{2N+3}\left[\frac{1}{H}\right] := \left\{\frac{z}{H} : z \in \mathbb{Z}^{2N+3}\right\}.$$

Observe that $(4N + 6)(\log_2 H + 1)$ is a bound for the number of binary digits required to write any point $y \in Y^H$ in a computer.

For any positive real number $H > 0$, we denote by $\mathcal{G}_{(d)}^H \subseteq \mathcal{G}_{(d)}$ the finite set of points given by the equality:

$$\mathcal{G}_{(d)}^H := \{(G_{(d)}(y), e_0) : y \in Y^H\}.$$

We consider $\mathcal{G}_{(d)}^H$ endowed with the pull–back probability distribution obtained from Y^H via $G_{(d)}$. Namely, in order to choose a random point $(g, e_0) \in \mathcal{G}_{(d)}^H$, we choose a random point (uniform distribution) $y \in Y^H$ and we compute the point $(G_{(d)}(y), e_0) \in \mathcal{G}_{(d)}$. Then, the following statement also holds.

Theorem 1.6 *(Beltrán & Pardo (2005)) There exists a universal constant $C > 0$ such that for every two positive real numbers $\varepsilon > 0, H > 0$ satisfying*

$$\log_2 H \geq CnN^3 \log_2 d + 2\log_2 \varepsilon^{-1},$$

the following properties hold.

- *The probability (uniform distribution) that a randomly chosen data $(G, e_0) \in \mathcal{G}_{(d)}^H$ is ε–efficient is greater than*

$$1 - 2\varepsilon.$$

- *For ε–efficient initial pairs $(G, e_0) \in \mathcal{G}_{(d)}^H$, the probability that a randomly chosen input $F \in \mathcal{H}_{(d)}$ is solved by NHD with initial data (G, e_0) performing $O(n^5 N^2 d^4 \varepsilon^{-2})$ steps is at least*

$$1 - \varepsilon.$$

Theorem 1.5 and its consequences represent a small step forward in the theory introduced by Shub and Smale. It simply shows the existence of a true, although probabilistic, algorithm that computes partial information about solution varieties for most homogeneous systems of polynomial equations in time which depends polynomially on the input length.

However, things are not as optimistic as they appear. First of all, the algorithm we propose here is probabilistic and, hence, "uniform" as demanded in Smale (2000). Obviously, a deterministic version is also desirable. Nevertheless, we consider this a minor drawback that time and some investment of scientific effort will probably overcome.

The second drawback is more important. The algorithm, its efficiency and its probability of success depends upon the generic (dense) encoding of input polynomials. Namely, it is based on the full space $\mathcal{H}_{(d)}$ of input systems. This seems to be an overly mathematical working hypothesis. Note that *the aim of such kind of result is essentially that of explaining why computational numerical analysis methods run efficiently in real life computing, even when there is no well–founded reason proving their efficiency.*

In real life computing, problems modelled by polynomial equations have some structure and are a subset of the generic class of polynomials in dense encoding. Namely, real life problems provide inputs that belong to some particular subclasses of polynomial data (as unirational families of input systems given by regular mappings $\Phi : W \subseteq \mathbb{C}^m \longrightarrow \mathcal{H}_{(d)}$).

Theorem 1.5 states that $\mathcal{G}_{(d)}$ is a questor set of initial pairs for generic input data. However, it does not mean that $\mathcal{G}_{(d)}$ is also a questor set for input systems F in a unirational family of data like $Im(\Phi)$. As we claimed in Section 1.3, source dimension is usually much smaller than the dimension $N + 1$ of the space of input systems. Hence, $Im(\Phi)$ is commonly a set of measure zero and it may be unfortunately contained in the class of systems for which $\mathcal{G}_{(d)}$ $\mathcal{G}_{(d)}$ does not apply. Hence the question is *whether this algorithmic scheme (or anything inspired by these ideas) can be adapted to particular classes of input data.*

In order to deal with this open question, we need to reconsider most of the studies done by Shub and Smale on the generic case $\mathcal{H}_{(d)}$, this time applied to special subsets $Im(\Phi)$ of $\mathcal{H}_{(d)}$.

Theorem 1.5 owes much of its strength to the good behavior of the probability distribution of a condition number μ_{norm} in the full space of generic inputs $\mathcal{H}_{(d)}$. This is the main semantic invariant involved in the complexity of numerical analysis polynomial equation solvers (as remarked in Shub & Smale (1993b)).

Any answer to the above adaptability question requires a preliminary study on the behavior of the probability distribution of the condition number for nonlinear systems when restricted to submanifolds, subvarieties and particular subclasses of the generic space $\mathcal{H}_{(d)}$.

In the next section we illustrate some of the main difficulties that may arise in such a study. We deal with the adaptability question of a simpler problem. We study the probability distribution of the condition number associated with the class of singular matrices. It contains some of the main results of Beltrán & Pardo (2004a, 2004b).

1.5 The Distribution of the Condition Number of Singular Complex Matrices

Condition numbers in linear algebra were introduced in Turing (1948). They were also studied in Goldstine & von Neumann (1947) and Wilkinson (1965). Variations of these condition numbers may be found in the literature of numerical linear algebra (cf. Demmel (1988), Golub & van Loan (1996), Higham (2002), Trefethen & Bau (1997) and references therein).

A relevant breakthrough was the study of the probability distribution of these condition numbers. The paper Smale (1985) and, mainly, Edelman (1988, 1992) showed the exact values of the probability distribution of the condition number of dense complex matrices.

From a computational point of view, these statements can be translated into the following terms. Let $\mathcal{P}$ be a numerical analysis procedure whose space of input data is the space of arbitrary square complex matrices $\mathcal{M}_n(\mathbb{C})$. Then, Edelman's statements mean that *the probability that a randomly chosen dense matrix in $\mathcal{M}_n(\mathbb{C})$ is a well–conditioned input for $\mathcal{P}$ is high.*

Sometimes, however, we deal with procedures $\mathcal{P}$ whose input space is a proper subset $\mathcal{C} \subseteq \mathcal{M}_n(\mathbb{C})$. Additionally such procedures with particular data lead to particular condition numbers $\kappa_{\mathcal{C}}$ adapted both for the procedure $\mathcal{P}$ and the input space $\mathcal{C}$. Edelman's and Smale's results do not apply with these constraints. In Beltrán & Pardo (2004a, 2004b) we introduced a new technique to study the probability distribution of condition numbers $\kappa_{\mathcal{C}}$. Namely, we introduce a technique to exhibit upper bound estimates for the quantity

$$\frac{vol[\{A \in \mathcal{C} : \kappa_{\mathcal{C}}(A) > \varepsilon^{-1}\}]}{vol[\mathcal{C}]}, \tag{1.3}$$

where $\varepsilon > 0$ is a positive real number, and $vol[\cdot]$ is some suitable measure on the space $\mathcal{C}$ of acceptable inputs for $\mathcal{P}$.

As an example of how our technique applies, let $\mathcal{C} := \Sigma^{n-1} \subseteq \mathcal{M}_n(\mathbb{C})$ be the class of all singular complex matrices. In Kahan (2000) and Stewart & Sun (1990), a condition number for singular matrices $A \in \mathcal{C}$ is considered. This condition number measures the precision required to perform kernel computations. For every singular matrix $A \in \Sigma^{n-1}$ of corank 1, the condition number $\kappa_D^{n-1}(A) \in \mathbb{R}$ is defined by the identity

$$\kappa_D^{n-1}(A) := \|A\|_F \|A^\dagger\|_2,$$

where $\|A\|_F$ is the Frobenius norm of the matrix A, $A^\dagger$ is the Moore–

Penrose pseudo–inverse of A and $\|A^\dagger\|_2$ is the norm of $A^\dagger$ as a linear operator.

As Σ^{n-1} is a complex homogeneous hypersurface in $\mathcal{M}_n(\mathbb{C})$ (i.e., a cone of complex codimension 1), it is endowed with a natural volume vol induced by the $2(n^2 - 1)-$dimensional Hausdorff measure of its intersection with the unit disk. In Beltrán & Pardo (2004b), we prove the following statement.

Theorem 1.7 *With the same notation and assumptions as above, the following inequality holds:*

$$\frac{vol[A \in \Sigma^{n-1} : \kappa_D^{n-1}(A) > \varepsilon^{-1}]}{vol[\Sigma^{n-1}]} \leq 18n^{20}\varepsilon^6,$$

This statement is an (almost) immediate consequences of a wider class of results that we state below.

First of all, most condition numbers are by nature projective functions. For example, the classical condition number κ of numerical linear algebra is naturally defined as a function on the complex projective space $\mathbb{P}(\mathcal{M}_n(\mathbb{C}))$ defined by the complex vector space $\mathcal{M}_n(\mathbb{C})$. Namely, we may see κ as a function

$$\kappa : \mathbb{P}(\mathcal{M}_n(\mathbb{C})) \longrightarrow \mathbb{R}_+ \cup \infty.$$

Secondly, statements like the Schmidt–Mirsky–Eckart–Young Theorem (cf. Schmidt (1907), Eckart & Young (1936), Mirsky (1963)) imply that Smale's and Edelman's estimates are, in fact, estimates of the volume of a tube about a concrete projective algebraic variety in $\mathbb{P}(\mathcal{M}_n(\mathbb{C}))$.

In Beltrán & Pardo (2004b), we prove a general upper bound for the volume of a tube about any (possibly singular) complex projective algebraic variety (see Theorem 1.8 below).

Estimates on volumes of tubes is a classic topic that began with Weyl's Tube Formula for tubes in the affine space (cf. Weyl (1939)). Formulae for the volumes of some tubes about analytic submanifolds of complex projective spaces are due to Gray (2004) and references therein. However, Gray's results do not apply even in Smale's and Edelman's case. Nor do they apply to particular classes $\mathcal{C}$, as above. Firstly, Gray's statements are only valid for smooth submanifolds and not for singular varieties (as, for instance, Σ^{n-1}). Secondly, Gray's theorems are only valid for tubes of small enough radius (depending on intrinsic features of the manifold under consideration) which may become dramatically small in the presence of singularities. These two drawbacks motivated us to search for a general statement that may be stated as follows.

Let $d\nu_n$ be the volume form associated with the complex Riemannian structure of $\mathbb{P}_n(\mathbb{C})$. Let $V \subseteq \mathbb{P}_n(\mathbb{C})$ be any subset of the complex projective space and let $\varepsilon > 0$ be a positive real number. We define *the tube of radius ε about V in $\mathbb{P}_n(\mathbb{C})$* as the subset $V_\varepsilon \subseteq \mathbb{P}_n(\mathbb{C})$ given by the identity:

$$V_\varepsilon := \{x \in \mathbb{P}_n(\mathbb{C}) : d_{\mathbf{P}}(x, V) < \varepsilon\},$$

where $d_{\mathbf{P}}(x, y) := \sin d_R(x, y)$ and $d_R : \mathbb{P}_n(\mathbb{C})^2 \longrightarrow \mathbb{R}$ is the Fubini–Study distance.

Theorem 1.8 *Let $V \subseteq \mathbb{P}_n(\mathbb{C})$ be a (possibly singular) equidimensional complex algebraic variety of (complex) codimension r in $\mathbb{P}_n(\mathbb{C})$. Let $0 < \varepsilon \leq 1$ be a positive real number. Then, the following inequality holds*

$$\frac{\nu_n[V_\varepsilon]}{\nu_n[\mathbb{P}_n(\mathbb{C})]} \leq 2 \deg(V) \left(\frac{e\, n\, \varepsilon}{r}\right)^{2r},$$

where $\deg(V)$ is the degree of V.

This theorem can be applied to Edelman's conditions to obtain the estimate:

$$\frac{vol[\{A \in \mathcal{M}_n(\mathbb{C}) : \kappa_D(A) > \varepsilon^{-1}\}]}{vol[\mathcal{M}_n(\mathbb{C})]} \leq 2e^2 n^5 \varepsilon^2,$$

where $\kappa_D(A) := \|A\|_F \|A^{-1}\|_2$, and *vol* is the standard Gaussian measure in $\mathbb{C}^{n^2}$. We also prove that the constants on the right–hand side of the inequality in Theorem 1.8 are essentially optimal.

The reader will observe that our bound is less sharp than Edelman's or Smale's bounds, although it is a particular instance of a more general statement.

Next, observe that neither Smale's nor Edelman's results nor Theorem 1.8 exhibit upper bounds on the probability distribution described in equation (1.3). In particular, they do not apply to prove Theorem 1.7. In order to deal with this kind of estimate, *we need an upper bound for the volume of the intersection of an extrinsic tube with a proper subvariety*. This is our main result from Beltrán & Pardo (2004b) and is contained in the following statement.

Theorem 1.9 *Let $V, V' \subseteq \mathbb{P}_n(\mathbb{C})$ be two projective equidimensional algebraic varieties of respective dimensions $m > m' \geq 1$. Let $0 < \varepsilon \leq 1$ be a positive real number. With the same notation as in Theorem 1.8, the* *following inequality holds:*

$$\frac{\nu_m[V_\varepsilon' \cap V]}{\nu_m[V]} \le c \deg(V')n \binom{n}{m'}^2 \left[e\frac{n-m'}{m-m'}\,\varepsilon\right]^{2(m-m')},$$

where $c \le 4e^{1/3}\pi$, ν_m is the $2m-$dimensional natural measure in the algebraic variety V, and $\deg(V')$ is the degree of V'.

Now, observe that the projective point defined by a corank 1 matrix $A \in \Sigma^{n-1} \subseteq \mathbb{P}(\mathcal{M}_n(\mathbb{C}))$ satisfies:

$$\kappa_d^{n-1}(A) := \frac{1}{d_\mathbf{P}(A, \Sigma^{n-2})},$$

where $\Sigma^{n-2} \subseteq \mathbb{P}(\mathcal{M}_n(\mathbb{C}))$ is the projective algebraic variety of all complex matrices of corank at least 2, $d_\mathbf{P}(A, \Sigma^{n-2}) := \sin d_R(A, \Sigma^{n-2})$, and d_R is the Fubini–Study distance in the complex projective space $\mathbb{P}(\mathcal{M}_n(\mathbb{C}))$.

Hence, Theorem 1.7 becomes an immediate consequence of Theorem 1.9. In Beltrán & Pardo (2004b) other examples of applications of Theorem 1.9 are shown, including the stratification by corank of the space of complex matrices $\mathcal{M}_n(\mathbb{C})$ and the corresponding condition number.

This is just an example of how research on the adaptability question (discussed in Section 1.4) can be initiated. It is, however, far from being an appropriate treatment for the adaptability question in either linear algebra or in the nonlinear case. Future advances in this direction are required.

Acknowledgements This research was partially supported by the Spanish grant MTM2004-01167. The authors gratefully acknowledge the valuable suggestions for improving the manuscript which were made by the editors.

References

B. Bank, M. Giusti, J. Heintz and G. Mbakop (1997), 'Polar varieties, real equation solving and data structures: the hypersurface case', *J. of Complexity* **13**, 5–27.

B. Bank, M. Giusti, J. Heintz and G. Mbakop (2001), 'Polar varieties and efficient real elimination', *Math. Zeit.* **238**, 115–144.

B. Bank, M. Giusti, J. Heintz and L. M. Pardo (2004), 'Generalized polar varieties and an efficient real elimination procedure', *Kybernetika* **40**, 519–550.

B. Bank, M. Giusti, J. Heintz and L. M. Pardo (2005), 'Generalized polar varieties: Geometry and algorithms', *J. of Complexity* **21**, 377–412.

T. Becker and V. Weispfenning (1993), *Groebner Bases: A Computational Approach to Commutative Algebra*, Volume 141 of *Grad. Texts in Maths*, Springer Verlag.

C. Beltrán and L. M. Pardo (2004a), 'Upper bounds on the distribution of the condition number of singular matrices', *Comptes Rendus Mathématique* **340**, 915–919.

C. Beltrán and L. M. Pardo (2004b), 'Estimates on the probability distribution of the condition number of singular matrices', submitted to *Found. of Comput. Math.* .

C. Beltrán and L. M. Pardo (2005), 'On Smale's 17th Problem: probabilistic polynomial time', manuscript.

C. Berenstein and A. Yger (1991), 'Effective Bézout identities in $\mathbb{Q}[X_1, \ldots, X_n]$', *Acta Math.* **166**, 69–120.

C. Berenstein and A. Yger (1991a), 'Une formule de Jacobi et ses conséquences', *Ann. Sci. E.N.S.* **24**, 363–377.

L. Blum, F. Cucker, M. Shub and S. Smale (1998), *Complexity and Real Computation*, Springer Verlag, New York.

L. Blum, M. Shub and S. Smale (1989), 'On a theory of computation and complexity over the real numbers: **NP**-completeness, recursive functions and universal machines', *Bulletin of the Amer. Math. Soc.* **21**, 1–46.

J.-B. Bost, H. Gillet, and C. Soulé (1994), 'Heights of projective varieties and positive green forms', *J. Amer. Math. Soc.* **7**, 903–1027.

W. D. Brownawell (1987), 'Bounds for the degree in the Nullstellensatz', *Annals of Math.* **126**, 577–591.

B. Buchberger (1965), *Ein Algoritmus zum Auffinden der Basiselemente des Restklassenringes nach einen nulldimensionallen Polynomideal*, Leopold–Franzens Universität, Innsbruck, Ph.D. Thesis.

L. Caniglia, A. Galligo and J. Heintz (1988), 'Borne simplement exponentielle pour les degrés dans le théorème des zéros sur un corps de caractéristique quelconque', *C.R. Acad. Sci. Paris* **307**, Série I, 255–258.

D. Castro, K. Hägele, J. E. Morais and L. M. Pardo (2001), 'Kronecker's and Newton's approaches to solving: a first comparison', *J. Complexity* **17**, 212–303.

D. Castro, M. Giusti, J. Heintz, G. Matera and L. M. Pardo (2003), 'The hardness of polynomial equation solving', *Found. Comput. Math.* **3**, 347–420.

D. Castro, J.L. Montaña, J. San Martín and L. M. Pardo (2002), 'The distribution of condition numbers of rational data of bounded bit length', *Found. Comput. Math.* **2**, 1–52.

D. Castro, J. San Martín and L. M. Pardo (2003), 'Systems of rational polynomial equations have polynomial size approximate zeros on the average', *J. Complexity* **19**, 161–209.

D. Cox, J. Little, D. O'Shea (1997), *Ideals, Varieties, and Algorithms,*Springer Verlag, New York.

J. P. Dedieu (2001), 'Newton's method and some complexity aspects of the zero–finding problem', in *Foundations of Computational Mathematics (Oxford, 1999)*, Volume 284 of *London Math. Soc. Lecture Note Ser.*, 45–67, Cambridge Univ. Press, Cambridge.

J. P. Dedieu (2003), *Points Fixes, Zéros et la Méthode de Newton,* manuscript, Univ. Paul Sabatier.

J. W. Demmel (1988), 'The probability that a numerical analysis problem is difficult', *Math. Comp.* **50**, 449–480.

C. Eckart, G. Young (1936), 'The approximation of one matrix by another of lower rank', *Psychometrika* **1**, 211–218.

A. Edelman (1988), 'Eigenvalues and condition numbers of random matrices', *SIAM J. Matrix Anal. Appl.* **9**, 543–560.

A. Edelman (1992), 'On the distribution of a scaled condition number', *Math. Comp.* **58**, 185–190.

M. R. Garey and D. S. Johnson (1979), *Computers and Intractability: A Guide to the Theory of NP–Completness,* W.H. Freeman, San Francisco, California.

M. Giusti, K. Hägele, J. Heintz, J. L. Montaña, J. E. Morais and L. M. Pardo (1997), 'Lower bounds for diophantine approximations', *J. Pure Appl. Algebra* **117& 118**, 277–317.

M. Giusti and J. Heintz (1991), 'Algorithmes – disons rapides – pour la décomposition d' une variété algébrique en composantes irréductibles et équidimensionelles', in *Proceedings of MEGA'90,* Volume 94 of Progress in Mathematics, 169–194, ed. T. Mora and C. Traverso, Birkhäuser Verlag, Basel.

M. Giusti and J. Heintz (1993), 'La détermination des points isolés et de la dimension d'une variété algébrique peut se faire en temps polynomial', in *Computational Algebraic Geometry and Commutative Algebra*, volume XXXIV of *Symposia Matematica*, 216–256, ed. D. Eisenbud and L. Robbiano, Cambridge University Press, Cambridge.

M. Giusti, J. Heintz, J. E. Morais, J. Morgenstern and L. M. Pardo (1998), 'Straight–line programs in geometric elimination theory', *J. Pure Appl. Algebra* **124**, 101–146.

M. Giusti, J. Heintz, J. E. Morais and L. M. Pardo (1995). 'When polynomial equation systems can be "solved" fast?', in *Applied Algebra, Algebraic Algorithms and Error-Correcting Codes (Paris, 1995)*, Volume 948 of *Lecture Notes in Comput. Sci.*, 205–231, ed. A. Cohen, M. Giusti and T. Mora, Springer Verlag, Berlin.

M. Giusti, J. Heintz, J. E. Morais and L. M. Pardo (1997), 'Le rôle des structures de données dans les problèmes d'élimination', *C. R. Acad. Sci. Paris, Sér. I Math.* **325**, 1223–1228.

M. Giusti, G. Lecerf and B. Salvy (2001), 'A Gröbner free alternative for polynomial system solving', *J. of Complexity* **17**, 154–211.

M. Giusti and E. Schost (1999), 'Solving some over–determined systems', in *Proceedings of the 1999 International Symposium on Symbolic and Algebraic Computation, ISSAC'99*, 1–8, ed. S. Dooley, ACM Press, New York.

H. H. Goldstine and J. von Neumann (1947), 'Numerical inverting of matrices of high order', *Bull. Amer. Math. Soc.* **53**, 1021–1099.

G. H. Golub and C. F. Van Loan (1996), *Matrix Computations*, 3rd Edition, *Johns Hopkins Studies in the Mathematical Sciences*, Johns Hopkins University Press, Baltimore, MD.

A. Gray (2004), *Tubes*, 2dn Edition, Volume 221 of *Progress in Mathematics* , Birkhäuser Verlag, Basel.

K. Hägele, J. E. Morais, L. M. Pardo and M. Sombra (2000), 'The intrinsic complexity of the Arithmetic Nullstellensatz', *J. of Pure and App. Algebra* **146**, 103–183.

J. Heintz (1983), 'Definability and fast quantifier elimination in algebraically closed fields'. *Theoret. Comput. Sci.* **24**, 239–27.

J. Heintz, G. Matera, L. M. Pardo and R. Wachenchauzer (1998), 'The intrinsic complexity of parametric elimination methods', *Electron. J. SADIO* **1**, 37–51.

J. Heintz, G. Matera and A. Weissbein (2001), 'On the time–space complexity of geometric elimination procedures', *App. Algebra Eng. Comm. Comput.* **11**, 239–296.

J. Heintz and J. Morgenstern (1993), 'On the intrinsic complexity of elimination theory', *J. of Complexity* **9**, 471–498.

J. Heintz and C. P. Schnorr (1980), 'Testing Polynomials which are easy to compute', in *Logic and Algorithmic (an International Symposium in honour of Ernst Specker)*, Monographie n. 30 de *l'Enseignement Mathématique*, 237–254. A preliminary version appeared in *Proc. 12th Ann. ACM Symp. on Computing* (1980) 262–268.

G. Hermann (1926). 'Die Frage der endlich vielen Schritte in der Theorie der Polynomideale', *Math. Ann.* **95**, 736–788.

N. J. Higham (2002), *Accuracy and Stability of Numerical Algorithms*, 2nd Edition, SIAM, Philadelphia, PA.

D. Hilbert (1890), 'Über theorie der Algebraischen Formen', *Math. Ann.* **36**, 473–534.

H. Hironaka (1964), 'Resolution of singularities of an algebraic variety over a field of characteristic zero', *Annals of Math.* **79**, I : 109–203, II : 205–326.

W. Kahan (2000), *Huge Generalized Inverses of Rank-Deficient Matrices,* unpublished manuscript.

M. H. Kim (1988), 'On approximate zeros and root finding algorithms for a complex polynomial', *Math. Comp.* **51**, 707–719.

M. H. Kim (1989), 'Topological complexity of a root finding algorithm', *J. Complexity* **5**, 331–344.

P. Koiran (1996), 'Hilbert's Nullstellensatz is in the Polynomial Hierarchy', *J. of Complexity* **12**, 273–286.

J. Kollár (1988), 'Sharp Effective Nullstellensatz', *J. of Amer. Math. Soc.* **1**, 963–975.

J. König (1903), *Einleitung in die allgemeine Theorie der algebraischen Grözen*, Druck und Verlag von B.G. Teubner, Leipzig.

T. Krick and L. M. Pardo (1996), 'A computational method for diophantine approximation', in *Algorithms in Algebraic Geometry and Applications, Proceedings of MEGA'94*, Volume 143 of *Progress in Mathematics*, 193–254, ed. T. Recio, L. Gonzalez–Vega, Birkhäuser Verlag, Basel.

T. Krick, L. M. Pardo and M. Sombra (2001), 'Sharp estimates for the Arithmetic Nullstellensatz', *Duke Math. Journal* **109**, 521–598.

L. Kronecker (1882), 'Grundzüge einer arithmetischen theorie de algebraischen grössen', *J. Reine Angew. Math.* **92**, 1–122.

G. Lecerf (2001), *Une Alternative aux Méthodes de Réécriture pour la Résolution des Systèmes Algébriques*, École Polytechnique, Thèse.

F. S. Macaulay (1916), *The Algebraic Theory of Modular Systems*, Cambridge University Press, Cambridge.

G. Malajovich (1994), 'On generalized Newton algorithms: quadratic convergence, path–following and error analysis', *Theoret. Comput. Sci.* **133**, 65–84.

G. Malajovich and M. J. Rojas (2002), 'Polynomial systems and the momentum map', in *Foundations of Computational Mathematics (Hong Kong, 2000)*, 251–266, World Sci. Publishing, River Edge, NJ.

E. Mayr and A. Meyer (1982) 'The complexity of the word problem for commutative semigroups', *Advances in Math.* **46**, 305–329.

L. Mirsky (1963), 'Results and problems in the theory of doubly-stochastic matrices', *Z. Wahrscheinlichkeitstheorie und Verw. Gebiete* **1**, 319–334.

J. L. Montaña, J. E. Morais and L. M. Pardo (1996), 'Lower bounds for arithmetic networks II : Sum of Betti numbers', *Appl. Algebra Engrg. Comm. Comput.* **7**, 41–51.

T. Mora (2003), *Solving Polynomial Equation Systems. I. The Kronecker-Duval Philosophy*, Volume 88 of *Encyclopedia of Mathematics and its Applications*, Cambridge University Press, Cambridge.

P. Philippon (1991), 'Sur des hauteurs alternatives, I', *Math. Ann.* **289**, 255–283.

P. Philippon (1994), 'Sur des hauteurs alternatives, II', *Ann. Inst. Fourier, Grenoble* **44**, 1043–1065.

P. Philippon (1995), 'Sur des hauteurs alternatives, III', *J. Math. Pures Appl.* **74**, 345–365.

L. M. Pardo (1995), 'How lower and upper complexity bounds meet in elimination theory', in *Applied Algebra, Algebraic Algorithms and Error–Correcting Codes (Paris, 1995)*, Volume 948 of *Lecture Notes in Comput. Sci.*, 33–69, ed. A. Cohen, M. Giusti and T. Mora, Springer Verlag, Berlin.

L. M. Pardo (2000), 'Universal Elimination requires exponential running time', in *Actas EACA'2000*, 25–50, ed. A. Montes, UPC, Barcelona.

M. Rojas (2001), 'Computational arithmetic geometry I: Diophantine sentences nearly within the Polynomial Hierarchy', *J. of Comput and Syst. Sci.* **62**, 216–235.

M. J. Rojas (2003), 'Dedekind Zeta functions and the complexity of Hilbert's Nullstellensatz', Math ArXiV preprint math.NT/0301111.

J. San Martín and L. M. Pardo (2004), 'Deformation techniques that solve generalized Pham systems', *Theoret. Comput. Sci.* **315**, 593–625.

E. Schmidt (1907), 'Zur Theorie der linearen und nichtlinearen Integralgleichungen. I Tiel. Entwicklung willkÜrlichen Funktionen nach System vorgeschriebener', *Math. Annalen* **63**, 433–476.

M. Shub (1993), 'Some remarks on Bezout's theorem and complexity theory', in *From Topology to Computation: Proceedings of the Smalefest (Berkeley, CA, 1990)*, 443–455, Springer Verlag, New York.

M. Shub and S. Smale (1993a), 'Complexity of Bézout's theorem I: Geometric aspects', *J. Am. Math. Soc.* **6**, 459–501.

M. Shub and S. Smale (1993b), 'Complexity of Bezout's theorem. II. Volumes and probabilities', in *Computational Algebraic Geometry, Proc. MEGA'92*, Volume 109 of *Progress in Mathematics*, 267–285, Birkhäuser Verlag, Basel.

M. Shub and S. Smale (1993c), 'Complexity of Bezout's theorem. III. Condition number and packing', *J. Complexity* **9**, 4–14.

M. Shub and S. Smale (1994), 'Complexity of Bezout's theorem. V. Polynomial time', *Theoret. Comput. Sci.* **133**, 141–164.

M. Shub and S. Smale (1996), 'Complexity of Bezout's theorem. IV. Probability of success; extensions', *SIAM J. Numer. Anal.* **33**, 128–148.

S. Smale (1985), 'On the efficiency of algorithms of analysis', *Bull. Amer. Math. Soc. (N.S.)* **13**, 87–121.

S. Smale (2000), 'Mathematical problems for the next century', in *Mathematics: Frontiers and Perspectives*, 271–294, ed. V. Arnold, M. Atiyah, P. Lax and B. Mazur, Amer. Math. Soc., Providence, RI.

G. W. Stewart and J. G. Sun (1990), *Matrix Perturbation Theory*, Computer Science and Scientific Computing, Academic Press, Boston, MA.

B. Sturmfels (1996), *Gröbner Bases and Convex Polytopes* , Volume 8 of *University Lecture Series* Amer. Math. Soc., Providence, RI.

J. Traub and A. G. Werschultz (1998), *Complexity and Information*, Lezioni Lincee, Cambridge University Press, Cambridge.

L. N. Trefethen and D. Bau (1997), *Numerical Linear Algebra*, SIAM, Philadelphia, PA.

A. M. Turing (1948), 'Rounding–off errors in matrix processes', *Quart. J. Mech. Appl. Math.* **1**, 287–308.

W. V. Vasconcelos (1998), *Computational Methods in Commutative Algebra and Algebraic Geometry*, Volume 2 of *Algorithms and Computation in Mathematics*, Springer Verlag, Berlin.

J. Verschelde (2000), 'Toric Newton method for polynomial homotopies', *J. of Symb. Comput.* **29**, 1265–1287.

A. Weyl (1939), 'On the volume of tubes', *Amer. J. Math* **61**, 461–472.

J. H. Wilkinson (1965), *The Algebraic Eigenvalue Problem*, Clarendon Press, Oxford.

J. C. Yakoubsohn (1995), 'A universal constant for the convergence of Newton's method and an application to the classical homotopy method', *Numer. Algorithms* **9**, 223–244.

O. Zariski (1995), *Algebraic Surfaces*, Springer Verlag, Berlin.

2

Toward Accurate Polynomial Evaluation in Rounded Arithmetic

James Demmel[a]

Mathematics Department and CS Division
University of California, Berkeley, CA
e-mail: demmel@cs.berkeley.edu

Ioana Dumitriu[b]

Mathematics Department
University of California, Berkeley, CA
e-mail: dumitriu@math.berkeley.edu

Olga Holtz

Mathematics Department
University of California, Berkeley, CA
e-mail: holtz@math.berkeley.edu

[a] The author acknowledges support of NSF under grants CCF-0444486, ACI-00090127, CNS-0325873 and of DOE under grant DE-FC02-01ER25478.
[b] The author acknowledges support of the Miller Institute for Basic Research in Science.

Abstract

Given a multivariate real (or complex) polynomial p and a domain $\mathcal{D}$, we would like to decide whether an algorithm exists to evaluate $p(x)$ accurately for all $x \in \mathcal{D}$ using rounded real (or complex) arithmetic. Here "accurately" means with relative error less than 1, i.e., with some correct leading digits. The answer depends on the model of rounded arithmetic: We assume that for any arithmetic operator $op(a, b)$, for example $a + b$ or $a \cdot b$, its computed value is $op(a, b) \cdot (1 + \delta)$, where $|\delta|$ is bounded by some constant ϵ where $0 < \epsilon \ll 1$, but δ is otherwise arbitrary. This model is the traditional one used to analyze the accuracy of floating point algorithms.

Our ultimate goal is to establish a decision procedure that, for any p and $\mathcal{D}$, either exhibits an accurate algorithm or proves that none exists. In contrast to the case where numbers are stored and manipulated as finite bit strings (e.g., as floating point numbers or rational numbers) we show that some polynomials p are impossible to evaluate accurately. The existence of an accurate algorithm will depend not just on p and $\mathcal{D}$, but on which arithmetic operators are available (perhaps beyond $+$, $-$, and $\cdot$), which constants are available to the algorithm (integers, algebraic numbers, ...), and whether branching is permitted in the algorithm. For floating point computation, our model can be used to identify which accurate operators beyond $+$, $-$ and $\cdot$ (e.g., dot products, 3x3 determinants, ...) are necessary to evaluate a particular $p(x)$.

Toward this goal, we present necessary conditions on p for it to be accurately evaluable on open real or complex domains $\mathcal{D}$. We also give sufficient conditions, and describe progress toward a complete decision procedure. We do present a complete decision procedure for homogeneous polynomials p with integer coefficients, $\mathcal{D} = \mathbb{C}^n$, and using only the arithmetic operations $+$, $-$ and $\cdot$.

2.1 Introduction

Let $x = (x_1, \ldots, x_n)$ be a vector of real (or complex) numbers, let $p(x)$ denote a multivariate polynomial, and let $\mathcal{D}$ be a subset of $\mathbb{R}^n$ (or $\mathbb{C}^n$). We would ideally like to evaluate $p(x)$ *accurately* for all $x \in \mathcal{D}$, despite any rounding errors in arithmetic operations. The nature of the problem depends on how we measure accuracy, what kinds of rounding errors we consider, the class of polynomials $p(x)$, the domain $\mathcal{D}$, and what operations and constants our evaluation algorithms may use. Depending on these choices, an accurate algorithm for evaluating $p(x)$ may or may not exist. Our ultimate goal is a decision procedure that will either exhibit an algorithm that evaluates $p(x)$ accurately for all $x \in \mathcal{D}$, or else exhibit a proof that no such algorithm exists.

By *accuracy*, we mean that we compute an approximation $p_{comp}(x)$ to $p(x)$ that has small *relative* error: $|p(x) - p_{comp}(x)| \leq \eta |p(x)|$ for some desired $0 < \eta < 1$. In particular, $\eta < 1$ implies that $p(x) = 0$ if and only if $p_{comp}(x) = 0$. This requirement that p and p_{comp} define the same variety will be crucial in our development. We justify this definition of accuracy in more detail in Section 2.

Our motivation for this work is two-fold. First, it is common for numerical analysts to seek accurate formulas for particularly important or common expressions. For example, in computational geometry and mesh generation, certain geometric predicates like "Is point x inside, on or outside circle C?" are expressed as multivariate polynomials $p(\cdot)$ whose signs determine the answer; the correctness of the algorithms depends critically on the correctness of the sign, which is in turn guaranteed by having a relative error less than 1 in the value of $p(\cdot)$ [Shewchuk (1997)]. We would like to automate the process of finding such formulas.

The second motivation is based on recent work of Koev and one of the authors [Demmel and Koev (2004a)] which identified several classes of structured matrices (e.g., Vandermonde, Cauchy, totally positive, certain discretized elliptic partial differential equations, ...) for which algorithms exist to accurately perform some (or all) computations from linear algebra: determinants, inversion, Gaussian elimination, computing singular values, computing eigenvalues, and so on. The proliferation of these classes of structured matrices led us to ask what common algebraic structure these matrix classes possess that made these accurate algorithms possible. This paper gives a partial answer to this question; see Section 2.6.

Now we consider our model of rounded arithmetic. Let $op(\cdot)$ denote a *basic arithmetic operation*, for example $op(x, y) = x + y$ or $op(x, y, z) = x + y \cdot z$. Then we assume that the rounded value of $op(\cdot)$, which we denote $rnd(op(\cdot))$, satisfies

$$rnd(op(\cdot)) = op(\cdot)(1 + \delta) \tag{2.1}$$

where we call δ the *rounding error*. We assume only that $|\delta|$ is tiny, $|\delta| \leq \epsilon$, where $0 < \epsilon < 1$ and typically $\epsilon \ll 1$; otherwise δ is an arbitrary real (or complex) number. The constant ϵ is called the *machine precision*, by analogy to floating point computation, since this model is the traditional one used to analyze the accuracy of floating point algorithms [Higham (2002), Wilkinson (1963)].

To illustrate the obstacles to accurate evaluation that this model poses, consider evaluating $p(x) = x_1 + x_2 + x_3$ in the most straightforward way: add (and round) x_1 and x_2, and then add (and round) x_3. If we let δ_1 be the first rounding error and δ_2 be the second rounding error, we get the computed value $p_{comp}(x) = ((x_1 + x_2)(1 + \delta_1) + x_3)(1 + \delta_2)$. To see that this algorithm is not accurate, simply choose $x_1 = x_2 = 1$ and $x_3 = -2$ (so $p(x) = 0$) and $\delta_1 \neq 0$. Then $p_{comp}(1, 1, -2) = 2\delta_1(1 + \delta_2) \neq 0$, so the relative error is infinite. Indeed, it can be shown that there is an open set of x and of (δ_1, δ_2) where the relative error is large, so that this loss of accuracy occurs on a "large" set. We will see that unless $x_1 + x_2 + x_3$ is itself a basic arithmetic operation, or unless the variety $\{x_1 + x_2 + x_3 = 0\}$ is otherwise constructible from varieties derived from basic operations as described in Theorem 2.9, then *no* algorithm exists to evaluate $x_1 + x_2 + x_3$ accurately for all arguments.

In contrast, if we were to assume that the x_i and coefficients of p were given as exact rational numbers (e.g., as floating point numbers), then by performing integer arithmetic with sufficiently large integers it would clearly be a straightforward matter to evaluate any $p(x)$ as an exact rational number. (One could also use floating point arithmetic to accomplish this; see Sections 2.2.3 and 2.2.8.) In other words, accurate evaluation is always possible, and the only question is cost. Our model addresses this by identifying which composite operations have to be provided with high precision in order to evaluate $p(x)$ accurately. For further discussion of the challenge of evaluating a simple polynomial like $x_1 + x_2 + x_3$ accurately, see Section 2.2.8.

We give some examples to illustrate our results. Consider the family of homogeneous polynomials

$$M_{jk}(x) = j \cdot x_3^6 + x_1^2 \cdot x_2^2 \cdot (j \cdot x_1^2 + j \cdot x_2^2 - k \cdot x_3^2)$$

where j and k are positive integers, $\mathcal{D} = \mathbb{R}^n$, and we allow only addition, subtraction and multiplication of two arguments as basic arithmetic operations, along with comparisons and branching.

- When $k/j < 3$, $M_{jk}(x)$ is *positive definite*, i.e., zero only at the origin and positive elsewhere. This will mean that $M_{jk}(x)$ is easy to evaluate accurately using a simple method discussed in Section 2.3.
- When $k/j > 3$, then we will show that $M_{jk}(x)$ cannot be evaluated accurately by *any* algorithm using only addition, subtraction and multiplication of two arguments. This will follow from a simple necessary condition on the real variety $V_{\mathbb{R}}(M_{jk})$, the set of real x where $M_{jk}(x) = 0$, see Theorem 2.3. This theorem requires the real variety $V_{\mathbb{R}}(p)$ (or the complex variety $V_{\mathbb{C}}(p)$) to lie in a certain explicitly given finite set of varieties called *allowable varieties* in order to be able to evaluate $p(x)$ accurately in real arithmetic (or in complex arithmetic, resp.).
- When $k/j = 3$, i.e., on the boundary between the above two cases, $M_{jk}(x)$ is a multiple of the Motzkin polynomial [Reznick (2000)]. Its real variety $V_{\mathbb{R}}(M_{jk}) = \{x : |x_1| = |x_2| = |x_3|\}$ satisfies the necessary condition of Theorem 2.3, and the simplest accurate algorithm to evaluate it that we know is shown (in part) below:

$$\text{if} \quad |x_1 - x_3| \le |x_1 + x_3| \wedge |x_2 - x_3| \le |x_2 + x_3| \quad \text{then}$$

$$\begin{aligned}
p = {} & x_3^4 \cdot [4((x_1 - x_3)^2 + (x_2 - x_3)^2 + (x_1 - x_3)(x_2 - x_3))] \\
& + x_3^3 \cdot [2(2(x_1 - x_3)^3 + 5(x_2 - x_3)(x_1 - x_3)^2 \\
& \quad + 5(x_2 - x_3)^2(x_1 - x_3) + 2(x_2 - x_3)^3)] \\
& + x_3^2 \cdot [(x_1 - x_3)^4 + 8(x_2 - x_3)(x_1 - x_3)^3 \\
& \quad + 9(x_2 - x_3)^2(x_1 - x_3)^2 \\
& \quad + 8(x_2 - x_3)^3(x_1 - x_3) + (x_2 - x_3)^4] \\
& + x_3 \cdot [2(x_2 - x_3)(x_1 - x_3)((x_1 - x_3)^3 \\
& \quad + 2(x_2 - x_3)(x_1 - x_3)^2 \\
& \quad + 2(x_2 - x_3)^2(x_1 - x_3) + (x_2 - x_3)^3)] \\
& + (x_2 - x_3)^2(x_1 - x_3)^2((x_1 - x_3)^2 + (x_2 - x_3)^2)
\end{aligned}$$

$$p = j \cdot p$$

$$\text{else} \quad \dots \text{ 7 more analogous cases.}$$

In general, for a Motzkin polynomial in n real variables ($n = 3$ above), the algorithm has 2^n separate cases. Just n tests and branches are needed to choose the correct case for any input x, so that the cost of running the algorithm is still just a polynomial function of n for any particular x.

In contrast to the real case, when $\mathcal{D} = \mathbb{C}^n$ then Theorem 2.3 will show that $M_{jk}(x)$ is not accurately evaluable using only addition, subtraction and multiplication.

If we still want to evaluate $M_{jk}(x)$ accurately in one of the cases where addition, subtraction and multiplication alone do not suffice, it is natural to ask which composite or "black-box" operations we would need to implement accurately to do so. Section 2.5 addresses this question.

The necessary condition for accurate evaluability of $p(x)$ in Theorem 2.3 depends only on the variety of $p(x)$. The next example shows that the variety alone is not enough to determine accurate evaluability, at least in the real case. Consider the two irreducible, homogeneous, degree $2d$, real polynomials

$$p_i(x) = \left(x_1^{2d} + x_2^{2d}\right) + \left(x_1^2 + x_2^2\right)(q_i(x_3, \ldots, x_n))^2 \quad \text{for } i = 1,\, 2 \qquad (2.2)$$

where $q_i(\cdot)$ is a homogeneous polynomial of degree $d - 1$. Both $p_1(x)$ and $p_2(x)$ have the same real variety $V_{\mathbb{R}}(p_1) = V_{\mathbb{R}}(p_2) = \{x : x_1 = x_2 = 0\}$, which is allowable, i.e., satisfies the necessary condition for accurate evaluability in Theorem 2.3. However, near $x_1 = x_2 = 0$, $p_i(x)$ is "dominated" by $(x_1^2 + x_2^2)(q_i(x_3, \ldots, x_n))^2$, so accurate evaluability of $p_i(x)$ in turn depends on accurate evaluability of $q_i(x_3, \ldots, x_n)$. Since $q_1(\cdot)$ may be accurately evaluable while $q_2(\cdot)$ is not, we see that $V_{\mathbb{R}}(p_i)$ alone cannot determine whether $p_i(x)$ is accurately evaluable. Applying the same principle to $q_i(\cdot)$, we see that any decision procedure must be recursive, expanding $p_i(x)$ near the components of its variety and so on. We show current progress toward a decision procedure in Section 2.4.3. In particular, Theorem 2.7 shows that, at least for algorithms without branching, being able to compute dominant terms of p (suitably defined) accurately on $\mathbb{R}^n$ is a necessary condition for computing p accurately on $\mathbb{R}^n$. Furthermore, Theorem 2.8 shows that accurate evaluability of the dominant terms, along with branching, is sufficient to evaluate p accurately.

In contrast to the real case, Theorem 2.5 shows that for the complex case, with $\mathcal{D} = \mathbb{C}^n$, and using only addition, subtraction and multiplication of two arguments, a homogeneous polynomial $p(x)$ with integer coefficients is accurately evaluable if and only if it satisfies the necessary condition of Theorem 2.3. More concretely, $p(x)$ is accurately evaluable for all $x \in \mathbb{C}^n$

if and only if $p(x)$ can be completely factored into a product of factors of the form x_i, $x_i + x_j$ and $x_i - x_j$.

The results described so far from Section 2.4 consider only addition, subtraction, multiplication and (exact) negation (which we call *classical arithmetic*). Section 2.5 considers the same questions when accurate *black-box* operations beyond addition, subtraction and multiplication are permitted, such as fused-multiply-add [Montoye, Hokenek, and Runyon (1990)], or indeed any collection of polynomials at all (e.g., dot products, 3x3 determinants, ...). The necessary condition on the variety of p from Theorem 2.3 is generalized to black-boxes in Theorem 2.9, and the sufficient conditions Theorem 2.5 in the complex case are generalized in Theorems 2.12 and 2.13.

The rest of this paper is organized as follows. Section 2.2 discusses further details of our algorithmic model, explains why it is a useful model of floating point computation, and otherwise justifies the choices we have made in this paper. Section 2.3 discusses the evaluation of positive polynomials. Section 2.4 discusses necessary conditions (for real and complex data) and sufficient conditions (for complex data) for accurate evaluability, when using only classical arithmetic. Section 2.4.3 describes progress toward devising a decision procedure for accurate evaluability in the real case using classical arithmetic. Section 2.5 extends Section 2.4's necessary conditions to arbitrary black-box arithmetic operations, and gives sufficient conditions in the complex case. Section 2.6 describes implications for accurate linear algebra on structured matrices.

2.2 Models of Algorithms and Related Work

Now we state more formally our decision question. We write the output of our algorithm as $p_{comp}(x, \delta)$, where $\delta = (\delta_1, \delta_2, \ldots \delta_k)$ is the vector of rounding errors made during the algorithm.

Definition 2.1 *We say that $p_{comp}(x, \delta)$ is an* accurate algorithm *for the evaluation of $p(x)$ for $x \in \mathcal{D}$ if*

$\forall\, 0 < \eta < 1 \quad \ldots$ *for any $\eta = $ desired relative error*

$\exists\, 0 < \epsilon < 1 \quad \ldots$ *there is an $\epsilon = $ machine precision*

$\forall\, x \in \mathcal{D} \quad \ldots$ *so that for all x in the domain*

$\forall\, |\delta_i| \leq \epsilon \quad \ldots$ *and for all rounding errors bounded by ϵ*

$|p_{comp}(x, \delta) - p(x)| \leq \eta \cdot |p(x)| \ldots$ *the relative error is at most η.*

Our ultimate goal is a decision procedure (a "compiler") that takes $p(\cdot)$ and $\mathcal{D}$ as input, and either produces an accurate algorithm p_{comp} (including how to choose the machine precision ϵ given the desired relative error η) or exhibits a proof that none exists.

To be more precise, we must say what our set of possible algorithms includes. The above decision question is apparently not Tarski-decidable [Renegar (1992), Tarski (1951)] despite its appearance, because we see no way to express "there exists an algorithm" in that format.

The basic decisions about algorithms that we make are as follows, with details given in the indicated sections:

Sec. 2.2.1: We insist that the inputs x are given exactly, rather than approximately.

Sec. 2.2.2: We insist that the algorithm compute the exact value of $p(x)$ in finitely many steps when all rounding errors $\delta = 0$. In particular, we exclude iterative algorithms which might produce an approximate value of $p(x)$ even when $\delta = 0$.

Sec. 2.2.3: We describe the basic arithmetic operations we consider, beyond addition, subtraction and multiplication. We also describe the constants available to our algorithms.

Sec. 2.2.4: We consider algorithms both with and without comparisons and branching, since this choice may change the set of polynomials that we can accurately evaluate.

Sec. 2.2.5: If the computed value of an operation depends only the values of its operands, i.e., if the same operands x and y of $op(x, y)$ always yield the same δ in $rnd(op(x, y)) = op(x, y) \cdot (1 + \delta)$, then we call our model *deterministic,* else it is *nondeterministic.* We show that comparisons and branching let a nondeterministic machine simulate a deterministic one, and subsequently restrict our investigation to the easier nondeterministic model.

Sec. 2.2.6: What domains of evaluation $\mathcal{D}$ do we consider? In principle, any semialgebraic set $\mathcal{D}$ is a possibility, but for simplicity we mostly consider open $\mathcal{D}$, especially $\mathcal{D} = \mathbb{R}^n$ or $\mathcal{D} = \mathbb{C}^n$. We point out issues in extending results to other $\mathcal{D}$.

Finally, Section 2.2.7 summarizes the axioms our model satisfies, and Section 2.2.8 compares our model to other models of arithmetic, and explains the advantages of our model.

2.2.1 Exact or Rounded Inputs

We must decide whether we assume that the arguments are given exactly [Blum, Cucker, Shub, and Smale (1996b), Cucker and Grigoriev (1999),

Renegar (1992), Tarski (1951)] or are known only approximately [Cucker and Smale (1999), Edelat and Sünderhauf (1998), Ko (1991), Pour-El and Richards (1989)]. Not knowing the input x exactly means that at best (i.e., in the absence of any further error) we could only hope to compute the exact value of $p(\hat{x})$ for some $\hat{x} \approx x$, an algorithmic property known as *backward stability* [Demmel (1997), Higham (2002)]. Since we insist that zero outputs be computed exactly in order to have bounded relative error, this means there is no way to guarantee that $p(\hat{x}) = 0$ when $p(x) = 0$, for nonconstant p. This is true even for simple addition $x_1 + x_2$. So we insist on exact inputs in our model.

2.2.2 Finite Convergence

Do we consider algorithms that take a bounded amount of time for all inputs $x \in \mathcal{D}$, and return $p_{comp}(x, 0) = p(x)$, i.e., the exact answer when all rounding errors are zero? Or do we consider possibly iterative algorithms that might take arbitrarily long on some inputs to produce an adequately accurate answer? We consider only the former, because (1) it seems natural to use a finite algorithm to evaluate a finite object like a polynomial, (2) we have seen no situations where an iterative algorithm offers any advantage to obtaining guaranteed *relative* accuracy and (3) this lets us write any algorithm as a piecewise polynomial function and so use tools from algebraic geometry.

2.2.3 Basic Arithmetic Operations and Constants

What are the basic arithmetic operations? For most of the paper we consider addition, subtraction and multiplication of two arguments, since this is necessary and sufficient for polynomial evaluation in the absence of rounding error. Furthermore, we consider negation as a basic operation that is always exact (since this mimics all implementations of rounded arithmetic). Sometimes we will also use (rounded) multiplication by a constant $op(x) = c \cdot x$. We also show how to extend our results to include additional basic arithmetic operations like $op(x, y, z) = x + y \cdot z$. The motivations for considering such additional "black-box" operations are as follows:

1. By considering operations like $x + c$, $x - c$ and $c \cdot x$ for any c in a set C of constants, we may investigate how the the choice of C affects accurate evaluability. For example, if C includes the roots

of a polynomial like $p(x) = x^2 - 2$, then we can accurately evaluate $p(x)$ with the algorithm $(x - \sqrt{2}) \cdot (x + \sqrt{2})$, but otherwise it may be impossible. We note that having C include all algebraic numbers would in principle let us evaluate any univariate polynomial $p(x)$ accurately by using its factored form $p(x) = c \prod_{i=1}^{d}(x - r_i)$.

In the complex case, it is natural to consider multiplication by $\sqrt{-1}$ as an exact operation, since it only involves "swapping" and possibly negating the real and imaginary parts. We can accommodate this by introducing operations like $x + \sqrt{-1} \cdot y$ and $x - \sqrt{-1} \cdot y$.

The necessary conditions in Theorem 2.9 and sufficient conditions in Theorems 2.12 and 2.13 do not depend on how one chooses an operation $x - r_i$ from a possibly infinite set, just whether that operation exists in the set. On the other hand, a decision procedure must effectively choose that operation, so our decision procedures will restrict themselves to enumerable (typically finite) sets of possible operations.[1]

2. Many computers now supply operations like $x + y \cdot z$ in hardware with the accuracy we demand (the *fused-multiply-add* instruction [Montoye, Hokenek, and Runyon (1990)]). It is natural to ask how this operation extends the class of polynomials that we can accurately evaluate.

3. It is natural to build a library (in software or perhaps even hardware) containing several such accurate operations, and ask how much this extends the class of polynomials that can be evaluated accurately. This approach is taken in computational geometry, where the library of accurate operations is chosen to implement certain geometric predicates precisely (e.g., "is point x inside, outside or on circle C?" written as a polynomial whose sign determines the answer). These precise geometric predicates are critical to performing reliable mesh generation [Shewchuk (1997)].

4. A common technique for extending floating point precision is to simulate and manipulate extra precision numbers by representing a high precision number y as a sum $y = \sum_{i=1}^{k} y_i$ of numbers satisfying $|y_i| \gg |y_{i+1}|$, the idea being that each y_i represents (nearly) disjoint parts of the binary expansion of y (see Bailey (1993), Bailey (1995), Dekker (1971), Demmel and Hida (2003), Khachiyan (1984), Møller (1965), Pichat (1972), Priest (1991)) and the references therein; similar techniques were used by Gill as early as 1951). This technique can be modeled by the correct choice of black-box operations as we

---

[1] We could in principle deal with the set of all instructions $x - r$ for r an arbitrary algebraic number, because the algebraic numbers are enumerable.

now illustrate. Suppose we include the enumerable set of black-box operations $\sum_{i=1}^{n} p_i$, where n is any finite number, and each p_i is the product of 1 to d arguments. In other words, we include the accurate evaluation of arbitrary multivariate polynomials in $\mathbb{Z}[x]$ of degree at most d among our black-box operations. Then the following sequence of operations produces as accurate an approximation of any such polynomial

$$\sum_{i=1}^{n} p_i = y_1 + y_2 + \cdots + y_k$$

as desired:

$$y_1 = rnd\left(\sum_{i=1}^{n} p_i\right) = (1 + \delta_1)\left(\sum_{i=1}^{n} p_i\right)$$

$$y_2 = rnd\left(\sum_{i=1}^{n} p_i - y_1\right) = (1 + \delta_2)\left(\sum_{i=1}^{n} p_i - y_1\right)$$

$$\cdots$$

$$y_k = rnd\left(\sum_{i=1}^{n} p_i - \sum_{j=1}^{k-1} y_j\right) = (1 + \delta_k)\left(\sum_{i=1}^{n} p_i - \sum_{j=1}^{k-1} y_j\right)$$

Induction shows that

$$\sum_{j=1}^{k} y_j = \left(1 - (-1)^k \left(\prod_{j=1}^{k} \delta_j\right)\right) \cdot \left(\sum_{i=1}^{n} p_i\right)$$

so that $y = \sum_{j=1}^{k} y_j$ approximates the desired quantity with relative error at most ϵ^k. Despite this apparent power, our necessary conditions in Theorem 2.9 and Section 2.6.1 will still show limits on what can be evaluated accurately. For example, no irreducible polynomial of degree ≥ 3 can be accurately evaluable over $\mathbb{C}^n$ if only dot products (degree $d = 2$) are available.

5. Another standard technique for extending floating point precision is to split a floating point number x with b bits in its fraction into the exact sum $x = x_{hi} + x_{lo}$, where x_{hi} and x_{lo} each have only $b/2$ bits in their fractions. Then products like

$$x \cdot y = (x_{hi} + x_{lo}) \cdot (y_{hi} + y_{lo}) = x_{hi} \cdot y_{hi} + x_{hi} \cdot y_{lo} + x_{lo} \cdot y_{hi} + x_{lo} \cdot y_{lo}$$

can be represented exactly as a sum of 4 floating point numbers, since each product like $x_{hi} \cdot y_{hi}$ has at most b bits in its fraction and so can be computed without error. Arbitrary products may be computed accurately by applying this technique repeatedly, which is the basis of some extra-precise software floating point libraries like Shewchuk (1997). The proof of accuracy of the algorithm for splitting $x = x_{hi} + x_{lo}$ is intrinsically discrete, and depends on a sequence of classical operations some of which can be proven to be free of error [Shewchuk (1997), Thm 17], and similarly for the exactness of $x_{hi} \cdot y_{hi}$. Therefore this exact multiplication operation cannot be built from simpler, inexact operations in our classical model. But we may still model this approach as follows: We imagine an exact multiplication operation $x \cdot y$, and note that all we can do with it is feed it into the inputs of other operations. This means that from the operation $rnd(z + w)$ we also get $rnd(x \cdot y + w)$, $rnd(x \cdot y + r \cdot s)$, $rnd(x \cdot y \cdot z + w)$, and so on. In other words, we take the other operations in our model and from each create an enumerable set of other black-boxes, to which we can apply our necessary and sufficient conditions.

2.2.4 Comparisons and Branching

Are we permitted to do comparisons and then branch based on their results? Are comparisons *exact*, i.e., are the computed values of $x > y$, $x = y$ and $x < y$ (true or false) always correct for real x and y? (For complex x and y we consider only the comparison $x = y$.) We consider algorithms both without comparisons (in which case $p_{comp}(x, \delta)$ is simply a polynomial), and with exact comparisons and branching (in which case $p_{comp}(x, \delta)$ is a piecewise polynomial, on semialgebraic sets determined by inequalities among other polynomials in x and δ). We conjecture that using comparisons and branching strictly enlarges the set of polynomials that we can evaluate accurately.

We note that by comparing $x - r_i$ to zero for selected constants r_i, we could extract part of the bit representation of x. Since we are limiting ourselves to a finite number of operations, we could at most approximate x this way, and as stated in Section 2.2.1, this means we could not exploit this to get high relative accuracy near $p(x) = 0$. We note that the model of arithmetic in Cucker and Smale (1999) excludes real→integer conversion instructions.

2.2.5 Nondeterminism

As currently described, our model is *nondeterministic*, e.g., the rounded result of $1 + 1$ is not necessarily identical if it is performed more than once. This is certainly different behavior than the deterministic computers whose behavior we are modeling. However, it turns out that this is not a limitation, because we can always simulate a deterministic machine with a nondeterministic one using comparisons and branching. The idea is simple: The first addition instruction (say) records its input arguments and computed sum in a list. Every subsequent addition instruction compares its arguments to the ones in the list (which it can do exactly), and either just uses the precomputed sum if it finds them, or else does the addition and appends the results to the list. In other words, the existence (or nonexistence) of an accurate algorithm in a model with comparisons and branching does not depend on whether the machine is deterministic. So for simplicity, we will henceforth assume that our machines are nondeterministic.

2.2.6 Choice of Domain $\mathcal{D}$

As mentioned in the introduction, it seems natural to consider any semialgebraic set as a possible domain of evaluation for $p(x)$. While some choices, like $\mathcal{D} = \{x : p(x) = 0\}$ make evaluating $p(x)$ trivial, they beg the question of how one would know whether $x \in \mathcal{D}$. Similarly, if $\mathcal{D}$ includes a discrete set of points, then $p(x)$ can be evaluated at these points by looking up the answers in a table. To avoid these pathologies, it may seem adequate restrict $\mathcal{D}$ to be a sufficiently "fat" set, say open. But this still leads to interesting complications; for example the algorithm

$$p_{comp}(x, \delta) = ((x_1 + x_2)(1 + \delta_1) + x_3)(1 + \delta_2)$$

for $p(x) = x_1 + x_2 + x_3$, which is inaccurate on $\mathbb{R}^n$, is accurate on the open set $\{|x_1 + x_2| > 2|x_3|\}$, whose closure intersects the variety $V_{\mathbb{R}}(p)$ on $\{x_1 + x_2 = 0 \wedge x_3 = 0\}$.

In this paper we will mostly deal with open $\mathcal{D}$, especially $\mathcal{D} = \mathbb{R}^n$ or $\mathcal{D} = \mathbb{C}^n$, and comment on when our results apply to smaller $\mathcal{D}$.

2.2.7 Summary of Arithmetic and Algorithmic Models

We summarize the axioms our arithmetic and algorithms must satisfy. We start with the axioms all arithmetic operations and algorithms satisfy:

Exact Inputs. Our algorithm will be given the input exactly.

Finite Convergence. An accurate algorithm must, when all roundoff errors $\delta = 0$, compute the exact value of the polynomial in a finite number of steps.

Roundoff Model. Except for negation, which is exact, the rounded value of any arithmetic operation $op(x_1, \ldots, x_k)$ satisfies

$$rnd(op(x_1, \ldots, x_k)) = op(x_1, \ldots, x_k) \cdot (1 + \delta)$$

where δ is arbitrary number satisfying $|\delta| \leq \epsilon$, where ϵ is a nonzero value called the *machine precision*. If the data x is real (or complex), then δ is also real (resp. complex).

Nondeterminism. Every arithmetic operation produces an independent roundoff error δ, even if the arguments to different operations are identical.

Domain $\mathcal{D}$. Unless otherwise specified, the domain of evaluation $\mathcal{D}$ is assumed to be all of $\mathbb{R}^n$ (or all of $\mathbb{C}^n$).

We now list the alternative axioms our algorithms may satisfy. In each category, an algorithm must satisfy one set of axioms or the other.

Branching or Not. Some of our algorithms will permit exact comparisons of intermediate quantities ($<$, $=$ and $>$ for real data), and subsequent branching based on the result of the comparison. Other algorithms will not permit branching. In the complex case, we will see that branching does not matter (see Sections 2.4.2 and 2.5.2).

Classical or "black-box" operations. Some of our algorithms will use only "classical" arithmetic operations, namely addition, subtraction and multiplication. Others will use a set of arbitrary polynomial "black-box" operations, like $op(x, y, z) = x + y \cdot z$ or $op(x) = x - \sqrt{2}$, of our choice. In particular, we omit division.

2.2.8 Other Models of Error and Arithmetic

Our goal in this paper is to model rounded, finite precision computation, i.e., arithmetic with numbers represented in scientific notation, and rounded to their leading k digits, for some fixed k. It is natural to ask about models related to ours.

First, we point out some positive attributes of our model:

1. The model $rnd(op(a, b)) = op(a, b)(1 + \delta)$ has been the most widely used model for floating point error analysis [Higham (2002)] since the early papers of von Neumann and Goldstine (1947), Turing (1948) and Wilkinson (1963).
2. The extension to include black-boxes includes widely used floating point techniques for extending the precision.

3. Though the model is for real (or complex) arithmetic, it can be efficiently simulated on a conventional Turing machine by using a simple variation of floating point numbers $m \cdot 2^e$, stored as the pair of integers (m, e), where m is of fixed length, and $|e|$ grows as necessary. In particular, any sequence of n addition, subtraction, multiplication or division (by nonzero) operations can increase the largest exponent e by at most $O(n)$ bits, and so can be done in time polynomial in the input size. See Demmel and Koev (2001) for further discussion. This is in contrast to repeated squaring in the BSS model [Blum (2004)] which can lead to exponential time simulations.

Models of arithmetic may be categorized according to several criteria (the references below are not exhaustive, but illustrative):

- Are numbers (and any errors) represented discretely (e.g., as bit strings such as floating point numbers) [Demmel and Koev (2001), Higham (2002), Wilkinson (1963)], or as a (real or complex) continuum [Blum, Cucker, Shub, and Smale (1997), Cucker and Dedieu (2001)]?
- Is arithmetic exact [Blum, Cucker, Shub, and Smale (1997), Blum, Cucker, Shub, and Smale (1996a)] or rounded [Cucker and Grigoriev (1999), Cucker and Smale (1999), Higham (2002), Wilkinson (1963)]? If it is rounded, is the error bounded in a relative sense [Higham (2002)], absolute sense [Blum, Cucker, Shub, and Smale (1997)], or something else [Lozier and Olver (1990), Demmel (1987), Demmel (1984)] [Higham (2002), Sec. 2.9]?
- In which of these metrics is the final error assessed?
- Is the input data exact [Blum, Cucker, Shub, and Smale (1997)] or considered "rounded" from its true value [Chatelin and Frayssé (1996), Edelat and Sünderhauf (1998), Ko (1991), Pour-El and Richards (1989), Renegar (1994)] (and if rounded, again how is the error bounded)?
- Do we want a "worst case" error analysis [Higham (2002), Wilkinson (1963)], or by modeling rounding errors as random variables, a statistical analysis [Vignes (1993), Kahan (1998), Spielman and Teng (2002)] [Higham (2002), Sec. 2.8]? Does a condition number appear explicitly in the complexity of the problem [Cucker and Smale (1999)]?

First we consider floating point arithmetic itself, i.e., where real numbers are represented by a pair of integers (m, n) representing the real number $m \cdot r^n$, where r is a fixed number called the *radix* (typically $r = 2$ or $r = 10$). Either by using one of many techniques in the literature for using an array $(x_1, ..., x_s)$ of floating point numbers to represent $x = \sum_{i=1}^{s} x_i$ to very high accuracy and to perform arithmetic on such high precision numbers (e.g., Bailey (1993), Bailey (1995), Priest (1991)), or by converting $m \cdot r^n$ to an exact rational number and performing exact rational arithmetic, one can clearly evaluate *any* polynomial $p(x)$ without error, and the only question is cost. In light of this, our results on classical vs black-box arithmetic can be interpreted as saying when such high precision techniques are necessary, and which black-box operations must be implemented this way, in order to evaluate p accurately.

Let us revisit the accurate evaluation of the simple polynomial $y_1 + y_2 + y_3$. The obvious algorithm is to carry enough digits so that the sum is computed exactly, and then rounded at the end. But then to compute $(2^e + 1) - 2^e$ accurately would require carrying at least e bits, which is *exponential* in the size of the input ($\log_2 e$ bits to represent e). Instead, most practical algorithms rely on the technique in the above paragraph, repeatedly replacing partial sums like $y_1 + y_2$ by $x_1 + x_2$ where $|x_1| \gg |x_2|$ and in fact the bits of x_1 and x_2 do not "overlap." These techniques depend intrinsically on the discreteness of the number representation to prove that certain intermediate additions and subtractions are in fact exact. Our model treats this by modeling the entire operation as a black-box (see Section 2.2.3).

Second, consider our goal of guaranteed high *relative* accuracy. One might propose that *absolute* accuracy is a more tractable goal, i.e., guaranteeing $|p_{comp}(x, \delta) - p(x)| \leq \eta$ instead of $|p_{comp}(x, \delta) - p(x)| \leq \eta |p(x)|$. However, we claim that as long as our basic arithmetic operations are defined to have bounded relative error ϵ, then trying to attain relative error in p_{comp} is the most natural goal.

Indeed, we claim that tiny absolute accuracy is impossible to attain for *any* nonconstant polynomial $p(x)$ when $\mathcal{D} = \mathbb{R}^n$ or $\mathcal{D} = \mathbb{C}^n$. For example, consider $p(x) = x_1 + x_2$, for which the obvious algorithm is $p_{comp}(x, \delta) = (x_1 + x_2)(1 + \delta)$. Thus the absolute error $|p_{comp}(x, \delta) - p(x)| = |x_1 + x_2|\delta \leq |x_1 + x_2|\epsilon$. This absolute error is at most η precisely when $|x_1 + x_2| \leq \eta/\epsilon$, i.e., for x in a diagonal strip in the (x_1, x_2) plane. For $p(x) = x_1 \cdot x_2$ we analogously get accuracy only for x in a region bounded by hyperbolas. In other words, even for the simplest possible polynomials that take one operation to evaluate, they cannot be evaluated to high absolute accuracy on most of $\mathcal{D} = \mathbb{R}^n$ or $\mathbb{C}^n$. The natural error model to consider when trying to attain low absolute error in $p(x)$ is to have low absolute error in the basic arithmetic operations, and this is indeed the approach taken in Cucker and Smale (1999)(though as stated before, repeated squaring can lead to an exponential growth in the number of bits a real number represents [Blum (2004)]).

One could also consider more complicated error models, for example *mixed absolute/relative error*: $|p_{comp}(x, \delta) - p(x)| \leq \eta \cdot \max(|p(x)|, 1)$. Similar models have been used to model underflow error in floating point arithmetic [Demmel (1984)]. A small mixed error implies that either the relative error or the absolute error must be small, and so may be easier to attain than either small absolute error or small relative error alone. But we argue that, at least for the class of homogeneous polynomials evaluated on homogeneous $\mathcal{D}$, the question of whether $p(x)$ is accurately evaluable yields the same answer whether we mean accuracy in the relative sense or mixed sense. To see why, note that $x \in \mathcal{D}$ if and only if $\alpha x \in \mathcal{D}$ for any scalar α, since $\mathcal{D}$ is homogeneous, and that $p(\alpha x) = \alpha^d p(x)$, where $d = degree(p)$. Thus for any nonzero $p(x)$, scaling x to αx will make $\eta \cdot \max(|p(\alpha x)|, 1) = \eta |p(\alpha x)|$ once α is large enough, i.e., relative error η must be attained. By results in Section 2.4.3, this will mean that $p_{comp}(x, \delta)$ must also be homogeneous in x of the same degree, i.e., $p_{comp}(\alpha x, \delta) = \alpha^d p_{comp}(x, \delta)$. Thus for any $x \in \mathcal{D}$ at which we can evaluate $p(x)$ with high mixed accuracy, we can choose α large enough so that

$$\alpha^d |p_{comp}(x, \delta) - p(x)| = |p_{comp}(\alpha x, \delta) - p(\alpha x)| \leq \eta \cdot \max(|p(\alpha x)|, 1)$$
$$= \eta \cdot |p(\alpha x)| = \alpha^d \cdot \eta \cdot |p(x)|$$

implying that $p(\alpha x)$ can be evaluated with high relative accuracy for all α. In summary, changing our goal from relative accuracy to mixed relative/absolute accuracy will not change any of our results, for the case of homogeneous p and homogeneous $\mathcal{D}$.

Yet another model is to assume that the input x is given only approximately, instead of exactly as we assume. This corresponds to the approach taken in Edelat and Sünderhauf (1998), Ko (1991), Pour-El and Richards (1989), in which one can imagine reading as many leading bits as desired of each input x_i from an infinite tape, after which one tries to compute the answer using a conventional Turing machine model. This gives yet different results, since, for example, the difference $x_1 - x_2$ cannot be computed with small relative error in a bounded amount of time, since x_1 and x_2 may agree in arbitrarily many leading digits. Absolute error is more appropriate for this model.

It is worth commenting on why high accuracy of the sort we want is desirable in light of inevitable uncertainties in the inputs. Indeed, many numerical algorithms are successfully analyzed using *backward error analysis* [Higham (2002), Demmel (1997)], where the computed results are shown to be the exact result for a slightly perturbed value of the input. This is the case, for example, for polynomial evaluation using Horner's rule where one shows that one gets the exact value of a polynomial at x but with slightly perturbed coefficients. Why is this not always accurate enough?

We already mentioned mesh generation [Shewchuk (1997)], where the inputs are approximately known physical coordinates of some physical object to be triangulated, but where geometric predicates about the vertices defining the triangulation must be answered consistently; this means the signs of certain polynomials must be computed exactly, which is in turn guaranteed by guaranteeing any relative accuracy $\eta < 1$.

More generally, in many physical simulations, the parameters describing the physical system to be simulated are often known to only a few digits, if that many. Nonetheless, intermediate computations must be performed to much higher accuracy than the input data is known, for example to make sure the computed system conserves energy (which it should to high accuracy for the results to be meaningful, even if the initial conditions are uncertain).

Another example where high accuracy is important are the trigonometric functions: When x is very large and slightly uncertain, the value of $\sin x$ may be completely uncertain. Still, we want the computed trigonometric functions to (nearly) satisfy identities like $\sin^2 x + \cos^2 x = 1$ and $\sin 2x = 2 \sin x \cos x$ so that we can reason about program correctness. Many other examples of this sort can be found in articles posted at Kahan (webpage).

In the spirit of backward error analysis, one could consider the polynomial p fixed, but settle for accurately computing $p(\hat{x})$ where $\hat{x}$ differs from x by only a small relative change in each component x_i. This is not guaranteed by Horner's rule, which is equivalent to changing the polynomial p slightly but not x. Would it be easier to compute $p(\hat{x})$ accurately than $p(x)$ itself? This is the case for some polynomials, like $x_1 + x_2 + x_3$ or $c_1 x_2^2 x_3^3 + c_2 x_1^2 x_3^3 + c_3 x_1 x_2^4$, where there is a unique x_i that we can associate with each monomial to "absorb" the rounding error from Horner's rule. In particular, with Horner's rule, the number of monomials in $p(x)$ may at most be equal to the number of x_i. In analogy to this paper, one could ask for a decision procedure to identify polynomials that permit accurate evaluation of $p(\hat{x})$ using any algorithm. This is a possible topic for future work.

Another possibility is to consider error probabilistically [Higham (2002), Sec. 2.8]. This has been implemented in a practical system [Vignes (1993)], where a program is automatically executed repeatedly with slightly different rounding errors made at each step in order to assess the distribution of the final error. This approach is criticized in Kahan (1998) for improperly modeling the discrete, non-random behavior of roundoff, and for possibly invalidating (near) identities like $\sin 2x = 2 \sin x \cos x$ upon which correctness may depend.

In *smoothed analysis* [Spielman and Teng (2002)], one considers complexity (or for us, relative error) by averaging over a Gaussian distribution around each input. For us, input could mean either the argument x of a fixed polynomial p, or the polynomial itself, or both. First consider the case of a fixed polynomial p with a randomly perturbed x. This case is analogous to the previous paragraph, because the inputs can be thought of as slightly perturbed before starting the algorithm. Indeed, one could imagine rounding the inputs slightly to nearby rational or floating point numbers, and then computing exactly. But in this case, it is easy to see that, at least for codimension 1 varieties of p, the "smoothed" relative error is finite or infinite precisely when the worst case relative error is finite or infinite. So smoothing does not change our basic analysis.[1] Now suppose one smooths over the polynomial p, i.e., over its coefficients. If we smooth using a Gaussian distribution, then as we will see, the genericity of "bad" p will make the smoothed relative error infinite for all polynomials. Changing the distribution from Gaussian to one with a small support would only distinguish between positive definite polynomials, the easy case discussed in Section 2.3, and polynomials that are not positive definite.

In *interval arithmetic* [Moore (1979), Neumaier (1990), Alefeld and Herzberger (1983)] one represents each number by a floating point interval guaranteed to contain it. To do this one rounds interval endpoints "outward" to ensure that, for example, the sum $c = a + b$ of two intervals yields an interval c guaranteed to contain the sum of any two numbers in a and b. It is intuitive that if an interval algorithm existed to evaluate $p(x)$ for $x \in \mathcal{D}$ that always computed an interval whose width was small compared to the number of smallest magnitude in the interval, and if the algorithm obeyed the rules in Section 2.2.7, then it would satisfy our accuracy requirements. Conversely, one might conjecture that an algorithm accurate by our criteria would straightforwardly provide an accurate interval algorithm, where one would simply replace all arithmetic operation by interval operations. The issue of interpreting comparisons and branches using possibly overlapping intervals makes this question interesting, and a possible subject for future work.

Finally, many authors use condition numbers in their analysis of the complexity of solving certain problems. This is classical in numerical analysis [Higham (2002)]; more recent references are Chatelin and Frayssé (1996),

---

[1] The logarithm of the relative error, like the logarithm of many condition numbers, does however have a finite average.

Cucker and Smale (1999), Cucker and Dedieu (2001). In this approach, one is willing to do more and more work to get an adequate answer as the condition number grows, perhaps without bound. Such a conditioning question appears in our approach, if we ask how small the machine precision ϵ must be as a function of the desired relative error η, as well as p, $\mathcal{D}$, and allowed operations. Computing this condition number (outside the easy case described in Section 2.3) is an open question.

2.3 Evaluating positive polynomials accurately

Here we address the simpler case where the polynomial $p(x)$ to be evaluated has no zeros in the domain of evaluation $\mathcal{D}$. It turns out that we need more than this to guarantee accurate evaluability: we will require that $|p(x)|$ be bounded both above and below in an appropriate manner on $\mathcal{D}$.

We let $\bar{\mathcal{D}}$ denote the closure of $\mathcal{D}$.

Theorem 2.1 *Let $p_{comp}(x, \delta)$ be any algorithm for $p(x)$ satisfying the condition $p_{comp}(x, 0) = p(x)$, i.e., it computes the correct value in the absence of rounding error. Let $p_{min} := \inf_{x \in \bar{\mathcal{D}}} |p(x)|$. Suppose $\bar{\mathcal{D}}$ is compact and $p_{min} > 0$. Then $p_{comp}(x, \delta)$ is an accurate algorithm for $p(x)$ on $\mathcal{D}$.*

Proof Since the relative error on $\mathcal{D}$ is $|p_{comp}(x, \delta) - p(x)|/|p(x)| \leq |p_{comp}(x, \delta) - p(x)|/p_{min}$, it suffices to show that the numerator approaches 0 uniformly as $\delta \to 0$. This follows by writing the value of $p_{comp}(x, \delta)$ along any branch of the algorithm as $p_{comp}(x, \delta) = p(x) + \sum_{\alpha > 0} p_\alpha(x) \delta^\alpha$, where $\alpha > 0$ is a multi-index $(\alpha_1, \ldots, \alpha_k)$ with at least one component exceeding zero and $\delta^\alpha := \delta_1^{\alpha_1} \cdots \delta_k^{\alpha_k}$. By compactness of $\bar{\mathcal{D}}$, $|\sum_{\alpha > 0} p_\alpha(x) \delta^\alpha| \leq C \sum_{\alpha > 0} |\delta|^\alpha$ for some constant C, which goes to 0 uniformly as the upper bound ϵ on each $|\delta_i|$ goes to zero. $\square$

Next we consider domains $\mathcal{D}$ whose closure is not compact. To see that merely requiring $p_{min} > 0$ is not enough, consider evaluating $p(x) = 1 + (x_1 + x_2 + x_3)^2$ on $\mathbb{R}^3$. Intuitively, $p(x)$ can only be accurate if its "dominant term" $(x_1 + x_2 + x_3)^2$ is accurate, once it is large enough, and this is not possible using only addition, subtraction and multiplication. (These observations will be formalized in Sections 2.4.)

Instead, we consider a homogeneous polynomial $p(x)$ evaluated on a homogeneous $\mathcal{D}$, i.e., one where $x \in \mathcal{D}$ implies $\gamma x \in \mathcal{D}$ for any scalar γ. Even though such $\mathcal{D}$ are unbounded, homogeneity of p will let us consider just the behavior of $p(x)$ on $\mathcal{D}$ intersected with the unit ball S^{n-1} in $\mathbb{R}^n$ (or S^{2n-1} in $\mathbb{C}^n$). On this intersection we can use the same compactness argument as above:

Theorem 2.2 *Let $p(x)$ be a homogeneous polynomial, let $\mathcal{D}$ be a homogeneous domain, and let S denote the unit ball in $\mathbb{R}^n$ (or $\mathbb{C}^n$). Let*

$$p_{min,homo} := \inf_{x \in \mathcal{D} \cap S} |p(x)|$$

Then $p(x)$ can be evaluated accurately if $p_{min,homo} > 0$.

Proof We describe an algorithm $p_{comp}(x, \delta)$ for evaluating $p(x)$. There are many such algorithms, but we only describe a simple one. (Indeed, we will see that the set of all accurate algorithms for this situation can be characterized completely by Definition 2.11 and Lemma 2.3.) Write $p(x) = \sum_\alpha c_\alpha x^\alpha$, where α is a multi-index $(\alpha_1, \ldots, \alpha_n)$ and $c_\alpha \neq 0$ is a scalar. Homogeneity implies $|\alpha| = \sum_i \alpha_i$ is constant. Then the algorithm simply

1. computes each x^α term by repeated multiplication by x_is,
2. computes each $c_\alpha x^\alpha$ either by multiplication by c_α or by repeated addition if c_α is an integer, and
3. sums the $c_\alpha x^\alpha$ terms.

Since each multiplication, addition and subtraction contributes a $(1 + \delta_i)$ term, it is easy to see that

$$p_{comp}(x, \delta) = \sum_\alpha c_\alpha x^\alpha \Delta_\alpha$$

where each Δ_α is the product of at most some number f of factors of the form $1 + \delta_i$.

Now let $\|x\|_2 = (\sum_i |x_i|^2)^{1/2}$, so $\hat{x} = x/\|x\|_2$ is in the unit ball S. Then the relative error may be bounded by

$$
\begin{aligned}
\left| \frac{p_{comp}(x, \delta) - p(x)}{p(x)} \right| &= \left| \frac{\sum_\alpha c_\alpha x^\alpha \Delta_\alpha - \sum_\alpha c_\alpha x^\alpha}{\sum_\alpha c_\alpha x^\alpha} \right| \\
&= \left| \frac{\sum_\alpha c_\alpha \hat{x}^\alpha (\Delta_\alpha - 1)}{\sum_\alpha c_\alpha \hat{x}^\alpha} \right| \\
&\leq \frac{\sum_\alpha |c_\alpha| \cdot |\Delta_\alpha - 1|}{p_{min}} \\
&\leq \frac{\sum_\alpha |c_\alpha| \cdot ((1 + \epsilon)^f - 1)}{p_{min}}
\end{aligned}
$$

which goes to zero uniformly in ϵ. $\qquad\square$

2.4 Classical arithmetic

In this section we consider the simple or classical arithmetic over the real or complex fields, with the three basic operations $\{+, -, \cdot\}$, to which we add negation. The model of arithmetic is governed by the laws in Section 2.2.7. We remind the reader that this arithmetic model *does not allow* the use of constants.

In Section 2.4.1 we find a necesary condition for accurate evaluability over either field, and in Section 2.4.2 we prove that this condition is also sufficient for the complex case.

Throughout this section, we will make use of the following definition of allowability.

Definition 2.2 *Let p be a polynomial over $\mathbb{R}^n$ or $\mathbb{C}^n$, with variety $V(p) := \{x \ : \ p(x) = 0\}$. We call $V(p)$ allowable if it can be represented as a union of intersections of sets of the form*

$$1. \qquad Z_i := \{x : x_i = 0\}, \tag{2.3}$$

$$2. \qquad S_{ij} := \{x : x_i + x_j = 0\}, \tag{2.4}$$

$$3. \qquad D_{ij} := \{x : x_i - x_j = 0\}. \tag{2.5}$$

If $V(p)$ is not allowable, we call it unallowable.

Remark 2.1 *For a polynomial p, having an allowable variety $V(p)$ is obviously a Tarski-decidable property (following Tarski (1951)), since the number of unions of intersections of hyperplanes (2.3)-(2.5) is finite.*

2.4.1 Necessity: real and complex

All the statements and proofs in this section work equally well for both the real and the complex case, and thus we may treat them together. At the end of the section we use the necessity condition to obtain a partial result relating to domains.

Definition 2.3 *From now we will refer to the space of variables as $\mathcal{S} \in \{\mathbb{R}^n, \mathbb{C}^n\}$.*

To state and prove the main result of this section, we need to introduce some additional notions and notation.

Definition 2.4 *Given a polynomial p over $\mathcal{S}$ with unallowable variety $V(p)$, consider all sets W that are finite intersections of allowable hyperplanes defined by (2.3), (2.4), (2.5), and subtract from $V(p)$ those W for which $W \subset V(p)$. We call the remaining subset of the variety points in general position and denote it by $G(p)$.*

Remark 2.2 *If $V(p)$ is not allowable, then from definition 2.4 it follows that $G(p) \neq \emptyset$. One may also think of points in $G(p)$ as "unallowable" or "problematic", because, as we will see, we necessarily get large relative errors in their vicinity.*

Definition 2.5 *Given $x \in S$, define the set* Allow(x) *as the intersection of all allowable hyperplanes going through x:*

$$\text{Allow}(x) := \left(\cap_{x \in Z_i} Z_i \right) \cap \left(\cap_{x \in S_{ij}} S_{ij} \right) \cap \left(\cap_{x \in D_{ij}} D_{ij} \right),$$

with the understanding that

$$\text{Allow}(x) := S \qquad \text{whenever} \qquad x \notin Z_i,\ S_{ij},\ D_{ij} \quad \text{for all} \quad i, j.$$

Note that Allow(x) *is a linear subspace of S.*

We will be interested in the sets Allow(x) primarily when $x \in G(p)$. For such cases we make the following observation.

Remark 2.3 *For each $x \in G(p)$, the set* Allow(x) *is not a subset of $V(p)$:*

$$\text{Allow}(x) \not\subseteq V(p),$$

which follows directly from the definition of $G(p)$.

We can now state the main result of this section, which is a necessity condition for the evaluability of polynomials over domains.

Theorem 2.3 *Let p be a polynomial over a domain $\mathcal{D} \in S$. Let $G(p)$ be the set of points in general position on the variety $V(p)$. If there exists $x \in \mathcal{D} \cap G(p)$ such that* Allow$(x) \cap \text{Int}(\mathcal{D}) \neq \emptyset$*, then p is not accurately evaluable on $\mathcal{D}$.*

To prove Theorem 2.3, we need to recall the notion of Zariski topology (see, e.g., Hulek (2003)).

Definition 2.6 *A subset $Y \subseteq \mathbb{R}^n$ (or $\mathbb{C}^n$) is called a Zariski closed set if there a subset T of the polynomial ring $\mathbb{R}[x_1, \ldots, x_n]$ (or $\mathbb{C}[x_1, \ldots, x_n]$) such that Y is the variety of T: $Y = V(T) := \cap_{p \in T} V(p)$. A complement of a Zariski closed set is said to be Zariski open. The class of Zariski open sets defines the Zariski topology on S.*

In this paper, we consider the Zariski topology not on S, but on a hypercube centered at the origin in δ-space (the space in which the vector of error variables δ lies). This topology is defined in exactly the same fashion.

Note that a Zariski closed set has measure zero unless it is defined by the zero polynomial only; then the set is the whole space. In the coming proof we will deal with nonempty Zariski open sets, which are all of full measure. Finally, it is worth noting that the Zariski sets we will work with are algorithm-dependent.

Finally, we represent any algorithm as in Blum, Shub, and Smale (1989) and in Aho, Hopcroft, and Ullman (1974) by a directed acyclic graph (DAG) with input nodes, branching nodes, and output nodes. For simplicity in dealing with negation (given that negation is an *exact* operation), we define a special type of edge which indicates that the value carried along the edge is negated. We call these special edges *dotted,* to distinguish them from the regular *solid* ones.

Every computational node has two inputs (which may both come from a single other computational node); depending on the source of these inputs we have computational nodes with inputs from two distinct nodes and computational nodes with inputs from the same node. The latter type correspond either to

1. doubling $((x, x) \overset{+}{\mapsto} 2x)$,
2. doubling and negating $((-x, -x) \overset{+}{\mapsto} -2x)$,
3. computing zero exactly $((-x, x) \overset{+}{\mapsto} 0, (-x, -x) \overset{\cdot}{\mapsto} 0,$ or $(x, x) \overset{\cdot}{\mapsto} 0)$,
4. squaring $((x, x) \overset{\cdot}{\mapsto} x^2$ or $(-x, -x) \overset{\cdot}{\mapsto} x^2)$,
5. squaring and negating $((-x, x) \overset{\cdot}{\mapsto} -x^2)$.

All nodes are labeled by $(op(\cdot), \delta_i)$ with $op(\cdot)$ representing the operation that takes place at that node. It means that at each node, the algorithm takes in two inputs, executes the operation, and multiplies the result by $(1 + \delta_i)$.

Finally, for each branch, there is a single destination node, with one input and no output, whose input value is the result of the algorithm.

Throughout the rest of this section, unless specified, we consider only non-branching algorithms.

Definition 2.7 *For a given $x \in \mathcal{S}$, we say that a computational node N is* of non-trivial type *if its output is a nonzero polynomial in the variables δ when the algorithm is run on the given x and with symbolic δs.*

Definition 2.8 *For a fixed x, let N be any non-trivial computational node in an algorithm. We denote by $L(N)$ (resp., $R(N)$) the set of computational nodes in the left (resp., right) subgraphs of N. If both inputs come* *from the same node, i.e. $L(N)$ and $R(N)$ overlap, we will only talk about $L(N)$.*

Definition 2.9 *For a given $\epsilon > 0$, we denote by H_ϵ the hypercube of edge length 2ϵ centered at the origin, in δ-space.*

We will need the following Proposition.

Proposition 2.10 *Given any algorithm, any $\epsilon > 0$, and a point $x \in G(p)$, there exists a Zariski open set Δ in H_ϵ such that no non-trivial computational node has a zero output on the input x for all $\delta \in \Delta$.*

Proof The proof follows from the definition of the non-trivial computational node.

Since every non-trivial computational node outputs a non-trivial polynomial in δ, it follows that each non-trivial computational node is nonzero on a Zariski open set (corresponding to the output polynomial in δ) in H_ϵ. Intersecting this finite number of Zariski open sets we obtain a Zariski open set which we denote by Δ; for any $\delta \in \Delta$ the output of any non-trivial computational node is nonzero. $\qquad \square$

We can now state and prove the following crucial lemma.

Lemma 2.1 *For a given algorithm, any $x \in G(p)$, and $\epsilon > 0$, exactly one of the following holds:*

1. *there exists a Zariski open set $\Delta \subseteq H_\epsilon$ such that the value $p_{comp}(x, \delta)$ computed by the algorithm is not zero when the algorithm is run with source input x and $\delta \in \Delta$;*
2. *$p_{comp}(y, \delta) = 0$ for all $y \in \mathrm{Allow}(x)$ and all δ in H_ϵ.*

Proof of Lemma 2.1. We recall that the algorithm can be represented as a DAG, as described in the paragraphs preceding Definition 2.7.

Fix a point $x \in G(p)$. Once x is fixed, the result of each computation is a polynomial expression in the δs. Consider the Zariski open set Δ whose existence is guaranteed by Proposition 2.10. There are now two possibilities: either the output node is of non-trivial type, in which case $p_{comp}(x, \delta) \neq 0$ for all $\delta \in \Delta$, or the output node is not of non-trivial type, in which case $p_{comp}(x, \delta_0) = 0$ for some $\delta_0 \in \Delta$.

In the latter case the output of the computation is zero; we trace back this zero to its origin, by marking in descending order all computational nodes that produced a zero (and thus we get a set of paths in the DAG, all of whose nodes produced exact zeros). Note that we are not interested in *all* nodes that produced a 0; only those which are on paths of zeros to the output node.

We will examine the last occurrences of zeros on paths of marked vertices, i.e. the zeros that are farthest from the output on such paths.

Lemma 2.2 *The* last *zero on such a path must be either*

1. *a source;*
2. *the output of a node where* $(-x, x) \overset{+}{\mapsto} 0$, $(-x, -x) \overset{-}{\mapsto} 0$, *or* $(x, x) \overset{-}{\mapsto} 0$ *are performed;*
3. *the output of an addition or subtraction node with two nonzero source inputs.*

Proof of Lemma 2.2. Note that a nonzero non-source output will be a non-constant polynomial in the δ specific to that node.

Clearly the last zero output cannot happen at a multiplication node; we have thus to show that the last occurrence of a zero output cannot happen at an addition or subtraction node which has two nonzero inputs from different nodes, at least one of which is a non-source. We prove the last statement by reductio ad absurdum.

Assume we could have a zero output at a node N with two nonzero inputs, at least one of which is not a source. Let $R(N)$ and $L(N)$ be as in Definition 2.8. Let $\delta(L(N))$ and $\delta(R(N))$ be the sets of errors δ_i corresponding to the left, respectively the right subtrees of N.

By assumption, $\delta(R(N)) \cup \delta(L(N)) \neq \emptyset$ (since at least one of the two input nodes is a non-source). Let δ_l (δ_r) denote the δ associated to the left (right) input node of N. Then we claim that either $\delta_l \notin \delta(R(N))$ or $\delta_r \notin \delta(L(N))$. (There is also the possibility that one of the two input nodes *is* a source and does not have a δ, but in that case the argument in the next paragraph becomes trivial.)

Indeed, since each δ is specific to a node, if δ_l were in $\delta(R(N))$, there would be a path from the left input node to the right input node. Similarly, if δ_r were in $\delta(L(N))$, then there would be a path from the right input node of N to the left input node of N. So if both events were to happen at the same time, there would be a cycle in the DAG. This cannot happen, hence either $\delta_l \notin \delta(R(N))$ or $\delta_r \notin \delta(L(N))$.

Assume w.l.o.g. $\delta_l \notin \delta(R(N))$. Then the left input of N is a non-trivial polynomial in δ_l, while the right input does not depend on δ_l at all. Hence their sum or difference is still a non-trivial polynomial in δ_l. Contradiction. □

Now that Lemma 2.2 has been proven, we can state the crucial fact of the proof of Lemma 2.1: *all last occurrences of a zero appear at nodes which either correspond to allowable constraints (i.e., zero sources, or sums and differences of sources,) or are addition/subtraction nodes with both inputs from the same node, which always, on any source inputs, produce a zero.*

Take now any point $y \in \text{Allow}(x)$; then y produces the same chains of consecutive zeros constructed (marked) in Lemma 2.1 as x does, with errors given by $\delta_0 \in \Delta$. Indeed, any node on such a chain that has a zero output at x when the error variables are δ_0 can trace this zero back to an allowable constraint (which is satisfied by both x and y) or to an addition/subtraction node with both inputs from the same node; hence the node will also have a zero output at y with errors δ_0. In particular, if $p_{comp}(x, \delta_0) = 0$ for $\delta_0 \in \Delta$, then $p_{comp}(y, \delta_0) = 0$. Moreover, changing δ_0 can only introduce additional zeros, but cannot eliminate zeros on the zero paths that we traced for x (by the choice of Δ). Therefore, $p_{comp}(y, \delta) = 0$ for all $y \in \text{Allow}(x)$ and $\delta \in H_\epsilon$. This completes the proof of Lemma 2.1. □

From Lemma 2.1 we obtain the following corollary.

Corollary 2.1 *For any $\epsilon > 0$ and any $x \in G(p)$, exactly one of the following holds: the relative error of computation, $|p_{comp} - p|/|p|$, is either infinity at x for all δ in a Zariski open set or 1 at all points $y \in (\text{Allow}(x) \setminus V(p))$ and all $\delta \in H_\epsilon$.*

We now consider algorithms with or without branches.

Theorem 2.4 *Given a (branching or non-branching) algorithm with output function $p_{comp}(\cdot)$, $x \in G(p)$, and $\epsilon > 0$, then one of the following is true:*

1. *there exists a set Δ_1 of positive measure in H_ϵ such that $p_{comp}(x, \delta)$ is nonzero whenever the algorithm is run with errors $\delta \in \Delta_1$, or*
2. *there exists a set Δ_2 of positive measure in H_ϵ such that for every $\delta \in \Delta_2$, there exists a neighborhood $N_\delta(x)$ of x such that for every $y \in N_\delta(x) \cap (\text{Allow}(x) \setminus V(p))$, $p_{comp}(y, \delta) = 0$ when the algorithm is run with errors δ.*

Remark 2.4 *This implies that, on a set of positive measure in H_ϵ, the relative accuracy of any given algorithm is either ∞ or 1.*

Proof With $p_{comp}(\cdot)$ the output function and x a fixed point in general position, we keep the δs symbolic. Depending on the results of the comparisons, the algorithm splits into a finite number of non-branching algorithms, which all start in the same way (with the input nodes) and then differ in accordance with a finite set of polynomial constraints on the δs and xs.

Some of these branches will be chosen by sets of δs of measure zero; at least one of the branches will have to be chosen by a set of δs of positive measure whose interior is nonempty (all constraints being polynomials). Call that branch B, and let the set of δs that choose it be called Δ_B.

By Proposition 2.10, there exists a Zariski open set $\Delta \in H_\epsilon$ such that, for all $\delta \in \Delta$, no non-trivial node in the subgraph representing our branch B has a zero output. In particular, this includes all quantities computed for comparisons that define B. Let $\Delta_2 := \text{Int}(\Delta_B \cap \Delta)$, where Int denotes the interior of a set. By the choice of Δ_B and Δ, the obtained set Δ_2 is non-empty.

Suppose the algorithm is run with errors $\delta_0 \in \Delta_2$ and $p_{comp}(x, \delta_0) \neq 0$. Then, by continuity, there must be a neighborhood Δ_1 in the set Δ_2 on which the computation will still be directed to branch B and $p_{comp}(x, \cdot)$ will still be nonzero, so we are in Case 1.

Assume now that we are not in Case 1, i.e. there is no $\delta \in \Delta_2$ such that $p_{comp}(x, \delta) \neq 0$. In this case we show by contradiction that $p_{comp}(y, \delta) = 0$ for all $y \in \text{Allow}(x)$ if y is sufficiently close to x (since $\text{Allow}(x)$ is a linear subspace containing x, there exist points in $\text{Allow}(x)$ which are arbitrarily close to x), thus, that Case 2 must be fulfilled.

If this claim is not true, then there is no neighborhood $N_\delta(x)$ of x such that when $y \in N_\delta(x) \cap \text{Allow}(x)$, the algorithm is directed to branch B on δ. In that case, there must be a sequence $\{y_n\} \in (\text{Allow}(x) \setminus V(p))$ such that $y_n \to x$ and y_n is always directed elsewhere for this choice of δ. The reason for this is that $\text{Allow}(x)$ is a linear subspace which is *not* contained in $V(p)$; hence no neighborhood of x in $\text{Allow}(x)$ can be contained in $V(p)$, and then such a sequence y_n must exist.

Since there is a finite number of branches, we might as well assume that all y_n will be directed to the same branch B' for this δ and that they split off at the same branching node (pigeonhole principle).

Now consider the branching node where the splitting occurs, and let $r(z, \delta)$ be the quantity to be compared to 0 at that node. Since we always go to B' with y_n but to B with x, it follows that we *necessarily must have* $r(y_n, \delta) \neq 0$ whereas $r(x, \delta) = 0$. On the other hand, until that splitting point the algorithm followed the same path with y_n and with x, computing with the same errors δ. Applying then case 2 of Lemma 2.1 (which can be read to state that any algorithm computing r, and obtaining $r(x, \delta) = 0$, will also obtain $r(y_n, \delta) = 0$), we get a contradiction.

This completes the proof of Theorem 2.4. $\square$

Corollary 2.2 *Let p be a polynomial over S with unallowable variety $V(p)$. Choose any algorithm with output function $p_{comp}(\cdot)$, any point $x \in G(p)$, $\epsilon > 0$, and $\eta < 1$. Then there exists a set Δ_x of positive measure arbitrarily close to x and a set Δ of positive measure in H_ϵ, such that $|p_{comp} - p|/|p|$ is strictly larger than η when computed at a point $y \in \Delta_x$ using any vector of relative errors $\delta \in \Delta$.*

Proof On symbolic input x and with symbolic δ, the algorithm will have m branches $B_1, \ldots, B_m$ that correspond to constraints yielding (semi-algebraic) sets of positive measure $S_1, \ldots, S_m$ in (x, δ)-space. Choose $x \in G(p)$, and let $(x, 0)$ be a point in (x, δ)-space.

1. If $(x, 0)$ is in $\text{Int}(S_i)$ (the interior of some region S_i), then by Lemma 2.1 and Corollary 2.1 there exists either

   (a) a δ_0 in δ-space sufficiently small such that (x, δ_0) is in $\text{Int}(S_i)$ and $p_{comp}(x, \delta_0) \neq 0$. The relative error at (x, δ_0) is in this case ∞, and (by continuity) there must be a small ball around (x, δ_0) which is still in $\text{Int}(S_i)$, on which the minimum relative error is arbitrarily large, certainly larger than 1;

   (b) a δ_0 in δ-space sufficiently small and a $y \in \text{Allow}(x) \backslash V(p)$ sufficiently close to x such that (y, δ_0) is in $\text{Int}(S_i)$ and $p_{comp}(y, \delta_0) = 0$. In this case the relative error at (y, δ_0) is 1, and (by continuity) there must be a small ball around (y, δ_0) which is still in $\text{Int}(S_i)$, on which the relative error is strictly larger than our $\eta < 1$.

2. Otherwise, $(x, 0)$ must be on the boundary of some of the regions S_i; assume w.l.o.g. that it is on the boundary of the regions $S_1, \ldots, S_l$. In this case, we choose a small hyperdisk $B_{\bar\epsilon}((x, 0))$ in the linear subspace $(x, \cdot)$ such that $B_{\bar\epsilon}((x, 0))$ intersects the closures of $S_1, \ldots, S_l$ (and no other S_is). We can do this because the sets S_i are all semi-algebraic.

   (a) If there exists a δ_0 in δ-space such that $(x, \delta_0) \in B_{\bar\epsilon}((x, 0))$ and $(x, \delta_0) \in \text{Int}(S_i)$ for some $i \in \{1, \ldots, l\}$, then by the same argument as in case 1. we obtain a small ball included in $\text{Int}(S_i)$ on which the relative error is greater than η;

   (b) Otherwise, if there exists a δ_0 such that $(x, \delta_0) \in B_{\bar\epsilon}((x, 0))$ is on the boundary of some region S_i for which the local algorithm corresponding to it would yield $p_{comp}(x, \delta_0) \neq 0$, then (by continuity) there exists a small ball around (x, δ_0) such that the intersection of that small ball with S_i is of positive measure, and the relative error on that small ball as computed by the algorithm corresponding to S_i is greater than 1;

(c) Finally, otherwise, choose some point $(x, \delta_1) \in B_{\bar{\varepsilon}}((x, 0))$, so that (x, δ_1) is on the boundary of a subset of regions $S \subset \{S_1, \ldots, S_l\}$. We must have that $p_{comp}(x, \delta_1) = 0$ when computed using any of the algorithms that correspond to any $S_i \in S$.

Let now $B(x)$ be a small ball around x in x-space, and consider $\tilde{B}(x) := B(x) \cap (\text{Allow}(x) \setminus V(p))$.

There exists some $y \in \tilde{B}(x)$, close enough to x, such that (y, δ_1) is either in the interior or on the boundary of some $S_k \in S$.

By Lemma 2.1, since we must have $p_{comp}(y, \delta_1) = 0$ as computed by the algorithm corresponding to S_k, if follows (by continuity) that there is a small ball around (y, δ_1) on which the relative error, when computed using the algorithm corresponding to S_k, is greater than η. The intersection of that small ball with S_k must have positive measure.

From the above analysis, it follows that there is always a set of positive measure, arbitrarily close to $(x, 0)$, on which the algorithm will produce a relative error larger than η. $\qquad \square$

Proof of Theorem 2.3. Follows immediately from Theorem 2.4 and Corollary 2.2. $\qquad \square$

Remark 2.5 *Consider the polynomial $p(x, y) = (1 - xy)^2 + x^2$, whose variety is at infinity. We believe that Theorem 2.3 can be extended to show that polynomials like $p(x, y)$ cannot be evaluated accurately on $\mathbb{R}$; this is future work.*

2.4.2 Sufficiency: the complex case

Suppose we now restrict input values to be complex numbers and use the same algorithm types and the notion of accurate evaluability from the previous sections. By Theorem 2.3, for a polynomial p of n complex variables to be accurately evaluable over $\mathbb{C}^n$ it is necessary that its variety $V(p) := \{z \in \mathbb{C}^n : p(z) = 0\}$ be allowable.

The goal of this section is that this condition is also sufficient, as stated in the following theorem.

Theorem 2.5 *Let $p : \mathbb{C}^n \to \mathbb{C}$ be a polynomial with integer coefficients and zero constant term. Then p is accurately evaluable on $\mathcal{D} = \mathbb{C}^n$ if and only if the variety $V(p)$ is allowable.*

To prove this we first investigate what allowable complex varieties can look like. We start by recalling a basic fact about complex polynomial varieties, which can for example be deduced from Theorem 3.7.4 on page 53 of Taylor (2004). Let V denote any complex variety. To say that $\dim_{\mathbb{C}}(V) = k$ means that, for each $z \in V$ and each $\delta > 0$, there exists $w \in V \cap B(z, \delta)$ such that w has a V-neighborhood that is homeomorphic to a real $2k$-dimensional ball.

Theorem 2.6 *Let p be a non-constant polynomial over $\mathbb{C}^n$. Then*

$$\dim_{\mathbb{C}}(V(p)) = n - 1.$$

Corollary 2.3 *Let $p : \mathbb{C}^n \to \mathbb{C}$ be a nonconstant polynomial whose variety $V(p)$ is allowable. Then $V(p)$ is a union of allowable hyperplanes.*

Proof Suppose $V(p) = \cup_j S_j$, where each S_j is an intersection of the sets in Definition 2.2 and, for some j_0, S_{j_0} is not a hyperplane but an irredundant intersection of hyperplanes. Let $z \in S_{j_0} \setminus \cup_{j \neq j_0} S_j$. Then, for some $\delta > 0$, $B(z, \delta) \cap V(p) \subset S_{j_0}$. Since $\dim_C(S_{j_0}) < n-1$, no point in $B(z, \delta) \cap V(p)$ has a $V(p)$-neighborhood that is homeomorphic to a real $2(n-1)$-dimensional ball. Contradiction. $\qquad\square$

Corollary 2.4 *If $p : \mathbb{C}^n \to \mathbb{C}$ is a polynomial whose variety $V(p)$ is allowable, then it is a product $p = c \prod_j p_j$, where each p_j is a power of x_i, $(x_i - x_j)$, or $(x_i + x_j)$.*

Proof By Corollary 2.3, the variety $V(p)$ is a union of allowable hyperplanes. Choose a hyperplane H in that union. If $H = Z_{j_0}$ for some J_0, expand p into a Taylor series in x_{j_0}. If $H = D_{i_0 j_0}$ (or $H = S_{i_0 j_0}$) for some i_0, j_0, expand p into a Taylor series in $(x_{i_0} - x_{j_0})$ (or $(x_{i_0} + x_{j_0})$). In either case, in this expansion, the zeroth coefficient of p must be the zero polynomial in x_j, $j \neq j_0$ (or $j \notin \{i_0, j_0\}$). Hence there is a k such that $p(x) = x_{j_0}^k \, \widetilde{p}(x)$ in the first case, or $p(x) = (x_{i_0} \pm x_{j_0})^k \, \widetilde{p}(x)$ in the second (third) one. In any case, we choose k maximal, so that the variety $V(\widetilde{p})$ is the closure of the set $V(p) \setminus Z_{j_0}$ in the first case, or $V(p) \setminus D_{i_0 j_0}$ $(V(p) \setminus S_{i_0, j_0})$ in the second (third) case. Then proceed by factoring $\widetilde{p}$ in the same fashion. $\qquad\square$

Proof of Theorem 2.5. By Corollary 2.4, $p = c \prod_j p_j$, with each p_j a power of x_k or $(x_k \pm x_l)$. It also follows that c must be an integer since all coefficients of p are integers.

Since each of the factors is accurately evaluable, and we can get any integer constant c in front of p by repeated addition (followed, if need be, by negation), which are again accurate operations, the algorithm that forms their product and then adds/negates to obtain c evaluates p accurately.   □

Remark 2.6 *From Theorem 2.5, it follows that only homogeneous polynomials are accurately evaluable over $\mathbb{C}^n$.*

2.4.3 Toward a necessary and sufficient condition in the real case

In this section we show that accurate evaluability of a polynomial over $\mathbb{R}^n$ is ultimately related to accurate evaluability of its "dominant terms". This latter notion is formally defined later in this section. Informally, it describes the terms of the polynomial that dominate the remaining terms in a particular semialgebraic set close to a particular component of its variety; thus it depends on how we "approach" the variety of a polynomial.

For reasons outlined in Section 2.3, we consider here only homogeneous polynomials. Furthermore, most of this section is devoted to non-branching algorithms, but we do need branching for our statements at the end of the section. The reader will be alerted to any change in our basic assumptions.

Here is a short walk through this section:

- In Section **2.4.3.1. Homogeneity**, we discuss an expansion of the relative error $|p_{comp}(x, \delta) - p(x)|/|p(x)|$ as a function of x and δ, and prove a result about accurate evaluability of homogeneous polynomials that will be used in Section **2.4.3.3. Pruning**.
- In Section **2.4.3.2. Dominance**, we introduce the notion of dominance and present different ways of looking at an irreducible component of the variety $V(p)$ using various simple linear changes of variables. These changes of variables allow us to identify all the dominant terms of the polynomial, together with the "slices" of space where they dominate.
- In Section **2.4.3.3. Pruning**, we explain how to "prune" an algorithm to manufacture an algorithm that evaluates one of its dominant terms, and prove a necessary condition, Theorem 2.7, for the accurate evaluation of a homogeneous polynomial by a non-branching algorithm. Roughly speaking, this condition says that accurate evaluation

of the dominant terms we identified in Section **2.4.3.2. Dominance**, is necessary.

- In Section **2.4.3.4. Sufficiency of evaluating dominant terms**, we identify a special collection of dominant terms, together with the slices of space where they dominate. If accurately evaluable by (branching or non-branching) algorithms, these dominant terms allow us to construct a branching algorithm for the evaluation of the polynomial over the entire space (Theorem 2.8). These are just some of the terms present in the statement of Theorem 2.7.

2.4.3.1 Homogeneity

We begin by establishing some basic facts about non-branching algorithms that evaluate homogeneous polynomials.

Definition 2.11 *We call an algorithm $p_{comp}(x, \delta)$ with error set δ for computing $p(x)$ homogeneous of degree d if*

1. *the final output is of degree d in x;*
2. *no output of a computational node exceeds degree d in x;*
3. *the output of every computational node is homogeneous in x.*

Lemma 2.3 *If $p(x)$ is a homogeneous polynomial of degree d and if a non-branching algorithm evaluates $p(x)$ accurately by computing $p_{comp}(x, \delta)$, the algorithm must itself be homogeneous of degree d.*

Proof First note that the output of the algorithm must be of degree at least d in x, since $p_{comp}(x, \delta) = p(x)$ when $\delta = 0$. Let us now write the overall relative error as

$$rel_{err}(x, \delta) = \frac{p_{comp}(x, \delta) - p(x)}{p(x)} = \sum_{\alpha} \frac{p_\alpha(x)}{p(x)} \delta^\alpha$$

where α is a multi-index. If $p_{comp}(x, \delta)$ is accurate then $p_\alpha(x)/p(x)$ must be a bounded rational function on the domain ($\mathbb{R}^n$, or in the homogeneous case the sphere $S^{(n-1)}$). This implies, in particular, that the output cannot be of degree higher than d in x. So, Condition 1 of Definition 2.11 must be satisfied.

Now suppose Condition 3. of Definition 2.11 is violated. We would like to show that the final output is also inhomogeneous. We can assume without loss of generality that the algorithm does not contain nodes that do operations like $x - x$ or $0 \cdot x$ (these can be "pre-pruned" and replaced with a 0 source). There exists a highest node $(op(\cdot), \delta_i)$ whose output is not homogeneous. If it is the output node, we are done. Otherwise, look at the next node $(op(\cdot), \delta_j)$ on the path toward the output node. The output of $op(\cdot, \delta_j)$ is homogeneous. On the other hand, the output of $(op(\cdot), \delta_i)$ (which is one of the two inputs to $(op(\cdot), \delta_j)$) must be inhomogeneous in x and must contain a term $\delta_i r(x)$ with $r(x)$ an inhomogeneous polynomial in x.

If $op(\cdot, \delta_i)$ is the only input to $op(\cdot, \delta_j)$, then inhomogeneity will be present in both outputs, since neither doubling nor squaring can cancel it; contradiction. Otherwise there is another input to $op(\cdot, \delta_j)$ (call it $op(\cdot, \delta_k)$). The output of $op(\cdot, \delta_k)$ must therefore also be inhomogeneous to cancel the inhomogeneous $r(x)$. Since the DAG is acyclic, δ_i is not present in the output of $op(\cdot, \delta_k)$ or δ_k is not present in the output $op(\cdot, \delta_i)$. Without loss of generality, assume the former case. Then the term $\delta_i r(x)$ will create inhomogeneity in the output of $(op(\cdot), \delta_j)$, and hence $(op(\cdot), \delta_i)$ is *not* a highest node with inhomogeneous output, contradiction. Hence $p_{comp}(x, \delta)$ is not homogeneous in x, thus one of the $p_\alpha(x)$'s has to contain terms in x of higher or smaller degree than d.

Similarly, if Condition 2. of Definition 2.11 were violated, then for some δs the final output would be a polynomial of higher degree in x, and that would also mean some $p_\alpha(x)$ would be of higher degree in x.

In either of these cases, if some $p_\alpha(x)$ contained terms of smaller degree than d, by scaling the variables appropriately and letting some of them go to 0, we would deduce that $p_\alpha(x)/p(x)$ could not be bounded. If some $p_\alpha(x)$ contained terms of higher degree than d, by scaling the variables appropriately and letting some of them go to ∞, we would once again obtain that $p_\alpha(x)/p(x)$ could not be bounded. $\qquad\square$

This proof shows that an algorithm evaluates a homogeneous polynomial p accurately on $\mathbb{R}^n$ if and only if each fraction p_α/p is bounded on $\mathbb{R}^n$. It also shows each p_α has to be homogeneous of the same degree as p. Therefore, each fraction p_α/p is bounded on $\mathbb{R}^n$ if and only if it is bounded on the unit sphere $S^{(n-1)}$. We record this as a corollary.

Corollary 2.5 *A non-branching homogeneous algorithm is accurate on $\mathbb{R}^n$ if and only if it is accurate on $S^{(n-1)}$.*

2.4.3.2 Dominance

Now we begin our description of "dominant terms" of a polynomial. Given a polynomial p with an allowable variety $V(p)$, let us fix an irreducible component of $V(p)$. Any such component is described by linear allowable constraints, which, after reordering variables, can be grouped into l groups as

$$x_1 = \cdots = x_{k_1} = 0, \quad x_{k_1+1} = \cdots = \pm x_{k_2}, \ldots, \quad x_{k_{l-1}+1} = \cdots = \pm x_{k_l}.$$

To consider terms of p that "dominate" in a neighborhood of that component, we will change variables to map any component of a variety to a set of the form

$$\tilde{x}_1 = \cdots = \tilde{x}_{k_1} = 0, \quad \tilde{x}_{k_1+2} = \cdots = \tilde{x}_{k_2} = 0, \quad \ldots, \\ \tilde{x}_{k_{l-1}+2} = \cdots = \tilde{x}_{k_l} = 0. \tag{2.6}$$

The changes of variables we will use are defined inductively as follows.

Definition 2.12 *We call a change of variables associated with a set of the form*

$$\sigma_1 x_1 = \sigma_2 x_2 = \cdots = \sigma_k x_k, \qquad \sigma_l = \pm 1, \quad l = 1, \ldots, k,$$

basic *if it leaves one of the variables unchanged, which we will refer to as the* representative *of the group, and replaces the remaining variables by their sums (or differences) with the representative of the group. In other words,*

$$\tilde{x}_j := x_j, \quad \tilde{x}_l := x_l - \sigma_j \sigma_l x_j \qquad for \quad l \neq j,$$

where x_j is the representative of the group $x_1, \ldots, x_k$. A change of variables associated with a set of all x satisfying conditions

$$x_1 = \cdots = x_{k_1} = 0, \\ \sigma_{k_1+1} x_{k_1+1} = \sigma_{k_1+2} x_{k_1+2} = \cdots = \sigma_{k_2} x_{k_2}, \\ \cdots\cdots\cdots\cdots\cdots\cdots\cdots\cdots\cdots\cdots\cdots\cdots\cdots\cdots \\ \sigma_{k_{l-1}+1} x_{k_{l-1}+1} = \sigma_{k_{l-1}+2} x_{k_{l-1}+2} = \cdots = \sigma_{k_l} x_{k_l}, \tag{2.7}$$

$$\sigma_j = \pm 1 \ for \ all \ pertinent \ j$$

is basic *if it is a composition of the identity map on the first k_1 variables and $(l-1)$ basic changes of variables associated with each set $\sigma_{k_1+1} x_{k_1+1} = \cdots = \sigma_{k_2} x_{k_2}$ through $\sigma_{k_{l-1}+1} x_{k_{l-1}+1} = \cdots = \sigma_{k_l} x_{k_l}$.*

Finally, a change of variables associated with a set S of type (2.7) is standard *if it is a basic change of variables associated with some allowable irreducible superset $\widetilde{S} \supseteq S$ and it maps S to (2.6).*

Thus, a standard change of variables amounts to splitting the group $x_1, \ldots, x_{k_1}$ into smaller groups and either keeping the conditions $x_r = \cdots = x_q = 0$ or assigning arbitrary signs to members of each group so as to obtain a set $\sigma_r x_r = \cdots = \sigma_q x_q$. It may also involve splitting the chains of conditions $\sigma_{k_m+1} x_{k_m+1} = \cdots = \sigma_{k_{m+1}} x_{k_{m+1}}$ into several subchains.

The standard change of variables is then just one of the basic changes of variables associated with the obtained set.

Example 1 *There are 5×3 standard changes of variables associated with the set*

$$x_1 = x_2 = 0, \qquad x_3 = -x_4 = x_5 :$$

$\tilde{x}_1 = x_1,\ \tilde{x}_2 = x_2,\qquad \tilde{x}_3 = x_3,\ \tilde{x}_4 = x_4 + x_3,\ \tilde{x}_5 = x_5 - x_3,$ or

$\tilde{x}_1 = x_1,\ \tilde{x}_2 = x_2,\qquad \tilde{x}_3 = x_3 + x_4,\ \tilde{x}_4 = x_4,\ \tilde{x}_5 = x_5 + x_4,$ or

$\tilde{x}_1 = x_1,\ \tilde{x}_2 = x_2,\qquad \tilde{x}_3 = x_3 - x_5,\ \tilde{x}_4 = x_4 + x_5,\ \tilde{x}_5 = x_5,$ or

$\tilde{x}_1 = x_1,\ \tilde{x}_2 = x_2 - x_1,\ \tilde{x}_3 = x_3,\ \tilde{x}_4 = x_4 + x_3,\ \tilde{x}_5 = x_5 - x_3,$ or

$\tilde{x}_1 = x_1,\ \tilde{x}_2 = x_2 - x_1,\ \tilde{x}_3 = x_3 + x_4,\ \tilde{x}_4 = x_4,\ \tilde{x}_5 = x_5 + x_4,$ or

$\tilde{x}_1 = x_1,\ \tilde{x}_2 = x_2 - x_1,\ \tilde{x}_3 = x_3 - x_5,\ \tilde{x}_4 = x_4 + x_5,\ \tilde{x}_5 = x_5,$ or

$\tilde{x}_1 = x_1,\ \tilde{x}_2 = x_2 + x_1,\ \tilde{x}_3 = x_3,\ \tilde{x}_4 = x_4 + x_3,\ \tilde{x}_5 = x_5 - x_3,$ or

$\tilde{x}_1 = x_1,\ \tilde{x}_2 = x_2 + x_1,\ \tilde{x}_3 = x_3 + x_4,\ \tilde{x}_4 = x_4,\ \tilde{x}_5 = x_5 + x_4,$ or

$\tilde{x}_1 = x_1,\ \tilde{x}_2 = x_2 + x_1,\ \tilde{x}_3 = x_3 - x_5,\ \tilde{x}_4 = x_4 + x_5,\ \tilde{x}_5 = x_5,$ or

$\tilde{x}_1 = x_1 - x_2,\ \tilde{x}_2 = x_2,\ \tilde{x}_3 = x_3,\ \tilde{x}_4 = x_4 + x_3,\ \tilde{x}_5 = x_5 - x_3,$ or

$\tilde{x}_1 = x_1 - x_2,\ \tilde{x}_2 = x_2,\ \tilde{x}_3 = x_3 + x_4,\ \tilde{x}_4 = x_4,\ \tilde{x}_5 = x_5 + x_4,$ or

$\tilde{x}_1 = x_1 - x_2,\ \tilde{x}_2 = x_2,\ \tilde{x}_3 = x_3 - x_5,\ \tilde{x}_4 = x_4 + x_5,\ \tilde{x}_5 = x_5,$ or

$\tilde{x}_1 = x_1 + x_2,\ \tilde{x}_2 = x_2,\ \tilde{x}_3 = x_3,\ \tilde{x}_4 = x_4 + x_3,\ \tilde{x}_5 = x_5 - x_3,$ or

$\tilde{x}_1 = x_1 + x_2,\ \tilde{x}_2 = x_2,\ \tilde{x}_3 = x_3 + x_4,\ \tilde{x}_4 = x_4,\ \tilde{x}_5 = x_5 + x_4,$ or

$\tilde{x}_1 = x_1 + x_2,\ \tilde{x}_2 = x_2,\ \tilde{x}_3 = x_3 - x_5,\ \tilde{x}_4 = x_4 + x_5,\ \tilde{x}_5 = x_5,$

The supersets $\widetilde{S}$ for this example are the set S itself together with the set $\{x : x_1 + x_2 = 0, x_3 = -x_4 = x_5\}$ and the set $\{x : x_1 - x_2 = 0, x_3 = -x_4 = x_5\}$.

Note that we can write the vector of new variables $\tilde{x}$ as Cx where C is a matrix, so can label the change of variables by the matrix C.

Now let us consider components of the variety $V(p)$. We have seen that any given component of $V(p)$ can be put into the form $x_1 = x_2 = \cdots = x_k = 0$ using a standard change of variables, provided $V(p)$ is allowable. (To avoid cumbersome notation, we renumber all the variables set to zero as x_1 through x_k for our discussion that follows. We will return to the original description to introduce the notion of pruning.)

Write the polynomial $p(x)$ in the form

$$p(x) = \sum_{\lambda \in \Lambda} c_\lambda x_{[1:k]}^\lambda q_\lambda \left(x_{[k+1:n]} \right), \tag{2.8}$$

where, almost following MATLAB notation, we write $x_{[1:k]} := (x_1, \ldots, x_k)$, $x_{[k+1:n]} := (x_{k+1}, \ldots, x_n)$. Also, we let Λ be the set of all multi-indices $\lambda := (\lambda_1, \ldots, \lambda_k)$ occuring in the monomials of $p(x)$.

To determine all dominant terms associated with the component $x_1 = x_2 = \cdots = x_k = 0$, consider the Newton polytope P of the polynomial p with respect to the variables x_1 through x_k only, i.e., the convex hull of the exponent vectors $\lambda \in \Lambda$ (see, e.g., p. 71 of Miller and Sturmfels (2005)). Next, consider the normal fan $N(P)$ of P (see pp. 192–193 of Ziegler (1995)) consisting of the cones of all row vectors η from the dual space $(\mathbb{R}^k)^*$ whose dot products with $x \in P$ are maximal for x on a fixed face of P. That means that for every nonempty face F of P we take

$$N_F := \{\eta = (n_1, \ldots, n_k) \in (\mathbb{R}^k)^* :$$
$$F \subseteq \{x \in P : \eta x (:= \textstyle\sum_{j=1}^{k} n_j x_j) = \max_{y \in P} \eta y\}\}$$

and

$$N(P) := \{N_F : \ F \text{ is a face of } P\}.$$

Finally, consider the intersection of the negative of the normal fan $-N(P)$ and the nonnegative quadrant $(\mathbb{R}^k)^*_+$. This splits the first quadrant $(R^k)^*_+$ into several regions S_{Λ_j} according to which subsets Λ_j of exponents λ "dominate" close to the considered component of the variety $V(p)$, in the following sense:

Definition 2.13 *Let Λ_j be a subset of Λ that determines a face of the Newton polytope P of p such that the negative of its normal cone $-N(P)$ intersects $(\mathbb{R}^k)^*_+$ nontrivially (not only at the origin). Define $S_{\Lambda_j} \in (\mathbb{R}^k)^*_+$ to be the set of all nonnegative row vectors η such that*

$$\eta\lambda_1 = \eta\lambda_2 < \eta\lambda, \quad \forall \lambda_1, \lambda_2 \in \Lambda_j, \quad \text{and } \lambda \in \Lambda \setminus \Lambda_j.$$

Note that if x_1 through x_k are small, then the exponential change of variables $x_j \mapsto -\log|x_j|$ gives rise to a correspondence between the nonnegative part of $-N(P)$ and the space of original variables $x_{[1:k]}$. We map back the sets S_{Λ_j} into a neighborhood of 0 in $\mathbb{R}^k$ by lifting.[1]

Definition 2.14 *Let $F_{\Lambda_j} \subseteq [-1, 1]^k$ be the set of all points $x_{[1:k]} \in \mathbb{R}^k$ such that*

$$\eta := (-\log|x_1|, \ldots, -\log|x_k|) \in S_{\Lambda_j}.$$

Remark 2.7 *For any j, the closure of F_{Λ_j} contains the origin in $\mathbb{R}^k$.*

Remark 2.8 *Given a point $x_{[1:k]} \in F_{\Lambda_j}$, and given $\eta = (n_1, n_2, \ldots, n_k) \in S_{\Lambda_j}$, for any $t \in (0, 1)$, the vector $(x_1 t^{n_1}, \ldots, x_k t^{n_k})$ is in F_{Λ_j}. Indeed,*

---

[1] This is reminiscent of the concept of an amoeba introduced in Gelfand, Kapranov, and Zelevinsky (1994).

if $(-\log|x_1|, \ldots, -\log|x_k|) \in S_{\Lambda_j}$, *then so is* $(-\log|x_1|, \ldots, -\log|x_k|) -$ $\log|t|\eta$, *since all equalities and inequalities that define* S_{Λ_j} *will be preserved, the latter because* $\log|t| < 0$.

Example 2 *Consider the following polynomial*

$$p(x_1, x_2, x_3) = x_2^8 x_3^{12} + x_1^2 x_2^2 x_3^{16} + x_1^8 x_3^{12} + x_1^6 x_2^{14} + x_1^{10} x_2^6 x_3^4.$$

We show below the Newton polytope P *of* p *with respect to the variables* x_1, x_2, *its normal fan* $N(P)$, *the intersection* $-N(P) \cap R_+^2$, *the regions* S_{Λ_j}, *and the regions* F_{Λ_j}.

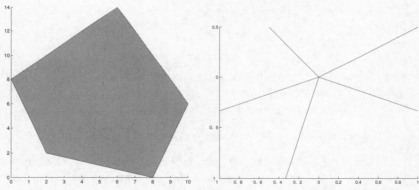

Figure 1.   A Newton polytope P.   Figure 2.   Its normal fan $N(P)$.

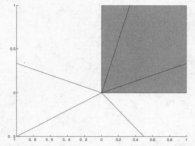
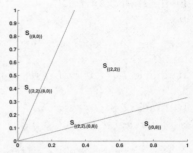

Figure 3. The intersection $-N(P) \cap \mathbb{R}_+^k$.   Figure 4.   The regions S_{Λ_j}.

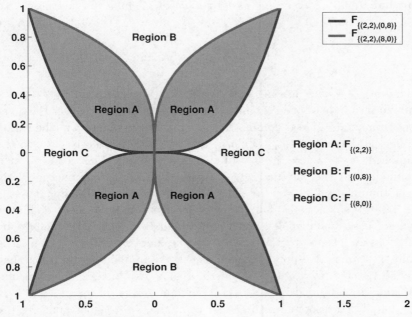

Figure 5. The regions F_{Λ_j}.

Definition 2.15 *We define the* dominant term *of $p(x)$ corresponding to the component $x_1 = \cdots = x_k = 0$ and the region F_{Λ_j} by*

$$p_{dom_j}(x) := \sum_{\lambda \in \Lambda_j} c_\lambda x_{[1:k]}^\lambda q_\lambda \big(x_{[k+1:n]} \big).$$

The following observations about dominant terms are immediate.

Lemma 2.4 *Let $\eta = (n_1, \ldots, n_k) \in S_{\Lambda_j}$ and let $d_j := \sum_{\lambda_i \in \Lambda_j} \lambda_i n_i$. Let x^0 be fixed and let*

$$x(t) := (x_1(t), \ldots, x_n(t)), \qquad x_j(t) := \begin{cases} t^{n_j} x_j^0 & j = 1, \ldots, k, \\ x_j^0, & j = k+1, \ldots, n. \end{cases}$$

Then $p_{dom_j}(x(t))$ has degree d_j in t and is the lowest degree term of $p(x(t))$ in t, that is

$$p(x(t)) = p_{dom_j}(x(t)) + o(t^{d_j}) \quad \text{as } t \to 0, \qquad \deg_t p_{dom_j}(x(t)) = d_j.$$

Proof Follows directly from the definition of a dominant term. $\qquad\square$

Corollary 2.6 *Under the assumptions of Lemma 2.4 suppose that* $p_{dom_j}(x^0) \neq 0$. *Then*

$$\lim_{t \to 0} \frac{p_{dom_j}(x(t))}{p(x(t))} = 1.$$

Thus p_{dom_j} is the leading term along each curve traced by $x(t)$ as t tends to zero from above. An important question now is whether the dominant term p_{dom_j} indeed dominates the remaining terms of p in the region F_{Λ_j} in the sense that $p_{dom_j}(x)/p(x)$ is close to 1 sufficiently close to the component $x_1 = \cdots = x_k = 0$ of the variety $V(p)$. This requires, at a minimum, that the variety $V(p_{dom_j})$ does not have a component strictly larger than the set $x_1 = \cdots = x_k = 0$. Note that most dominant terms of a polynomial actually fail this requirement. Indeed, most dominant terms of p are monomials, which correspond to regions F_{Λ_j} indexed by singletons Λ_j, hence to the vertices of the Newton polytope of p. The dominant terms corresponding to larger sets Λ_j are more useful, since they pick up terms relevant not only in the region F_{Λ_j} but also in its neighborhood. In Example 2 above the dominant terms for $F_{\{(2,2),(8,0)\}}$ and $F_{\{(2,2),(0,8)\}}$, corresponding to the edges of the Newton polygon, are the useful ones. This points to the fact that we should be ultimately interested only in dominant terms corresponding to the facets, i.e., the highest-dimensional faces, of the Newton polytope of p. Note that the convex hull of Λ_j is a facet of the Newton polytope N if and only if the set S_{Λ_j} is a one-dimensional ray.

The next lemma will be instrumental for our results in Section 2.4.3.4. It shows that each dominant term p_{dom_j} such that the convex hull of Λ_j is a facet of the Newton polytope of p and whose variety $V(p_{dom_j})$ does not have a component strictly larger than the set $x_1 = \cdots = x_k = 0$ indeed dominates the remaining terms in p in a certain "slice" $\widetilde{F}_{\Lambda_j}$ around F_{Λ_j}.

Lemma 2.5 *Let* p_{dom_j} *be the dominant term of a homogeneous polynomial* p *corresponding to the component* $x_1 = \cdots = x_k = 0$ *of the variety* $V(p)$ *and to the set* Λ_j *whose convex hull is a facet of the Newton polytope* N.

Let $\widetilde{S}_{\Lambda_j}$ *be any closed pointed cone in* $(\mathbb{R}^k)^*_+$ *with vertex at 0 that does not intersect other one-dimensional rays* S_{Λ_l}, $l \neq j$, *and contains* $S_{\Lambda_j} \setminus \{0\}$ *in its interior. Let* $\widetilde{F}_{\Lambda_j}$ *be the closure of the set*

$$\left\{ x_{[1:k]} \in [-1,1]^k : (-\log|x_1|, \ldots, -\log|x_k|) \in \widetilde{S}_{\Lambda_j} \right\}. \qquad (2.9)$$

Suppose the variety $V(p_{dom_j})$ *of* p_{dom_j} *is allowable and intersects* $\widetilde{F}_{\Lambda_j}$ *only at 0. Let* $\| \cdot \|$ *be any norm. Then, for any* $\delta = \delta(j) > 0$, *there exists*

$\varepsilon = \varepsilon(j) > 0$ *such that*

$$\left| \frac{p_{dom_j}(x_{[1:k]}, x_{[k+1:n]})}{p(x_{[1:k]}, x_{[k+1:n]})} - 1 \right| < \delta \quad \text{whenever} \quad \frac{\|x_{[1:k]}\|}{\|x_{[k+1:n]}\|} \leq \varepsilon \tag{2.10}$$
$$\text{and} \quad x_{[1:k]} \in \widetilde{F}_{\Lambda_j}.$$

Proof We prove the lemma in the case $\widetilde{F}_{\Lambda_j}$ does not intersect nontrivially any of the coordinate planes (the proof extends to the other case via limiting arguments). Let $\|x_{[k+1:n]}\| = 1$ and let x_1 through x_k be ± 1. If $\eta = (n_1, \ldots, n_k) \in \widetilde{S}_{\Lambda_j}$, then, directly from the definition of the set $\widetilde{F}_{\Lambda_j}$, the curve $(t^{n_1} x_1, \ldots, t^{n_k} x_k)$, $t \in (0, 1]$, lies in $\widetilde{F}_{\Lambda_j}$ (and every point in $\widetilde{F}_{\Lambda_j}$ lies on such a curve). Denote $(t^{n_1} x_1, \ldots, t^{n_k} x_k, x_{[k+1:n]})$ by $x(t)$ and let t decrease from 1 to 0, keeping the x_m, $m = 1, \ldots, n$, fixed. By the assumption of the Lemma, $p_{dom_j}(x(t))$ does not vanish for sufficiently small $t > 0$. Moreover, by Lemma 2.4, p_{dom_j} is the leading term of p in F_{Λ_j}. Since the cone $\widetilde{S}_{\Lambda_j}$ around S_{Λ_j} does not intersect any other one-dimensional rays S_{Λ_l}, $l \neq j$, all the monomials present in any term that dominates in $\widetilde{F}_{\Lambda_j} \setminus F_{\Lambda_j}$ are already present in p_{dom_j}. Thus p_{dom_j} contains all terms that dominate in $\widetilde{F}_{\Lambda_j}$. Therefore, there exists $\varepsilon(x) > 0$ such that $|p_{dom_j}(x(t))/p(x(t)) - 1| < \delta$ whenever $t < \varepsilon(x)$. The function $f : x \to \varepsilon(x)$ is lower semicontinuous. Since the set $S := \{x : x_m = \pm 1, \ m = 1, \ldots, k, \ \|x_{[k+1:n]}\| = 1\}$ is compact, the minimum $\varepsilon := \min f(S)$ is necessarily positive and satisfies (2.10). $\quad\square$

The above discussion of dominance was based on the transformation of a given irreducible component of the variety to the form $x_1 = \cdots = x_k = 0$. We must reiterate that the identification of dominant terms becomes possible only after a suitable change of variables C is used to put a given irreducible component into the standard form $x_1 = \cdots = x_k = 0$ and then the sets Λ_j are determined. Note however that the polynomial p_{dom_j} is given in terms of the original variables, i.e., as a sum of monomials in the original variables x_q and sums/differences $x_q \pm x_r$. We will therefore use the more precise notation $p_{dom_j, C}$ in the sequel.

Without loss of generality we can assume that any standard change of variables has the form

$$\begin{aligned}
x &= (x_{[1:k_1]}, x_{[k_1+1:k_2]}, \ldots, x_{[k_{l-1}+1:k_l]}) \longmapsto \\
\widetilde{x} &= (\widetilde{x}_{[1:k_1]}, \widetilde{x}_{[k_1+1:k_2]}, \ldots, \widetilde{x}_{[k_{l-1}+1:k_l]}), \quad \text{where} \\
\widetilde{x}_{k_m+1} &:= x_{k_m+1}, \quad \widetilde{x}_{k_m+2} := x_{k_m+2} - \sigma_{k_m+2} x_{k_m+1}, \ \ldots, \\
\widetilde{x}_{k_{m+1}} &:= x_{k_{m+1}} - \sigma_{k_{m+1}} x_{k_m+1}, \\
k_0 &:= 0, \quad \sigma_r = \pm 1 \quad \text{for all pertinent } r
\end{aligned} \tag{2.11}$$

Note also that we can think of the vectors $\eta \in S_{\Lambda_j}$ as being indexed by integers 1 through k_l, i.e., $\eta = (n_1, \ldots, n_{k_l})$. Moreover, to define pruning in the next subsection we will assume that

$$n_{k_m+1} \le n_r \quad \text{for all} \quad r = k_m + 2, \ldots, k_{m+1}$$
$$\text{and for all} \quad m = 0, \ldots, l-1. \tag{2.12}$$

Remark 2.9 *This condition is trivially satisfied if $n_{k_m+1} = 0$, as is the case for any group $x_{k_m+1} = \sigma_{k_m+2} x_{k_m+2} = \cdots = \sigma_{k_{m+1}} x_{k_{m+1}}$ of original conditions that define the given irreducible component of $V(p)$, since x_{k_m+1} does not have to be close to 0 in the neighborhood of that component of $V(p)$. If, however, the same group of equalities was created from the original conditions $x_{k_m+1} = x_{k_m+2} = \cdots = x_{k_{m+1}} = 0$ due to the particular change of variables C, the condition (2.12) is no longer forced upon us. Yet (2.12) can be assumed without loss of generality. Indeed, if, say, $n_{k_m+2} < n_{k_m+1}$, then we can always switch to another standard change of variables by taking x_{k_m+2} to be the representative of the group $x_{k_m+1}, \ldots, x_{k_{m+1}}$ and taking the sums/differences with x_{k_m+2} as the other new variables. Also note that (2.12) is satisfied either by all or by no vectors in S_{Λ_j}. In other words, (2.12) is a property of the entire set S_{Λ_j}. So, with a slight abuse of terminology we will say that a set S_{Λ_j} satisfies or fails (2.12).*

Finally note that the curves $(x(t))$ corresponding to the change of variables (2.11) are described as follows:

$$x(t) := (x_{[1:k_1]}(t), x_{[k_1+1:k_2]}(t), \ldots, x_{[k_{l-1}+1:k_l]}(t), x_{[k_l+1:n]}), \quad \text{where}$$
$$x_{[k_m+1:k_{m+1}]}(t) := (t^{n_{k_m+1}} x_{k_m+1}, t^{n_{k_m}+2} x_{k_m+2} + \sigma_{k_m+2} t^{n_{k_m+1}} x_{k_m+1}, \ldots,$$
$$t^{n_{k_m+1}} x_{k_{m+1}} + \sigma_{k_{m+1}} t^{n_{k_m+1}} x_{k_m+1}) \quad \text{where} \quad k_0 := 1, \quad m = 0, \ldots, l.$$
$$\tag{2.13}$$

This description will be instrumental in our discussion of pruning, which follows immediately.

2.4.3.3 Pruning

Now we discuss how to convert an accurate algorithm that evaluates a polynomial p into an accurate algorithm that evaluates a selected dominant term $p_{dom_j,C}$. This process, which we will refer to as *pruning*, will consist of deleting some vertices and edges and redirecting certain other edges in the DAG that represents the algorithm.

Definition 2.16 (Pruning) *Given a non-branching algorithm represented by a DAG for computing $p_{comp}(x, \delta)$, a standard change of variables C of the form (2.11) and a subset $\Lambda_j \in \Lambda$ satisfying (2.12), we choose any $\eta \in S_{\Lambda_j}$, we input (formally) the expression (2.13), and then perform the following process.*

We can perform one of two actions: redirection *or* deletion. *By* redirection *(of sources) we mean replacing an edge from a source node corresponding to a variable x_j to a computational node i by an edge from the representative x_{rep} of x_j followed by exact negation if $\sigma_j = -1$. This corresponds to replacing x_j by the product $\sigma_j x_{rep}$. To define* deletion, *consider a node i with distinct input nodes j and k. Then deletion of node i from node j means deleting the out-edge to node i from node j, changing the origin of all out-edges from node i to input node k, and deleting node i.*

Starting at the sources, we process each node as follows, provided that both its inputs have already been processed (this can be done because of acyclicity). Let the node being processed be node i, i.e., $(op(\cdot), \delta_i)$, and assume it has input nodes k and l. Both inputs being polynomials in t, we determine the lowest degree terms in t present in either of them and denote these degrees by $\deg(k)$ and $\deg(l)$.

if $op(\cdot) = \cdot$ **and one or both inputs are sources,** *then*

  redirect each source input.

if $op(\cdot) = \pm$, *then*

  if $\deg(k) \neq \deg(l)$, **say** $\deg(k) > \deg(l)$, *delete input node i from node k.*

  else *If nodes k and l are sources and the operation $op(\cdot)$ leads to cancellation of their lowest degree terms in t, examine their second-lowest degree terms. If those degrees coincide or if one second-lowest term is missing, we change nothing. If one is bigger than the other, we do not change the source containing the lower degree term in t, but redirect the other source.*

  If only one of nodes k and l is a source or if both inputs are sources but there is no cancellation of lowest degree terms, redirect each source.

We then delete inductively all nodes which no longer are on any path to the output.

We call this process pruning, *and denote the output of the pruned algorithm by $p_{dom_j, C, comp}(x, \delta)$.*

Remark 2.10 *Note that the outcome of pruning does not depend on the choice of $\eta \in S_{\Lambda_j}$. Since each region S_{Λ_j} is determined by linear homogeneous equalities and inequalities with integer coefficients, the vector η can always be chosen to have all integer entries.*

Example 3 *Figure 6 shows an example of pruning an algorithm that evaluates the polynomial*

$$x_1^2 x_2^2 + (x_2 - x_3)^4 + (x_3 - x_4)^2 x_5^2$$

using the substitution

$$(tx_1, x_2, tx_3 + x_2, tx_4 + x_2, x_5)$$

near the component

$$x_1 = 0, \quad x_2 = x_3 = x_4.$$

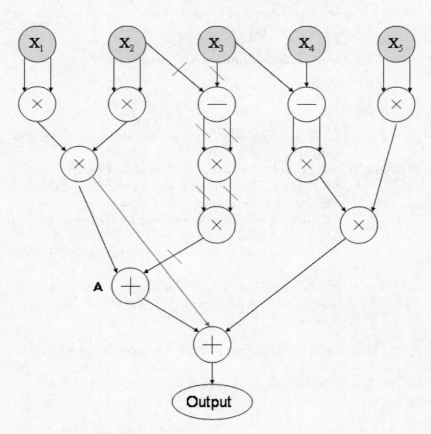

Figure 6.   Pruning an algorithm for $p(x) = x_1^2 x_2^2 + (x_2 - x_3)^4$
              $+(x_3 - x_4)^2 x_5^2$.

The result of pruning is an algorithm that evaluates the dominant term

$$x_1^2 x_2^2 + (x_3 - x_4)^2 x_5^2.$$

One of two branches leading to the node A is pruned due to the fact that it computes a quantity of order $O(t^4)$ whereas the other branch produces a quantity of order $O(t^2)$.

The output of the original algorithm is given by

$$
\begin{aligned}
\big(\big(& (x_1^2(1 + \delta_1)x_2^2(1 + \delta_2)(1 + \delta_3) \\
& + (x_2 - x_3)^4(1 + \delta_4)^4(1 + \delta_5)^2(1 + \delta_6)\big)\,(1 + \delta_7) \\
& + (x_3 - x_4)^2(1 + \delta_8)^2(1 + \delta_9)x_5^2(1 + \delta_{10})(1 + \delta_{11})\big)\,(1 + \delta_{12}).
\end{aligned}
$$

The output of the pruned algorithm is

$$
\begin{aligned}
\big(& x_1^2(1 + \delta_1)x_2^2(1 + \delta_2)(1 + \delta_3) \\
& + (x_3 - x_4)^2(1 + \delta_8)^2(1 + \delta_9)x_5^2(1 + \delta_{10})(1 + \delta_{11})\big)\,(1 + \delta_{12}).
\end{aligned}
$$

Let us prove that this process will indeed produce an algorithm that accurately evaluates the corresponding dominant term.

Theorem 2.7 *Suppose a non-branching algorithm evaluates a polynomial p accurately on $\mathbb{R}^n$ by computing $p_{comp}(x, \delta)$. Suppose C is a standard change of variables (2.11) associated with an irreducible component of $V(p)$. Let $p_{dom_j, C}$ be one of the corresponding dominant terms of p and let S_{Λ_j} satisfy (2.12). Then the pruned algorithm defined in Definition 2.16 with output $p_{dom_j, C, comp}(x, \delta)$ evaluates $p_{dom_j, C}$ accurately on $\mathbb{R}^n$. In other words, being able to compute all such $p_{dom_j, C}$ for all components of the variety $V(p)$ and all standard changes of variables C accurately is a condition necessary to compute p accurately.*

Proof Directly from the definition of pruning it can be seen that the output of each computational node is a homogeneous polynomial in t. This can be checked inductively starting from computational nodes operating on two sources, using the pruning rules. Moreover, the pruning rules are equivalent to taking the lowest degree terms in t (as well as setting some δs to zero). This can be checked inductively as well, once we rule out the situation when a $\pm$ node in the original algorithm leads to exact cancellation of lowest degree terms of the inputs, and at least one of the inputs is not a source. Indeed, in that case one of the inputs contains a factor $(1 + \delta)$ and that δ by acyclicity is not present in the other input. Therefore no exact cancellation of lowest degree terms can occur.

Thus, the final output $p_{dom_j,C,comp}(x,\delta)$ of the pruned algorithm takes the lowest degree terms in t of the final output of the original algorithm, so $p_{dom_j,C,comp}(x,\delta)$ is homogeneous in t (of degree $d_j = \eta\lambda = \sum \lambda_i n_i$). We write

$$\frac{p_{comp}(x,\delta) - p(x)}{p(x)} = \sum_\alpha \frac{p_\alpha(x)}{p(x)} \delta^\alpha,$$

$$\frac{p_{dom_j,C,comp}(x) - p_{dom_j,C}(x)}{p_{dom_j,C}(x)} = \sum_\alpha \frac{p_{\alpha,dom_j,C}(x)}{p_{dom_j,C}(x)} \delta^\alpha.$$

Note that, in eliminating nodes and redirecting the out-edges in the pruning process, we do the equivalent of setting some of the δs to 0, and hence the set of monomials δ^α present in the second sum is a subset of the set of monomials δ^α present in the first sum.

Indeed, first of all, we can focus on the effect of deleting some nodes, since redirection does not affect δs at all, because only sources can be redirected. So, if δ^α appears in the second sum, there is a path which yields the corresponding multi-index, on which some term in $p_{\alpha,dom_j,C}(x)$ is computed. But since this path survived the pruning, there is a corresponding path in the original DAG which perhaps has a few more nodes that have been deleted in the pruning process and a few source nodes that were redirected. In the computation, the effect of deleting a node was to set that δ equal to 0 (and make some terms of higher degree disappear). So the surviving term was also present in the computation of $p_{comp}(x,\delta)$, with the same multi-index: just choose the 1 in the $(1 + \delta)$ each time when you hit a node that will be deleted (i.e., whose δ will be set to 0).

Now note that $p_{\alpha,dom_j,C}(x)$ is the leading term of $p_\alpha(x)$, i.e., the term of smallest degree d_j in t. This happens since each term of degree d_j in $p_\alpha(x)$ must survive on the same path in the DAG, with the same choices of 1 in $(1 + \delta)$ each time we hit a deleted node.

We can now prove that $p_{dom_j,C,comp}(x)$ is accurate. To do that, it is enough to show that each $p_{\alpha,dom_j,C}(x)/p_{dom_j,C}(x)$ is bounded, provided that there is some constant M such that $|p_\alpha(x)/p(x)| \leq M$ for all x.

Choose a point $x = (x_1, \ldots, x_n)$ not on the variety of $p_{dom_j,C}$ and consider the curve traced by the associated point $x(t)$ from (2.13) as t tends to 0. Since both $p_{\alpha,dom_j,C}(x(t))$ and $p_{dom_j,C}(x(t))$ are homogeneous of degree d_j in t, we have

$$\left| \frac{p_{\alpha,dom_j,C}(x)}{p_{dom_j,C}(x)} \right| = \left| \frac{p_{\alpha,dom_j,C}(x(t))}{p_{dom_j,C}(x(t))} \right| = \lim_{t \to 0} \left| \frac{p_{\alpha,dom_j,C}(x(t))}{p_{dom_j,C}(x(t))} \right|.$$

Since both $p_{\alpha,dom_j,C}(x(t))$ *and* $p_{dom_j,C}(x(t))$ are the dominant terms in

$p_\alpha(x)$, respectively $p(x)$, along the curve $\{x(t) : t \to 0\}$, we conclude that

$$\left| \frac{p_{\alpha,dom_j,C}(x)}{p_{dom_j,C}(x)} \right| = \lim_{t\to 0} \left| \frac{p_{\alpha,dom_j,C}(x(t))}{p_{dom_j,C}(x(t))} \right| = \lim_{t\to 0} \left| \frac{p_\alpha(x(t))}{p(x(t))} \right| \leq M .$$

Invoking the density of the Zariski open set $\{x : p_{dom_j,C}(x) \neq 0\}$, we are done. $\qquad\square$

2.4.3.4 Sufficiency of evaluating dominant terms

Our next goal is to prove a converse of a sort to Theorem 2.7. Strictly speaking, the results that follow do not provide a true converse, since branching is needed to construct an algorithm that evaluates a polynomial p accurately from algorithms that evaluate its dominant terms accurately.

For the rest of this section, we make two assumptions, viz., that our polynomial p is homogeneous and irreducible. The latter assumption effectively reduces the problem to that of accurate evaluation of a nonnegative polynomial, due to the following lemma.

Lemma 2.6 *If a polynomial p is irreducible and has an allowable variety $V(p)$, then it is either a constant multiple of a linear form that defines an allowable hyperplane or it does not change its sign in $\mathbb{R}^n$.*

Proof Suppose that p changes its sign. Then the sets $\{x : p(x) > 0\}$ and $\{x : p(x) < 0\}$ are both open, hence the part of the variety $V(p)$ whose neighborhood contains points from both sets must have dimension $n-1$. The only allowable sets of dimension $n-1$ are allowable hyperplanes. Therefore, the linear polynomial that defines an allowable hyperplane must divide p. As p is assumed to be irreducible, p must be a constant multiple of the linear polynomial.

Thus, unless p is a constant multiple of a linear factor of the type x_i or $x_i \pm x_j$, it must satisfy $p(x) \geq 0$ for all $x \in \mathbb{R}^n$ or $p(x) \leq 0$ for all $x \in \mathbb{R}^n$. $\qquad\square$

From now on we therefore restrict ourselves to the nontrivial case when a (homogeneous and irreducible) polynomial p is nonnegative everywhere in $\mathbb{R}^n$.

Theorem 2.8 *Let p be a homogeneous nonnegative polynomial whose variety $V(p)$ is allowable. Suppose that all dominant terms $p_{dom_j,C}$ for all components of the variety $V(p)$, all standard changes of variables C and all subsets Λ_j satisfying (2.12) are accurately evaluable. Then there exists a branching algorithm that evaluates p accurately over $\mathbb{R}^n$.*

Proof We first show how to evaluate p accurately in a neighborhood of each irreducible component of its variety $V(p)$. We next evaluate p accurately off these neighborhoods of $V(p)$. The final algorithm will involve branching depending on which region the input belongs to, and the subsequent execution of the corresponding subroutine. We fix the relative accuracy η that we want to achieve.

Consider a particular irreducible component V_0 of the variety $V(p)$. Using any standard change of variables C, say, a basic change of variables associated with V_0, we map V_0 to a set of the form $\widetilde{x}_1 = \cdots = \widetilde{x}_k = 0$. Our goal is to create an ε-neighborhood of V_0 where we can evaluate p accurately. It will be built up from semialgebraic ε-neighborhoods. We begin with any set S containing a neighborhood V_0, say, we let S coincide with $[-1,1]^k \times \mathbb{R}^{n-k}$ after the change of variables C. We partition the cube $[-1,1]^k$ into sets $\widetilde{F}_{\Lambda_j}$ of type (2.9) as follows: We consider, as in Section 2.4.3.2 above, the Newton polytope P of p, form the intersection of the negative of its normal fan $-N(P)$ with the nonnegative quadrant $\mathbb{R}_+^k$ (in the new coordinate system), and determine the sets S_{Λ_j}. If condition (2.12) fails for some of the sets S_{Λ_j}, we transform them using a suitable standard change of variables as described in Remark 2.9 so as to meet condition (2.12). For the rest of the argument, we can assume that all sets S_{Λ_j} satisfy (2.12).

To form conic neighborhoods of one-dimensional rays of $-N(P) \cap \mathbb{R}_+^k$ (which are normal to facets of the Newton polytope), we intersect $-N(P) \cap \mathbb{R}_+^k$ with, say, the hyperplane $\widetilde{x}_1 + \cdots + \widetilde{x}_k = 1$. Perform the Voronoi tesselation (see, e.g., Ziegler (1995)) of the simplex $\widetilde{x}_1 + \cdots + \widetilde{x}_k = 1$, $\widetilde{x}_j \geq 0$, $j = 1, \ldots, k$ relative to the intersection points of $-N(P) \cap \mathbb{R}_+^k$ with the hyperplane $\widetilde{x}_1 + \cdots + \widetilde{x}_k = 1$. Connecting each Voronoi cell of the tesselation to the origin $\widetilde{x}_1 = \cdots \widetilde{x}_k = 0$ by straight rays, we obtain cones $\widetilde{S}_{\Lambda_j}$ and the corresponding sets $\widetilde{F}_{\Lambda_j}$ of type (2.9). Note that the Voronoi cells and therefore the cones $\widetilde{S}_{\Lambda_j}$ are determined by rational inequalities since the tesselation centers have rational coordinates. Hence the sets $\widetilde{F}_{\Lambda_j}$ are semialgebraic, and moreover are determined by polynomial inequalities with integer coefficients. Indeed, even though the sets $\widetilde{F}_{\Lambda_j}$ are defined using logarithms, the resulting inequalities are among powers of absolute values of the variables and/or their sums and differences. For example, if a particular set $\widetilde{F}_{\Lambda_j}$ is described by the requirement that $(-\log|x_1|, -\log|x_2|)$ lie between two lines through the origin, with slopes $1/2$ and $2/3$, respectively, this translates into the condition $|x_2|^4 \leq |x_1|^2 \leq |x_2|^3$.

Now consider a particular "slice" $\widetilde{F}_{\Lambda_j}$ and the dominant term $p_{dom_j,C}$. By Theorem 2.7, the dominant term $p_{dom_j,C}$ must be accurately evaluable everywhere. Hence, in particular, its variety $V(p_{dom_j,C})$ must be allowable.

Since the polynomial p vanishes on V_0, the dominant term $p_{dom_j,C}$ must vanish on V_0 as well. So, there are two possibilities: $V(p_{dom_j,C}) \cap \widetilde{F}_{\Lambda_j}$ either coincides with V_0 or is strictly larger.

In the first case we apply Lemma 2.5 to show that we can evaluate p accurately in $\widetilde{F}_{\Lambda_j}$ sufficiently close to V_0, as follows: Since the polynomial $p_{dom_j,C}$ is accurately evaluable everywhere, for any number $\eta_j > 0$ there exists an algorithm with output $p_{dom_j,C,comp}$ such that $|p_{dom_j,C,comp} - p_{dom_j,C}|/|p_{dom_j,C}| < \eta_j$ everywhere. Next, by Lemma 2.5, for any $\delta_j > 0$ there exists $\varepsilon_j > 0$ such that (2.10) holds. Choose η_j and δ_j so that

$$\eta_j(1 + \delta_j) + \delta_j < \eta.$$

Then we have

$$\left| \frac{p_{dom_j,C,comp} - p}{p} \right| \leq \frac{|p_{dom_j,C,comp} - p_{dom_j,C}| + |p_{dom_j,C} - p|}{|p|}$$

$$= \frac{|p_{dom_j,C,comp} - p_{dom_j,C}|}{|p_{dom_j,C}|} \cdot \frac{|p_{dom_j,C}|}{|p|} + \frac{|p_{dom_j,C} - p|}{|p|}$$

$$\leq \eta_j(1 + \delta_j) + \delta_j < \eta$$

in the ε_j-neighborhood of V_0 within the set $\widetilde{F}_{\Lambda_j}$. Therefore, p can be evaluated by computing $p_{dom_j,C,comp}$ to accuracy η in the ε_j-neighborhood of V_0 within $\widetilde{F}_{\Lambda_j}$.

In the second case $V(p_{dom_j,C})$ has an irreducible component, say, W_0, that is strictly larger than V_0 and intersects $\widetilde{F}_{\Lambda_j}$ nontrivially. Since $V(p_{dom_j,C})$ is allowable, it follows from Definition 2.12 that there exists a standard change of variables C_1 associated with V_0 that maps W_0 to a set $\widetilde{x}_1 = \cdots = \widetilde{x}_l = 0$, $l < k$. Use that change of variables and consider the new Newton polytope and dominant terms. Since the polynomial p is positive in $S \setminus V_0$, there are terms in p that do not contain variables $\widetilde{x}_1$ through $\widetilde{x}_l$. Therefore each new p_{dom_r,C_1} picks up some of those non-vanishing terms. Hence the (allowable) varieties $V(p_{dom_r,C_1})$ have irreducible components strictly smaller than W_0 (but still containing V_0). So, we can subdivide the set $\widetilde{F}_{\Lambda_j}$ further using the sets $\widetilde{F}_{\Lambda_r}$ coming from the change of variables C_1 and the resulting dominant terms will vanish on a set strictly smaller than W_0. In this fashion, we refine our subdivision repeatedly until we obtain a subdivision of the original set S into semialgebraic pieces (S_j) such that the associated dominant terms p_{dom_j} vanish in S_j only on V_0. Applying Lemma 2.5 in each such situation, we conclude that p can be evaluated accurately sufficiently close to V_0 within each piece S_j.

For each V_0, we therefore can find a collection (S_j) of semialgebraic sets, all determined by polynomial inequalities with integer coefficients, and the corresponding numbers ε_j, so that the polynomial p can be evaluated with accuracy η in each ε_j-neighborhood of V_0 within the piece S_j. Note that we can assume that each ε_j is a reciprocal of an integer, so that testing whether a particular point x is within ε_j of V_0 within S_j can be done by branching based on polynomial inequalities with integer coefficients.

The final algorithm will be organized as follows. Given an input x, determine by branching whether x is in S_j and within the corresponding ε_j of a component V_0. If that is the case, evaluate $p(x)$ using the algorithm that is accurate in S_j in that neighborhood of V_0. For x not in any of the neighborhoods, evaluate p by Horner's rule. Since the polynomial p is strictly positive off the neighborhoods of the components of its variety, the reasoning of Section 2.3 applies, showing that the Horner's rule algorithm is accurate. If x is on the boundary of a set S_j, any applicable algorithm will do, since the inequalities we use are not strict. Thus the resulting algorithm for evaluating p will have accuracy η as required. $\qquad\square$

2.4.3.5 Obstacles to full induction

The reasoning above suggests that there could be an inductive decision procedure that would allow us to determine whether or not a given polynomial is accurately evaluable by reducing the problem for the original polynomial p to the same problem for its dominant terms, then their dominant terms, and so forth, going all the way to monomials or other polynomials that are easy to analyze. However, this idea would only work if the dominant terms were somehow "simpler" than the original polynomial itself. This approach would require an induction variable that would decrease at each step.

Two possible choices are the number of variables or the degree of the polynomial under consideration. Sometimes, however, neither of the two goes down, and moreover, the dominant term may even coincide with the polynomial itself. For example, if

$$p(x) = A(x_{[3:n]})x_1^2 + B(x_{[3:n]})x_1x_2 + C(x_{[3:n]})x_2^2$$

where A, B, C are nonnegative polynomials in x_3 through x_n, then the only useful dominant term of p in the neighborhood of the set $x_1 = x_2 = 0$ is the polynomial p itself. Thus no progress whatsoever is made in this situation.

Another possibility is induction on domains but we do not yet envision how to make this idea precise, since we do not know exactly when a given polynomial is accurately evaluable on a given domain.

Further work to establish a full decision procedure is therefore highly desirable.

2.5 "Black-box" arithmetic

In this section we prove a necessary condition (for both the real and the complex cases) for a more general type of arithmetic, which allows for "black-box" polynomial operations. We describe the type of operations below.

Definition 2.17 *We call a black-box operation any type of operation that takes a number of inputs (real or complex) $x_1, \ldots, x_k$ and produces an output q such that q is a polynomial in $x_1, \ldots, x_k$.*

Example 4 $q(x_1, x_2, x_3) = x_1 + x_2 x_3$.

Remark 2.11 *Note that $+, -,$ and $\cdot$ are all black-box operations.*

Consider a fixed set of multivariate polynomials $\{q_j : j \in J\}$ with real or complex inputs (this set may be infinite). In our model under consideration, the arithmetic operations allowed are given by the black-box operations $q_1, \ldots, q_k$, and negation (which will be dealt with by way of dotted edges, as in Section 2.4). With the exception of negation, which is exact, all the others yield a $rnd(op(a_1, \ldots, a_l)) = op(a_1, \ldots, a_l)(1 + \delta)$, with $|\delta| < \epsilon$ (ϵ here is the machine precision). All arithmetic operations have unit cost. We consider the same arithmetical models as in Section 2.2.7, with this larger class of operations.

2.5.1 Necessity: real and complex

In order to see how the statement of the necessity Theorem 2.3 changes, we need to introduce a different notion of allowability.

Recall that we denote by $\mathcal{S}$ the space of variables (which may be either $\mathbb{R}^n$ or $\mathbb{C}^n$). From now on we will denote the set $\{1, \ldots, n\}$ by $\mathcal{A}$.

Definition 2.18 *Let $p(x_1, \ldots, x_n)$ be a multivariate polynomial over $\mathcal{S}$ with variety $V(p)$. Let $\mathcal{A}_Z \subseteq \mathcal{A}$, and let $\mathcal{A}_D, \mathcal{A}_S \subseteq \mathcal{A} \times \mathcal{A}$. Modify p as follows: impose conditions of the type Z_i for each $i \in \mathcal{A}_Z$, and of type D_{ij}, respectively S_{ij}, on all pairs of variables in $\mathcal{A}_D$, respectively $\mathcal{A}_S$. Rewrite p subject to those conditions (e.g. set $X_i = 0$ for all $i \in \mathcal{A}_Z$), and denote it by $\tilde{p}$, and denote by $\mathcal{A}_R$ the set of remaining independent variables (use* *the convention which eliminates the second variable in each pair in $\mathcal{A}_D$ or $\mathcal{A}_S$).*

Choose a set $T \subseteq \mathcal{A}_R$, and let

$$V_{T,\mathcal{A}_Z,\mathcal{A}_D,\mathcal{A}_S}(p) = \cap_\alpha V(q_\alpha) \ ,$$

where the polynomials q_α are the coefficients of the expansion of $\tilde{p}$ in the variables x_T:

$$\tilde{p}(x_1,\ldots,x_n) = \sum_\alpha q_\alpha x_T^\alpha \ ,$$

with q_α being polynomials in $x_{\mathcal{A}_R \setminus T}$ only.

Finally, let $\mathcal{A}_N$ be a subset of $\mathcal{A}_R \setminus T$. We negate each variable in $\mathcal{A}_N$, and let $V_{T,\mathcal{A}_Z,\mathcal{A}_D,\mathcal{A}_S,\mathcal{A}_N}(p)$ be the variety obtained from $V_{T,\mathcal{A}_Z,\mathcal{A}_D,\mathcal{A}_S}(p)$, with each variable in $\mathcal{A}_N$ negated.

Remark 2.12 $V_{\emptyset,\emptyset,\emptyset,\emptyset,\emptyset}(p) = V(p)$. *We also note that, if we have a black-box computing p, then the set of all polynomials $\tilde{p}$ that can be obtained from p by permuting, repeating, and negating the variables (as in the definition above) is exactly the set of all polynomials that can be evaluated with a single rounding error, using that black box.*

Definition 2.19 *For simplicity, we denote a set $(T, \mathcal{A}_Z, \mathcal{A}_D, \mathcal{A}_S, \mathcal{A}_N)$ by $\mathcal{I}$, and a set $(T, \mathcal{A}_Z, \mathcal{A}_D, \mathcal{A}_S)$ by $\mathcal{I}_+$.*

Example 5 *Let $p(x, y, z) = x + y \cdot z$ (the fused multiply-add). We record below all possibilities for $\mathcal{I} = (T, \mathcal{A}_Z, \mathcal{A}_D, \mathcal{A}_S, \mathcal{A}_N)$, together with the obtained subvariety $V_\mathcal{I}(p)$.*

Without loss of generality, assume that we have eliminated all redundant or complicated conditions, like $(x, y) \in \mathcal{A}_D$ and $(x, y) \in \mathcal{A}_S$ (which immediately leads to $x = y = 0$, that is, $x, y \in \mathcal{A}_Z$). We assume thus that all variables not present in $\mathcal{A}_Z$ cannot be deduced to be 0 from conditions imposed by $\mathcal{A}_D$ or/and $\mathcal{A}_S$.

We obtain that all possibilities for $V_\mathcal{I}(p)$ are, up to a permutation of the variables,

  ⋄ $\{x = 0\}$, $\{x = 1\}$, $\{x = -1\}$,
  ⋄ $\{x = 0\} \cup \{x = 1\}$, $\{x = 0\} \cup \{x = -1\}$,
  ⋄ $\{x = 0\} \cup \{y = 0\}$, $\{x = 0\} \cup \{y = 1\}$, $\{x = 0\} \cup \{y = -1\}$,
  ⋄ $\{x = -y^2\}$, $\{x = y^2\}$, $\{x - y \cdot z = 0\}$, *and* $\{x + y \cdot z = 0\}$.

Definition 2.20 *We define $q_{-2}(x_1, x_2) = x_1 x_2$, $q_{-1}(x_1, x_2) = x_1 + x_2$, and $q_0(x_1, x_2) = x_1 - x_2$.*

Remark 2.13 *The sets*

$$1. \quad Z_i = \{x : x_i = 0\}, \tag{2.14}$$

$$2. \quad S_{ij} = \{x : x_i + x_j = 0\}, \tag{2.15}$$

$$3. \quad D_{ij} = \{x : x_i - x_j = 0\}, \tag{2.16}$$

and unions thereof, describe all non-trivial (neither $\emptyset$ nor $\mathcal{S}$) sets of type $V_{\mathcal{I}}$, for q_{-2}, q_{-1}, and q_0.

We will assume from now on that the black-box operations q_{-2}, q_{-1}, q_0 defined in 2.20, and some arbitrary extra operations q_j, with $j \in J$ (J may be infinite) are given and fixed.

Definition 2.21 *We call any set $V_{\mathcal{I}}(q_j)$ with $\mathcal{I} = (T, \mathcal{A}_Z, \mathcal{A}_D, \mathcal{A}_S, \mathcal{A}_N)$ as defined above and q_j a black-box operation basic q-allowable.*

We call any set R irreducible q-allowable if it is an irreducible component of a (finite) intersection of basic q-allowable sets, i.e., when R is irreducible and

$$R \subseteq \cap_l Q_l \ ,$$

where each Q_l is a basic q-allowable set.

We call any set Q q-allowable if it is a (finite) union of irreducible q-allowable sets, i.e.,

$$Q = \cup_j R_j \ ,$$

where each R_j is an irreducible q-allowable set.

Any set R which is not q-allowable we call q-unallowable.

Remark 2.14 *Note that the above definition of q-allowability is closed under taking union, intersection, and irreducible components. This parallels the definition of allowability for the classical arithmetic case – in the classical case, every allowable set was already irreducible (being an intersection of hyperplanes).*

Once again, we need to build the setup to state and prove our new necessity condition. To do this, we will modify the statements of the definitions and the statements and proofs of the lemmas from Section 2.4.1. Since most proofs just follow in the footsteps of those from Section 2.4.1, instead of repeating them, we will only point out the places where they differ and show how we modified them to work in the new context.

Definition 2.22 *Given a polynomial p with q-unallowable variety $V(p)$, consider all sets W that are q-allowable (as in Definition 2.21), and subtract from $V(p)$ those W for which $W \subset V(p)$. We call the remaining subset of the variety* points in general position *and denote it by $\mathcal{G}(p)$.*

Remark 2.15 *Since $V(p)$ is q-unallowable, $\mathcal{G}(p)$ is non-empty.*

Definition 2.23 *Given $x \in \mathcal{S}$, define the set* q$-$Allow(x) *as the intersection of all basic q-allowable sets going through x:*

$$\text{q}-\text{Allow}(x) := \cap_{j \in J \cup \{-2,-1,0\}} \left(\cap_{\mathcal{I} \ : \ x \in V_i(q_j)} \ V_{\mathcal{I}}(q_j) \right),$$

for all possible choices of $T, \mathcal{A}_Z, \mathcal{A}_D, \mathcal{A}_S, \mathcal{A}_N$.

    The intersection in parentheses is $\mathcal{S}$ whenever $x \notin V_{\mathcal{I}}(q_j)$ for all possible $\mathcal{I}$.

Remark 2.16 *When $x \in \mathcal{G}(p)$,* q$-$Allow$(x) \not\subseteq \mathcal{G}(p)$.

We can now state our necessity condition.

Theorem 2.9 *Given the black-box operations q_{-2}, q_{-1}, q_0, and $\{q_j : j \in J\}$, and the model of arithmetic described above, let p be a polynomial defined over a domain $\mathcal{D} \subset \mathcal{S}$. Let $\mathcal{G}(p)$ be the set of points in general position on the variety $V(p)$. If there exists $x \in \mathcal{D} \cap \mathcal{G}(p)$ such that* q$-$Allow$(x) \cap$ Int$(\mathcal{D}) \neq \emptyset$, *then p is not accurately evaluable on $\mathcal{D}$.*

We proceed to the construction of the elements of the proof of Theorem 2.9. The algorithm will once again be represented by a DAG with input nodes, branching nodes, and output nodes. As in Section 2.4, for simplicity in dealing with negation (since negation is exact), we will work with *solid* edges, which convey a value unchanged, and *dotted* edges, which indicate that negation of the conveyed quantity has occurred.

From now on, unless specified, we will consider only non-branching algorithms.

We will continue to use the definition of a Zariski set (Definition 2.6) on a hypercube in δ-space, and we work with the same definition of a non-trivial computational node (recalled below).

Definition 2.24 *For a given $x \in \mathcal{S}$, we say that a computational node N is of q-non-trivial type if its output is a nonconstant polynomial of δ.*

Recall the notation H_ϵ from Definition 2.9.

The equivalent of Proposition 2.10 becomes the following.

Proposition 2.25 *Given any algorithm, any $\epsilon > 0$, and a point x in $\mathcal{G}(p)$, there exists a Zariski open set $\Delta \subset H_\epsilon$ such that no q-non-trivial computational node has a zero output on the input x for all $\delta \in \Delta$.*

Proof The proof of Proposition 2.25 follows the same path as that of Proposition 2.10. To each q-non-trivial node corresponds a Zariski open set in δ-space; there is a finite number of them, and their intersection provides us with the Zariski open set we are looking for. $\square$

We will now state and sketch the proof of the equivalent of Lemma 2.1.

Lemma 2.7 *For a given algorithm, $x \in \mathcal{G}(p)$, and $\epsilon > 0$, exactly one of the following holds:*

1. *there exists a Zariski open set $\Delta \in H_\epsilon$ such that the value $p_{comp}(x, \delta)$ computed by the algorithm is not zero when the algorithm is run with source input x and errors $\delta \in \Delta$;*
2. *$p_{comp}(y, \delta) = 0$ for all $y \in$ q$-$Allow(x) and all $\delta \in H_\epsilon$.*

Proof of Lemma 2.7. Give $x \in \mathcal{G}(p)$, choose δ from the Zariski open set Δ whose existence is given by Proposition 2.25. Either the output node is of q-non-trivial type (in which case $p_{comp}(x, \cdot) \neq 0$ and we are done) or the output is a nonzero constant polynomial in δ (and again we are done) or the output is the zero polynomial in δ. In this latter case, we trace back all the zeros again, as in the proof of Lemma 2.1, and get a set of paths of nodes that produced all 0.

Let us start from the last nodes on these paths and work our way up, level after level. The last node on such a path is either a source, or a node with all inputs from sources, or a node with at least one input which is not a source and not 0 (and hence a polynomial in δ). In the former two cases, we have traced back the zeros to basic q-allowable conditions. We will show that this is also true for the latter case.

Lemma 2.8 *If the last zero occurs at a node which computes a black-box operation q_j and which has some source inputs and some (nontrivial) polynomial inputs, then the sources lie in some $V_\mathcal{I}(q_j)$ (and this constraint causes the zero output of this node).*

Proof of Lemma 2.8. Label the node $\tilde{N}$ (assume it corresponds to the black-box q_j) with output 0; then some inputs are sources, and some are polynomials of δ (since this was the last node on the path, it has no zero inputs). By the choice of δ, it follows that the output has to be the 0 polynomial in δ.

Some of the non-source inputs to $\tilde{N}$ might come from the same nodes; assume you have a total of l distinct nodes which input to $\tilde{N}$ (nonconstant) polynomials of δ, and let $I_1(\delta), I_2(\delta), \ldots, I_l(\delta)$ denote these inputs. We will need the following lemma.

Lemma 2.9 *Since the DAG is acyclic, $I_1(\delta), I_2(\delta), \ldots, I_l(\delta)$ are algebraically independent polynomials in δ.*

Proof of Lemma 2.9. Suppose there is some polynomial dependence among $I_1(\delta), \ldots, I_l(\delta)$. Let $N_1, \ldots, N_l$ be the nodes which have computed $I_1(\delta), \ldots, I_l(\delta)$, and let $\delta_1, \delta_2, \ldots, \delta_l$ be the specific δs at these nodes. Let $D_1, \ldots, D_l$ be the set of δs present (non-trivially) in the output of each node, e.g., $\delta_i \in D_i$. At least one δ_i is not present in $\cup_{i \neq j} D_j$; otherwise we get a cycle. But then $I_i(\delta)$ is algebraically independent from the other inputs, i.e., there is some dependence among the inputs $I_j(\delta)$ with $j \neq i$. We use induction on the number of remaining inputs, and exclude one input at a time, until we're left with a contradiction. This proves Lemma 2.9. $\square$

Replace each $I_i(\delta)$ (or $-I_i(\delta)$) in Y by the same *dummy* variable z_i (respectively $-z_i$), for each $i \in \{1, \ldots, l\}$. The variables z_i are algebraically independent by Lemma 2.9.

Denote by z the new vector of inputs to $\tilde{N}$ (both values *and* variables). The value $q_j(z) = 0$, regardless of the z_is (since it was 0 regardless of the δs and the z_i are algebraically independent variables). It follows that the constraints which place z on the variety of q_j are twofold: they come from constraints of the type D_{ij} and S_{ij} which describe the places where we inputted the values z_i, and they come from imposing conditions on the *other* inputs, which are *sources*. Thus the constraints on z are of the form $V_{T,\emptyset,\mathcal{A}_D,\mathcal{A}_S,\mathcal{A}_N}(q_j) =: V_{\mathcal{I}}$. This concludes the proof of Lemma 2.8. $\square$

Now that we have shown that the last marked vertices all provide basic q-allowable conditions, we proceed by induction: we look at a "next" marked vertex (here "next" means that all its marked ancestors have been examined already). It has some zero inputs, some source inputs, some of the inputs satisfy constraints of type D_{ij} and S_{ij}, and some of the inputs are polynomial. From here on we proceed as in Lemma 2.8, and obtain a set of new constraints to be imposed on the sources, of the type $V_{\mathcal{I}}(q_j)$, which we will intersect with the rest of the constraints obtained so far.

At the end of the examinations, we have found a set of basic q-allowable constraints which the sources must satisfy, i.e., a list of basic q-allowable sets with the property that the sources lie in their intersection; the fact that the sources satisfy these constraints is responsible for the zero output at the end of the computation.

It is not hard to see that in this case, once again, it follows that for all y in $\mathrm{q-Allow}(x)$ and any $\delta \in \Delta$, the output is 0 (just as in Lemma 2.1). Thus we have proved Lemma 2.7. □

From Lemma 2.7 we obtain the following corollary.

Corollary 2.7 *For any algorithm, for any $\epsilon > 0$, and any $x \in \mathcal{G}(p)$, exactly one of the following holds: the relative error of computation, $|p_{comp} - p|/|p|$, is either infinity at x for all δ in a Zariski open set or 1 at all points $y \in (\mathrm{q-Allow}(x) \setminus V(p))$ and all $\delta \in H_\epsilon$.*

We will now consider algorithms with or without branches.

Theorem 2.10 *Given a (branching or non-branching) algorithm with output function $p_{comp}(\cdot)$, $x \in \mathcal{G}(p)$, and $\epsilon > 0$, then one of the following is true:*

1. *there exists a set $\Delta_1 \in H_\epsilon$ of positive measure such that $p_{comp}(x, \delta)$ is nonzero whenever the algorithm is run with errors $\delta \in \Delta_1$, or*
2. *there exists a set $\Delta_2 \in H_\epsilon$ of positive measure such that for every $\delta \in \Delta_2$, there exists a neighborhood $N_\delta(x)$ of x such that for every $y \in N_\delta(x) \cap (\mathrm{q-Allow}(x) \setminus V(p))$, $p_{comp}(y, \delta) = 0$ when the algorithm is run with errors δ.*

Remark 2.17 *Just as before, this implies that, on a set of positive measure in H_ϵ, the relative accuracy of any given algorithm is either ∞ or 1.*

Proof The proof is essentially the same as in Theorem 2.4; the only thing that needs to be examined is the existence in $\mathrm{q-Allow}(x) \setminus V(p)$ of an infinite sequence $\{y_n\}$ with $y_n \to x$.

We will make use of the following basic result in the theory of algebraic varieties, which can for example be found as Theorem 1 in Section 6.1 of Shafarevich (1994).

Result 2.11 *If X and Y are polynomial varieties such that $X \subseteq Y$, then $dim(X) \leq dim(Y)$. If Y is irreducible and $X \subseteq Y$ is a (closed) subvariety with $dim(X) = dim(Y)$, then $X = Y$.*

We write $\mathrm{q-Allow}(x)$ as a union of irreducible q-allowable components. By the way we defined $\mathcal{G}(p)$, it follows that none of these components is included in $V(p)$; by Result 2.11, it follows that the intersection of any irreducible q-allowable component P of $\mathrm{q-Allow}(x)$ with $V(p)$ has a smaller dimension than P.

Choose the (unique) irreducible component P that contains x; this component must have dimension at least 1 (since if it contained only x, the set $\{x\}$ would be q-allowable, and hence we would have extracted it from $V(p)$, which is a contradiction with the fact that $x \in \mathcal{G}(p)$). Since $P \setminus V(p)$ has a smaller dimension than P, there *must* be some infinite sequence $\{y_n\}$ in $P \setminus V(p)$, i.e., in $q-\text{Allow}(x) \setminus V(p)$, such that $y_n \to x$.

The rest of the argument goes through just as in Theorem 2.4. □

Finally, as in Section 2.4.1, we have a corollary.

Corollary 2.8 *Let p be a polynomial over $\mathcal{S}$ with unallowable variety $V(p)$. Given any algorithm with output function $p_{comp}(\cdot)$, a point $x \in \mathcal{G}(p)$, $\epsilon > 0$, and $\eta < 1$, there exists a set Δ_x of positive measure arbitrarily close to x and a set $\Delta \in H_\epsilon$ of positive measure, such that $|p_{comp} - p|/|p|$ is strictly larger than η when computed at a point $y \in \Delta_x$ with errors $\delta \in \Delta$.*

The proof is based on the topology of $\mathcal{S}$, and is identical to the proof of Corollary 2.2; we choose not to repeat it.

Proof of Theorem 2.9. Follows immediately from Theorem 2.10 and Corollary 2.8. □

2.5.2 Sufficiency: the complex case

In this section we obtain a sufficiency condition for the accurate evaluability of a complex polynomial, given a black-box arithmetic with operations q_{-2}, q_{-1}, q_0 and $\{q_j | j \in J\}$ (J may be an infinite set).

Throughout this section, we assume our black-box operations include q^c, which consists of multiplication by a complex constant: $q^c(x) = c \cdot x$. Note that this operation is natural, and that most computers perform it with relative accuracy.

We believe that the sufficiency condition we obtain here is sub-optimal in general, but it subsumes the sufficiency condition we found for the basic complex case with classical arithmetic $\{+, -, \cdot\}$.

We assume that the black-box polynomials defining the operations q_j with $j \in J$ are *irreducible*.

Lemma 2.10 *The varieties $V_{\mathcal{I}}(q_j)$ are irreducible for any $j \in J$ and any $\mathcal{I}$ as in Definition 2.18 if and only if all q_j, $j \in J$ are affine polynomials.*

Proof If q_j is an affine polynomial then any $V(q_j)$ is also affine, hence irreducible over $\mathbb{C}^n$. Conversely, if q_j is not an affine polynomial, then by inputting a single value x for all the variables, we obtain a one-variable polynomial of degree at least 2, which is necessarily reducible over $\mathbb{C}^n$.   □

We state here the best sufficiency condition for the accurate evaluability of a polynomial we were able to find in the general case, and a necessary and sufficient condition for the all-affine black-box operations case.

Theorem 2.12 (General case) *Given a polynomial* $p : \mathbb{C}^n \to \mathbb{C}$ *with* $V(p)$ *a finite union of intersections of hyperplanes* Z_i, S_{ij}, D_{ij}, *and varieties* $V(q_j)$, *for* $j \in J$, *then* p *is accurately evaluable.*

Theorem 2.13 (Affine case) *If all black-box operations* q_j, $j \in J$ *are affine, then a polynomial* $p : \mathbb{C}^n \to \mathbb{C}$ *is accurately evaluable iff* $V(p)$ *is a union of intersections of hyperplanes* Z_i, S_{ij}, D_{ij}, *and varieties* $V_{\mathcal{I}}(q_j)$, *for* $j \in J$ *and* $\mathcal{I}$ *as in Definition 2.18.*

We will begin by proving Theorem 2.12. We will once again make use of Theorem 2.6 and of Theorem 2.11.

Lemma 2.11 *If* $V(p)$ *is as in Theorem 2.12, then* $V(p)$ *is a simple finite union of hyperplanes* Z_i, S_{ij}, D_{ij} *and varieties* $V(q_j)$ *(with no intersections).*

Proof Indeed, if that were not the case, then some irreducible q-allowable component P of $V(p)$ would be an intersection of two or more sets described in Theorem 2.12. If P were contained in the intersection of two or more (distinct) hyperplanes, its dimension would be smaller than $n - 2$, and we would get a contradiction to Theorem 2.6.

Suppose now that P was contained in the intersection of a $V(q_j)$ with some other variety or hyperplane. All such varieties, by Theorem 2.6, must have dimension $n - 1$, and since all such varieties and hyperplanes are irreducible, by Result 2.11, their intersection must have dimension strictly smaller than $n - 2$. Contradiction; we have thus proved that the variety $V(p)$ is a *simple* union of hyperplanes Z_i, D_{ij}, S_{ij}, and varieties $V(q_j)$.   □

Corollary 2.9 *If* $p : \mathbb{C}^n \to \mathbb{C}$ *is a polynomial whose variety* $V(p)$ *is* q-*allowable, then it is a product* $p = c \prod_j p_j$, *where each* p_j *is a power of* x_i, $(x_i - x_j)$, $(x_i + x_j)$, *or* q_j, *and* c *is a complex constant.*

Proof By Lemma 2.11, the variety $V(p)$ is a union of basic q-allowable hyperplanes and varieties $V(q_j)$.

Choose an irreducible q-allowable set in the union. If this set is a hyperplane, then by following the same argument as in Corollary 2.4, we obtain that p factors into some $\widetilde{p}$ and some power of either x_i, $(x_i - x_j)$, or $(x_i + x_j)$.

Suppose now that the irreducible q-allowable set were a variety $V(q_j)$; since p is 0 whenever q_j is 0 and q_j is irreducible, it follows that q_j divides p. We factor then p into the largest power of q_j which divides it, and some other polynomial $\widetilde{p}$.

In either of the two cases, we proceed by factoring $\widetilde{p}$ in the same fashion, until we encounter a polynomial $\widetilde{p}$ of degree 0. That polynomial is the constant c. $\qquad\square$

Proof of Theorem 2.12. By Corollary 2.4, $p = c \prod_j p_j$, with each p_j a power of x_k, $(x_k \pm x_l)$, or q_l.

Since each of the factors is accurately evaluable, the algorithm that forms their product evaluates p accurately. Multiplication by c (corresponding to the black-box q^c) is also accurate, hence p is accurately evaluable. $\qquad\square$

The proof of Theorem 2.13 follows the path described above; we sketch it here.

Proof of Theorem 2.13. The key fact in obtaining this condition is the irreducibility of all sets $V_{\mathcal{I}}(q_j)$, which is guaranteed by Lemma 2.10, together with the result of Lemma 2.11. Once again, we can write the polynomial as a product of powers of x_i, $(x_i \pm x_j)$, or $V_{\mathcal{I}}(q_j)$, times a constant; this takes care of the sufficiency part, while the necessity follows from Theorem 2.9. $\qquad\square$

Remark 2.18 *Note that Theorem 2.13 is a more general necessary and sufficient condition than Theorem 2.5, which only considered having q_{-2}, q_{-1}, and q_0 as operations, and restricted the polynomials to have integer coefficients (thus eliminating the need for q^c).*

2.6 Accurate Linear Algebra in Rounded Arithmetic

Now we describe implications of our results to the question of whether we can accurately do numerical linear algebra on structured matrices. By a *structured matrix* we mean a family of n-by-n matrices M whose entries $M_{ij}(x)$ are simple polynomial or rational functions of parameters x. Typically there are only $O(n)$ parameters, and the polynomials $M_{ij}(x)$ are closely related (for otherwise little can be said). Typical examples include Cauchy matrices $(M_{ij}(x, y) = 1/(x_i + y_j))$, Vandermonde matrices $(M_{ij}(x) = x_i^{j-1})$, generalized Vandermonde matrices $(M_{ij}(x) = x_i^{j-1+\lambda_j}$, where the λ_j are a nondecreasing sequence of nonnegative integers), Toeplitz matrices $(M_{ij}(x) = x_{i-j})$, totally positive matrices (where M is expressed as a product of simple nonnegative bidiagonal matrices arising from its Neville factorization), acyclic matrices, suitably discretized elliptic partial differential operations, and so on [Demmel (2002), Demmel and Koev (2004a), Demmel and Koev (2001), Demmel, Gu, Eisenstat, Slapničar, Veselić, and Z. Drmač (1999), Demmel (1999), Demmel and Koev (2004b), Demmel and Koev (2005b), Demmel and Koev (2005a), Koev (2004), Koev (2005)].

It has been recently shown that all the matrices on the above list (except Toeplitz and non-totally-positive generalized Vandermonde matrices) admit accurate algorithms in rounded arithmetic for many or all of the problems of numerical linear algebra:

- computing the determinant
- computing all the minors
- computing the inverse
- computing the triangular factorization from Gaussian elimination, with various kinds of pivoting
- computing eigenvalues
- computing singular values

We have gathered together these results in Table 2.1.

[1]: Demmel, Gu, Eisenstat, Slapničar, Veselić, and Z. Drmač (1999)
[2]: Dopico, Molera, and Moro (2003)
[3]: O'Cinneide (1996)
[4]: Attahiru, Junggong, and Ye (2002a)
[5]: Attahiru, Junggong, and Ye (2002b)
[6]: Demmel and Koev (2001)
[7]: Demmel and Koev (2004b)
[8]: Demmel (1999)
[9]: Demmel and Koev (2005b)
[10]: Higham (1990)

The proliferation of these accurate algorithms for some but not all matrix structures motivates us to ask for which structures they exist.

To convert this to a question about polynomials, we begin by noting that being able to compute the determinant accurately is a *necessary* condition for most of the above computations. For example, if the diagonal entries of a triangular factorization of A, or its eigenvalues, are computable with small relative error, then so is their product, the determinant.

Table 2.1. *General Structured Matrices*

Type of matrix		$\det A$	A^{-1}	Any minor	LDU	SVD	Sym EVD
Acyclic		n	n^2	n	$\leq n^2$	n^3	N/A
(e.g., bidiagonal)		[1]	[1]	[1]	[1]	[1]	
Total Sign		n	n^3	n	n^4	n^4	n^4
Compound		[1]	[1]	[1]	[1]	[1]	[2]
Diagonally Scaled		n^3	n^5	n^3	n^3	n^3	n^3
Totally Unimodular		[1]		[1]	[1]	[1]	[2]
Weakly diagonally		n^3	n^3	No	n^3	n^3	n^3
dominant M-matrix		[3]	[4,5]	[6]	[7]	[7]	[2]
Dis- place- ment Rank One	Cauchy	n^2	n^2	n^2	$\leq n^3$ [8]	n^3 [8]	n^3 [2]
	Vandermonde	n^2	No [6]	No [6]	No [6]	n^3 [8,9]	n^3 [2]
	Polynomial Vandermonde	n^2 [10]	**No** 2.6.1	**No** 2.6.1	**No** 2.6.1	$*$ [9]	$*$ [2]

It is also true that being able to compute all the minors accurately is a *sufficient* condition for many of the above computations. For the inverse and triangular factorization, this follows from Cramer's rule and Sylvester's theorem, resp., and for the singular values an algorithm is described in Demmel, Gu, Eisenstat, Slapničar, Veselić, and Z. Drmač (1999), Demmel and Koev (2001).

Thus, if the determinants $p_n(x) = \det M^{n \times n}(x)$ of a class of n-by-n structured matrices M do not satisfy the necessary conditions described in Theorem 2.9 for *any* enumerable set of black-box operations (perhaps with other properties, like bounded degree), then we can conclude that accurate algorithms of the sort described in the above citations are impossible.

In particular, to satisfy these necessary conditions would require that the varieties $V(p_n)$ be allowable (or q-allowable). For example, if V is a Vandermonde matrix, then $\det(V) = \prod_{i<j}(x_i - x_j)$ satisfies this condition, using only subtraction and multiplication.

The following theorem states a condition which guarantees the impossibility of an algorithm using *any* enumerable set of black-box operations of bounded degree:

Theorem 2.14 *Let $M(x)$ be an n-by-n structured complex matrix with determinant $p_n(x)$ as described above. Suppose $p_n(x)$ has an irreducible* *factor $\hat{p}_n(x)$ whose degree goes to infinity as n goes to infinity. Then for any enumerable set of black-box arithmetic operations of bounded degree, for sufficiently large n it is impossible to accurately evaluate $p_n(x)$ over the complex numbers.*

Proof Let $q_1, \ldots, q_m$ be any finite set of black-box operations. To obtain a contradiction, suppose the complex variety $V(p_n)$ satisfies the necessary conditions of Theorem 2.9, i.e., that $V(p_n)$ is allowable. This means that $V(p_n)$, which includes the hypersurface $V(\hat{p}_n)$ as an irreducible component, can be written as the union of irreducible q-allowable sets (by Def. 2.21). This means that $V(\hat{p}_n)$ must itself be equal to an irreducible q-allowable set (a hypersurface), since representations as unions of irreducible sets are unique. The irreducible q-allowable sets of codimension 1 are defined by single irreducible polynomials, which are in turn derived by the process of setting variables equal to one another, to one another's negation, or zero (as described in Defs. 2.18 and 2.21), and so have bounded degree. This contradicts the unboundedness of the degree of $V(\hat{p}_n)$. □

In the next theorem we apply this result to the set of complex Toeplitz matrices. We use the following notation. Let T be an n-by-n Toeplitz matrix, with x_j on the j-th diagonal, so x_0 is on the main diagonal, x_{n-1} is in the top right corner, and x_{1-n} is in the bottom left corner.

Theorem 2.15 *The determinant of a Toeplitz matrix T is irreducible over any field.*

Corollary 2.10 *The determinants of the set of complex Toeplitz matrices cannot be evaluated accurately using any enumerable set of bounded-degree black-box operations.*

Proof of Theorem 2.15. We use induction on n. We note that $\det T$ depends on every variable x_j, because $\det T$ includes the monomials $\pm x_j^{n-j} x_{j-n}^j$ for $j > 0$, as well as x_0^n, and these monomials contain the maximum powers of x_j and x_{j-n} appearing in the determinant. Now x_{n-1} appears exactly once in T, so $\det T$ must be an affine function of x_{n-1}, say $\det T = x_{n-1} \cdot p_{1n} + p_{2n}$. By expanding $\det T$ along the first row or column, we see that p_{1n} is itself the determinant of a Toeplitz matrix with diagonals $x_{1-n}, \ldots, x_{n-3}$, and p_{2n} depends on $x_{1-n}, \ldots, x_{n-2}$ but not x_{n-1}. If $\det T = x_{n-1} \cdot p_{1n} + p_{2n}$ were reducible, its factorization would have to look like $x_{n-1} \cdot p_{1n} + p_{2n} = (x_{n-1} \cdot p_{3n} + p_{4n}) p_{5n}$, where all the subscripted p polynomials are independent of x_{n-1}, implying either that $p_{1n} = p_{3n} p_{5n}$

were reducible, a contradiction by our induction hypothesis, or $p_{3n} = 1$ and so $p_{1n}|p_{2n}$. Now we can write $p_{2n} = x_{n-2}^2 q_{1n} + x_{n-2}q_{2n} + q_{3n}$ where $q_{1n} \neq 0$, since $\det T$ includes the monomial $\pm x_{n-2}^2 x_{-2}^{n-2}$ and no higher powers of x_{n-2}. Furthermore q_{1n} is independent of x_{n-1}, x_{n-2} and x_{n-3}, and q_{2n} and q_{3n} are independent of x_{n-1} and x_{n-2}. Since p_{1n} is independent of x_{n-2}, the only way we could have $p_{1n}|p_{2n}$ is to have $p_{1n}|q_{1n}$, $p_{1n}|q_{2n}$, and $p_{1n}|q_{3n}$. But since p_{1n} depends on x_{n-3} and q_{1n} is independent of x_{n-3}, this is a contradiction. So the determinant of a Toeplitz matrix must be irreducible.   □

In the real case, irreducibility of p_n is not enough to conclude that p_n cannot be evaluated accurately, because $V_{\mathbb{R}}(p_n)$ may still be allowable (and even vanish). So we consider another necessary condition for allowability: Since all black-boxes have a finite number of arguments, their associated codimension-1 irreducible components must have the property that whether $x \in V_{\mathcal{I}}(q_j)$ depends on only a finite number of components of x. Thus to prove that the hypersurface $V_{\mathbb{R}}(p_n)$ is not allowable, it suffices to find at least one regular point x^* in $V_{\mathbb{R}}(p_n)$ such that the tangent hyperplane at x^* is not parallel to sufficiently many coordinate directions, i.e., membership in $V_{\mathbb{R}}(p_n)$ depends on more variables than any $V_{\mathcal{I}}(q_j)$. This is easy to do for real Toeplitz matrices.

Theorem 2.16 *Let V be the variety of the determinant of real singular Toeplitz matrices. Then V has codimension 1, and at almost all regular points, its tangent hyperplane is parallel to no coordinate directions.*

Corollary 2.11 *The determinants of the set of real Toeplitz matrices cannot be evaluated accurately using any enumerable set of bounded-degree black-box operations.*

Proof of Theorem 2.16. Let $Toep(i, j)$ denote the Toeplitz matrix with diagonal entries x_i through x_j; thus $Toep(i, j)$ has dimension $(j - i)/2 + 1$. Let U be the Zariski open set where $\det Toep(i, j) \neq 0$ for all $1 - n \leq i \leq j < n - 1$ and $j - i$ even. Then $\det T$ is a nonconstant affine function of x_{n-1}, and so for any choice of $x_{1-n}, \ldots, x_{n-2}$ in U, $\det T$ is zero for a unique choice of x_{n-1}. This shows that $V_{\mathbb{R}}(\det T)$ has real codimension 1.

Furthermore, $\det T$ has highest order term in each x_i, $0 < i \leq n - 1$, equal to $\pm Toep(1 - n, 2i - n - 1)x_i^{n-i}$, i.e., with nonzero coefficient on U. It also has the highest order term in each x_i, $1 - n \leq i < 0$, equal to $\pm Toep(n + 2i + 1, n - 1)x_i^{n+i}$, i.e., with nonzero coefficient on U. Finally, the highest order term in x_0 is x_0^n, with coefficient 1. Thus the gradient of det T has all nonzero components on a Zariski open set, and whether det $T = 0$ depends on all variables. $\square$

2.6.1 Vandermonde matrices and generalizations

In this section we will explain the entries filled "**No**" in Table 2.1. First we will show that polynomial Vandermonde matrices do not have algorithms for computing accurate inverses, by proving that certain minors needed in the expression cannot be accurately computed (this will also explain the "**No**" in the *Any minor* column). Finally, we will show that the LDU factorization for polynomial Vandermonde matrices cannot be computed accurately.

First we consider the class of generalized Vandermonde matrices V, where $V_{ij} = P_{j-1}(x_i)$ is a polynomial function of x_i, with $1 \leq i, j \leq n$. This class includes the standard Vandermonde (where $P_{j-1}(x_i) = x_i^{j-1}$) and many others.

Consider a generalized Vandermonde matrix where $P_{j-1}(x_i) = x_i^{j-1+\lambda_j}$ with $0 \leq \lambda_1 \leq \lambda_2 \leq \cdots \leq \lambda_n$. The tuple $\lambda = (\lambda_1, \lambda_2, \ldots, \lambda_n)$ is called a *partition*. Any square submatrix of such a generalized Vandermonde matrix is also a generalized Vandermonde matrix. A generalized Vandermonde matrix is known to have determinant of the form $s_\lambda(x) \prod_{i<j}(x_i - x_j)$ where $s_\lambda(x)$ is a polynomial of degree $|\lambda| = \sum_i \lambda_i$, and called a Schur function [MacDonald (1995)]. In infinitely many variables (not our situation) the Schur function is irreducible [Farahat (1958)], but in finitely many variables, the Schur function is sometimes irreducible and sometimes not [Stanley (1999), Exer. 7.30]. But there are irreducible Schur functions of arbitrarily high degree. Thus we conclude by Theorem 2.14 that no enumerable set of black-box operations of bounded degree can compute all Schur functions accurately when the x_i are complex.

If we restrict the domain $\mathcal{D}$ to be nonnegative real numbers, then the situation changes: The nonnegativity of the coefficients of the Schur functions shows that they are positive in $\mathcal{D}$, and indeed the generalized Vandermonde matrix is totally positive [Karlin (1968)]. Combined with the homogeneity of the Schur function, Theorem 2.2 implies that the Schur function, and so determinants (and minors) of totally positive generalized Vandermonde matrices can be evaluated accurately in classical arithmetic. For accurate algorithms that are more efficient than the one in Theorem 2.2, see Demmel and Koev (2005c).

Now consider a polynomial Vandermonde matrix V_P defined by a family $\{P_k(x)\}_{k \in \mathbb{N}}$ of polynomials such that $\deg(P_k) = k$, and $V_P(i, j) = P_{j-1}(x_i)$.

Note that any V_P can be written as $V_P = VC$, with V being a regular Vandermonde matrix, and C being an upper triangular matrix of coefficients of the polynomials P_k, i.e.,

$$P_{j-1}(x) = \sum_{i=1}^{j} C(i,j)x^{i-1}, \quad \forall 1 \le j \le n .$$

Denote by $c_{i-1} := \tilde{D}(i,i)$, for all $1 \le i \le n$ the highest-order coefficients of the polynomials $P_0(x), \ldots, P_{n-1}(x)$.

To compute the inverse of the matrix V_P, we need to compute the minors that result from deleting a row and a column, i.e., (in MATLAB notation) $\det(V_P([1\!:\!i-1, i+1\!:\!n], [1\!:\!j-1, j+1\!:\!n]))$. We will focus our attention on the computation of the $(i, n-1)$ minors, i.e., the ones that result from deleting any of the rows and the $(n-1)$st column.

The resulting matrices look like

$$M_{P,i} := V_P([1\!:\!i-1, i+1\!:\!n], [1\!:\!n-2, n])$$

$$= \begin{bmatrix}
c_0 & P_1(x_1) & \cdots & P_{n-3}(x_1) & P_{n-1}(x_1) \\
\vdots & \vdots & \ddots & \vdots & \vdots \\
c_0 & P_1(x_{i-1}) & \cdots & P_{n-3}(x_{i-1}) & P_{n-1}(x_{i-1}) \\
c_0 & P_1(x_{i+1}) & \cdots & P_{n-3}(x_{i+1}) & P_{n-1}(x_{i+1}) \\
\vdots & \vdots & \ddots & \vdots & \vdots \\
c_0 & P_1(x_n) & \cdots & P_{n-3}(x_n) & P_{n-1}(x_n)
\end{bmatrix} .$$

Hence, we can manipulate the columns of $\det(M_{P,i})$ by subtracting from them linear combinations of other columns, to obtain

$$\det(M_{P,i})$$

$$= \begin{vmatrix}
c_0 & c_1 x_1 & \cdots & c_{n-3} x_1^{n-3} & c_{n-1} x_1^{n-1} + C(n-1,n)x_1^{n-2} \\
\vdots & \vdots & \ddots & \vdots & \\
c_0 & c_1 x_{i-1} & \cdots & c_{n-3} x_{i-1}^{n-3} & c_{n-1} x_{i-1}^{n-1} + C(n-1,n)x_{i-1}^{n-2} \\
c_0 & c_1 x_{i+1} & \cdots & c_{n-3} x_{i+1}^{n-3} & c_{n-1} x_{i+1}^{n-1} + C(n-1,n)x_{i+1}^{n-2} \\
\vdots & \vdots & \ddots & \vdots & \\
c_0 & c_1 x_n & \cdots & c_{n-3} x_n^{n-3} & c_{n-1} x_n^{n-1} + C(n-1,n)x_n^{n-2}
\end{vmatrix}$$

By expanding on the last column, and using the results from Demmel and Koev (2001), we obtain that there are constants E and F, specifically $E = C(n-1,n) \prod_{i=1}^{n-2} c_{i-1}$ and $F = E \frac{c_{n-1}}{C(n-1,n)}$, such that

$$\det(M_{P,i}) = \prod_{\substack{k < j \\ k,j \ne i}} (x_j - x_k) \left[E + F \cdot s_{[1]}(x_1, \ldots, x_{i-1}, x_{i+1}, \ldots, x_n) \right] ,$$

with $s_{[1]}$ being the Schur function corresponding to the partition $\lambda = (1, 0, \ldots, 0)$, i.e.,

$$\det(M_{P,i}) = \prod_{\substack{k < j \\ k, j \neq i}} (x_j - x_k) \left[E + F \cdot (x_1 + \ldots + x_{i-1} + x_{i+1} + \ldots + x_n) \right] ,$$

and it is not hard to see that for any $n \geq 4$, the above polynomial in $x_1, \ldots, x_{i-1}, x_{i+1}, \ldots, x_n$ *does not have an allowable variety,* and hence the inverse *cannot be evaluated accurately* in classical arithmetic.

Denote $\vec{x}_{\neq i} := (x_1, \ldots, x_{i-1}, x_{i+1}, \ldots, x_n)$.

Similarly, one can prove that the $(i, n - k)$ minor $\det M_{P,i,k}$ can be obtained as

$$\det(M_{P,i,k}) = \prod_{\substack{m < j \\ m, m \neq i}} (x_j - x_m) \left[A_1 + A_2 s_{[1]}(\vec{x}_{\neq i}) + \ldots + A_k s_{[1^k]}(\vec{x}_{\neq i}) \right] ,$$

where $[1^l] = (1, 1, 1, \ldots, 0)$, the right side containing exactly l ones and the rest 0; $A_1, \ldots, A_k$ are constants which can be computed easily in terms of the entries of the matrix C.

Since for any $l < n - 1$, $s_{[1^l]}$ is a homogeneous *irreducible* polynomial of degree l in $n - 1$ variables, the factor in the square brackets has degree k. Appropriate choices of n and the matrix C are likely to make this factor irreducible (for example, by making $|A_k| \gg |A_l|$ for all $l \neq k$). If this is the case, then by Theorem 2.14, this family of matrices has inverses that cannot be evaluated accurately even with the addition of any enumerable set of bounded-degree black-boxes.

This explains why we have filled in with "**No**" the entries corresponding to columns "A^{-1}" and "*Any minor*" in the Polynomial Vandermonde row of Table 2.1. Below we explain the "**No**" in the Polynomial Vandermonde row, in the column "*LDU*".

We can write $C = \tilde{D}\tilde{C}$, with $\tilde{D}$ being the diagonal matrix of highest-order coefficients, i.e., $\tilde{D}(i, i) = C(i, i)$ for all $1 \leq i \leq n$. We will assume that the matrices C and $\tilde{D}$ are given to us exactly.

If we let $V_P = L_P D_P U_P$ and $V = LDU$, it follows that

$$L_P = L \ ;$$
$$D_P = D\tilde{D} \ ;$$
$$U_P = \tilde{D}^{-1} U C \ .$$

Since we cannot compute L accurately in the general Vandermonde case, it follows that we cannot compute L_P accurately in the polynomial Vandermonde case.

Finally, we explain the "*" entries in the polynomial Vandermonde row. These depend on special properties of the polynomial. In general, neither the SVD nor the symmetric eigenvalue decomposition (EVD) are computable accurately, but if the polynomials are certain orthogonal polynomials, then the accurate SVD is possible [Demmel and Koev (2005b)], and an accurate symmetric EVD may also be possible [Dopico, Molera, and Moro (2003)].

Acknowledgments

We are grateful to Bernd Sturmfels, Gautam Bharali, Plamen Koev, William (Velvel) Kahan, and Gregorio Malajovich for interesting discussions, to Jonathan Dorfman for his help with the graphs, and to the anonymous referee for helpful critique and suggestions.

References

A. Aho, J. Hopcroft, and J. Ullman (1974), *The Design and Analysis of Computer Algorithms*, Addison-Wesley.

G. Alefeld and J. Herzberger (1983), *Introduction to Interval Computations*, Academic Press.

S. A. Attahiru, X. Junggong, and Q. Ye (2002a), Accurate computation of the smallest eigenvalue of a diagonally dominant M-matrix, preprint.

S. A. Attahiru, X. Junggong, and Q. Ye (2002b), Entrywise perturbation theory for diagonally dominant M-matrices with applications, *Numer. Math.* **90**, 401–414.

D. H. Bailey (1993), Multiprecision translation and execution of Fortran programs, *ACM Trans. Math. Soft.* **19**, 288–319.

D. H. Bailey (1995), A Fortran-90 based multiprecision system, *ACM Trans. Math. Soft.* **21**, 379–387.

L. Blum (2004), Computing over the reals: Where Turing meets Newton, *Notices of the AMS* **51**, October 2004.

L. Blum, F. Cucker, M. Shub, and S. Smale (1996a), Complexity and real computation: A manifesto, *Intern. J. Bifurcation and Chaos* **6**, 3–26.

L. Blum, F. Cucker, M. Shub, and S. Smale (1996b), *The Mathematics of Numerical Analysis*, volume 32 of *Lecture Notes in Applied Mathematics*, chapter Algebraic Settings for the Problem "P ≠ NP?". AMS.

L. Blum, F. Cucker, M. Shub, and S. Smale (1997), *Complexity and Real Computation*, Springer.

L. Blum, M. Shub, and S. Smale (1989), On a theory of computation and complexity over the real numbers: NP-completeness, recursive functions and universal machines, *AMS Bulletin (New Series)* **21**, 1–46.

F. Chatelin and V. Frayssé (1996), *Lectures on Finite Precision Computations*, SIAM, Philadelphia, PA.

F. Cucker and J. P. Dedieu (2001), Decision problems and round-off machines, *Theory of Computing Systems* **34**, 433–452.

F. Cucker and D. Grigoriev (1999), Complexity lower bounds for approximate algebraic computation trees, *J. Complexity* **15**, 499–512.

F. Cucker and S. Smale (1999), Complexity estimates depending on condition and roundoff error, *J. ACM* **46**, 113–184.

T. Dekker (1971), A floating point technique for extending the available precision, *Numer. Math.* **18**, 224–242.

J. Demmel (1984), Underflow and the reliability of numerical software, *SIAM J. Sci. Stat. Comput.* **5**, 887–919.

J. Demmel (1987), On error analysis in arithmetic with varying relative precision, In *Proceedings of the 8th Symposium on Computer Arithmetic*, Como, Italy. IEEE Computer Society Press.

J. Demmel (1997), *Applied Numerical Linear Algebra*, SIAM.

J. Demmel (1999), Accurate SVDs of structured matrices, *SIAM J. Mat. Anal. Appl.* **21**, 562–580.

J. Demmel (2002), The complexity of accurate floating point computation, In *Proceedings of the 2002 International Congress of Mathematicians*, Beijing.

J. Demmel, M. Gu, S. Eisenstat, I. Slapničar, K. Veselić, and Z. Drmač (1999), Computing the singular value decomposition with high relative accuracy, *Lin. Alg. Appl.* **299**, 21–80.

J. Demmel and Y. Hida (2003), Accurate and efficient floating point summation, *SIAM J. Sci. Comp.* **25**, 1214–1248.

J. Demmel and P. Koev (2001), Necessary and sufficient conditions for accurate and efficient rational function evaluation and factorizations of rational matrices, In V. Olshevsky, editor, *Special Issue on Structured Matrices in Mathematics, Computer Science and Engineering*, volume 281 of *Contemporary Mathematics*, pages 117–145. AMS.

J. Demmel and P. Koev (2004a), Accurate and efficient algorithms for floating point computation, In *Proceedings of the 2003 International Congress of Industrial and Applied Mathematics*, Sydney.

J. Demmel and P. Koev (2004b), Accurate SVDs of weakly diagonally dominant M-matrices, *Numer. Math.* **98**, 99–104.

J. Demmel and P. Koev (2005a), The accurate and efficient solution of a totally positive generalized Vandermonde linear system, *SIAM J. Matrix Anal. Appl.* **27**, 142–152.

J. Demmel and P. Koev (2005b), Accurate SVDs of polynomial Vandermonde matrices involving orthonormal polynomials, to appear in SIAM J. Matrix Anal. Appl.; `www-math.mit.edu/~plamen/files/chebvand.ps`.

J. Demmel and P. Koev (2005c), Efficient and accurate evaluation of Schur and Jack polynomials, to appear in Mathematics of Computation, `math.mit.edu/~plamen/files/s2.pdf`.

F. M. Dopico, J. M. Molera, and J. Moro (2003), An orthogonal high relative accuracy algorithm for the symmetric eigenproblem, *SIAM J. Matrix Anal. Appl.* **25**, 301–351 (electronic).

A. Edelat and P. Sünderhauf (1998), A domain-theoretic approach to real number computation, *Theoretical Computer Science* **210**, 73–98.

H. K. Farahat (1958), On Schur functions, *Proc. London Math. Soc.* **8**, 621–630.

I. M. Gelfand, M. M. Kapranov, and A. V. Zelevinsky (1994), *Discriminants, Resultants, and Multidimensional Determinants*, Birkhäuser, Boston.

N. J. Higham (1990), Stability analysis of algorithms for solving confluent Vandermonde-like systems, *SIAM J. Mat. Anal. Appl.* **11**, 23–41.

N. J. Higham (2002), *Accuracy and Stability of Numerical Algorithms*, SIAM, Philadelphia, PA, 2nd edition.

K. Hulek (2003), *Elementary Algebraic Geometry*, AMS, translated by H. Verrill.

W. Kahan (webpage), www.cs.berkeley.edu/~wkahan.

W. Kahan (1998), The Improbability of Probabilistic Error Analysis, UC Berkeley, www.cs.berkeley.edu/~wkahan/improber.pdf.

S. Karlin (1968), *Total Positivity*, Stanford University Press.

L. G. Khachiyan (1984), Convexity and complexity in polynomial programming, In *Proceedings of the 1984 International Congress of Mathematicians*, pages 1569–1577, Warsaw.

K.-I. Ko (1991), *Complexity Theory of Real Functions*, Birkhäuser.

P. Koev (2004), Accurate computations with totally nonnegative matrices, submitted to SIAM Journal on Matrix Analysis and Applications, www-math.mit.edu/~plamen/files/acctp.pdf.

P. Koev (2005), Accurate eigenvalues and SVDs of totally nonnegative matrices, *SIAM J. Matrix Anal. Appl.* **27**, 1–23.

D. Lozier and F. Olver (1990), Closure and precision in level-index arithmetic, *SIAM J. Num. Anal.* **27**, 1295–1304.

I. G. MacDonald (1995), *Symmetric Functions and Hall Polynomials*, Oxford University Press, 2nd edition.

E. Miller and B. Sturmfels (2005), *Combinatorial Commutative Algebra*, Springer-Verlag.

O. Møller (1965), Quasi double precision in floating-point arithmetic, *BIT* **5**, 37–50.

R. K. Montoye, E. Hokenek, and S. L. Runyon (1990), Design of the IBM RISC System 6000 floating point execution unit, *IBM Journal of Research and Development* **34**, 59–70.

R. E. Moore (1979), *Methods and Applications of Interval Analysis*, SIAM, Philadelphia.

A. Neumaier (1990), *Interval Methods for Systems of Equations*, Cambridge University Press, Cambridge, England.

C. O'Cinneide (1996), Relative-error for the LU decomposition via the GTH algorithm, *Numer. Math.* **73**, 507–519.

M. Pichat (1972), Correction d'une somme en arithmetique à virgule flottante, *Numer. Math.* **19**, 400–406.

M. Pour-El and J. Richards (1989), *Computability in Analysis and Physics*, Springer-Verlag.

D. Priest (1991), Algorithms for arbitrary precision floating point arithmetic, In P. Kornerup and D. Matula, editors, *Proceedings of the 10th Symposium on Computer Arithmetic*, pages 132–145, Grenoble, France, June 26-28, 1991, IEEE Computer Society Press.

J. Renegar (1992), On the computational complexity and geometry of the first-order theory of the reals: Parts I, II and III, *J. Symb. Comp.* **13**.

J. Renegar (1994), Is it possible to know a problem instance is ill-posed? *J. Complexity* **10**, 1–56.

B. Reznick (2000), *Some Concrete Aspects of Hilbert's 17th Problem*, volume 253 of *Contemporary Mathematics*, AMS.

I. R. Shafarevich (1994), *Basic Algebraic Geometry*, Springer, 2nd edition.

J. R. Shewchuk (1997), Adaptive Precision Floating-Point Arithmetic and Fast Robust Geometric Predicates, *Discrete & Computational Geometry* **18**, 305–363.

D. Spielman and S.-H. Teng (2002), Smoothed analysis of algorithms, In *Proceedings of the 2002 International Congress of Mathematicians*, Beijing.

R. Stanley (1999), *Enumerative Combinatorics*, volume 2, Cambridge Univ. Press.

A. Tarski (1951), *A Decision Method for Elementary Algebra and Geometry*, University of California Press, Berkeley.

J. Taylor (2004), *Several Complex Variables with Connections to Algebraic Geometry and Lie Groups*, AMS Series on Graduate Studies in Mathematics. American Math Society.

A. Turing (1948), Rounding-off errors in matrix processes. *Quart. J. Mech. Appl. Math.* **1**, 287–308.

J. Vignes (1993), A stochastic arithmetic for reliable computation, *Math. and Comp. in Sim.* **35**, 233–261.

J. von Neumann and H. Goldstine (1947), Numerical inverting matrices of high order, *Bull. AMS* **53**, 1021–1099.

J. H. Wilkinson (1963), *Rounding Errors in Algebraic Processes*, Prentice Hall, Englewood Cliffs.

G. Ziegler (1995), *Lectures on Polytopes*, Springer-Verlag.

3

Sparse Grids and Related Approximation Schemes for Higher Dimensional Problems

Michael Griebel

Institut für Numerische Simulation
Universität Bonn
53113 Bonn, Germany
e-mail: griebel@iam.uni-bonn.de

Abstract

The efficient numerical treatment of high-dimensional problems is hampered by the curse of dimensionality. We review approximation techniques which overcome this problem to some extent. Here, we focus on methods stemming from Kolmogorov's theorem, the ANOVA decomposition and the sparse grid approach and discuss their prerequisites and properties. Moreover, we present energy-norm based sparse grids and demonstrate that, for functions with bounded mixed derivatives on the unit hypercube, the associated approximation rate in terms of the involved degrees of freedom shows no dependence on the dimension at all, neither in the approximation order nor in the order constant.

3.1 Introduction

The discretization of PDEs by conventional methods is limited to problems with up to three or four dimensions due to storage requirements and computational complexity. The reason is the so-called curse of dimensionality, a term coined in (Bellmann 1961). Here, the cost to compute and represent an approximation with a prescribed accuracy ε depends exponentially on the dimensionality d of the problem considered. We encounter complexities of the order $O(\varepsilon^{-d/r})$ with $r > 0$ depending on the respective approach, the smoothness of the function under consideration, the polynomial degree of the ansatz functions and the details of the implementation. If we consider simple uniform grids with piecewise d-polynomial functions over a bounded domain in a finite element or finite difference approach, this complexity estimate translates to $O(N^d)$ grid points or degrees of freedom for which approximation accuracies of the order $O(N^{-r})$ are achieved.[1] Thus, the computational cost and storage requirements grow exponentially with the dimensionality of the problem, which is the reason for the dimensional restrictions mentioned above, even on the most powerful machines presently available.

The curse of dimensionality can be circumvented to some extent by restricting the class of functions under consideration. If we make a stronger assumption on the smoothness of the solution such that the order of accuracy depends on d as $O(N^{-c \cdot d})$ with $r = c \cdot d$, we directly see that the cost complexity is independent of d and that it is of the order $O(\varepsilon^{-d/(c \cdot d)}) = O(\varepsilon^{-1/c})$, for some c independent of d. This way, the curse of dimensionality can be broken easily.[2] In any case, such a smoothness assumption is somewhat unrealistic.

Nevertheless for practical applications in high(er) dimensions often a certain smoothness assumption on the function is implicitly present (e.g., in the data of the problem) which, in some way, relates to its dimensionality. Then, the curse of dimensionality is weakened or can even be broken completely. The problem on the one hand is to detect and classify applications where this may happen and on the other hand to develop and implement numerical schemes which exploit such a situation. This is the subject of this article. We intend to give an overview on recent approaches and results in this direction from the view of function approximation and solution of partial differential equations.

The remainder of this paper is organized as follows: In Section 3.2, we briefly consider applications in which high-dimensional partial differential equations appear. We then discuss the breaking of the curse of dimensionality from the theoretical point of view. Here, we collect known approaches for getting rid of the exponential dependence on d. Furthermore we consider the theorem of Kolmogorov in more detail and give a survey of approximation schemes which are related to it.

In Section 3.3 we consider dimension-wise decompositions of high-dimensional functions. Here we resort to ANOVA-type decompositions where a function is split into its contributions from different groups of subdimensions, an approach which is widely used in statistics. It basically involves the splitting of a one-dimensional function space into the constant subspace and the remainder space. A product construction then gives the

---

[1] If the solution is not smooth but possesses singularities, the order r of accuracy deteriorates. Adaptive refinement/nonlinear approximation is employed with success. In the best case, the cost-benefit ratio of a smooth solution can be recovered.

[2] An example would be the p-version of the finite element method if we couple the polynomial degree p to the dimension d and consider functions from the Sobolev space H^{p+1}.

associated splitting for the d-dimensional case. This reveals the relative importance of different dimensions as well as their interactions and correlations. For certain applications it can be observed that an apparently high-dimensional function possesses a low effective dimension or that there is a certain decay for the component functions with their dimension. Then the curse of dimensionality may be avoided. We formalize this with the help of reproducing kernel Hilbert spaces. The importance of the different contributions in an ANOVA splitting can then be expressed by certain weights.

For practical computations the associated subspaces need to be further discretized. This leads to so-called sparse grids which are discussed in detail in Section 3.4. To this end, we refine the remainder space of the one-dimensional splitting, i.e., we equip it with a basis. We use the standard piecewise linear hierarchical basis in one dimension (Faber 1909, Yserentant 1986) as the simplest example of a one-dimensional multiscale series expansion which involves interpolation by piecewise linears. Then the tensor product construction generates a basis for the d-dimensional case. A proper truncation – that can be formally derived by solving an optimization problem closely related to M-term approximation which involves the error norm and the smoothness assumption – results in sparse grids. For functions with bounded mixed second derivatives, approximation schemes are gained which exhibit cost complexities of the order $O(N(\log N)^{d-1})$ and give an accuracy of $O(N^{-2}(\log N)^{d-1})$ if we measure the error in the L_2-norm. However if we consider the energy norm, optimality leads us to an energy-based sparse grid with cost complexity $O(N)$ and accuracy $O(N^{-1})$ only. Thus, the exponential dependence of the logarithmic terms on d is completely removed (but is still present in the constants). Finally we discuss the order constants in more detail. In one special case we are able to show that, for the best approximation $v_M^{(E)}$ in the energy-norm based sparse grid space with dimension M, the following error estimate holds:

$$\|u - v_M^{(E)}\|_E \leq c \cdot d^2 \cdot 0.97515^d \cdot M^{-1} \cdot |u|_{2,\infty}$$

where the regularity term $|u|_{2,\infty}$ involves mixed second derivatives of u.

The concluding remarks of Section 3.5 summarize the discussion and give an outlook on current developments with sparse grids.

3.2 High dimensional problems and the curse of dimensionality

Usually in classical physics most problems are formulated as systems of (nonlinear) partial differential equations in three space dimension and one time dimension.[1] Here, the geometry of the object under consideration can be quite complicated. As examples consider flow around a car or an airplane, combustion in an engine, structural analysis of mechanical machines and buildings in civil engineering or related multiphysics applications which involve coupled systems of partial differential equations. These problems can nowadays be well treated with parallel adaptive finite element methods involving multilevel solvers on large parallel computers. They are at the edge of today's applications of numerical simulation in science and engineering. An efficient geometry description, subsequent parallel mesh generation, reliable a-posteriori error estimators for adaptive finite element discretizations, robust parallel multilevel solvers and load-balancing techniques for the overall approach are subjects of current research and form the mainstream in scientific computing.

However there are also problems which involve substantially more than just three spatial dimensions. Then, high-dimensionality often results from mathematical modelling. Besides pure integration problems stemming from physics and finance, typically models from the stochastics and data analysis world show up. For example, high-dimensional Laplace/diffusion problems and high-dimensional convection diffusion problems result from diffusion approximation techniques or the Fokker-Planck equation. Examples are the description of queueing networks (Mitzlaff 1997, Shen, Chen, Dai and Dai 2002), random excitations of mechanical structures (Johnson, Wojtkiewicz, Bergman and Spencer 1997, McWilliam, Knappett and Fox 2000, Wojtkiewicz and Bergman 2000), reaction mechanisms in molecular biology (Sjöberg 2002, Elf, Lötstedt and Sjöberg 2001), the viscoelasticity in polymer fluids (Rouse 1953, Prakash and Öttinger 1999, Prakash 2000, Venktiteswaran and Junk 2005a, Venktiteswaran and Junk 2005b, Lozinski and Chauviere 2003, Lozinski, Chauviere, Fang and Owens 2003, Chauviere and Lozinski 2004, Süli 2006), or various models for the pricing of financial derivatives (Duffie 1996, Kwok 1998, Wilmott 1998, Reisinger 2003, Schwab 2003, Escobar and Seco 2005). Furthermore, homogenization with multiple scales (Allaire 1992, Cioranescu, Damlamian and Griso 2002, Matache 2002, Hoang and Schwab 2003) as well as stochastic elliptic equations (Schwab and Todor 2003a, Schwab and Todor 2003b) result in high-dimensional PDEs. Next, we find quite high-dimensional problems in quantum mechanics and particle physics. Here, the dimensionality of the Schrödinger equation (Messiah 2000) grows with the number of considered

---

[1] Of course physical theories may involve more than just three spatial dimensions. For example the equations related to superstring theories which can be regarded as limits of the M-theory or the theory of supergravitation are formulated in 10 or 11 dimensions, respectively, see (Green, Schwarz and Witten 1998) and the references cited therein.

electrons and nuclei. Then, problems in statistical mechanics lead to the Liouville equation or the Langevin equation and related phase space models where the dimension depends on the number of particles (Balescu 1997). Furthermore, reinforcement learning and stochastic optimal control in continuous time give raise to the Hamilton-Jacobi-Bellman equation in high dimensions (Sutton and Barto 1998, Munos 2000, Munos and Moore 2002). Finally data mining problems involve differential operators as smoothing or regularization terms (priors) whose dimension grows with the number of features of the data (Girosi, Jones and Poggio 1995, Garcke, Griebel and Thess 2001, Schölkopf and Smola 2002, Hegland 2003, Garcke 2004).

Now, in higher dimensions, the question of the shape of the domain is not as important as in the two- and three-dimensional case, since complicated domains typically do not appear in applications. Conceptually, besides $\mathbb{R}^d$ itself, we use mainly hypercubes like $[-a, a]^d$, $a \in \mathbb{R}$, and their straightforward generalizations using different values of a for each coordinate direction as well as the corresponding structures in polar coordinates. These domains are of tensor product structure. This is an important prerequisite for numerical methods for higher-dimensional partial differential equations as we will see later.

3.2.1 Curse of dimensionality

Classical approximation schemes exhibit the curse of dimensionality (Bellmann 1961). We then have

$$||f - f_M|| = O(M^{-r/d}),$$

where r and d denote the isotropic smoothness of the function f and the problem's dimensionality, respectively. This is one of the main obstacles in the numerical treatment of high-dimensional problems. Therefore, the question is whether we can find situations, i.e., either function spaces or error norms, for which the curse of dimensionality can be broken. At first glance, there is an easy way out: if we make a stronger assumption on the smoothness of the function f such that $r = O(d)$, then, we directly obtain $||f - f_M|| = O(M^{-c})$ with constant $c > 0$. Of course, such an assumption is quite unrealistic.

However, about thirteen years ago, (Barron 1993) found an interesting result: Denote by $\mathcal{F}L_1$ the class of functions with Fourier transforms in L_1. Then, consider the class of functions of $\mathbb{R}^d$ with $\nabla f \in \mathcal{F}L_1$. We expect for the best M-term approximation f_M an approximation rate

$$||f - f_M|| = O(M^{-1/d})$$

since $\nabla f \in \mathcal{F}L_1 \approx r = 1$. However Barron was able to show that

$$||f - f_M|| = O(M^{-1/2}),$$

independent of d. Meanwhile, other function classes have been introduced with such properties. They comprise certain radial basis schemes, stochastic sampling techniques and approaches that work with spaces of functions with bounded mixed derivatives.

A better understanding of these results is possible with the help of harmonic analysis (Donoho 2000). Here, we resort to the approach of the L_1-combination of L_∞-atoms, see also (Triebel 1992, DeVore 1998). Consider the class of functions $\mathcal{F}(K)$ with integral representation

$$f(x) = \int A(x,t)d\mu(t) \quad \text{with} \quad \int d|\mu|(t) \le K, \tag{3.1}$$

where for fixed t we call $A(x,t) = A_t(x)$ an L_∞-atom, if $|A_t(x)| \le 1$ holds. Then, there are results from Maurey for Banach spaces and Stechkin in Fourier analysis which state that there exists an M-term sum

$$f_M(x) = \sum_{j=1}^{M} a_j A_{t_j}(x)$$

where

$$||f - f_M||_\infty \le C \cdot M^{-1/2}$$

with C independent of d.

As a first example we consider superpositions of Gaussian bumps (radial basis schemes). These resemble the space $\mathcal{F}(K, Gaussians)$ with $t := (x_0, s)$ and Gaussian atoms $A(x,t) = \exp(-||x - x_0||^2/s^2)$. Now, if the sum of the height of all Gaussians is bounded by K, (Niyogi and Girosi 1998) showed that the resulting approximation rate is independent of d for the corresponding radial basis schemes. There is no further condition on the widths or positions of the bumps. Note that this corresponds to a ball in Besov space $B_{1,1}^d(\mathbb{R}^d)$ which is just the bump algebra in (Meyer 1992). Thus, we have nothing but a restriction to smoother functions for higher dimensions such that the ratio r/d stays constant and, consequently, $M^{-r/d}$ does again not grow with d.

Another class of functions with an approximation rate independent of d is $\mathcal{F}(K, Orthant)$ which uses the parameter set of shifted orthants. Now $t = (x_0, k)$, and k is the orthant indicator. Furthermore, $A(x,t)$ is the indicator of orthant k with apex at x_0. Again, if the integral (3.1) is at most K, the resulting approximation rate is of order $O(M^{-1/2})$ independent of d. A typical and well-known example for such a construction is the cumulative distribution function in $\mathbb{R}^d$. This just results in the Monte Carlo method.

A more general class are the functions which are formed by any super-position of 2^d functions, each orthantwise monotone for a different orthant. Now, the condition $\int d|\mu|(t) \leq 1$ is the same as

$$\frac{\partial^d f}{\partial x_1 \cdots \partial x_d} \in L_1 \,, \tag{3.2}$$

i.e., we obtain the space of bounded mixed first variation. Again, this means to consider only functions which get smoother as the dimensionality increases, but, in contrast to the examples mentioned above, now, only an anisotropic smoothness assumption is involved. Note that this is just the prerequisite for sparse grids with the piecewise constant hierarchical basis.

Further results on high-dimensional (and even infinite-dimensional) prob-lems and their tractability were given by (Wasilkovski and Woźniakowski 1995, Sloan and Woźniakowski 1998, Wasilkovski and Woźniakowski 1999, Sloan 2001, Hickernell, Sloan and Wasilkowski 2004, Dick, Sloan, Wang and Woźniakowski 2004, Sloan, Wang and Woźniakowski 2004). Here, es-pecially in the context of numerical integration, the notion of *weighted* Sobolev spaces was introduced. Following the observation that for some problems the integrand becomes less and less variable in successive co-ordinate directions, a sequence of positive weights $\{\gamma_j\}$ with decreasing values is used, with the weight γ_j being associated with coordinate direc-tion j. Then it can be shown that the integration problem in a particular Sobolev space setting becomes *strongly tractable* (Traub and Woźniakowski 1980, Traub, Wasilkowski and Woźniakowski 1983, Traub, Wasilkowski and Woźniakowski 1988), i.e., that the worst-case error for all functions in the unit ball of the weighted Sobolev space is bounded independently of d and tends polynomially to zero if and only if the sum of the weights is asymptotically bounded from above. This corresponds to a decay of the kernel contributions in a reproducing kernel Hilbert space with in-creasing d. The original paper (Sloan and Woźniakowski 1998) assumes that the integrand belongs to a Sobolev space of functions with square-integrable mixed first derivatives with the weights built into the definition of the associated inner product. Note that this assumption is closely re-lated to that of (3.2) above. Since then, more general assumptions on the weights and, thus, on the induced weighted function spaces have been found (Dick et al. 2004, Hickernell et al. 2004, Sloan et al. 2004, Hickernell and Woźniakowski 2000, Wasilkowski and Woźniakowski 2004).

In any case, we observe that a certain smoothness assumption on the function under consideration changes with d and leads to approximation rates that no longer depend exponentially on d. This raises the question what smoothness for changing d and smoothness for $d \rightarrow \infty$ mean at all.

To this end, let us note an interesting aspect, namely the concentration of measure phenomenon (Milman 1988, Milman and Schechtman 2001, Talagrand 1995, Gromov 1999, Ledoux 2001) for probabilities in normed spaces in high dimensions (also known as the geometric law of large numbers). This is an important development in modern analysis and geometry, manifesting itself across a wide range of mathematical sciences, particularly geometric functional analysis, probability theory, graph theory, diverse fields of computer science, and statistical physics. In the statistical setting it states the following: Let f be a Lipschitz function with Lipschitz constant L on the d-sphere. Let P be a normalized Lebesgue measure on the sphere and let X be a random variable uniformly distributed with respect to P. Then,

$$P\{|f(X) - Ef(X)| > t\} \leq c_1 \exp(-c_2 t^2 / L^2)$$

with constants c_1, c_2 independent of f and d. In its simplest form, the phenomenon of concentration of measure just says that every Lipschitz function on a sufficiently high-dimensional domain Ω is well approximated by a constant function (Hegland and Pestov 1999, Baxter and Iserles 2003). Thus, there is some chance to treat high-dimensional problems despite the curse of dimensionality.

The relation of the concentration of measure phenomenon to approximation estimates was further elaborated upon in (Hegland and Pozzi 2005). There, with the help of a concentration function which expresses the concentration effect of the underlying metric space, new inequalities for the error of function approximations have been derived. Besides estimates for the above mentioned approximation of a function by a constant (e.g., by its mean or by the evaluation at a random point) and radial basis functions, also piecewise constant approximation schemes and piecewise approximations of higher order by Hermite polynomials have been studied. The resulting approximation rates were the same as with conventional estimates based on finite elements. However the constants in the estimates were substantially better. They are independent of the dimension and in addition allow realistic bounds for multimodal distributions which is not the case for classical approaches based on interpolation theory. These techniques may be employed to obtain better estimates for the constants in the order estimates with respect to the dimension in e.g., sparse grid approximation schemes. This however is future work.

3.2.2 The theorem of Komogorov and related approximation schemes

One approach to develop efficient approximations which allow one to overcome the curse of dimensionality is to describe multivariate continuous functions as a superposition (Rassias and Simsa 1995, Khavinson 1997) of a number of continuous functions with fewer variables. This question is related to Hilbert's 13th problem, see (Vitushkin 2004) and the references cited therein. It was answered in (Kolmogorov 1957) who found that every continuous function of several variables can be represented by the superposition of continuous functions with only two variables. Kolmogorov even showed that every continuous function of several variables can be represented by the superposition of continuous functions with only one variable,[1] see also (Sprecher 1965, Lorentz, v. Golitschek and Makovoz 1996, Khavinson 1997) for improved versions.

Kolmogorov's famous result can be expressed as follows: Let f be a multivariate continuous function on the unit cube, i.e., $f(x_1, \ldots, x_d) : [0,1]^d \to \mathbb{R}$. Each function $f \in C([0,1]^d)$ has a representation

$$f(x_1, \ldots, x_d) = \sum_{i=1}^{2d+1} f_i \left(\sum_{j=1}^{d} \phi_{i,j}(x_j) \right) \tag{3.3}$$

where all $\{f_i\}$ and $\{\phi_{i,j}\}$ are one-dimensional continuous functions defined on $\mathbb{R}$ and all $\{\phi_{i,j}\}$ are independent of the choice of f. An improvement was given in (Fridman 1967) where it was shown that the inner functions $\{\phi_{i,j}\}$ can be chosen to be Lipschitz continuous with exponent one.

There have been various refinements of this result. A version with just one outer function and $2d + 1$ inner functions, see (Lorentz et al. 1996) Chapter 17, reads: There exist $2d + 1$ continuous, strictly increasing functions $\phi_i : [0,1] \to [0,1]$ and d positive constants λ_i with $\sum_{i=1}^{d} \lambda_i \leq 1$ with the property that each function $f \in C([0,1]^d)$ has a representation

$$f(x_1, \ldots, x_d) = \sum_{i=1}^{2d+1} g \left(\sum_{j=1}^{d} \lambda_j \phi_i(x_j) \right) \tag{3.4}$$

for some non-smooth $g \in C([0,1])$ depending on f. Here, the functions ϕ_i together with their summation provide a one-to-one embedding of the unit cube $[0,1]^d$ into $\mathbb{R}^{2d+1}$, i.e., we have with $X_i := \sum_{j=1}^{d} \lambda_j \phi_i(x_j)$ the representation $f(x_1, \ldots, x_d) = \sum_{i=1}^{2d+1} g(X_i)$. Note that there is a close relation to d-dimensional topology: The theorem of Menger and Nöbeling, see (Hurewicz and Wallman 1948), page 84, tells us that any d-dimensional

---

[1] Kolmogorov's student Arnold showed even before in (Arnold 1957, Arnold 1958, Arnold 1959) that any $f \in C([0,1]^3)$ can be represented as a superposition of continuous functions in two variables, and thus refuted Hilbert's conjecture.

compact set can be homeomorphically embedded into $[0,1]^{2d+1}$. Thus, Kolmogorov's theorem (3.4) can be seen as just a special case of it.

Another version with $2d + 1$ outer functions and one inner function is due to (Sprecher 1965). It reads

$$f(x_1,\ldots,x_d) = \sum_{i=1}^{2d+1} f_i \left(\sum_{j=1}^{d} \lambda_j \phi(x_j + i \cdot \alpha) \right) \qquad (3.5)$$

with suitable constants λ_i, α and a continuous one-dimensional function ϕ. It also gives an embedding of $[0,1]^d$ into $\mathbb{R}^{2d+1}$ by $f(x_1,\ldots,x_d) = \sum_{i=1}^{2d+1} f_i(X_i)$ with $X_i := \sum_{j=1}^{d} \lambda_j \phi(x_j + i \cdot \alpha)$.

The proof of Kolmogorov's theorem is non-constructive and does not provide us with a way to choose the inner and outer functions in (3.3), (3.4) or (3.5), respectively. A first attempt to remedy this problem with merely an approximation of the inner function and an interpolation of the outer functions was made in (de Figueiredo 1980). Recently, however, algorithms were given to explicitly construct the functions in (3.5). The implementation of an inner function ϕ which does not depend on f was discussed in (Sprecher 1996). Here ϕ is pointwise defined on an everywhere dense set of rational numbers in $[0,1]$ from which it can be uniquely extended to a continuous function on $[0,1]$. The resulting ϕ is non-continuous. There is also a close relation of ϕ to space-filling curves, see (Sprecher and Draghici 2002, Sagan 1994). An implementation of the outer functions f_i by an iterative method was presented in (Sprecher 1997). This established the first constructive proof of Kolmogorov's theorem. It furthermore allows to realize (3.5) as a feedforward neural network with a hidden layer that computes the variables X_i and therefore involves the embedding mapping only, and a single output layer in which f is computed by means of the functions $f_i(X_i)$, see also (Hecht-Nielsen 1987a, Hecht-Nielsen 1987b). In (Köppen 2002) the construction was improved to give a continuous inner function ϕ.

Thus, in view of Kolmogorov's result, it seems that there are no high-dimensional functions and thus no high-dimensional problems at all. However, it turns out that the representing functions are quite bad, i.e., they are at best only continuous and highly non-smooth. This limits their practical use for approximation and interpolation purposes (Girosi and Poggio 1989), like e.g., for the discretization of PDEs within the Galerkin approach. In particular, the representing functions cannot be chosen to be differentiable. This even holds if one wants to represent an analytic function f only, see (Vitushkin 1964).

Nevertheless Kolmogorov's theorem inspired many linear and nonlinear approximation schemes and there have been various attempts to generalize Kolmogorov's formula. Moreover, in (Kurkova 1991) it was noticed that in the proof of Kolmogorov's superposition theorem the fixed number of $2d + 1$ basis functions can be replaced by a variable number m and the task of function representation can be replaced by the task of function approximation. In the following we give an (incomplete) list of approaches which are related to Kolmogorov's theorem.

- Popular approximation schemes in statistics are the so-called additive models, see (Hastie and Tibshirani 1986, Hastie and Tibshirani 1990). They resemble the approximation

$$f(x_1, \ldots, x_d) \approx \sum_{i=1}^{d} f_i(x_i). \tag{3.6}$$

This form can be derived from (3.3) by choosing d instead of $2d + 1$ and replacing the inner functions ϕ_{ij} trivially by the identity if $i = j$ and zero otherwise.

- The projection pursuit algorithm (Friedman and Stützle 1981, Stone 1985) approximates a function f by

$$f(x_1, \ldots, x_d) \approx f_0 + \sum_{i=1}^{m} f_i \left(\sum_{j=1}^{d} \beta_{ij} \cdot x_j \right)$$

with the so-called projection directions $\vec{\beta}_i := (\beta_{i1}, \ldots, \beta_{id})$ and with f_0 as the average of the function f. Here the parameter vectors $\vec{\beta}_i$ and the functions f_i are estimated from the data. This scheme can also be interpreted as a special case of (3.6) using linear combinations of the original coordinates.

- Closely related are multilayer perceptrons with a single hidden layer. They approximate f by

$$f(x_1, \ldots, x_d) \approx h \left(\sum_{i=1}^{m} \alpha_i g \left(\sum_{j=1}^{d} \beta_{ij} x_j \right) \right)$$

where h and g are arbitrary nonlinear functions. Here, the network is trained with a given set of input and output values and the approximation is then determined by the values of α_i and the vectors $\vec{\beta}_i$ which are found by least squares minimization. In (Kurkova 1991) it was demonstrated how to approximate a Hecht-Nielsen network which implements Komogorov's superposition approach by such traditional neural networks.

- Also radial basis schemes belong to the class of approximation schemes which can be derived from (3.3). Here the dimension embedding takes place by a distance function, i.e., the sum of the inner functions gets replaced by the Euclidean norm. They can under certain assumptions be written as

$$f(x) \approx \sum_{i=1}^{m} \beta_i f_i(\|x - y_i\|) + P(x)$$

where $x = (x_1, \ldots, x_d)$, the f_i are a chosen set of radial basis functions, $y_i \in \mathbb{R}^d$ are their centers, the β_i are constants and $P(x)$ is a polynomial. The coefficients are then fitted to the data by means of least-squares minimization.

- Starting from the representation (3.4), (Igelnik and Parikh 2003) introduced so-called Kolmogorov spline networks. There, $2d + 1$ is replaced by a general m and the outer function g and the inner functions ϕ_i are replaced by cubic splines $s(\cdot, \gamma_i)$ and $s(\cdot, \gamma_{i,j})$ where the parameters γ_i and $\gamma_{i,j}$ of the splines are adjusted to fit given data on f properly. The approximation scheme is defined as

$$f_m(x_1, \ldots, x_d) = \sum_{i=1}^{m} s \left(\sum_{j=1}^{d} \lambda_j s(x_j, \gamma_{i,j}), \gamma_i \right) \qquad (3.7)$$

with positive numbers $\lambda_1, \ldots, \lambda_d$ with $\sum \lambda_j \leq 1$ which can be chosen independent of f. It was shown that, for any function f from the class of continuously differentiable functions on $[0, 1]^d$ with bounded gradient, there exists a function f_m of the form (3.7) such that $\|f - f_m\| = O(1/m)$. The number of degrees of freedom involved in the network is of the order $O(m^{2/3})$. This result compares favorably with the approximation order $O(1/\sqrt{m})$ and the number of degrees of freedom $O(m^2)$ usually achieved for general one-hidden layer feedforward networks for this class of functions, compare also (Barron 1993, Igelnik and Parikh 2003) and the references cited therein. A similar approach was also presented in (Coppejans 2004).

- Finally, Kronecker-product type approximations of the form

$$f(x) \approx \sum_{i=1}^{m} c_i \prod_{j=1}^{d} f_{ij}(x_j), \quad c_i \in \mathbb{R},$$

possess a structure similar to Kolmogorov's theorem. Here however the inner sum is replaced by a product of one-dimensional functions. To see the relation to (3.3) choose there $f_i(\cdot) = c_i \cdot \exp(\cdot)$ and $\phi_{ij}(\cdot) = \log(f_{ij}(\cdot))$. For numerical purposes, the one-dimensional functions f_{ij} are further expanded in a series with a suitable multilevel basis which is then properly truncated or, more often, they are just simply discretized on a uniform grid. For details and applications see (Beylkin and Mohlenkamp 2002, Beylkin and Mohlenkamp 2005, Tyrtyshnikov 2004, Hackbusch and

Khoromskij 2004), the related developments on so-called H- and H^2-matrices (Hackbusch, Khoromskij and Sauter 2000, Grasedyck and Hackbusch 2003) and the references cited therein. A similar decomposition is used in the MCTDH approach (Beck, Jäckle, Worth and Meyer 2000).

The basic theory of this decomposition can be found in (Golomb 1959): In the case $d=2$ mainly the classical Hilbert-Schmidt theory appears, i.e., the functions f_{ij} are the unique solution of a system of two coupled linear integral equations which resemble the continuous analogue of the classical singular value decomposition. For the case $d > 2$ however, a system of nonlinear integral equations results for which the solution is no longer unique. Then heuristics must be used to obtain some solution. Nevertheless it is observed in applications that good approximations can be obtained with an already relatively small number m.

For most of these approximation schemes the parameters are obtained by some kind of (least-squares) minimization. Here, however the objective functional may not be globally convex and can have many minima which results in non-unique representations. Thus, the associated approximation rates for these schemes for increasing m are not always fully understood and, moreover, it is not clear which representation to prefer over another for a particular application. In the following we therefore study a simpler linear decomposition of a d-dimensional function into its contributions from different (groups of) subdimensions which can be seen as a multivariate generalization of (3.6).

3.3 Dimension-wise space decomposition

We consider a decomposition of the d-dimensional function f as

$$f(x_1,\ldots,x_d) = f_0 + \sum_{j_1}^{d} f_{j_1}(x_{j_1}) + \sum_{j_1<j_2}^{d} f_{j_1,j_2}(x_{j_1},x_{j_2}) \qquad (3.8)$$

$$+ \sum_{j_1<j_2<j_3}^{d} f_{j_1,j_2,j_3}(x_{j_1},x_{j_2},x_{j_3}) + \cdots + f_{j_1,\ldots,j_d}(x_{j_1},\ldots,x_{j_d}).$$

Here, f_0 is a constant function, f_{j_1} are one-dimensional functions, f_{j_1,j_2} are two-dimensional functions, and so on. This type of decomposition goes back to (Hoeffding 1948) and is well known in statistics under the name ANOVA (analysis of variance), see also (Efron and Stein 1981). Note that (3.8) is a finite expansion of f into 2^d different terms. Such a decomposition can be gained by a tensor product construction of a splitting of the one-dimensional function space into its constant subspace and its remainder. This will be explained in more detail in the following.

3.3.1 ANOVA-like decompositions

Let $V^{(d)}$ denote the underlying space of d-dimensional functions $f(x) = f(x_1, \ldots, x_d) : \bar{\Omega}^{(d)} \to \mathbb{R}$ with $\bar{\Omega}^{(d)} = [0,1]^d$. Let μ be a product measure with unit mass which has a density, i.e.,

$$d\mu(x) = \prod_{j=1}^{d} d\mu_j(x_j), \qquad \int_{\bar{\Omega}^{(1)}} d\mu(x_j) = 1, \tag{3.9}$$

$$d\mu(x) = h(x) = \prod_{j=1}^{d} h_j(x_j) dx_j, \tag{3.10}$$

where $h_j(x_j)$ is the marginal density of the input variable x_j. Furthermore, let $V^{(d)}$ be a Hilbert space equipped with the inner product $(f, g) = \int_{\bar{\Omega}^{(d)}} f(x)g(x) d\mu(x)$ and associated norm $\|\cdot\|$. First we will deal with the one-dimensional case. We decompose $V^{(1)}$ in a simple two-scale fashion by

$$V^{(1)} = \mathbf{1} \oplus W \tag{3.11}$$

where $\mathbf{1}$ denotes the one-dimensional subspace $span\{1\}$ which contains the constant functions. Associated to such a splitting is a mapping $P : V^{(1)} \to \mathbf{1}$ with

$$Pf(x) = \int_{\bar{\Omega}^{(1)}} f(x) d\mu(x). \tag{3.12}$$

Examples are the conventional Lebesque measure $d\mu(x) = dx$ which leads to the integral average

$$Pf(x) = \int_{\bar{\Omega}^{(1)}} f(x) dx$$

or the Dirac measure located at a point a, i.e., $d\mu(x) = \delta(x - a)dx$, which results in the simple evaluation at point a

$$Pf(x) = \int_{\bar{\Omega}^{(1)}} \delta(x - a) f(x) dx = f(a). \tag{3.13}$$

This introduces the decomposition

$$f(x) = f_0 + f_1(x) \tag{3.14}$$

with

$$f_0 = Pf(x) = \int_{\bar{\Omega}^{(1)}} f(x) d\mu(x) \in \mathbf{1} \text{ and} \tag{3.15}$$

$$f_1(x) = (I - P)f(x) = f(x) - \int_{\bar{\Omega}^{(1)}} f(x) d\mu(x) \in W. \tag{3.16}$$

Then W is the subspace of $V^{(1)}$ of functions which satisfy the relation $\int_{\bar{\Omega}^{(1)}} f(x) d\mu(x) = 0$. It is orthogonal to $\mathbf{1}$ and, with $\int_{\bar{\Omega}^{(1)}} f_1 d\mu(x) = 0$, it is easy to see that $\|f\| = \|f_0\| + \|f_1\|$ and $(f, g) = (f_0, g_0) + (f_1, g_1)$ with g split analogously to (3.14).

Note that for (3.13) with differentiable f there is a close relation to the Taylor expansion. The decomposition (3.14) is just the Taylor formula of first order and $f_1 = f(x) - f(a) = \int_a^x f'(t)dt$ is the remainder term.

Now we consider the d-dimensional case: The one-dimensional splitting introduces a natural decomposition of the d-dimensional function space $V^{(d)}$ by a tensor product construction

$$
\begin{aligned}
V^{(d)} &= \bigotimes_{j=1}^{d} (\mathbf{1}_j \oplus W_j) \\
&= \mathbf{1}_1 \otimes \cdots \otimes \mathbf{1}_d \\
&\oplus \bigoplus_{i=1}^{d} \mathbf{1}_1 \otimes \cdots \otimes W_i \otimes \cdots \otimes \mathbf{1}_d \\
&\oplus \bigoplus_{i=1}^{d} \bigoplus_{i<j} \mathbf{1}_1 \otimes \cdots \otimes W_i \otimes \cdots \otimes W_j \otimes \cdots \otimes \mathbf{1}_d \\
&\oplus \bigoplus_{i=1}^{d} \bigoplus_{i<j} \bigoplus_{i<j<k} \mathbf{1}_1 \otimes \cdots \otimes W_i \otimes \cdots \otimes W_j \otimes \cdots \otimes W_k \otimes \cdots \otimes \mathbf{1}_d \\
&\cdots \\
&\oplus W_1 \otimes \cdots \otimes W_d.
\end{aligned}
\tag{3.17}
$$

Here $\mathbf{1}_j = \mathbf{1}$ and $W_j = W$; we use the index j merely to indicate the respective coordinate direction for explanatory reasons. Another notation which involves the subsets of the index set $\{1, 2, \ldots, d\}$ is[1]

$$
V^{(d)} = \bigoplus_{\mathbf{u} \subset \{1,\ldots,d\}} \left(\bigotimes_{k \in \{1,\ldots,d\}/\mathbf{u}} \mathbf{1}_k \right) \otimes \left(\bigotimes_{j \in \mathbf{u}} W_j \right) =: \bigoplus_{\mathbf{u} \subset \{1,\ldots,d\}} W_{\mathbf{u}}. \tag{3.18}
$$

Then, a function $f \in V^{(d)}$ is decomposed accordingly as

$$
\begin{aligned}
f(x_1, \ldots, x_d) &= f_0 + \sum_{j_1}^{d} f_{j_1}(x_{j_1}) + \sum_{j_1 < j_2}^{d} f_{j_1, j_2}(x_{j_1}, x_{j_2}) \\
&\quad + \sum_{j_1 < j_2 < j_3}^{d} f_{j_1, j_2, j_3}(x_{j_1}, x_{j_2}, x_{j_3}) + \cdots + f_{j_1, \ldots, j_d}(x_{j_1}, \ldots, x_{j_d}) \\
&= \sum_{\mathbf{u} \subset \{1,\ldots,d\}} f_{\mathbf{u}}(x_{\mathbf{u}})
\end{aligned}
\tag{3.19}
$$

where $f_{\mathbf{u}} \in W_{\mathbf{u}}$ and $x_{\mathbf{u}}$ denotes the variables x_i of x with $i \in \mathbf{u}$. Note that

---

[1] Note the obvious identity $\prod_{j \in \mathbf{v}} (b_j + c_j) = \sum_{\mathbf{u} \subset \mathbf{v}} \prod_{k \in \mathbf{v}/\mathbf{u}} b_k \prod_{j \in \mathbf{u}} c_j, \forall b_j, c_j \in \mathbb{R}$. It can be applied for products of sums of functions and products of sums of subspaces in an analogous way.

due to the power set construct this is a finite expansion which involves 2^d different terms. The decomposition is unique for a fixed choice of the one-dimensional mapping $P : V^{(1)} \to \mathbf{1}$.

Associated is the identity

$$I^{(d)} = \bigotimes_{j=1}^{d} (P_j + (I_j - P_j))$$

$$= \sum_{\mathbf{u} \subset \{1,\ldots,d\}} \left(\prod_{k \in \{1,\ldots,d\}/\mathbf{u}} P_k \right) \cdot \left(\prod_{j \in \mathbf{u}} (I_j - P_j) \right) =: \sum_{\mathbf{u} \subset \{1,\ldots,d\}} P_{\mathbf{u}}$$

where P_j and I_j denote the one-dimensional projection operator (3.12) and the identity for the j-th coordinate direction, respectively. Here the projection $\prod_{j=1}^{d} P_j f$ is the unconditional mean of f (with respect to the measure μ) and the partial projection $\prod_{k \in \{1,\ldots,d\}/\mathbf{u}} P_k f$ is the conditional mean $\int .. \int f(x) \prod_{k \in \{1,\ldots,d\}/\mathbf{u}} d\mu_k(x_k)$.

For the example of the conventional Lebesque measure $d\mu(x) = dx$ we obtain the functions in (3.19) as

$$f_0 = \prod_{j=1}^{d} P_j f = \int_{\bar{\Omega}^{(d)}} f(x) \prod_{i=1}^{d} dx_i,$$

$$f_{j_1}(x_{j_1}) = \int_{\bar{\Omega}^{(d-1)}} f(x) \prod_{i \neq j_1} dx_i - f_0,$$

$$f_{j_1,j_2}(x_{j_1,j_2}) = \int_{\bar{\Omega}^{(d-2)}} f(x) \prod_{i \notin \{j_1,j_2\}} dx_i - f_{j_1}(x_{j_1}) - f_{j_2}(x_{j_2}) - f_0,$$

$$\cdots \quad \cdots$$

$$f_{j_1,\ldots,j_k}(x_{j_1,\ldots,j_k}) = \int_{\bar{\Omega}^{(d-k)}} f(x) \prod_{i \notin \{j_1,\ldots,j_k\}} dx_i \qquad (3.20)$$

$$- \sum_{i_1 < \cdots < i_{k-1} \subset \{j_1,\ldots,j_k\}} f_{i_1,\ldots,i_{k-1}}(x_{i_1},\ldots,x_{i_{k-1}})$$

$$- \sum_{i_1 < \cdots < i_{k-2} \subset \{j_1,\ldots,j_k\}} f_{i_1,\ldots,i_{k-2}}(x_{i_1},\ldots,x_{i_{k-2}})$$

$$\cdots$$

$$- \sum_{j_1} f_{j_1}(x_{j_1}) - f_0,$$

$$\cdots \quad \cdots$$

This is just the well known ANOVA decomposition used in statistics, see (Efron and Stein 1981, Wahba 1990) and the references cited therein. There, if the input consists of independently distributed uniform random variables (with respect to the Lebesque measure) then the component functions are uncorrelated and the total variance D can be written as

$$D = E(f - f_0)^2 = \sum_{j_1} D_{j_1} + \sum_{j_1 < j_2} D_{j_1, j_2} + \sum_{j_1 < j_2 < j_3} D_{j_1, j_2, j_3} + \ldots D_{j_1, \ldots j_d}$$

with the partial variances[1]

$$D_{j_1, \ldots, j_k} = \int_{\bar{\Omega}^{(k)}} (f_{j_1, \ldots, j_k})^2 dx_{j_1} \ldots dx_{j_k}.$$

For the example of the Dirac measure located at a point a_j, i.e., with $d\mu(x_j) = \delta(x_j - a_j)dx_j$ and $d\mu(x) = \prod_{j=1}^{d} d\mu(x_j)$, we obtain the functions in (3.19) as

$$f_0 = f(x)|_{x=a}$$
$$f_{j_1}(x_{j_1}) = f(x)|_{x=a \setminus x_{j_1}} - f_0,$$
$$f_{j_1, j_2}(x_{j_1, j_2}) = f(x)|_{x=a \setminus \{x_{j_1}, x_{j_2}\}} - f_{j_1}(x_{j_1}) - f_{j_2}(x_{j_2}) - f_0,$$
$$\cdots \quad \cdots$$
$$f_{j_1, \ldots, j_k}(x_{j_1, \ldots, j_k}) = f(x)|_{x=a \setminus \{x_{j_1}, \ldots, x_{j_k}\}} \tag{3.21}$$
$$- \sum_{\{i_1, \ldots, i_{k-1}\} \subset \{j_1, \ldots, j_k\}} f_{i_1, \ldots, i_{k-1}}(x_{i_1}, \ldots, x_{i_{k-1}})$$
$$- \sum_{\{i_1, \ldots, i_{k-2}\} \subset \{j_1, \ldots, j_k\}} f_{i_1, \ldots, i_{k-2}}(x_{i_1}, \ldots, x_{i_{k-2}})$$
$$\cdots$$
$$- \sum_{j_1} f_{j_1}(x_{j_1}) - f_0,$$
$$\cdots \quad \cdots$$

where now only (partial) point evaluations in the point $a = (a_1, \ldots, a_d)$ are involved. Here we use the notation

$$f(x)|_{x=a \setminus x_i} = f(a_1, \ldots, a_{i-1}, x_i, a_{i+1}, \ldots, a_d)$$

with its obvious generalization to $a \setminus \{x_{j_1}, \ldots, x_{j_k}\}$. This approach is considered in (Rabitz and Alis 1999) under the name cut-HDMR and is closely related to the anchor spaces of (Sloan et al. 2004, Dick et al. 2004, Hickernell and Woźniakowski 2000, Wasilkowski and Woźniakowski 2004). Note that a component function vanishes if the value of one of its input variables x_i is equal to the associated coordinate of the point a, i.e.,

$$f_{j_1, \ldots, j_k}(x_{j_1, \ldots, j_k})|_{x_i = a_i} = 0 \quad i \in \{j_1, \ldots, j_k\}.$$

Thus the decomposition (3.21) expresses f as a superposition of its values on lines, faces, hyperplanes etc. that pass through the point a. Also note

---

[1] The global sensitivity indices are then defined as $S_{j_1, \ldots, j_k} = D_{j_1, \ldots, j_k}/D$. They describe the contribution of the input $\{x_{j_1}, \ldots, x_{j_k}\}$ to the variance of the output.

that the component functions fulfill

$$f_{i_1,\ldots,i_p}(x_{i_1,\ldots,i_p})f_{j_1,\ldots,j_q}(x_{j_1,\ldots,j_q})|_{x_k=a_k} = 0 \quad k \in \{i_1,\ldots,i_p\}\cup\{j_1,\ldots,j_p\},$$

which is a direct consequence of the orthogonality

$$\int_{\bar{\Omega}^{(d)}} f_{i_1,\ldots,i_p}(x_{i_1,\ldots,i_p})f_{j_1,\ldots,j_q}(x_{j_1,\ldots,j_q})d\mu(x) = 0$$

with $d\mu(x) = \prod_{j=1}^d \delta(x_j - a_j)dx_j$.

Note furthermore that for differentiable f there is a close relation to the multivariate Taylor expansion. The decomposition (3.21) is just the multivariate Taylor formula of first order in each coordinate direction with partial remainder terms. Moreover, the multivariate Taylor expansion of f around a (provided that f is sufficiently many times differentiable of course) and a short calculation shows the following: The component functions of first order, i.e., f_{j_1}, are the sum of all terms in the Taylor series which depend only on x_{j_1}, the component functions of second order, i.e., f_{j_1,j_2}, are the sum of all terms in the Taylor series which depend on x_{j_1} and x_{j_2}, and so on. Thus (3.21) resembles a rearrangement of the infinite number of terms in the full Taylor series into a finite number, i.e., 2^d, of different groups where each group corresponds to one component function (which still contains as series an infinite number of terms).

Note finally that there are various generalizations of (3.21). Instead of the Dirac measure at one point a we could also take the average of the Dirac measures at m different points and build an ANOVA-type decomposition on it. This approach is closely related to the multi-cut-HDMR method of (Li, Rosenthal and Rabitz 2001b, Li, Schoendorf, Ho and Rabitz 2004). Other variants (mp-cut-HDMR and lp-RS) can be found in (Li, Wang, Rosenthal and Rabitz 2001c) and (Li, Atramonov, Rabitz, Wang, Georgopoulos and Demiralp 2001a), respectively.

In summary an ANOVA-type decomposition of f into component functions reveals the relative importance of the different dimensions as well as their interactions and correlations. In general, an arbitrary function f may result in zero components except of its highest order term $f_{j_1,\ldots,j_d}(x_{j_1},\ldots,x_{j_d})$ or might have all its components beeing relevant. Then, nothing is gained with respect to the curse of dimensionality when switching from f to its ANOVA decomposition. However in many practical applications it can be observed that the finite series (3.19) decays rapidly. In some cases it is even of *finite order* q, i.e., for the components $f_{\mathbf{u}}$ of the decomposition (3.19) there holds

$$f_{\mathbf{u}} = 0 \text{ with } |\mathbf{u}| > q$$

with $q << d$. This usually expresses the fact that reasonable, meaningful (observable) coordinates of the physical system under consideration had been chosen. Alternatively it may happen that the different dimensions are not of equal importance and we find a decay in the contribution of the dimensions (after sorting according to their relevance) and their associated higher order interactions.

Examples with such types of behavior of the expansion (3.19) are:

- In most molecular dynamics simulation codes only two-body (bonds), three-body (angle) and four-body (dihedral) potential functions are used to describe molecules, i.e., it holds that the finite order of the associated ANOVA decomposition is trivially $q \leq 4$. It seems that this mostly gives a sufficient representation of the potential energy hypersurface of a system, especially when macroscopic variables are sought.

- Closely related is the Mayer cluster expansion in statistical mechanics. Here, for pair potentials $U_{ij}(x_i, x_j)$ which express the interaction between two particles, the term

$$\exp\left(-\sum_{i<j} U_{ij}(x_i, x_j)\right) = \prod_{i<j} \exp(-U_{ij}(x_i, x_j))$$

gets transformed by $\exp(-U_{ij}(x_i, x_j)) =: 1 + \phi_{ij}(x_i, x_j)$ into

$$1 + \sum_{i<j} \phi_{ij} + \sum_{i<j}\sum_{k<l} \phi_{ij}\phi_{kl} + \sum_{i<j}\sum_{k<l}\sum_{m<n} \phi_{ij}\phi_{kl}\phi_{mn} + \cdots$$

which allows one to write the partition function of the canonical ensemble of a particle system by means of cluster integrals, for details see (Hill 1956), page 123 ff.

- In many statistical applications a statistics of second order is sufficient, i.e., the covariances of the input variables play an important role, but higher-order correlations are neglected. Again this means that we have finite order $q \leq 2$ in the associated ANOVA decomposition.

- In data mining it is found from multivariate adaptive regression splines (MARS), see (Friedman 1991), that even for really high-dimensional data there appear at most 5-7 dimensional interactions, i.e., $q \leq 7$, and higher-order interactions are practically not significant.

- The Brownian bridge representation of a Markov process results in a concentration of the total variation in the first few levels of the discretization since the variance decays with the factor $2^{-1/2}$ from level to level, see also (Caflisch, Morokoff and Owen 1997, Morokoff 1998, Gerstner and Griebel 2003, Gerstner and Griebel 1998) where the Brownian bridge was used in high-dimensional integration problems. A further analysis in view of reproducing kernel Hilbert spaces with weights is given in (Leobacher, Scheicher and Larcher 2003). Note that for the Karhunen-Loewe decomposition an even better decay may result than for the Brownian bridge.

- Many problems in mathematical finance can be formulated as high-dimensional integrals, where the large number of dimensions arises from small time steps in time discretization and/or a large number of state variables. Examples are option pricing, bond valuation or the pricing of collateral mortgage backed securities. There, it turns out that for the ANOVA decomposition of the integrand the importance of each dimension is naturally weighted by certain hidden weights where with the increase of dimension the lower-order terms continue to play a significant role and the higher-order terms tend to be negligible, see (Caflisch et al. 1997, Sloan and Wang 2005). This is a reason why Quasi-Monte-Carlo performs better than expected, especially for high-dimensional integrands.

- There is also a counterexample. In quantum chemistry, the solution of Schrödinger's equation for d fermions has to obey the antisymmetry condition due to Pauli's principle. It can be shown that for an ANOVA decomposition of an antisymmetric f all terms $f_{\mathbf{u}}$ with $|\mathbf{u}| < d - 1$ are identically zero and all information of f is contained in the terms with order $d - 1$ and d.[1]

In summary, in certain applications, i.e., for f from special function spaces, we know a-priori how the ANOVA-components decay and may resort to a truncation of (3.19) after the q-th order terms, where q is related to the needed accuracy. Or we may a-priori know if ANOVA-terms of order higher than some q are present at all or not. The question is how these situations and the associated function spaces can be characterized. A possibility are reproducing kernel Hilbert spaces. The associated multi-dimensional kernel function can be decomposed analogously to the ANOVA expansion into a sum of kernels. Then these partial kernels can be equipped with different individual weights. These weights allow one to model various behaviors of decay for the different contributions in the ANOVA decomposition as well as truncations to finite order. This will be dealt with in the following.

---

[1] For bosonic systems a symmetry condition is needed instead. Then all functions $f_{\mathbf{u}}$ with the same order $|\mathbf{u}|$ are the same.

3.3.2 Reproducing kernel Hilbert spaces

The theory of reproducing kernel Hilbert spaces (RKHS) was introduced in (Aronzaijn 1950). It allows one to describe function spaces in a concise and elegant way by means of so-called reproducing kernel functions. To this end, we assume that $f : [0,1]^d \to \mathbb{R}$ belongs to a Hilbert space H with associated inner product $\langle \cdot, \cdot \rangle_H$ and norm $\|f\|_H = \langle f, f \rangle_H^{1/2}$. We assume that H is continuously embedded into $L_2([0,1]^d)$. Thus, we consider integrable functions f with respect to the Lebesgue measure for which $\|f\|_{L_2} := \left(\int_{[0,1]^d} f^2(t)dt \right)^{1/2} < \infty$. Furthermore, there is a non-negative number $c(H)$ depending on the space H such that

$$\|f\|_{L_2} \leq c(H) \|f\|_H \qquad \text{for all } f \in H. \tag{3.22}$$

Finally we assume that the evaluation of the function f is well-defined and continuous, i.e., that the linear functional $f \in H \mapsto f(x)$ is continuous for any $x \in [0,1]^d$. These assumptions are equivalent to the requirement that H is a reproducing kernel Hilbert space, see (Aronzaijn 1950). Hence, H has an associated kernel $K^{(d)} : [0,1]^d \times [0,1]^d \to \mathbb{R}$ which is uniquely defined by the following three conditions:

- $K^{(d)}(\cdot, t) \in H$ for all $t \in [0,1]^d$,
- $\left(K^{(d)}(x_i, x_j) \right)_{i,j=1}^n$ is a symmetric and non-negative definite matrix for all n and points x_i from $[0,1]^d$,
- $f(t) = \langle f, K^{(d)}(\cdot, t) \rangle_H$ for all $f \in H$ and all $t \in [0,1]^d$ (reproducing kernel property).

Thus it is sufficient to give $K^{(d)}$ to uniquely characterize the associated function space H. The theory of reproducing kernel Hilbert spaces can be found in detail in (Aronzaijn 1950); further aspects are discussed in (Wahba 1990, Ritter 2000).

From the three properties of reproducing kernels it easily follows that

$$K^{(d)}(t, x) = \left\langle K^{(d)}(\cdot, x), K^{(d)}(\cdot, t) \right\rangle_H \qquad \text{for all } t, x \in [0,1]^d,$$

$$\sqrt{K^{(d)}(t, t)} = \|K^{(d)}(\cdot, t)\|_H \qquad \text{for all } t \in [0,1]^d,$$

$$|f(t)| \leq \|f\|_H \sqrt{K^{(d)}(t, t)} \qquad \text{for all } f \in H, t \in [0,1]^d.$$

If H is separable, then for an arbitrary orthonormal basis $\{\eta_i\}$, we have $K^{(d)}(\cdot, x) = \sum_{i=1}^{\dim(H)} c_i \eta_i$ with $c_i = \langle \eta_i, K^{(d)}(\cdot, x) \rangle_H = \eta_i(x)$. Therefore

$$K^{(d)}(x, t) = \sum_{i=1}^{\dim(H)} \eta_i(x)\eta_i(t) \qquad \text{for all } x, t \in [0,1]^d. \tag{3.23}$$

In a way, the reverse of this argument is also true, see (Wahba 1990). To this end, let $\{\eta_i\}_{i=1}^{\infty}$ be a given arbitrary sequence of linearly independent functions defined on $[0,1]^d$ such that $\sum_{i=1}^{\infty} \eta_i^2(t) < \infty$ for all $t \in [0,1]^d$. Consider the space $H = \text{span}\{\eta_1, \eta_2, \dots\}$ of functions $f(t) = \sum_{i=1}^{\infty} f_i \eta_i(t)$ with real numbers f_i such that $\sum_{i=1}^{\infty} f_i^2 < \infty$. Observe that $f(t)$ is well-defined. For $f \in H$ the coefficients f_i are uniquely determined since the η_i's are linearly independent. The inner product in H is given by requiring that the η_i's be orthonormal, $\langle \eta_i, \eta_j \rangle_H = \delta_{i,j}$. Hence, for $f, g \in H$ we have $\langle f, g \rangle_H = \sum_{i=1}^{\infty} f_i g_i$ with f_i and g_i being the coefficients of f and g, respectively. Then H is a Hilbert space. Furthermore, it can be easily shown that

$$K^{(d)}(x,t) = \sum_{i=1}^{\infty} \eta_i(x) \eta_i(t)$$

is its reproducing kernel.

Note that the Hilbert space $L_2([0,1]^d)$ does not have a reproducing kernel, since point evaluation $t \in [0,1]^d \mapsto f(t)$ is not well-defined for $L_2([0,1]^d)$ and thus can not be continuous. It is easy to see that H is continuously embedded in L_2 if we assume that

$$\int_{[0,1]^d} K^{(d)}(t,t)\, dt < \infty. \tag{3.24}$$

Indeed, $f^2(t) \leq \|f\|_H^2 \cdot K^{(d)}(t,t)$, and therefore (3.22) holds with

$$c(H) = \left(\int_{[0,1]^d} K^{(d)}(t,t)\, dt \right)^{1/2}. \tag{3.25}$$

In this case, H is a proper subset of L_2, and $K^{(d)}(\cdot, t) \in L_2$ for arbitrary $t \in [0,1]^d$. Many examples of reproducing kernel Hilbert spaces can be found in the literature, see for example (Wahba 1990, Ritter 2000).

Remember now our approach in Subsection 3.3.1: We first split $V^{(1)} = \mathbf{1} \oplus W$ in (3.11) and then used the tensor product construction in (3.17) to gain the decomposition $V^{(d)} = \bigoplus_{\mathbf{u} \subset \{1,\dots,d\}} W_{\mathbf{u}}$ in (3.18) for the d-dimensional case. From (Aronzaijn 1950) we know the following facts:

- The reproducing kernel for the direct sum of two orthogonal subspaces is the sum of the single reproducing kernels.
- The reproducing kernel for a tensor product of two RKHS is the product of the single reproducing kernels.

This allows us, depending on the one-dimensional splitting and the associated orthogonal norm, to build a d-dimensional RKHS as a product of the sum of one-dimensional RKHSs. If we have for the orthogonal splitting (3.11) the associated sum of reproducing kernels $K(x,y) = K^{\mathbf{1}}(x,y) + K^{W}(x,y)$ we obtain for the splitting (3.18) the corresponding kernel

$$K^{(d)}(x,y) = \prod_{j=1}^{d} \left(K_j^{\mathbf{1}}(x_j,y_j) + K_j^{W}(x_j,y_j) \right)$$

$$= \sum_{\mathbf{u}\subset\{1,\ldots,d\}} \prod_{k\in\{1,\ldots,d\}\setminus\mathbf{u}} K_k^{\mathbf{1}}(x_k,y_k) \cdot \prod_{j\in\mathbf{u}} K_j^{W}(x_j,y_j)$$

$$=: \sum_{\mathbf{u}\subset\{1,\ldots,d\}} K_{\mathbf{u}}(x,y)$$

with $K_{\mathbf{u}}(x,y) = \prod_{k\in\{1,\ldots,d\}\setminus\mathbf{u}} K_k^{\mathbf{1}}(x_k,y_k) \cdot \prod_{j\in\mathbf{u}} K_j^{W}(x_j,y_j)$. Here we again use the indices j and k to indicate the respective coordinate directions. In the special case of $K_j^{\mathbf{1}}(x_j,y_j) = 1$ we directly have

$$K^{(d)}(x,y) = \sum_{\mathbf{u}\subset\{1,\ldots,d\}} \prod_{j\in\mathbf{u}} K_j^{W}(x_j,y_j).$$

3.3.3 Weighted spaces

Now we are in the position to introduce weights into the splittings. We follow (Sloan et al. 2004, Dick et al. 2004, Wasilkowski and Woźniakowski 2004), see also (Kuo and Sloan 2005). First we consider the simple case where each dimension gets its own weight $\gamma_j \in \mathbb{R}_0^+, j = 1, \ldots, d$, i.e., where d different non-negative weights are involved. We then have $K_j^{\mathbf{1}} = 1$ and replace K_j^{W} by $\gamma_j \cdot K_j^{W}$.[1] We obtain with $\prod_{k\in\{1,\ldots,d\}\setminus\mathbf{u}} K_k^{\mathbf{1}} = 1$

$$K^{(d)}(x,y) = \sum_{\mathbf{u}\subset\{1,\ldots,d\}} \prod_{j\in\mathbf{u}} \gamma_j \cdot \prod_{j\in\mathbf{u}} K_j^{W}(x_j,y_j)$$

$$=: \sum_{\mathbf{u}\subset\{1,\ldots,d\}} \gamma_{d,\mathbf{u}} \cdot K_{d,\mathbf{u}}(x_{\mathbf{u}},y_{\mathbf{u}})$$

with $\gamma_{d,\mathbf{u}} = \prod_{j\in\mathbf{u}} \gamma_j$ and $K_{d,\mathbf{u}}(x_{\mathbf{u}},y_{\mathbf{u}}) = \prod_{j\in\mathbf{u}} K_j^{W}(x_j,y_j)$. The resulting weights $\gamma_{d,\mathbf{u}}$ are just products of the γ_j.

We can generalize this approach as follows: We set

$$K^{(d)}(x,y) = \sum_{\mathbf{u}\subset\{1,\ldots,d\}} \gamma_{d,\mathbf{u}} \cdot \prod_{j\in\mathbf{u}} K_j^{W}(x_j,y_j)$$

$$=: \sum_{\mathbf{u}\subset\{1,\ldots,d\}} \gamma_{d,\mathbf{u}} \cdot K_{d,\mathbf{u}}(x_{\mathbf{u}},y_{\mathbf{u}}) \qquad (3.26)$$

where we now allow 2^d general non-negative weights $\gamma_{d,\mathbf{u}} \in \mathbb{R}_0^+$ which

---

[1] Without loss of generality we choose here the weight one in front of the kernel $K^{\mathbf{1}}$ which is associated to the subspace $\mathbf{1}$ of constants.

need no longer be formed as products of one-dimensional weights γ_i but may be chosen arbitrarily. Here we use the convention $\gamma_{d,\{\}} = 1$ and $\prod_{j \in \{\}} K_j^W = 1$.

As an example we consider the reproducing kernels

$$K_j^W(x,y) = \frac{1}{2}B_2(x-y) + \left(x - \frac{1}{2}\right)\left(y - \frac{1}{2}\right) + \mu_j(x) + \mu_j(y) + m_j$$

where $B_2(x) := x^2 - x + 1/6$ denotes the Bernoulli polynomial of degree 2, μ_j is a function with bounded derivative in $[0,1]$ such that

$$\int_0^1 \mu_j(x)dx = 0, \quad m_j := \int_0^1 (\mu'(x))^2 dx.$$

This kernel was presented for example in (Sloan et al. 2004). It allows to capture the two types of ANOVA-decompositions introduced in (3.20) and (3.21) as special cases and it also allows one to generalize them by means of the weights $\gamma_{d,\mathbf{u}}$.

The choice $\mu_j(x) = 0, j = 1, \ldots, d$, gives $m_j = 0$ and thus $K_j^W(x,y) = \frac{1}{2}B_2(x-y) + (x - \frac{1}{2})(y - \frac{1}{2})$. Note that $\int_0^1 K_j^W(x,y)dy = 0, \forall x \in [0,1]$. Then, the associated kernel (3.26) is called the ANOVA Sobolev kernel with general weights $\gamma_{d,\mathbf{u}}$. It can be shown that the associated inner product in $V^{(d)}$ is now

$$\langle f,g \rangle_{V^{(d)}} = \sum_{\mathbf{u} \subset \{1,\ldots,d\}} \frac{1}{\gamma_{d,\mathbf{u}}} \int_{[0,1]^{|\mathbf{u}|}} \left(\int_{[0,1]^{d-|\mathbf{u}|}} \frac{\partial^{|\mathbf{u}|} f(x)}{\partial x_{\mathbf{u}}} dx_{-\mathbf{u}} \right.$$
$$\left. \int_{[0,1]^{d-|\mathbf{u}|}} \frac{\partial^{|\mathbf{u}|} g(x)}{\partial x_{\mathbf{u}}} dx_{-\mathbf{u}} \right) dx_{\mathbf{u}} \qquad (3.27)$$

where we interpret the term associated to $\mathbf{u} = \{\}$ as the product of integrals $\int_{[0,1]^d} f(x)dx \int_{[0,1]^d} g(x)dx$. Here, $x_{\mathbf{u}}$ denotes the $|\mathbf{u}|$-dimensional vector of the components x_j with $j \in \mathbf{u}$ and $x_{-\mathbf{u}}$ denotes $x_{\{1,\ldots,d\}\setminus\mathbf{u}}$.

The choice

$$K_j^W(x,y) = \begin{cases} \min(|x - a_j|, |y - a_j|), & \text{if } (x - a_j)(y - a_j) > 0, \\ 0, & \text{else,} \end{cases}$$

leads for (3.26) to the so-called anchored ANOVA Sobolev kernel with point $a = (a_1, \ldots, a_d)$ and general weights $\gamma_{d,\mathbf{u}}$. The associated inner product is

$$\langle f,g \rangle_{V^{(d)}} = \sum_{\mathbf{u} \subset \{1,\ldots,d\}} \frac{1}{\gamma_{d,\mathbf{u}}} \int_{[0,1]^{|\mathbf{u}|}} \frac{\partial^{|\mathbf{u}|} f(x_{\mathbf{u}}, a_{-\mathbf{u}})}{\partial x_{\mathbf{u}}} \frac{\partial^{|\mathbf{u}|} g(x_{\mathbf{u}}, a_{-\mathbf{u}})}{\partial x_{\mathbf{u}}} dx_{\mathbf{u}} \qquad (3.28)$$

where $(x_{\mathbf{u}}, a_{-\mathbf{u}})$ denotes the d-dimensional vector whose j-th component is equal to x_j if $j \in \mathbf{u}$ and to a_j if $j \notin \mathbf{u}$, respectively. For the case $\mathbf{u} = \{\}$ we set $\int_{[0,1]^{|\{\}|}} f(x_{\{\}}, a_{-\{\}})dx_{\{\}} := f(a)$.

In both cases the associated weighted inner product can be written as

$$\langle f, g \rangle_{V^{(d)}} = \sum_{\mathbf{u} \subset \{1,\dots,d\}} \frac{1}{\gamma_{d,\mathbf{u}}} (f_{\mathbf{u}}, g_{\mathbf{u}})_{V_{\mathbf{u}}} \tag{3.29}$$

where

$$(f_{\mathbf{u}}, g_{\mathbf{u}})_{V_{\mathbf{u}}} = \int_{[0,1]^{|\mathbf{u}|}} \frac{\partial^{|\mathbf{u}|} f_{\mathbf{u}}(x_{\mathbf{u}})}{\partial x_{\mathbf{u}}} \frac{\partial^{|\mathbf{u}|} g_{\mathbf{u}}(x_{\mathbf{u}})}{\partial x_{\mathbf{u}}} d\mu(x_{\mathbf{u}}) \tag{3.30}$$

and $f = \sum_{\mathbf{u}} f_{\mathbf{u}}$ and $g = \sum_{\mathbf{u}} g_{\mathbf{u}}$ are the ANOVA decompositions of f and g with respect to the chosen measure $d\mu(x)$. A straightforward calculation which uses the fact that $f_{\mathbf{u}}$ and $g_{\mathbf{u}}$ are the components of an ANOVA decomposition and thus possess orthogonality properties shows that (3.30) and the integrals in the sum (3.27) and (3.28) are indeed equivalent.[1]

From (3.29) we see the effect of the weights on the inner product: The $\{\gamma_{d,\mathbf{u}}\}$ are non-negative numbers which measure the influence of the associated partial derivative of the function and, consequently, also the influence of the corresponding terms $f_{\mathbf{u}}$ of the decomposition (3.14). Note that for positive weights the associated weighted norm

$$\|f\|_{V^{(d)}} = \sqrt{\langle f, f \rangle_{V^{(d)}}} = \sqrt{\sum_{\mathbf{u}} \frac{1}{\gamma_{d,\mathbf{u}}} (f_{\mathbf{u}}, f_{\mathbf{u}})_{V_{\mathbf{u}}}}$$

is equivalent (up to a constant) to the conventional norm in $V^{(d)}$ (with just a weighting of the contributions to the overall norm). However, for any $\mathbf{u}$ with $\gamma_{d,\mathbf{u}} \to 0$ the associated contribution to the norm is forced to zero since

$$\frac{1}{\gamma_{d,\mathbf{u}}} (f_{\mathbf{u}}, f_{\mathbf{u}})_{V_{\mathbf{u}}} \le \text{const.} \Rightarrow (f_{\mathbf{u}}, f_{\mathbf{u}})_{V_{\mathbf{u}}} \le \text{const.} \cdot \gamma_{d,\mathbf{u}} \to 0.$$

Thus $f_{\mathbf{u}} = 0$, the associated subspace $W_{\mathbf{u}}$ is switched off and we obtain a true subspace of the overall space.

The weights $\{\gamma_{d,\mathbf{u}}\}$ therefore allow to explicitly prescribe the importance of different dimensions and of the correlations and interactions between (groups of) dimensions and thus allow one to characterize the associated function spaces and the possibly low, hidden dimensionality of nominally high-dimensional functions.

An attempt in this direction was the concept of effective dimension introduced in (Caflisch et al. 1997). There, based on the ANOVA

---

[1] Plug the ANOVA decompositions $f = \sum_{\mathbf{v}} f_{\mathbf{v}}$ and $g = \sum_{\mathbf{v}} g_{\mathbf{v}}$ into (3.27), change the order of the sum, the integral and the derivative and use the orthogonality of the ANOVA decomposition and the fact that the partial derivative is non-zero only if $\mathbf{u} \subset \mathbf{v}$. An analogous argument holds for (3.28).

decomposition of a function, the distribution of the overall variance to the ANOVA components was considered. This leads to the definitions of the truncation dimension d_t and the superposition dimension d_s of a function. There, f has truncation dimension d_t if the sum of the partial variances of the ANOVA terms $f_{\mathbf{u}}$ with $\mathbf{u} \subset \{1, \ldots, d_t\}$ exceeds 99 percent of the total variance $\sigma(f)$. Alternatively, f has superposition dimension d_s if the sum of the partial variances of the ANOVA terms $f_{\mathbf{u}}$ with order $|\mathbf{u}| \leq d_s$ exceeds 99 percent of the total variance. It was argued that the success of Quasi Monte Carlo methods for high-dimensional problems from finance is due to the relatively low effective dimensions of the integrands involved. In particular, the example of a mortgage backed security with nominal 360 dimensions from (Paskov and Traub 1995) showed an effective dimension of only about 50 in the truncation sense and about 32 in the superposition sense, see (Caflisch et al. 1997) for details and (Sloan and Wang 2005) for a further discussion on this subject. With the help of the general weights $\{\gamma_{d,\mathbf{u}}\}$, besides these two simple situations, more general situations can now be modeled and analyzed. In addition to the product weights mentioned above, also the case of order-dependent weights, i.e., the interaction between the variables in $x_{\mathbf{u}}$ depends only on $|\mathbf{u}|$, and the case of finite-order weights, i.e., there exists $q \in \mathbb{N}$ such that $\gamma_{d,\mathbf{u}} = 0$ for all $|\mathbf{u}| > q$, has been studied, see (Sloan and Woźniakowski 1998, Dick et al. 2004, Hickernell et al. 2004, Sloan et al. 2004, Hickernell and Woźniakowski 2000, Wasilkowski and Woźniakowski 2004) and the references cited therein. This was mainly done for the analysis of Quasi Monte Carlo methods and lattice rules for the numerical integration of high-dimensional functions.

A closely related approach with weighted kernels can be found in the area of data analysis, where the weights are called rescaling parameters. There, for so-called interaction spline models (Wahba 1990), page 129 ff., strategies are discussed to delete ANOVA-component subspaces driven by data fitting methods. The weights $\gamma_{d,\mathbf{u}}$ are not given a-priori but are determined in an adaptive fashion by statistical tests. Alternative techniques are the l_1-penalty method or the structured Multicategory Support Vector Machine where an updating algorithm is used for the tuning of the weights, see (Lee, Lin and Wahba 2004b, Lee, Kim, Lee and Koo 2004a).

Now, with a-priori knowledge, i.e., for the case of weights with finite order q, where $\gamma_{d,\mathbf{u}} = 0$ for $|\mathbf{u}| > q$, only a proper discretization of the remaining component functions is needed. Then, the curse of dimensionality is no longer present with respect to d but only with respect to q, as shown for quadrature and more general linear problems in (Wasilkowski and Woźniakowski 2004). In a similar fashion, for the case of sufficiently fast-decaying weights, the series expansion (3.19) may be truncated

accordingly, which then results in an approximation to the true function.[1]
Then, again, for a proper discretization of the remaining component functions, the curse of dimensionality may no longer be present with respect
to d but only with respect to a smaller intrinsic dimension of the overall function. In this sense, the curse of dimensionality can be broken and
nominally high-dimensional problems can be tackled.

Note that a truncation of the ANOVA series introduces a modelling error
whereas the subsequent discretization of the remaining subspaces relates to
a discretization error. This unnatural distinction between modelling error
and subsequent discretization error can be overcome if we intertwine the
truncation of the ANOVA series and the discretization. However, how these
two types of errors may be balanced, how this may be done in a purely adaptive fashion, what the smoothness assumptions, a-posteriori error indicators
and refinement procedures, first, for the ANOVA parts and, second, within
the discretization of the ANOVA parts have to be and how they relate
to each other is presently not completely clear, especially for PDE-based
problems.[2] Nevertheless, such a type of approach needs to be developed
and applied in the area of partial differential equations in the future.

3.4 Sparse grids

So far we have seen how an ANOVA-type decomposition may be used to
detect important and unimportant correlations and interactions between
(groups of) dimensions. However the components in the decomposition
(3.19) are still continuous functions and the corresponding subspaces $W_{\mathbf{u}}$
are in general infinite-dimensional. Moreover, for practical computations a
choice of basis and a further discretization is needed for each of the subspaces. To this end, we can follow the same principle as in Subsection 3.3.1.
First we equip the space W in the one-dimensional splitting (3.11) with a
proper (infinite) basis $\{\phi_k\}$ (the constant, i.e., $\phi_0 = 1$ is excluded). Then we
apply the tensor product construction (3.17) to come to the d-dimensional
case. Here we just form the products of the respective one-dimensional
basis functions. This results in an induced basis $\{\phi_{\mathbf{u},\mathbf{k}}\}$ for each of the

---

[1] Alternatively a closure by an approximation of the higher order terms by means of
products of lower order terms or similar may be sought.

[2] If we assume, for example, $f_{\mathbf{u}} \in C^{|\mathbf{u}| \cdot k}([0,1]^{|\mathbf{u}|}), \forall \mathbf{u}$, we directly see that $f \in
C^{k \cdot d}([0,1]^d)$. However there exist more partial derivatives like, for example, the
$|\mathbf{u}|$-th mixed derivative of $f_{\mathbf{u}}$ and, therefore, f belongs to the space of bounded k-th
mixed derivatives, compare also (Novak and Ritter 1998). Note that the inverse di-
rection of this implication is in general not valid. Now, for functions whose ANOVA
decomposition is of finite order q, only a prerequisite $f_{\mathbf{u}} \in C^{q \cdot k}([0,1]^{|\mathbf{u}|})$ is at most
necessary to have f from the space of bounded k-th mixed derivatives.

ANOVA subspaces $W_\mathbf{u}$ with multi-index $\mathbf{k} = (k_{j_1}, \ldots, k_{j_{|\mathbf{u}|}}) \in \mathbb{N}^{|\mathbf{u}|}$ where $\phi_{\mathbf{u},\mathbf{k}}(x_\mathbf{u}) = \prod_{j \in \mathbf{u}} \phi_{k_j}(x_j) \cdot \prod_{j \in \{1,\ldots,d\} \setminus \mathbf{u}} 1$. This may be seen as an (infinite) refinement of the ANOVA decomposition by a further decomposition of the space W (e.g., by means of the span of certain basis functions).[1] Then, we can write each component function as linear combination of these basis functions. This overall expansion of the sum of component functions of the ANOVA decomposition must be truncated properly in each of its components to obtain a finite dimensional approximation to f and to its component functions f_{j_1}, $f_{j_1,j_2}, \ldots,$ $f_{j_1,\ldots j_d}$. This leads to so-called sparse grids. Depending on the smoothness assumed – here usually certain mixed derivatives have to be bounded – and depending on the specific one-dimensional basis chosen, such an approach allows to get rid of the curse of dimensionality to some extent.

This concept works for quite general systems of one-dimensional basis functions. Candidates are the eigenbasis of an associated one-dimensional differential operator (which may be chosen depending on the respective higher dimensional problem under consideration), classical Fourier bases, (hierarchical) global polynomial systems (Boyd 2000, Karniadakis and Sherwin 1999, Szabo and Babuska 1991, Bungartz 1998, Bungartz and Griebel 2004) or function families with localization properties like wavelets (Daubechies 1992), prewavelets (Chui and Wang 1992, Griebel and Oswald 1995b) or interpolets (Deslauriers and Dubuc 1989, Donoho and Yu 1999) and related wavelet-like constructs, see (Cohen 2003, Bungartz and Griebel 2004) for a survey. But also multiscale finite element systems and frames (Oswald 1994, Griebel 1994, Griebel and Oswald 1994, Griebel and Oswald 1995a, Griebel and Oswald 1995b) may be used.

In the following, we restrict ourselves for reasons of simplicity to the standard hat function and the associated hierarchical Faber basis. It is closely related to piecewise linear finite elements and thus suited for the discretization of elliptic PDEs of second order in weak form. This choice allows in a straightforward way to derive approximation orders and cost complexities by simple geometric series arguments and triangle inequalities. Moreover, for this special choice of basis we are able to also derive estimates of the constants involved and their dependence on the dimension d. We

---

[1] Note the close relation to the work in (Lemieux and Owen 2002, Liu and Owen 2005), to MARS (Friedman 1991), to the WARNAX model (Wei and Billings 2004) and to tensor product space ANOVA models (Gu and Wahba 1993, Lin 2000). For example in (Wahba 1990), page 130, the one-dimensional two-scale splitting is extended to more terms which correspond to higher derivatives. Orthogonality is then given by a proper eigenbasis associated to this splitting. This results in so-called interaction splines.

now closely follow (Bungartz 1992, Bungartz and Griebel 1999, Bungartz 1998, Bungartz and Griebel 2004).

3.4.1 Hierarchical multilevel subspace splitting

3.4.1.1 Subspace decomposition

Let $\bar{\Omega} := [0,1]^d$ denote the d-dimensional unit interval. We consider multivariate functions u, $u(\mathbf{x}) \in \mathbb{R}$, $\mathbf{x} := (x_1, \ldots, x_d) \in \bar{\Omega}$, with (in some sense) bounded weak mixed derivatives

$$D^{\boldsymbol{\alpha}} u := \frac{\partial^{|\boldsymbol{\alpha}|_1} u}{\partial x_1^{\alpha_1} \cdots \partial x_d^{\alpha_d}} \tag{3.31}$$

up to some given order $r \in \mathbb{N}_0$. Here, $\boldsymbol{\alpha} \in \mathbb{N}_0^d$ denotes a d-dimensional multi-index with the norms $|\boldsymbol{\alpha}|_1 := \sum_{j=1}^d \alpha_j$ and $|\boldsymbol{\alpha}|_\infty := \max_{1 \le j \le d} \alpha_j$. Furthermore, we use for multi-indices component-wise arithmetic operations, for example $\boldsymbol{\alpha} \cdot \boldsymbol{\beta} := (\alpha_1 \beta_1, \ldots, \alpha_d \beta_d)$, $\gamma \cdot \boldsymbol{\alpha} := (\gamma \alpha_1, \ldots, \gamma \alpha_d)$, or $2^{\boldsymbol{\alpha}} := (2^{\alpha_1}, \ldots, 2^{\alpha_d})$, the relational operators $\boldsymbol{\alpha} \le \boldsymbol{\beta} :\Leftrightarrow \forall_{1 \le j \le d} \ \alpha_j \le \beta_j$ and $\boldsymbol{\alpha} < \boldsymbol{\beta} \quad :\Leftrightarrow \quad \boldsymbol{\alpha} \le \boldsymbol{\beta} \ \wedge \ \exists_{1 \le j \le d} \ \alpha_j < \beta_j$, and, finally, special multi-indices like $\mathbf{0} := (0, \ldots, 0)$ or $\mathbf{1} := (1, \ldots, 1)$, and so on.

In the following, for $q \in \{2, \infty\}$ and $r \in \mathbb{N}_0$, we study the spaces

$$\begin{aligned} X^{q,r}(\bar{\Omega}) &:= \left\{ u : \bar{\Omega} \to \mathbb{R} : D^{\boldsymbol{\alpha}} u \in L_q(\Omega), |\boldsymbol{\alpha}|_\infty \le r \right\}, \\ X_0^{q,r}(\bar{\Omega}) &:= \left\{ u \in X^{q,r}(\bar{\Omega}) : u|_{\partial\Omega} = 0 \right\}. \end{aligned} \tag{3.32}$$

Thus, $X^{q,r}(\bar{\Omega})$ denotes the space of all functions of bounded (with respect to the L_q-norm) mixed derivatives up to order r, and $X_0^{q,r}(\bar{\Omega})$ will be the subspace of $X^{q,r}(\bar{\Omega})$ consisting of those $u \in X^{q,r}(\bar{\Omega})$ which vanishes on the boundary $\partial\Omega$. Note that we first restrict ourselves to the case of homogeneous boundary conditions, i.e., to $X_0^{q,r}(\bar{\Omega})$. As smoothness parameter $r \in \mathbb{N}_0$, we need $r = 2$ for the case of piecewise linear approximations which will be in the following in the focus of our considerations. Finally, for functions $u \in X_0^{q,r}(\bar{\Omega})$ and multi-indices $\boldsymbol{\alpha}$ with $|\boldsymbol{\alpha}|_\infty \le r$, we introduce the seminorm

$$|u|_{\boldsymbol{\alpha},\infty} := \|D^{\boldsymbol{\alpha}} u\|_\infty. \tag{3.33}$$

Now, with the multi-index $\mathbf{l} = (l_1, \ldots, l_d) \in \mathbb{N}^d$ which indicates the level of refinement in a multivariate sense, we consider the family of d-dimensional standard rectangular grids

$$\{\Omega_{\mathbf{l}}, \mathbf{l} \in \mathbb{N}^d\} \tag{3.34}$$

on $\bar{\Omega}$ with mesh size

$$\mathbf{h_l} := (h_{l_1}, \ldots, h_{l_d}) := 2^{-\mathbf{l}}. \tag{3.35}$$

That is, the grid $\Omega_{\mathbf{l}}$ is equidistant with respect to each individual coordinate direction, but, in general, may have different mesh sizes in the different coordinate directions. The grid points $\mathbf{x}_{\mathbf{l},\mathbf{i}}$ of grid $\Omega_{\mathbf{l}}$ are just the points

$$\mathbf{x}_{\mathbf{l},\mathbf{i}} := (x_{l_1,i_1}, \ldots, x_{l_d,i_d}) := \mathbf{i} \cdot \mathbf{h}_{\mathbf{l}}, \qquad \mathbf{0} \leq \mathbf{i} \leq 2^{\mathbf{l}}. \tag{3.36}$$

Thus, here and in the following, the multi-index $\mathbf{l}$ indicates the level (of a grid, a point, or, later on, a basis function, respectively), whereas the multi-index $\mathbf{i}$ denotes the location of a given grid point $\mathbf{x}_{\mathbf{l},\mathbf{i}}$ in the respective grid $\Omega_{\mathbf{l}}$.

Next, we define discrete approximation spaces and sets of basis functions that span those discrete spaces. In a piecewise linear setting, the simplest choice of a 1 D basis function is the standard hat function $\phi(x)$,

$$\phi(x) := \begin{cases} 1 - |x|, & \text{if } x \in [-1,1], \\ 0, & \text{else.} \end{cases} \tag{3.37}$$

This function can be used to generate an arbitrary $\phi_{l_j,i_j}(x_j)$ with support $[x_{l_j,i_j} - h_{l_j}, x_{l_j,i_j} + h_{l_j}] = [(i_j-1)h_{l_j}, (i_j+1)h_{l_j}]$ by dilation and translation, that is

$$\phi_{l_j,i_j}(x_j) := \phi\left(\frac{x_j - i_j \cdot h_{l_j}}{h_{l_j}}\right). \tag{3.38}$$

The resulting 1 D basis functions are the input of the tensor product construction which provides a suitable piecewise d-linear basis function in each grid point $\mathbf{x}_{\mathbf{l},\mathbf{i}}$ (see Figure 3.1):

$$\phi_{\mathbf{l},\mathbf{i}}(\mathbf{x}) := \prod_{j=1}^{d} \phi_{l_j,i_j}(x_j). \tag{3.39}$$

Since we deal with homogeneous boundary conditions (i.e., with $X_0^{q,2}(\bar{\Omega})$), only those $\phi_{\mathbf{l},\mathbf{i}}(\mathbf{x})$ that correspond to *inner* grid points of $\Omega_{\mathbf{l}}$ are taken into

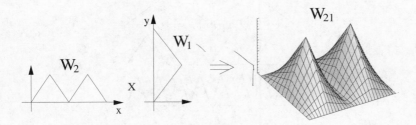

Fig. 3.1. Tensor product approach for piecewise bilinear basis functions.

account for the definition of

$$V_1 := \text{span}\left\{\phi_{1,i} : \mathbf{1} \le \mathbf{i} \le 2^1 - \mathbf{1}\right\}, \qquad (3.40)$$

the space of piecewise d-linear functions with respect to the interior of Ω_1. Obviously, the $\phi_{1,i}$ form a basis of V_1, with one basis function $\phi_{1,i}$ of a support of the fixed size $2 \cdot \mathbf{h}_1$ for each inner grid point $\mathbf{x}_{1,i}$ of Ω_1, and this basis $\{\phi_{1,i}\}$ is simply the standard nodal point basis of the finite dimensional space V_1.

Additionally, we introduce the hierarchical increments W_1,

$$W_1 := \text{span}\left\{\phi_{1,i} : \mathbf{1} \le \mathbf{i} \le 2^1 - \mathbf{1},\ i_j \text{ odd for all } 1 \le j \le d\right\}, \qquad (3.41)$$

for which the relation

$$V_1 = \bigoplus_{\mathbf{k} \le 1} W_{\mathbf{k}} \qquad (3.42)$$

can be easily seen. Note that the supports of all basis functions $\phi_{1,i}$ spanning W_1 are mutually disjoint. Thus, with the index set

$$\mathbf{I}_1 := \left\{\mathbf{i} \in \mathbb{N}^d : \mathbf{1} \le \mathbf{i} \le 2^1 - \mathbf{1},\ i_j \text{ odd for all } 1 \le j \le d\right\}, \qquad (3.43)$$

we get another basis of V_1, the *hierarchical basis*

$$\{\phi_{\mathbf{k},i} : \mathbf{i} \in \mathbf{I_k}, \mathbf{k} \le 1\} \qquad (3.44)$$

which generalizes the well-known $1\,\text{D}$ basis shown in Figure 3.2 to the d-dimensional case by means of a tensor product approach. With these

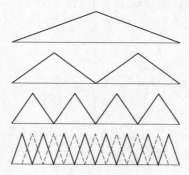

Fig. 3.2. Piecewise linear hierarchical basis (solid) vs. nodal point basis (dashed).

hierarchical difference spaces $W_\mathbf{l}$, we can define

$$V^{(d)} := \sum_{l_1=1}^{\infty} \cdots \sum_{l_d=1}^{\infty} W_{(l_1,\ldots,l_d)} = \bigoplus_{\mathbf{l}\in\mathbb{N}^d} W_\mathbf{l} \qquad (3.45)$$

with its natural hierarchical basis

$$\{\phi_{\mathbf{l},\mathbf{i}} : \mathbf{i} \in \mathbf{I}_\mathbf{l}, \mathbf{l} \in \mathbb{N}^d\}. \qquad (3.46)$$

Except for completion with respect to the H^1-norm, $V^{(d)}$ is just the underlying Sobolev space $H_0^1(\bar{\Omega})$, i.e., $\bar{V}^{(d)} = H_0^1(\bar{\Omega})$.

Now it is easy to see that any function $u \in H_0^1(\bar{\Omega})$ and, consequently, any $u \in X_0^{q,2}(\bar{\Omega})$ can be uniquely split by

$$u(\mathbf{x}) = \sum_{\mathbf{l}} u_\mathbf{l}(\mathbf{x}), \qquad u_\mathbf{l}(\mathbf{x}) = \sum_{\mathbf{i}\in\mathbf{I}_\mathbf{l}} v_{\mathbf{l},\mathbf{i}} \cdot \phi_{\mathbf{l},\mathbf{i}}(\mathbf{x}) \in W_\mathbf{l}, \qquad (3.47)$$

where the $v_{\mathbf{l},\mathbf{i}} \in \mathbb{R}$ are the coefficient values of the hierarchical product basis representation of u.

3.4.1.2 Basic properties of the subspaces

We summarize the most important properties of the hierarchical subspaces $W_\mathbf{l}$.

From (3.41) and (3.43), we immediately learn the dimension of $W_\mathbf{l}$, i. e. the number of degrees of freedom (grid points or basis functions, resp.) associated with $W_\mathbf{l}$:

$$|W_\mathbf{l}| = |\mathbf{I}_\mathbf{l}| = 2^{|\mathbf{l}-\mathbf{1}|_1}. \qquad (3.48)$$

The question is now how important $W_\mathbf{l}$ is for the interpolation of some given $u \in X_0^{q,2}(\bar{\Omega})$. In the following we will discuss the contribution of a subspace $W_\mathbf{l}$ to the overall interpolant according to (3.47). Here, besides the L_p-norms, $p \in \{2,\infty\}$ we will concentrate on the energy norm

$$\|u\|_E := \left(\int_\Omega \sum_{j=1}^{d} \left(\frac{\partial u(\mathbf{x})}{\partial x_j} \right)^2 d\mathbf{x} \right)^{1/2}, \qquad (3.49)$$

which is equivalent to the H^1-norm in $H_0^1(\bar{\Omega})$. For the Laplacian, (3.49) indeed indicates the energy norm in finite element terminology.[1]

---

[1] Note that analogous results for the maximum norm $\|\cdot\|_\infty$ and the L_p-norm $\|\cdot\|_p$ (in general $p = 2$) can be found in e.g., (Bungartz and Griebel 2004).

First, we look at the different hierarchical basis functions $\phi_{\mathbf{l},\mathbf{i}}(\mathbf{x})$. A straightforward calculation gives

$$\|\phi_{\mathbf{l},\mathbf{i}}\|_E \;=\; \sqrt{2}\cdot\left(\frac{2}{3}\right)^{(d-1)/2}\cdot 2^{-|\mathbf{l}|_1/2}\cdot\left(\sum_{j=1}^d 2^{2l_j}\right)^{1/2}. \tag{3.50}$$

Next, we consider the hierarchical coefficient values $v_{\mathbf{l},\mathbf{i}}$ in more detail. They can be computed from the function values $u(\mathbf{x}_{\mathbf{l},\mathbf{i}})$ in the following way:

$$v_{\mathbf{l},\mathbf{i}} \;=\; \left(\prod_{j=1}^d \left[-\tfrac{1}{2}\;1\;-\tfrac{1}{2}\right]_{x_{l_j,i_j},l_j}\right)u. \tag{3.51}$$

This is due to the definition of the spaces $W_{\mathbf{l}}$ and their basis functions (3.41), whose supports are mutually disjoint and do not contain coarse grid points $\mathbf{x}_{\mathbf{k},\mathbf{j}}$, $\mathbf{k} < \mathbf{l}$, in their interior. The right-hand side term in (3.51), as usual in multigrid terminology (see, for example, (Hackbusch 1985)), denotes a d-dimensional stencil which gives the coefficients for a linear combination of nodal values of its argument u.

A straightforward calculation using partial integration twice and the product structure of (3.51), see (Bungartz and Griebel 2004) for details, gives the integral representation

$$v_{\mathbf{l},\mathbf{i}} \;=\; \int_\Omega \psi_{\mathbf{l},\mathbf{i}}(\mathbf{x})\cdot D^2 u(\mathbf{x})\;\mathrm{d}\mathbf{x} \tag{3.52}$$

for any coefficient value $v_{\mathbf{l},\mathbf{i}}$ of the hierarchical representation (3.47) of $u \in X_0^{q,2}(\bar{\Omega})$. Here $\psi_{l_j,i_j}(x_j) := -2^{-(l_j+1)}\cdot\phi_{l_j,i_j}(x_j)$, and furthermore $\psi_{\mathbf{l},\mathbf{i}}(\mathbf{x}) := \prod_{j=1}^d \psi_{l_j,i_j}(x_j)$.

Starting from (3.52), we are now able to give bounds for the hierarchical coefficients with respect to the seminorm introduced in (3.33). For the detailed proof see again e.g., (Bungartz and Griebel 2004). We obtain

$$|v_{\mathbf{l},\mathbf{i}}| \;\leq\; 2^{-d}\cdot 2^{-2\cdot|\mathbf{l}|_1}\cdot|u|_{2,\infty}. \tag{3.53}$$

We are now ready to state the following Lemma.

Lemma 1 *Let $u \in X_0^{q,2}(\bar{\Omega})$ be given in its hierarchical representation (3.47). Then, the following estimate holds for its components $u_{\mathbf{l}} \in W_{\mathbf{l}}$:*

$$\|u_{\mathbf{l}}\|_E \;\leq\; \frac{1}{2\cdot 12^{(d-1)/2}}\cdot 2^{-2\cdot|\mathbf{l}|_1}\cdot\left(\sum_{j=1}^d 2^{2\cdot l_j}\right)^{1/2}\cdot|u|_{2,\infty}. \tag{3.54}$$

Proof Note that the supports of all $\phi_{\mathbf{l},\mathbf{i}}$ contributing to $u_{\mathbf{l}}$ according to (3.47) are mutually disjoint. Then

$$\|u_{\mathbf{l}}\|_E^2 = \left\|\sum_{\mathbf{i}\in\mathbf{I}_{\mathbf{l}}} v_{\mathbf{l},\mathbf{i}} \cdot \phi_{\mathbf{l},\mathbf{i}}\right\|_E^2 = \sum_{\mathbf{i}\in\mathbf{I}_{\mathbf{l}}} |v_{\mathbf{l},\mathbf{i}}|^2 \cdot \|\phi_{\mathbf{l},\mathbf{i}}\|_E^2$$

$$\leq \sum_{\mathbf{i}\in\mathbf{I}_{\mathbf{l}}} \frac{1}{4^d} \cdot 2^{-4\cdot|\mathbf{l}|_1} \cdot |u|_{\mathbf{2},\infty}^2 \cdot 2 \cdot \left(\frac{2}{3}\right)^{d-1} \cdot 2^{-|\mathbf{l}|_1} \cdot \left(\sum_{j=1}^d 2^{2\cdot l_j}\right)$$

$$= \frac{1}{2\cdot 6^{d-1}} \cdot 2^{-5\cdot|\mathbf{l}|_1} \cdot \left(\sum_{j=1}^d 2^{2\cdot l_j}\right) \cdot \sum_{\mathbf{i}\in\mathbf{I}_{\mathbf{l}}} |u|_{\mathbf{2},\infty}^2$$

$$= \frac{1}{4\cdot 12^{d-1}} \cdot 2^{-4\cdot|\mathbf{l}|_1} \cdot \left(\sum_{j=1}^d 2^{2\cdot l_j}\right) \cdot |u|_{\mathbf{2},\infty}^2.$$

This shows (3.54). $\qquad\qquad\square$

3.4.2 Energy-norm based sparse grids

We will now construct finite-dimensional approximation spaces U for $V^{(d)}$ or $X_0^{q,2}(\bar{\Omega})$, respectively. Such a U is based on a subspace selection $\mathbf{I} \subset \mathbb{N}^d$,

$$U := \bigoplus_{\mathbf{l}\in\mathbf{I}} W_{\mathbf{l}}, \qquad (3.55)$$

with corresponding interpolants

$$u_U := \sum_{\mathbf{l}\in\mathbf{I}} u_{\mathbf{l}}, \qquad u_{\mathbf{l}} \in W_{\mathbf{l}}. \qquad (3.56)$$

The estimate

$$\|u - u_U\| = \left\|\sum_{\mathbf{l}} u_{\mathbf{l}} - \sum_{\mathbf{l}\in\mathbf{I}} u_{\mathbf{l}}\right\| \leq \sum_{\mathbf{l}\notin\mathbf{I}} \|u_{\mathbf{l}}\| \leq \sum_{\mathbf{l}\notin\mathbf{I}} b(\mathbf{l}) \cdot |u| \quad (3.57)$$

will then allow the evaluation of the approximation space U with respect to a norm $\|\cdot\|$ and a corresponding seminorm $|\cdot|$ on the basis of the bounds from above indicating the *benefit* $b(\mathbf{l})$ of $W_{\mathbf{l}}$.

3.4.2.1 Construction of subspaces by optimization

We now address the question how to determine optimal subspace index sets $\mathbf{I}$ which optimize cost versus accuracy for the interpolation error for functions u from $X_0^{q,2}$. To this end, we look for an optimum $V^{(\text{opt})}$ by solving a restricted optimization problem of the type

$$\max_{u \in X_0^{q,2}:|u|=1} \|u - u_{V^{(\mathrm{opt})}}\| = \min_{U \subset V^{(d)}:|U|=w} \max_{u \in X_0^{q,2}:|u|=1} \|u - u_U\| \quad (3.58)$$

for some prescribed work count w. The aim is to profit from a given work count as much as possible.[1] Of course, any potential solution $V^{(\mathrm{opt})}$ of (3.58) has to be expected to depend on the norm $\|\cdot\|$ as well as on the seminorm $|\cdot|$ used to measure the error of u's interpolant $u_U \in U$ or the smoothness of u, respectively. Note that this *a-priori optimization* strategy depends only on the problem class (i.e., on the space u has to belong to – here $X_0^{q,2}(\bar{\Omega})$), but not on u itself.[2]

According to our hierarchical setting, we will allow discrete spaces of the type $U := \bigoplus_{l \in \mathbf{I}} W_l$ for an arbitrary finite index set $\mathbf{I} \subset \mathbb{N}^d$ as candidates for the optimization process only. Now, an approach like (3.58) selects certain W_l due to their importance and, thus, selects the respective grids and the underlying index sets $\mathbf{I} \subset \mathbb{N}^d$. This is done by using techniques known from combinatorial optimization as follows:

For the following, a grid and its representation $\mathbf{I}$ – formerly a finite set of multi-indices – is nothing but a bounded subset of $\mathbb{N}_+^d$, and a hierarchical subspace W_l just corresponds to a point $l \in \mathbb{N}_+^d$. First, we have to reformulate the optimization problem (3.58). We define the local functions $c(l)$ and $b(l)$, for the multi-indices $l \in \mathbb{N}^d$. According to (3.48), the local cost $c(l)$ is

$$c(l) := |W_l| = 2^{|l-1|_1}. \quad (3.59)$$

Obviously, $c(l) \in \mathbb{N}$ holds for all $l \in \mathbb{N}^d$. Concerning the local benefit $b(l)$, we define

$$b(l) := \gamma \cdot \beta(l), \quad (3.60)$$

where $\beta(l)$ is an upper bound for $\|u_l\|^2$ according to (3.54), and γ is a factor which depends on the problem's dimensionality d and on the smoothness of the data, i.e., of u, but which is constant with respect to l, such that $b(l) \in \mathbb{N}$. The bound in (3.54) shows that such a choice of γ is indeed possible. At the moment, we do not yet fix the norm to be used here.

Now, the search for an optimal grid $\mathbf{I} \subset \mathbb{N}^d$ can be restricted to all $\mathbf{I} \subset \mathbf{I}^{(\max)} := \{1, \ldots, N\}^d$ for a sufficiently large N without loss of generality.

---

[1] Note that an optimization the other way round could be done as well: Prescribe some desired accuracy ε and look for the discrete approximation scheme that achieves this with the smallest possible work count. This is in fact the point of view of computational complexity.

[2] This is in contrast to adaptive grid refinement which uses a-posteriori error estimators to approximate one given function u.

Next, global cost and benefit functions $C(\mathbf{I})$ and $B(\mathbf{I})$ are to be defined. For $C(\mathbf{I})$, we set

$$C(\mathbf{I}) := \sum_{\mathbf{l} \in \mathbf{I}} c(\mathbf{l}) = \sum_{\mathbf{l} \in \mathbf{I}^{(\max)}} x(\mathbf{l}) \cdot c(\mathbf{l}), \qquad (3.61)$$

where

$$x(\mathbf{l}) := \begin{cases} 0 & : \ \mathbf{l} \notin \mathbf{I}, \\ 1 & : \ \mathbf{l} \in \mathbf{I}. \end{cases} \qquad (3.62)$$

The interpolant to u on a grid $\mathbf{I}$ provides the global benefit $B(\mathbf{I})$:

$$\left\| u - \sum_{\mathbf{l} \in \mathbf{I}} u_{\mathbf{l}} \right\|^2 \approx \left\| \sum_{\mathbf{l} \in \mathbf{I}^{(\max)}} u_{\mathbf{l}} - \sum_{\mathbf{l} \in \mathbf{I}} u_{\mathbf{l}} \right\|^2$$

$$\leq \sum_{\mathbf{l} \in \mathbf{I}^{(\max)} \setminus \mathbf{I}} \| u_{\mathbf{l}} \|^2$$

$$\leq \sum_{\mathbf{l} \in \mathbf{I}^{(\max)}} (1 - x(\mathbf{l})) \cdot \gamma \cdot \beta(\mathbf{l}) \qquad (3.63)$$

$$= \sum_{\mathbf{l} \in \mathbf{I}^{(\max)}} \gamma \cdot \beta(\mathbf{l}) - \sum_{\mathbf{l} \in \mathbf{I}^{(\max)}} x(\mathbf{l}) \cdot \gamma \cdot \beta(\mathbf{l})$$

$$=: \sum_{\mathbf{l} \in \mathbf{I}^{(\max)}} \gamma \cdot \beta(\mathbf{l}) - B(\mathbf{I}).$$

Of course, (3.63) gives only an upper bound for an approximation to the (squared) interpolation error, because it does not take into account all $\mathbf{l} \notin \mathbf{I}^{(\max)}$. However, since N and, consequently, $\mathbf{I}^{(\max)}$ can be chosen to be as large as is appropriate, this is not a serious restriction. Altogether, we get the following reformulation of (3.58):

$$\max_{\mathbf{I} \subset \mathbf{I}^{(\max)}} B(\mathbf{I}) \quad \text{with} \quad C(\mathbf{I}) = w, \quad \text{i.e.,}$$

$$\max_{\mathbf{I} \subset \mathbf{I}^{(\max)}} \sum_{\mathbf{l} \in \mathbf{I}^{(\max)}} x(\mathbf{l}) \cdot \gamma \cdot \beta(\mathbf{l}) \quad \text{with} \quad \sum_{\mathbf{l} \in \mathbf{I}^{(\max)}} x(\mathbf{l}) \cdot c(\mathbf{l}) = w. \quad (3.64)$$

If we arrange the $\mathbf{l} \in \mathbf{I}^{(\max)}$ in some linear order (a lexicographical one, for instance) with local cost c_i and benefit b_i, $i = 1, \dots, N^d =: M$, (3.64) results in

$$\max_{\mathbf{x}} \mathbf{b}^T \mathbf{x} \quad \text{with} \quad \mathbf{c}^T \mathbf{x} = w, \qquad (3.65)$$

where $\mathbf{b} \in \mathbb{N}^M$, $\mathbf{c} \in \mathbb{N}^M$, $\mathbf{x} \in \{0, 1\}^M$, and, without loss of generality, $w \in \mathbb{N}$. In combinatorial optimization, a problem like (3.65) is called a *binary knapsack problem* (Martello and Toth 1990), which is known to be NP-hard. However, a slight change makes things much easier. If *rational* solutions, i.e., $\mathbf{x} \in ([0,1] \cap \mathbb{Q})^M$, are allowed, too, there exists a very simple algorithm that provides an optimal solution vector $\mathbf{x} \in ([0,1] \cap \mathbb{Q})^M$:

1. rearrange the order that $\frac{b_1}{c_1} \geq \frac{b_2}{c_2} \cdots \geq \frac{b_M}{c_M}$,

2. let $r := \max\left\{ j : \sum_{i=1}^{j} c_i \leq w \right\}$,

3. $x_1 := \cdots := x_r := 1$,

$$x_{r+1} := \left(w - \sum_{i=1}^{r} c_i \right) / c_{r+1},$$

$$x_{r+2} := \cdots := x_M := 0.$$

Although there is only one potential non-binary coefficient x_{r+1}, the rational solution vector $\mathbf{x}$, generally, has nothing to do with its binary counterpart. But, fortunately, our knapsack is of variable size, since the global work count w is an arbitrarily chosen natural number. Therefore, it is possible to force the solution of the *rational* problem to be a *binary* one which is, of course, also a solution of the corresponding *binary* problem. Consequently, the *global* optimization problem (3.58) or (3.65), respectively, can be reduced to the discussion of the *local* cost-benefit ratios b_i/c_i or $b(\mathbf{l})/c(\mathbf{l})$ of the underlying subspaces $W_{\mathbf{l}}$. Those subspaces with the best cost-benefit ratios are taken into account first, and the smaller these ratios become, the more negligible the underlying subspaces turn out to be.

Now, if our cost-benefit approach is based on the L_p-norms, with $p \in \{1, \infty\}$ we showed in (Bungartz and Griebel 2004) that this results in the regular sparse grid spaces

$$V_n^{(1)} := \bigoplus_{|\mathbf{l}|_1 \,\leq\, n+d-1} W_{\mathbf{l}} \tag{3.66}$$

which have been introduced in (Zenger 1991). Note that they are the finite element analogon of the well-known hyperbolic cross or Korobov spaces which are based on the Fourier series expansion instead of the hierarchical Faber basis. An example of a regular sparse grid is given for the two- and three-dimensional case in Figure 3.3. The basic concept can be traced back to (Smolyak 1963, Babenko 1960), see also (Gordon 1969, Gordon 1971, Delvos 1982, Delvos and Schempp 1989, DeVore, Konyagin and Temlyakov 1998).

The dimension of the space $V_n^{(1)}$ fulfills

$$\left| V_n^{(1)} \right| = O\big(h_n^{-1} \cdot |\log_2 h_n|^{d-1}\big) \tag{3.67}$$

with $h_n = 2^{-n}$, whereas, for the interpolation error of a function $u \in X_0^{q,2}(\bar{\Omega})$ in the sparse grid space $V_n^{(1)}$ there holds

$$\left\| u - u_n^{(1)} \right\|_p = O\big(h_n^2 \cdot n^{d-1}\big), \tag{3.68}$$

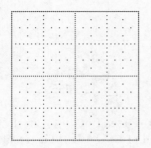
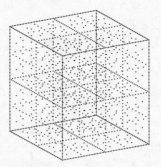

Fig. 3.3. L_p-norm based sparse grids: Two-dimensional example (left) and three-dimensional example (right), here including sparse grid points on the boundary.

for the L_p-norms, and

$$\left|\left|u - u_n^{(1)}\right|\right|_E \leq \frac{d \cdot |u|_{2,\infty}}{2 \cdot 3^{(d-1)/2} \cdot 4^{d-1}} \cdot 2^{-n} = O(h_n), \qquad (3.69)$$

for the energy-norm, see for example (Bungartz and Griebel 2004) for detailed proofs. Note that the conventional full grid space

$$V_n^{(\infty)} := \bigoplus_{|\mathbf{l}|_\infty \leq n} W_\mathbf{l}$$

results in an error in the L_p-norm of the order $O(h_n^2)$ and an error in the energy-norm of the order $O(h_n)$. It however possesses a dimension $|V_n^{(\infty)}| = O(h_n^{-d})$ and thus exhibits the curse of dimensionality with respect to h_n. In comparison to that we now see a crucial improvement for $V_n^{(1)}$: The number of degrees of freedom is reduced significantly, whereas the accuracy deteriorates only slightly for the L_p-norm and stays of the same order for the energy-norm. The curse of dimensionality is now present in the $\log(h_n)$-term only. Since this result is optimal with respect to the L_p-norms, a further improvement can only be expected if we change the setting. Therefore, in the following, we turn to the energy norm.

3.4.2.2 Energy-based sparse grids

We now base our cost-benefit approach on the energy norm. According to (3.48) and (3.54) and following the discussion in Section 3.4.2.1, we define

$$cbr_E(\mathbf{l}) := \frac{b_E(\mathbf{l})}{c(\mathbf{l})} := \frac{2^{-4 \cdot |\mathbf{l}|_1} \cdot |u|_{\mathbf{2},\infty}^2}{4 \cdot 12^{d-1} \cdot 2^{|\mathbf{l}-\mathbf{1}|_1}} \cdot \sum_{j=1}^{d} 4^{l_j}$$

$$= \frac{3}{6^d} \cdot 2^{-5 \cdot |\mathbf{l}|_1} \cdot \sum_{j=1}^{d} 4^{l_j} \cdot |u|_{\mathbf{2},\infty}^2 \qquad (3.70)$$

as the local cost-benefit ratio. Note that, instead of $\|u_\mathbf{l}\|_E$ itself, only an upper bound for the squared energy norm of $u_\mathbf{l}$ is used. The resulting optimal grid $\mathbf{I}^{(\mathrm{opt})}$ will consist of all those multi-indices $\mathbf{l}$ or their respective hierarchical subspaces $W_\mathbf{l}$ that fulfill $cbr_E(\mathbf{l}) \geq \sigma_E(n)$ for some given constant threshold $\sigma_E(n)$. Here, $\sigma_E(n)$ is defined via the cost-benefit ratio of $W_{\bar{\mathbf{l}}}$ with $\bar{\mathbf{l}} := (n, 1, \ldots, 1)$:

$$\sigma_E(n) := cbr_E(\bar{\mathbf{l}}) = \frac{3}{6^d} \cdot 2^{-5 \cdot (n+d-1)} \cdot \left(4^n + 4 \cdot (d-1)\right) \cdot |u|_{\mathbf{2},\infty}^2. \qquad (3.71)$$

Thus, applying the criterion $cbr_E(\mathbf{l}) \geq \sigma_E(n)$, we come to a sparse grid approximation space $V_n^{(E)}$ which is based on the energy norm:

$$V_n^{(E)} := \bigoplus_{|\mathbf{l}|_1 - \frac{1}{5} \cdot \log_2\left(\sum_{j=1}^d 4^{l_j}\right) \ \leq \ (n+d-1) - \frac{1}{5} \cdot \log_2(4^n + 4d - 4)} W_\mathbf{l}. \qquad (3.72)$$

For a comparison of the underlying subspace schemes of $V_n^{(1)}$ and $V_n^{(E)}$ in two dimensions, see Figure 3.4.

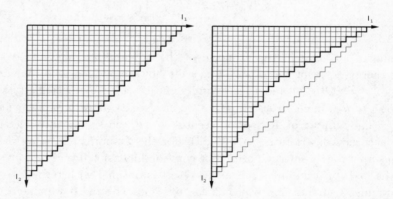

Fig. 3.4. Scheme of subspaces for $V_{30}^{(1)}$ (left) and $V_{30}^{(E)}$ (right), $d = 2$.

First, we look at the number of grid points of the underlying sparse grids.

Lemma 2 *The energy-based sparse grid space $V_n^{(E)}$ is a subspace of $V_n^{(1)}$, and its dimension fulfills*

$$|V_n^{(E)}| \ \leq \ 2^n \cdot \frac{d}{2} \cdot \left(1 - 2^{-\frac{2}{3}}\right)^{-d} \ \leq \ 2^n \cdot \frac{d}{2} \cdot e^d \ = \ O\big(h_n^{-1}\big). \qquad (3.73)$$

Proof For subspaces $W_\mathbf{l}$ with $|\mathbf{l}|_1 = n + d - 1 + i$, $i \in \mathbb{N}$, we have

$$|\mathbf{l}|_1 - \frac{1}{5} \cdot \log_2 \left(\sum_{j=1}^{d} 4^{l_j} \right) \geq n + d - 1 + i - \frac{1}{5} \cdot \log_2 \left(4^{n+i} + 4d - 4 \right)$$

$$\geq n + d - 1 + i - \frac{1}{5} \cdot \log_2 \left(4^i \left(4^n + 4d - 4 \right) \right)$$

$$> n + d - 1 - \frac{1}{5} \cdot \log_2 \left(4^n + 4d - 4 \right).$$

Therefore, no $W_\mathbf{l}$ with $|\mathbf{l}|_1 > n + d - 1$ can belong to $V_n^{(E)}$. Consequently, $V_n^{(E)}$ is a subspace of $V_n^{(1)}$ and $|V_n^{(E)}| \leq |V_n^{(1)}|$ for all $n \in \mathbb{N}$. Starting from that, (3.48) provides

$$|V_n^{(E)}| = \sum_{i=0}^{n-1} \sum_{\substack{|\mathbf{l}|_1 = n+d-1-i, \\ \sum_{j=1}^{d} 4^{l_j} \geq \frac{4^n + 4d - 4}{32^i}}} |W_\mathbf{l}|$$

$$= 2^n \cdot \frac{1}{2} \cdot \sum_{i=0}^{n-1} 2^{-i} \cdot \sum_{\substack{|\mathbf{l}|_1 = n+d-1-i, \\ \sum_{j=1}^{d} 4^{l_j} \geq \frac{4^n + 4d - 4}{32^i}}} 1$$

$$\leq 2^n \cdot \frac{1}{2} \cdot \lim_{n \to \infty} \sum_{i=0}^{n-1} 2^{-i} \cdot \sum_{\substack{|\mathbf{l}|_1 = n+d-1-i, \\ \sum_{j=1}^{d} 4^{l_j} \geq \frac{4^n + 4d - 4}{32^i}}} 1$$

$$= 2^n \cdot \frac{1}{2} \cdot \lim_{n \to \infty} \sum_{i=0}^{n-1} 2^{-i} \cdot d \cdot \binom{d - 1 - \lfloor 1.5i \rfloor}{d - 1},$$

since it can be shown that, for $n \to \infty$, our energy-based sparse grid and the grid resulting from the second condition $|\mathbf{l}|_\infty \geq n - \lfloor 2.5i \rfloor$ for the inner sum instead of

$$\sum_{j=1}^{d} 4^{l_j} \geq \frac{4^n + 4d - 4}{32^i}$$

are the same, and since there exist

$$\binom{d - 1 + \lfloor 1.5i \rfloor}{d - 1}$$

such subspaces $W_\mathbf{l}$ with $|\mathbf{l}|_\infty = l_1$. Consequently, we obtain

$$|V_n^{(E)}| \leq 2^n \cdot \frac{d}{2} \cdot \sum_{i=0}^{\infty} 2^{-\frac{2}{3}i} \cdot \binom{d - 1 + i}{d - 1}$$

$$= 2^n \cdot \frac{d}{2} \cdot \left(1 - 2^{-\frac{2}{3}} \right)^{-d}$$

$$\leq 2^n \cdot \frac{d}{2} \cdot e^d,$$

since $\sum_{i=0}^{\infty} x^i \cdot \binom{k+i}{k} = (1-x)^{-k-1}$ for $k \in \mathbb{N}_0$ and $0 < x < 1$. $\qquad\square$

Next, we have to deal with the interpolation accuracy of the energy-based sparse grid spaces $V_n^{(E)}$ and to study the sparse grid interpolant $u_n^{(E)} \in V_n^{(E)}$.

Theorem 1 *The energy norm of the interpolation error of some $u \in X_0^{q,2}(\bar{\Omega})$ in the energy-based sparse grid space $V_n^{(E)}$ is bounded by*

$$\|u - u_n^{(E)}\|_E \leq \frac{d \cdot |u|_{2,\infty}}{3^{(d-1)/2} \cdot 4^{d-1}} \cdot \left(\frac{1}{2} + \left(\frac{5}{2}\right)^{d-1}\right) \cdot 2^{-n} = O(h_n). \quad (3.74)$$

Proof First, since

$$\|u - u_n^{(E)}\|_E \leq \|u - u_n^{(1)}\|_E + \|u_n^{(1)} - u_n^{(E)}\|_E,$$

and since we know that $\|u - u_n^{(1)}\|_E$ is of the order $O(h_n)$, we can restrict ourselves to $\|u_n^{(1)} - u_n^{(E)}\|_E$. For that, it can be shown that, for $i \in \mathbb{N}_0$, each $W_\mathbf{l}$ with $|\mathbf{l}|_1 = n + d - 1 - i$ and $|\mathbf{l}|_\infty \geq n - 2.5i$ is a subspace of $V_n^{(E)}$. Therefore, we obtain with (3.54)

$$\|u_n^{(1)} - u_n^{(E)}\|_E \leq \sum_{W_\mathbf{l} \subseteq V_n^{(1)} \ominus V_n^{(E)}} \|u_\mathbf{l}\|_E \leq \sum_{i=0}^{i^*} \sum_{\substack{|\mathbf{l}|_1 = n+d-1-i, \\ |\mathbf{l}|_\infty < n-2.5i}} \|u_\mathbf{l}\|_E$$

$$\leq \frac{|u|_{2,\infty}}{2 \cdot 12^{(d-1)/2}} \cdot \sum_{i=0}^{i^*} \sum_{\substack{|\mathbf{l}|_1 = n+d-1-i, \\ |\mathbf{l}|_\infty < n-2.5i}} 4^{-|\mathbf{l}|_1} \cdot \left(\sum_{j=1}^{d} 4^{l_j}\right)^{1/2}$$

$$\leq \frac{|u|_{2,\infty}}{2 \cdot 12^{(d-1)/2}} \cdot 4^{-n-d+1} \cdot \sum_{i=0}^{i^*} 4^i \cdot \sum_{\substack{|\mathbf{l}|_1 = n+d-1-i, \\ |\mathbf{l}|_\infty < n-2.5i}} \left(\sum_{j=1}^{d} 2^{l_j}\right)$$

$$\leq \frac{|u|_{2,\infty}}{2 \cdot 12^{(d-1)/2}} \cdot 4^{-n-d+1} \cdot \sum_{i=0}^{i^*} 4^i \cdot \sum_{j=1}^{n-1-\lfloor 2.5i \rfloor} d \cdot \binom{n+d-2-i-j}{d-2} \cdot 2^j$$

$$= \frac{|u|_{2,\infty}}{2 \cdot 12^{(d-1)/2}} 4^{-n-d+1} \sum_{i=0}^{i^*} 4^i \sum_{k=1}^{n-1-\lfloor 2.5i \rfloor} d\binom{d-2+\lfloor 1.5i \rfloor+k}{d-2} 2^{n-\lfloor 2.5i \rfloor-k}$$

$$= \frac{d \cdot |u|_{2,\infty}}{2 \cdot 12^{(d-1)/2}} 4^{-(d-1)} 2^{-n} \sum_{i=0}^{i^*} 2^{-\lfloor \frac{i}{2} \rfloor} \sum_{k=1}^{n-1-\lfloor 2.5i \rfloor} \binom{d-2+\lfloor 1.5i \rfloor+k}{d-2} 2^{-k}$$

$$\leq \frac{d \cdot |u|_{2,\infty}}{2 \cdot 12^{(d-1)/2}} \cdot 4^{-(d-1)} \cdot 2^{-n} \cdot 2 \cdot 5^{d-1}$$

$$= \frac{d \cdot |u|_{2,\infty}}{3^{(d-1)/2} \cdot 4^{d-1}} \cdot \left(\frac{5}{2}\right)^{d-1} \cdot 2^{-n},$$

where $0 \leq i^* \leq n-1$ is the maximum value of i for which the set of indices $\mathbf{l}$ with $|\mathbf{l}|_1 = n+d-1-i$ and $|\mathbf{l}|_\infty < n - 2.5i$ is not empty. Together with (3.69) we get the result. □

The crucial result of this section is that, with the energy-based sparse grid spaces $V_n^{(E)}$, the curse of dimensionality can be overcome. In both (3.73) and (3.74), the n-dependent terms are free of any d-dependencies: There is an order of $O(2^n)$ for the dimension and $O(2^{-n})$ for the interpolation error. In particular, there is no longer any polynomial term in n like n^{d-1} as for the case of the space $V_1^{(1)}$. That is, apart from the factors that are constant with respect to n, there is no d-dependence in either $|V_n^{(E)}|$ or $||u - u_n^{(E)}||_E$ and, thus, no deterioration in complexity for higher dimensional problems. The curse of dimensionality has thus been completely overcome, at least with respect to n. However the constants in the order estimates are still dependent on the dimension d. This will be studied in more detail in the following section.

3.4.3 The constants and their dependence on d

So far, we derived the estimate

$$|V_n^{(E)}| \leq c_1(d) \cdot 2^n \tag{3.75}$$

for the degrees of freedom of the sparse grid spaces $V_n^{(E)}$ with the constant

$$c_1(d) = \frac{d}{2}(1 - 2^{-2/3})^{-d}$$

and the estimate

$$||u - u_n^{(E)}||_E \leq c_2(d) \cdot 2^{-n} \cdot |u|_{2,\infty} = O(h_n), \tag{3.76}$$

for the accuracy of the achieved interpolation error with the constant

$$c_2(d) = \frac{d}{3^{(d-1)/2} \cdot 4^{d-1}} \cdot \left(\frac{1}{2} + \left(\frac{5}{2}\right)^{d-1}\right).$$

Note that the upper bound (3.69) for the interpolant $u_n^{(E)}$ of u in $V_n^{(E)}$ also gives by virtue of Cea's lemma an upper bound for the best approximation in $V_n^{(E)}$. We thus have

$$\inf_{v_n \in V_n^{(E)}} ||u - v_n||_E \leq ||u - u_n^{(E)}||_E.$$

We are interested in casting these results in a form which is more common in approximation theory, i.e., we want to express the bound for the approximation error in terms of the amount of degrees of freedom involved.

To this end we define the number of degrees of freedom for the best approximation of u in $V_n^{(E)}$ as

$$M := |V_n^{(E)}|. \qquad (3.77)$$

We then express the estimate of the approximation error in terms of M.

Theorem 2 *For the best approximation of a function $u \in X_0^{q,2}(\bar{\Omega})$ in the space $V_n^{(E)}$ with respect to the energy norm, there holds*

$$\inf_{v_n \in V_n^{(E)}} \|u - v_n\|_E \le \|u - u_n^{(E)}\|_E \le c \cdot d^2 \cdot 0.97515^d \cdot M^{-1} \cdot |u|_{2,\infty}. \qquad (3.78)$$

Proof First, with the definition (3.77) we solve (3.75) for 2^n. Taking the inverse we obtain $2^{-n} \le c_1(d) \cdot M^{-1}$. Then

$$\|u - u_n^{(E)}\|_E \le c_2(d) \cdot 2^{-n} \cdot |u|_{2,\infty} \le c_1(d) \cdot c_2(d) \cdot M^{-1} \cdot |u|_{2,\infty}.$$

With

$$c_1(d) \cdot c_2(d) = \frac{d}{2}(1 - 2^{-2/3})^{-d} \cdot \frac{d}{3^{(d-1)/2} \cdot 4^{d-1}} \cdot \left(\frac{1}{2} + \left(\frac{5}{2}\right)^{d-1}\right)$$

$$= \sqrt{3} \cdot d^2 \left(\frac{1}{(1 - 2^{-2/3}) \cdot \sqrt{3} \cdot 4}\right)^d$$

$$+ \frac{4 \cdot \sqrt{3}}{5} \cdot d^2 \left(\frac{5}{(1 - 2^{-2/3}) \cdot \sqrt{3} \cdot 8}\right)^d$$

$$\le \sqrt{3} \cdot d^2 \cdot 0.39901^d + \frac{4\sqrt{3}}{5} \cdot d^2 \cdot 0.97515^d \qquad (3.79)$$

the estimate (3.78) results. $\qquad \square$

Thus we see that we obtain for the constant a decay for $d \to \infty$ to zero.[1] The term $|u|_{2,\infty}$ however is also dependent on d and may grow exponentially with it. Already for the simple example $u(\mathbf{x}) = \prod_{j=1}^d \sin(2\pi k x_j)$ we see that $|u|_{2,\infty} = (4\pi^2 k^2)^d$ grows faster than 0.97515^d decays. Obviously it is sufficient to restrict ourselves to the approximation of functions $u \in X_0^{q,2}$ with $|u|_{2,\infty} = o(1/(d^2 \cdot 0.97515^d))$ to insure that $\|u - u_n^{(E)}\|_E$ is bounded for all d. But it is not clear how interesting this function class for large d is in practice. Nevertheless, the facts we know from the concentration of measure phenomenon (i.e., that the best approximation in very high dimensions is nearly constant) give hope in this direction.

---

[1] Note that this holds for the asymptotics with respect to M, i.e., the estimates for c_1 and c_2 were done for asymptotically large n.

Note that if we use the seminorm

$$|u|_{2,2} := \left\| D^2 u \right\|_2 = \left(\int_{\bar{\Omega}} \left| D^2 u \right|^2 \, \mathrm{d}\mathbf{x} \right)^{1/2}$$

instead and rewrite all the above lemmata, theorems and their proofs in terms of the associated regularity assumption $|u|_{2,2}$ we are no longer able to derive a favorable estimate as in (3.79). We obtain slightly worse estimates where, for the c_1 and c_2 involved, we only get $c_1(d) \cdot c_2(d) \leq c \cdot \sqrt{3} \cdot d^2 \cdot 0.45041^d + \sqrt{3}\frac{4}{5} \cdot d^2 \cdot 1.12601^d$. Thus we see a blow-up to infinity for $d \to \infty$ for these estimates. Since the corresponding c_1 and c_2 are only upper bounds of the true d-dependent constants it is not clear if this also holds for them or not. Note furthermore that a rescaling of the size of the domain Ω of course also influences the constants c_1 and c_2 which has to be taken into account in the above discussion.

Let us finally consider the case of non-homogeneous boundary conditions, i.e., u from the space $X^{q,2}$. Now, to capture also functions living on the boundary of $\bar{\Omega}$, we generalize the two-scale splitting (3.11) to a three-scale decomposition $V^{(1)} = \mathbf{1} \oplus \mathbf{lin} \oplus \tilde{W}$ where $\mathbf{1}$ denotes the subspace of constants, $\mathbf{lin}$ denotes the subspace of linear functions (without the constants) and $\tilde{W}$ denotes the remainder, respectively. Note that $\mathbf{1} \oplus \mathbf{lin}$ is just the kernel of the second derivative. This augmented splitting is now used as the input of a tensor product construction to gain a splitting of the function space for the d-dimensional case. Analogously to (3.17) and (3.18) a decomposition of the d-dimensional space into now 3^d subspaces is introduced by $X^{q,2} \doteq V^{(d)} = \bigotimes_{j=1}^{d} (\mathbf{1}_j \oplus \mathbf{lin}_j \oplus \tilde{W}_j)$ and we can repeat the discussion of Section 3.3.1 in a similar way for the refined decomposition. Informally speaking, the space $X^{q,2}$ can then be decomposed into $X^{q,2} \setminus X_0^{q,2}$ and $X_0^{q,2}$ consistent with this refined decomposition, and we can split a function $u \in X^{q,2}$ accordingly into

$$u = \tilde{u} + v \quad \text{where} \quad \tilde{u} \in X^{q,2} \setminus X_0^{q,2} \quad \text{and} \quad v \in X_0^{q,2}.$$

The regularity condition $|u|_{2,\infty} \leq c < \infty$ translates to $\|D^2 u\|_\infty \leq c < \infty$, with some d-dependent constant c. For the term $D^2 u$ we obtain

$$D^2 u = \frac{\partial^{2d} u}{\partial x_1^2 \cdots \partial x_d^2} = \frac{\partial^{2d} (\tilde{u} + v)}{\partial x_1^2 \cdots \partial x_d^2} = \frac{\partial^{2d} v}{\partial x_1^2 \cdots \partial x_d^2}$$

since $\frac{\partial^{2d} \tilde{u}}{\partial x_1^2 \cdots \partial x_d^2}$ vanishes due to the involved constant and linear subspaces. We are thus just in the situation of homogeneous boundary conditions as treated previously and the lemmata and theorems above apply. Note that the more general assumption $D^\alpha u \in L_\infty, |\alpha|_\infty \leq 2$, from (3.32) relates to the (second) variation of Hardy and Krause (Owen 2004) and involves in a dimension-recursive way different partial (mixed) derivatives up to second order of the contributions of u from the various boundary manifolds of $\bar{\Omega}$.

In an analogous way, we enlarge our basis function set to also capture functions living on the boundary of $\bar{\Omega}$. To this end we introduce two more functions into the one-dimensional hierarchical basis from Figure 3.2 which are associated to the left and right boundary point of $[0,1]$ and number their associated level l by -1 for the left point and 0 for the right point, respectively. As basis functions we use the constant function $\phi_{-1,1}(x) := 1$ for the left boundary point and the linear function $\phi_{0,1}(x) = x$ for the right boundary point. Then, $\{\phi_{-1,1}, \phi_{0,1}\}$ just spans the subspace of constant and linear functions.[1] This augmented system of basis functions is now used as the input of the tensor product construction (3.39) to gain a function system for the d-dimensional case. Moreover, analogously to (3.45) and (3.47), a function $u \in X^{q,2}$ can now be represented as $u(\mathbf{x}) = \sum_{\mathbf{l} \in (\mathbb{N} \cup \{-1,0\})^d} u_{\mathbf{l}}(\mathbf{x})$ and the space $X^{q,2}$ gets decomposed as $\bigotimes_{\mathbf{l} \in (\mathbb{N} \cup \{-1,0\})^d} W_{\mathbf{l}}$.[2]

Our approach so far was focused on the space of bounded second mixed derivatives, the energy-norm as measure for the approximation error and (piecewise) linear hierarchical basis functions. It can be carried over to a more general setting where we measure the approximation error in the H^s-norm, $s \in (-\infty, \infty)$, assume a smoothness of the type $|u|_{H^{l,t}_{mix}}$, where l denotes isotropic and t mixed smoothness, see (Griebel and Knapek 2000), and use wavelet-type multilevel systems with sufficient primal and dual regularity. Then, depending on these additional parameters, we can again derive optimal discrete approximation spaces, we may study their cost complexities and approximation properties for different regimes of s, l, t and we can identify situations where the curse of dimensionality can be broken. The approach is based on norm-equivalences and associated wavelet-type multilevel systems. This is explained in more detail in (Griebel and Knapek 2000, Knapek 2000*a*), see also (Knapek 2000*b*) for a variant using Fourier bases. Since the constants in these norm-equivalences depend on d, the constants in the resulting error estimates also depend on d and cannot, in contrast to our approach in Section 3.4.3, be estimated explicitly.

Another generalization of the sparse grid concept uses optimization not with respect to a whole class of functions involving error norm and smoothness prerequisite (a-priori knowledge) but, in the interpolation context, with respect to one single given function or, alternatively, in the context of PDEs, with respect to a given right-hand side or other data in the partial

---

[1] We could also have used the linear function $1 - x$ at the left boundary point instead. Here, we use the constant one to be completely in sync with our splitting $V = \mathbf{1} \oplus W$ from (3.11).

[2] $X^{q,2} \setminus X^{q,2}_0$ is then (up to completion) just $\sum_{\mathbf{l} \in (\mathbb{N} \cup \{-1,0\})^d \setminus \mathbb{N}^d} W_{\mathbf{l}}$.

differential equation. This leads with proper a-posteriori error estimators to an adaptively refined sparse grid which adapts itself (hopefully in an optimal way) to the specific situation. The adaption and refinement process can be performed on the level of the subspaces W_l from (3.41). This results in a so-called dimension-adaptive method for sparse grids, see (Gerstner and Griebel 2003). This approach is well suited for high-dimensional functions and detects important and unimportant dimensions and groups of dimensions of an ANOVA-decomposition in an automatic way (provided that the error indicators are sound and no premature termination of the adaption process occurs). The method was developed and used so far for integration problems, its application to partial differential equations is future work. Alternatively, the adaption and refinement process can be performed on the level of the single basis functions $\phi_{l,i}$. We then obtain a method where, besides the detection of important and unimportant dimensions and groups of dimensions, singularities and similar local variations in a function are additionally found and resolved. Here, the development of sound local error estimators (via the dual problem), efficient refinement strategies and the associated complexities with respect to d are an area of active research (Griebel 1998, Bungartz and Griebel 2004, Bungartz 1998, Schneider 2000). Figure 3.5 gives examples of two- and three-dimensional adaptive sparse grids.

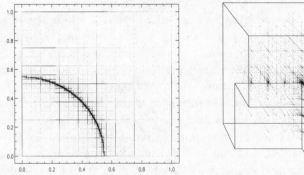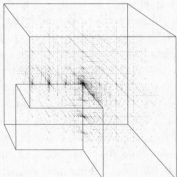

Fig. 3.5. Adaptively refined sparse grids: Two-dimensional example (left) and three-dimensional example (right).

3.5 Concluding remarks

We reviewed approximation techniques which have the potential to overcome the curse of dimensionality, which is a main obstacle in the numerical treatment of most high-dimensional problems. After a survey on methods stemming from Kolmogorov's theorem, we focused on the ANOVA decomposition and the sparse grid approach and discussed their properties and prerequisites. Moreover, we presented energy-norm based sparse grids and demonstrated that, for functions with bounded mixed second derivatives on the unit hypercube, the associated approximation rate in terms of the involved degrees of freedom shows no dependence on the dimension at all, neither in the approximation order nor in the order constant. Important ingredients were the product structure of the underlying domain in high dimensions, a one-dimensional splitting of the space into the constant and the remainder subspaces and, as a refinement of the remainder subspace, a one-dimensional multilevel basis. Then a tensor-product approach leads to both the ANOVA decomposition and a multilevel series expansion of the underlying function. Proper truncation may result in an ANOVA decomposition with finite order weights or, if the more elaborate one-dimensional multiscale splitting is employed, in sparse grids. In this sense sparse grids are closely related to the ANOVA approach and can be seen as a discretization and refined version of it. In the case of functions with low effective dimension or alternatively, bounded mixed second derivatives, the curse of dimensionality indeed can be broken.

These techniques can, together with a Galerkin approach or with a finite difference method, be applied successfully to higher-dimensional partial differential equations. Areas of actual research are here, besides elliptic partial differential equations, also parabolic problems, like e.g., the Fokker-Planck equation with many practical applications in the natural sciences and financial engineering, ranging from the modelling of mechanical systems with random oscillations to the pricing of financial derivatives. Also the Schrödinger equation is of utmost interest. Here, in (Griebel and Hamaekers 2006) we developed antisymmetric sparse grid spaces to cope with the antisymmetry condition stemming from Pauli's principle, see also (Yserentant 2004). Further actual work is the use of sparse grids in space-time (Griebel, Oeltz and Vassilevski 2005). There a product structure between space and time exists naturally and can be exploited.

To reach higher space dimensions the constants in the complexities must be kept as low as possible. Besides the theoretical results on the constants which we presented in the preceding section for approximations in the energy-norm, also the effect of the detailed implementation (data structures, fast solution of the discretized linear systems by e.g., multigrid) on the complexity constants has to be taken into account. Presently we are able to treat elliptic differential equations with up to about 120 dimensions on a modern workstation (provided that there are homogeneous boundary conditions and a product-type right hand side), see (Feuersänger 2005).

Further work has surely to be done to better relate ANOVA-type approaches from high-dimensional integration, data mining and statistics to the solution of partial differential equations by sparse grid techniques, especially with respect to adaptivity. Then certain classes of high-dimensional PDE problems with e.g., lower effective dimension or a decay in the interaction weights of the solution may be detected automatically and treated effectively. Finally there is hope to numerically deal with high-dimensional problems due to the concentration of measure phenomenon.

Acknowledgements

I am grateful to M. Arndt, H. Bungartz, P. Jones, F. Kuo, M. Mohlenkamp, A. Pinkus, M. Schweitzer, E. Süli, I. Sloan, G. Wasilkowski and H. Woźniakowski for valuable discussions and comments.

References

G. Allaire (1992), 'Homogenization and two-scale convergence', *SIAM J. Math. Anal.* **21**(6), 1482–1516.

V. Arnold (1957), 'On functions of three variables', *Dokl. Akad. Nauk SSSR* **114**, 679–681. English translation: American Math. Soc. Transl. (2), 28, 51-54, 1963.

V. Arnold (1958), 'On the representation of functions of several variables by superpositions of functions of fewer variables', *Mat. Prosveshchenie* **3**, 41–61.

V. Arnold (1959), 'On the representation of continuous functions of three variables by superpositions of continuous functions of two variables', *Mat. Sb.* **48**, 3–74. English translation: American Math. Soc. Transl. (2), 28, 61-147, 1963.

N. Aronzaijn (1950), 'Theory of reproducing kernels', *Trans. Amer. Math. Soc.* **68**, 337–404.

K. Babenko (1960), 'Approximation by trigonometric polynomials in a certain class of periodic functions of several variables', *Soviet Math. Dokl.* **1**, 672–675. Russian original in Dokl. Akad. Nauk SSSR, 132 (1960), pp. 982–985.

R. Balescu (1997), *Statistical Dynamics, Matter out of Equilibrium*, Imperial College Press, Imperial College, London.

A. Barron (1993), 'Universal approximation bounds for superpositions of a sigmoidal function', *IEEE Transactions on Information Theory* **39**(3), 930–945.

B. Baxter and A. Iserles (2003), On the foundations of computational mathematics, in *Foundations of Computational Mathematics, Special Volume of Handbook of Numerical Analysis Vol. XI* (F. Cucker, ed.), Elsevier Science, pp. 3–34.

M. Beck, A. Jäckle, G. Worth and H. Meyer (2000), 'The multiconfiguration time-dependent Hartree (MCTDH) method: A highly efficient algorithm for propagating wavepackets', *Phys. Reports* **324**, 1–105.

R. Bellmann (1961), *Adaptive Control Processes: A Guided Tour*, Princeton University Press.

G. Beylkin and M. Mohlenkamp (2002), 'Numerical operator calculus in higher dimensions', *PNAS* **99**(16), 10246–10251.

G. Beylkin and M. Mohlenkamp (2005), 'Algorithms for numerical analysis in high dimensions', *SIAM Journal on Scientific Computing* **26**(6), 2133–2159.

J. Boyd (2000), *Chebyshev and Fourier spectral methods*, Dover Publications, New York.

H. Bungartz (1992), *Dünne Gitter und deren Anwendung bei der adaptiven Lösung der dreidimensionalen Poisson-Gleichung*, Dissertation, Institut für Informatik, TU München.

H. Bungartz (1998), *Finite Elements of Higher Order on Sparse Grids*, Habilitationsschrift, Institut für Informatik, TU München, and Shaker Verlag, Aachen.

H. Bungartz and M. Griebel (1999), 'A note on the complexity of solving Poisson's equation for spaces of bounded mixed derivatives', *J. Complexity* **15**, 167–199.

H. Bungartz and M. Griebel (2004), 'Sparse grids', *Acta Numerica* **13**, 147–269.

R. Caflisch, W. Morokoff and A. Owen (1997), 'Valuation of mortgage backed securities using Brownian bridges to reduce effective dimension', *J. Comput. Finance* **1**(1), 27–46.

C. Chauviere and A. Lozinski (2004), 'Simulation of dilute polymer solutions using a Fokker-Planck equation', *Computers and Fluids* **33**, 687–696.

C. Chui and Y. Wang (1992), 'A general framework for compactly supported splines and wavelets', *J. Approx. Theory* **71**, 263–304.

D. Cioranescu, A. Damlamian and G. Griso (2002), 'Periodic unfolding and homogenization', *C. R. Acad. Sci. Paris, Ser. I* **335**, 99–104.

A. Cohen (2003), *Numerical Analysis of Wavelet Methods*, Studies in Mathematics and its Applications, Vol. 32, North Holland.

M. Coppejans (2004), 'On Kolmogorov's representation of functions of several variables by functions of one variable', *Journal of Econometrics* **123**, 1–31.

I. Daubechies (1992), *Ten Lectures on Wavelets*, CBMS-NSF Regional Conf. Series in Appl. Math. 61, SIAM.

R. de Figueiredo (1980), 'Implications and applications of Kolmogorov's superposition theorem', *IEEE Transactions on Automatic Control* **25**(6), 1227–1231.

F. Delvos (1982), '*d*-variate Boolean interpolation', *J. Approx. Theory* **34**, 99–114.

F. Delvos and W. Schempp (1989), *Boolean Methods in Interpolation and Approximation*, Vol. 230 of *Pitman Research Notes in Mathematics*, Longman Scientific and Technical, Harlow.

G. Deslauriers and S. Dubuc (1989), 'Symmetric iterative interpolation processes', *Constr. Approx.* **5**, 49–68.

R. DeVore (1998), 'Nonlinear approximation', *Acta Numerica* **7**,51–150.

R. DeVore, S. Konyagin and V. Temlyakov (1998), 'Hyperbolic wavelet approximation', *Constr. Approx.* **14**, 1–26.

J. Dick, I. Sloan, X. Wang and H. Woźniakowski (2004), 'Liberating the weights', *Journal of Complexity* **20**(5), 593–623.

D. Donoho (2000), 'High-dimensional data analysis: The curses and blessings of dimensionality'. Aide-Memoire.

D. Donoho and P. Yu (1999), Deslauriers-Dubuc: Ten years after, in *CRM Proceedings and Lecture Notes Vol. 18* (G. Deslauriers and S. Dubuc, eds).

D. Duffie (1996), *Dynamic Asset Pricing Theory*, Princeton University Press.

B. Efron and C. Stein (1981), 'The jackknife estimator of variance', *Annals of Statist.* **9**, 586–596.

J. Elf, P. Lötstedt and P. Sjöberg (2001), Problems of high dimension in molecular biology, in *17th Gamm Seminar Leipzig 2001* (W. Hackbusch, ed.), pp. 1–10.

M. Escobar and L. Seco (2005), A partial differential equation for credit derivatives pricing, in *Workshop High-dimensional Partial Differential Equations in Science and Engineering, August 7-12, CRM, Montreal.*

G. Faber (1909), 'Über stetige Funktionen', *Mathematische Annalen* **66**, 81–94.

C. Feuersänger (2005), Dünngitterverfahren für hochdimensionale elliptische partielle Differentialgleichungen, Diplomarbeit, Institut für Numerische Simulation, Universität Bonn.

B. Fridman (1967), 'An improvement in the smoothness of the functions in Kolmogorov's theorem on superpositions', *Dokl. Akad. Nauk SSSR* **177**, 1019–1022. English translation: Soviet Math. Dokl. (8), 1550-1553, 1967.

J. Friedman (1991), 'Multivariate adaptive regression splines', *Ann. Statist.* **19**(1), 1–141.

J. Friedman and W. Stützle (1981), 'Projection pursuit regression', *J. Amer. Statist. Assoc.* **76**, 817–823.

J. Garcke (2004), Maschinelles Lernen durch Funktionsrekonstruktion mit verallgemeinerten dünnen Gittern, Dissertation, Institut für Numerische Simulation, Universität Bonn.

J. Garcke, M. Griebel and M. Thess (2001), 'Data mining with sparse grids', *Computing* **67**(3), 225–253.

T. Gerstner and M. Griebel (1998), 'Numerical integration using sparse grids', *Numerical Algorithms* **18**, 209–232.

T. Gerstner and M. Griebel (2003), 'Dimension-adaptive tensor-product quadrature', *Computing* **71**, 65–87.

F. Girosi and T. Poggio (1989), 'Representation properties of networks: Kolmogorov's theorem is irrelevant', *Neural Comput.* **1**, 465–469.

F. Girosi, M. Jones and T. Poggio (1995), 'Regularization theory and neural networks architectures', *Neural Computation* **7**, 219–265.

M. Golomb (1959), Approximation by functions of fewer variables, in *On Numerical Approximation* (R. E. Langer, ed.), University of Wisconsin Press, Madison, pp. 275–327.

W. Gordon (1969), Distributive lattices and the approximation of multivariate functions, in *Approximation with Special Emphasis on Spline Functions* (I. Schoenberg, ed.), Academic Press, New York, pp. 223–277.

W. Gordon (1971), 'Blending function methods of bivariate and multivariate interpolation and approximation', *SIAM J. Numer. Anal.* **8**, 158–177.

L. Grasedyck and W. Hackbusch (2003), 'Construction and arithmetics of $\mathcal{H}$-matrices', *Computing* **70**, 295–334.

M. Green, J. Schwarz and E. Witten (1998), *Superstring Theory, Vol. 1 and 2*, Cambridge University Press.

M. Griebel (1994), 'Multilevel algorithms considered as iterative methods on semidefinite systems', *SIAM J. Sci. Stat. Comput.* **15**(3), 547–565.

M. Griebel (1998), 'Adaptive sparse grid multilevel methods for elliptic PDEs based on finite differences', *Computing* **61**(2), 151–179.

M. Griebel and J. Hamaekers (2006), 'Sparse grids for the Schrödinger equation', *Mathematical Modelling and Numerical Analysis.* to appear.

M. Griebel and S. Knapek (2000), 'Optimized tensor-product approximation spaces', *Constr. Approx.* **16**(4), 525–540.

M. Griebel and P. Oswald (1994), 'On additive Schwarz preconditioners for sparse grid discretizations', *Numer. Math.* **66**, 449–463.

M. Griebel and P. Oswald (1995*a*), 'On the abstract theory of additive and multiplicative Schwarz algorithms', *Numer. Math.* **70**(2), 161–180.

M. Griebel and P. Oswald (1995*b*), 'Tensor product type subspace splittings and multilevel iterative methods for anisotropic problems', *Adv. Comput. Math.* **4**, 171–206.

M. Griebel, D. Oeltz and P. Vassilevski (2005), Space-time approximation with sparse grids, Preprint UCRL-JRNL-211686, Lawrence Livermore National Laboratory. SIAM J. Sci. Comput. To appear. Also as Preprint No. 222, SFB 611, University of Bonn.

M. Gromov (1999), *Metric Structures for Riemannian and Non-Riemannian Spaces, Progress in Mathematics 152*, Birkhäuser Verlag.

C. Gu and G. Wahba (1993), 'Semiparametric analysis of variance with tensor product thin plate splines', *Journal of the Royal Statistical Society, Series B* **55**(2), 353–368.

W. Hackbusch (1985), *Multigrid Methods and Applications*, Springer, Berlin/Heidelberg.

W. Hackbusch and B. Khoromskij (2004), Kronecker tensor-product based approximation to certain matrix-valued functions in higher dimensions, Technical Report 16/2004, Max Planck Institute for Mathematics in the Sciences, Leipzig.

W. Hackbusch, B. Khoromskij and S. Sauter (2000), On $\mathcal{H}^2$-matrices, in *Lectures on Applied Mathematics* (H. Bungartz, R. Hoppe and C. Zenger, eds), Springer, Berlin, pp. 9–29.

T. Hastie and R. Tibshirani (1986), 'Generalized additive models (with discussion)', *Statist. Sci.* **1**, 297–318.

T. Hastie and R. Tibshirani (1990), *Generalized Additive Models*, Monographs on Statistics and Applied Probability 43, Chapman and Hall.

R. Hecht-Nielsen (1987*a*), 'Counter propagation networks', *Proceeedings IEEE Int. Conf. Neural Networks II* pp. 19–32.

R. Hecht-Nielsen (1987*b*), 'Kolmogorov's mapping neural network existence theorem', *Proceeedings IEEE Int. Conf. Neural Networks* pp. 11–13.

M. Hegland (2003), Adaptive sparse grids, in *Proc. of 10th Computational Techniques and Applications Conference CTAC-2001* (K. Burrage and R. Sidje, eds), Vol. 44 of *ANZIAM J.*, pp. C335–C353.

M. Hegland and V. Pestov (1999), Additive models in high dimensions, Technical Report 99-33, School of Mathematical and Computing Sciences, Victoria University of Wellington.

M. Hegland and P. Pozzi (2005), Concentration of measure and the approximation of functions of many variables, in *Mathematical Methods for Curves and Surfaces: Tromso 2004* (M. Daehlen, K. Morken and L. Schumaker, eds), Nashboro Press, Brentwood TN, pp. 101–114.

F. Hickernell and H. Woźniakowski (2000), 'Integration and approximation in arbitrary dimensions', *Adv. Comput. Math.* **12**, 25–58.

F. Hickernell, I. Sloan and G. Wasilkowski (2004), On tractability of weighted integration for certain Banach spaces of functions, in *Monte Carlo and Quasi-Monte Carlo Methods 2002* (H. Niederreiter, ed.), Springer, pp. 51–71.

T. Hill (1956), *Statistical Mechanics, Principles and Selected Applications*, McGraw-Hill, New York.

V. Hoang and C. Schwab (2003), High-dimensional finite elements for elliptic problems with multiple scales, Technical Report Nr. 2003-14, Seminar für Angewandte Mathematik, ETH Zürich.

W. Hoeffding (1948), 'A class of statistics with asympotically normal distributions', *Annals of Math. Statist.* **19**, 293–325.

W. Hurewicz and H. Wallman (1948), *Dimension Theory*, Princeton University Press, Princeton, NJ.

B. Igelnik and N. Parikh (2003), 'Kolmogorov's spline network', *IEEE Transactions on Neural Networks* **14**(4), 725–733.

E. Johnson, S. Wojtkiewicz, L. Bergman and B. Spencer (1997), Finite element and finite difference solutions to the transient Fokker-Planck equation, in *Nonlinear and Stochastic Beam Dynamics in Accelerators - A Challenge to Theoretical and Computational Physics* (A. Bazzani, J. Ellison, H. Mais and G. Turchetti, eds), pp. 290–306. Proceedings of a Workshop, Lüneburg, Germany, 29. Sep.- 3. Oct. 1997.

G. Karniadakis and S. Sherwin (1999), *Spectral/hp Element Methods for CFD*, Oxford University Press.

S. Khavinson (1997), *Best Approximation by Linear Superpositions*, Translations of Mathematical Monographs, Vol. 159, AMS.

S. Knapek (2000*a*), *Approximation and Kompression mit Tensorprodukt-Multiskalen-Approximationsräumen*, Dissertation, Institut für Angewandte Mathematik, Universität Bonn.

S. Knapek (2000*b*), Hyperbolic cross approximation of integral operators with smooth kernel, Technical Report 665, SFB 256, Univ. Bonn.

A. Kolmogorov (1957), 'On the representation of continuous functions of several variables by superpositions of continuous functions of several variables and addition', *Dokl. Akad. Nauk SSSR* **114**, 953–956. English translation: American Math. Soc. Transl. (2), 28, 55-59, 1963.

M. Köppen (2002), 'On the training of a Kolmogorov network', *Artificial neural networks, ICANN2002, LNCS 2415* pp. 474–479.

F. Kuo and I. Sloan (2005), 'Lifting the curse of dimensionality', *Notices of the AMS*, **52** (11), 1320–1328.

V. Kurkova (1991), 'Kolmogorov's theorem is relevant', *Neural Comput.* **3**, 617–622.

Y. Kwok (1998), *Mathematical Models of Financial Derivatives*, Springer.

M. Ledoux (2001), *The Concentration of Measure Phenomenon*, Math. surveys and monographs, Vol. 89, AMS.

Y. Lee, Y. Kim, S. Lee and J. Koo (2004*a*), Structured Multicategory Support Vector Machine with ANOVA decomposition, Technical Report 743, Department of Statistics, The Ohio State University.

Y. Lee, Y. Lin and G. Wahba (2004*b*), 'Multicategory Support Vector Machines, theory, and application to the classification of microarray data and satellite radiance data', *Journal of the American Statistical Association* **99**, 67–81.

C. Lemieux and A. Owen (2002), Quasi-regression and the relative importance of the ANOVA components of a function, in *Monte Carlo and Quasi-Monte Carlo Methods* (K. Fang, F. Hickernell and H. Niederreiter, eds), Springer-Verlag, Berlin, pp. 331–344.

G. Leobacher, K. Scheicher and G. Larcher (2003), 'On the tractability of the Brownian bridge algorithm', *Journal of Complexity* **19**, 511–528.

G. Li, M. Atramonov, H. Rabitz, S. Wang, P. Georgopoulos and M. Demiralp (2001*a*), 'Highdimensional model representations generated from low dimensional data samples: lp-RS-HDMR', *Journal of Comput. Chemistry* **24**, 647–656.

G. Li, C. Rosenthal and H. Rabitz (2001*b*), 'Highdimensional model representations', *Journal of Phys. Chem. A* **105**(33), 7765–7777.

G. Li, J. Schoendorf, T. Ho and H. Rabitz (2004), 'Multicut HDMR with an application to an ionospheric model', *Journal Comp. Chem.* **25**, 1149–1156.

G. Li, S. Wang, C. Rosenthal and H. Rabitz (2001*c*), 'Highdimensional model representations generated from low dimensional data samples: I. mp-cut-HDMR', *Journal of Mathematical Chemistry* **30**(1), 1–30.

B. Lin (2000), 'Tensor product space ANOVA models', *The Annals of Statistics* **28**(3), 734–755.

R. Liu and A. Owen (2005), Estimating mean dimensionality of ANOVA decompositions, Technical report, Stanford, Department of Statistics.

G. Lorentz, M. v. Golitschek and Y. Makovoz (1996), *Constructive Approximation*, Springer Verlag.

A. Lozinski and C. Chauviere (2003), 'A fast solver for Fokker-Planck equations applied to viscoelastic flow calculations: 2d FENE model', *Journal of Computational Physics* **189**, 607–625.

A. Lozinski, C. Chauviere, J. Fang and R. Owens (2003), 'Fokker-Planck simulations of fast flows of melts and concentrated polymer solutions in complex geometries', *Journal of Rheology* **47**(2), 535–561.

S. Martello and P. Toth (1990), *Knapsack Problems : Algorithms and Computer Implementations*, John Wiley & Sons, Chichester.

A. Matache (2002), 'Sparse two-scale FEM for homogenization problems', *SIAM Journal of Scientific Computing* **17**, 709–720.

S. McWilliam, D. Knappett and C. Fox (2000), 'Numerical solution of the stationary FPK equation using Shannon wavelets', *Journal of Sound and Vibration* **232**(2), 405–430.

A. Messiah (2000), *Quantum Mechanics*, Dover Publications, New York.

Y. Meyer (1992), *Wavelets and Operators*, Cambridge University Press.

V. Milman (1988), 'The heritage of P. Levy in geometrical functional analysis', *Asterisque* **157-158**, 273–301.

V. Milman and G. Schechtman (2001), *Asymptotic Theory of Finite Normed Spaces*, Lecture Notes in Mathematics Vol. 1200, Springer.

U. Mitzlaff (1997), *Diffusionsapproximation von Warteschlangensystemen*, Dissertation, Institut für Mathematik, Technische Universität Clausthal.

W. Morokoff (1998), 'Generating quasi-random paths for stochastic processes', *SIAM Review* **40**(4), 765–788.

R. Munos (2000), 'A study of reinforcement learning in the continuous case by the means of viscosity solutions', *Machine Learning Journal* **40**, 265–299.

R. Munos and A. Moore (2002), 'Variable resolution discretization in optimal control', *Machine Learning Journal* **49**, 291–323.

P. Niyogi and F. Girosi (1998), 'Generalization bounds for function approximation from scattered noisy data', *Advances in Computational Mathematics* **10**, 51–80.

E. Novak and K. Ritter (1998), The curse of dimension and a universal method for numerical integration, in *Multivariate Approximation and Splines* (G. Nürnberger, J. Schmidt and G. Walz, eds).

P. Oswald (1994), *Multilevel Finite Element Approximation*, Teubner Skripten zur Numerik, Teubner, Stuttgart.

A. Owen (2004), Multidimensional variation for quasi-Monte Carlo, Technical Report Nr. 2004-2, Dept. of Statistics, Stanford Univ.

S. Paskov and J. Traub (1995), 'Faster valuation of financial derivatives', *J. Portfolio Management* **22**, 113–120.

J. Prakash (2000), Rouse chains with excluded volume interactions: Linear viscoelasticity, Technical Report Nr. 221, Berichte der Arbeitsgruppe Technomathematik, Universität Kaiserslautern.

J. Prakash and H. Öttinger (1999), 'Viscometric functions for a dilute solution of polymers in a good solvent', *Macromolecules* **32**, 2028–2043.

H. Rabitz and O. Alis (1999), 'General foundations of high-dimensional model representations', *Journal of Mathematical Chemistry* **25**, 197–233.

T. Rassias and J. Simsa (1995), *Finite Sum Decompositions in Mathematical Analysis*, Pure and Applied Mathematics, Wiley.

C. Reisinger (2003), *Numerische Methoden für hochdimensionale parabolische Gleichungen am Beispiel von Optionspreisaufgaben*, Dissertation, Universität Heidelberg.

K. Ritter (2000), *Average-Case Analysis of Numerical Problems*, Lecture Notes in Mathematics, No. 1733, Springer, New York.

P. Rouse (1953), 'A theory of the linear viscoelastic properties of dilute solutions of coiling polymers', *J. Chem. Phys.* **21**(7), 1272–1280.

H. Sagan (1994), *Space-Filling Curves*, Springer Verlag.

S. Schneider (2000), *Adaptive solution of elliptic PDE by hierarchical tensor product finite elements*, Dissertation, Institut für Informatik, TU München.

B. Schölkopf and A. Smola (2002), *Learning with Kernels*, MIT Press.

C. Schwab (2003), Numerical solution of parabolic problems in high dimensions, in *Proceedings of the 19th GAMM-Seminar*, pp. 1–20.

C. Schwab and R. Todor (2003*a*), 'Sparse finite elements for stochastic elliptic problems', *Numerische Mathematik* **95**, 707–734.

C. Schwab and R. Todor (2003*b*), 'Sparse finite elements for stochastic elliptic problems – higher order moments', *Computing* **71**(1), 43–63.

X. Shen, H. Chen, J. Dai and W. Dai (2002), 'The finite element method for computing the stationary distribution on an SRBM in a hypercube with applications to finite buffer queueing networks', *Queuing Systems* **42**, 33–62.

P. Sjöberg (2002), *Numerical Solution of the Master Equation in Molecular Biology*, Master Thesis, Department for Scientific Computing, Uppsala Universität.

I. Sloan (2001), QMC integration – beating intractability by weighting the coordinate directions, Technical Report AMR01/12, University of New South Wales, Applied Mathematics report.

I. Sloan and X. Wang (2005), 'Why are high-dimensional finance problems often of low effective dimension', *SIAM Journal on Scientific Computing* **27**(1), 159–183.

I. Sloan and H. Woźniakowski (1998), 'When are quasi-Monte Carlo algorithms efficient for high dimensional integrals ?', *J. Complexity* **14**, 1–33.

I. Sloan, X. Wang and H. Woźniakowski (2004), 'Finite-order weights imply tractability of multivariate integration', *Journal of Complexity* **20**, 46–74.

S. Smolyak (1963), 'Quadrature and interpolation formulas for tensor products of certain classes of functions', *Soviet Math. Dokl.* **4**, 240–243. Russian original in Dokl. Akad. Nauk SSSR, 148 (1963), pp. 1042–1045.

D. Sprecher (1965), 'On the structure of continuous functions of several variables', *Transactions American Mathematical Society* **115**(3), 340–355.

D. Sprecher (1996), 'A numerical implementation of Kolmogorov's superpositions', *Neural Networks* **9**(5), 765–772.

D. Sprecher (1997), 'A numerical implementation of Kolmogorov's superpositions II', *Neural Networks* **10**(3), 447–457.

D. Sprecher and S. Draghici (2002), 'Space filling curves and Kolmogorov superposition-based neural networks', *Neural Networks* **15**, 57–67.

C. Stone (1985), 'Additive regression and other nonparametric models', *Ann. Statist.* **13**, 689–705.

E. Süli (2006), Finite element approximation of high-dimensional transport-dominated diffusion problems, in these proceedings.

R. Sutton and A. Barto (1998), *Reinforcement Learning. An Introduction*, MIT Press, Cambridge, London.

B. Szabo and I. Babuska (1991), *Finite Element Analysis*, Wiley.

M. Talagrand (1995), *Concentration of Measure and Isoperimetric Inequalities in Product Spaces*, Vol. 81, Publ. Math. IHES.

J. Traub and H. Woźniakowski (1980), *A General Theory of Optimal Algorithms*, Academic Press, New York.

J. Traub, G. Wasilkowski and H. Woźniakowski (1983), *Information, Uncertainty, Complexity*, Addison-Wesley, Reading.

J. Traub, G. Wasilkowski and H. Woźniakowski (1988), *Information-Based Complexity*, Academic Press, New York.

H. Triebel (1992), *Theory of function spaces II*, Birkhäuser Verlag.

E. Tyrtyshnikov (2004), 'Kronecker-product type approximations for some function-related matrices', *Linear Algebra and its Applications* **379**, 423–437.

G. Venktiteswaran and M. Junk (2005*a*), 'A QMC approach for high dimensional Fokker-Planck equations modelling polymeric liquids', *Math. Comp. Simul.* **68**, 23–41.

G. Venktiteswaran and M. Junk (2005*b*), 'Quasi-Monte Carlo algorithms for diffusion equations in high dimensions', *Math. Comp. Simul.* **68**, 43–56.

A. Vitushkin (1964), 'Proof of the existence of analytic functions of several variables not representable by linear superpositions of continuously differentiable functions of fewer variables', *Dokl. Akad. Nauk SSSR* **156**, 1258–1261. English translation: Soviet Math. Dokl. (5), 793-796, 1964.

A. Vitushkin (2004), 'On Hilbert's thirteenth problem and related questions', *Russian Math. Surveys* **51**(1), 11–25.

G. Wahba (1990), *Spline Models for Observational Data*, Vol. 59 of *CBMS-NSF Regional Conference Series in Applied Mathematics*, SIAM, Philadelphia.

G. Wasilkovski and H. Woźniakowski (1995), 'Explicit cost bounds of algorithms for multivariate tensor product problems', *J. Complexity* **11**, 1–56.

G. Wasilkovski and H. Woźniakowski (1999), 'Weighted tensor product algorithms for linear multivariate problems', *J. Complexity* **15**, 402–447.

G. Wasilkowski and H. Woźniakowski (2004), 'Finite-order weights imply tractability of linear multivariate problems', *Journal of Approximation Theory* **130**(1), 57–77.

H. Wei and A. Billings (2004), 'A unified wavelet-based modelling framework for non-linear system identification: The WARNARX model structure', *Int. J. Control* **77**(4), 351–366.

P. Wilmott (1998), *Derivatives, the Theory and Practice of Financial Engineering*, John Wiley & Sons.

S. Wojtkiewicz and L. Bergman (2000), Numerical solution of high-dimensional Fokker-Planck equations, in *8th ASCE Speciality Conference on Probabilistic Mechanics and Structural Reliability*. PCM2000-167.

H. Yserentant (1986), 'On the multi-level splitting of finite element spaces', *Numer. Math.* **49**, 379–412.

H. Yserentant (2004), 'On the regularity of the electronic Schrödinger equation in Hilbert spaces of mixed derivatives', *Numer. Math.* **98**(4), 731–759.

C. Zenger (1991), Sparse grids, in *Parallel Algorithms for Partial Differential Equations* (W. Hackbusch, ed.), Vol. 31 of *NNFM*, Vieweg, Braunschweig/Wiesbaden.

4

Long-time Energy Conservation of Numerical Integrators

Ernst Hairer

Section de Mathématiques
Université de Genève
Genève, Suisse
e-mail: Ernst.Hairer@math.unige.ch

Abstract

This article discusses the energy conservation of a wide class of numerical integrators applied to Hamiltonian systems. It surveys known material by various illustrations, and it also contains more recent and new results.

4.1 Introduction

In this introductory section we present the class of differential equations considered (Hamiltonian systems) together with properties of their flow, and we introduce numerical integration methods that can be expressed as B-series. We further discuss difficulties that can arise when one tries to conserve exactly the Hamiltonian.

4.1.1 Properties of Hamiltonian systems

We consider Hamiltonian systems

$$\begin{aligned} \dot{p} &= -\nabla_q H(p, q) \\ \dot{q} &= \nabla_p H(p, q) \end{aligned} \quad \text{or} \quad \dot{y} = J^{-1}\nabla H(y), \quad J = \begin{pmatrix} 0 & I \\ -I & 0 \end{pmatrix} \quad (4.1)$$

where $y = (p, q)^T$ and $H(y) = H(p, q)$ is a real-valued smooth function, and we emphasise the following two properties of such systems:

(P1) energy conservation, and
(P2) symplecticity.

Property (P1) just means that $H(y) = H(p, q)$ is constant along solutions of the differential equation. For classical mechanical systems, where $H(p, q) = \frac{1}{2}p^T M(q)^{-1}p + U(q)$ is the sum of kinetic and potential energy, this is equivalent to the conservation of the total energy.

162

Property (P2) – symplecticity – can be conveniently expressed in terms of the flow $\varphi_t(y_0)$ of the differential equation, which is the solution at time t for the initial value $y(0) = y_0$. Symplecticity then means that

$$\varphi_t'(y)^T J \varphi_t'(y) = J, \qquad (4.2)$$

where prime indicates the derivative with respect to y. It is interesting to mention that this property is characteristic for Hamiltonian systems.

4.1.2 B-series integrators

For a general differential equation $\dot{y} = f(y)$, the Taylor series of the exact solution with initial value $y(0) = y$ can be written as

$$\begin{aligned}
y(h) &= y + h f(y) + \frac{h^2}{2!} f'(y) f(y) \\
&\quad + \frac{h^3}{3!} \big(f''(y)\big(f(y), f(y)\big) + f'(y) f'(y) f(y) \big) + \dots
\end{aligned} \qquad (4.3)$$

We consider numerical integrators $y_{n+1} = \Phi_h(y_n)$, whose Taylor series have the same structure as (4.3) with additional real coefficients:

$$\begin{aligned}
\Phi_h(y) &= y + h a(\,\bullet\,) f(y) + h^2 a(\,\textbf{\textit{↑}}\,) f'(y) f(y) \\
&\quad + h^3 \Big(\frac{a(\textbf{V})}{2} f''(y)\big(f(y), f(y)\big) + a(\,\textbf{\textit{↑}}\,) f'(y) f'(y) f(y) \Big) + \dots
\end{aligned} \qquad (4.4)$$

The coefficients $a(\tau)$, which are in a one-to-one correspondence with rooted trees, characterise the integrator. Properties like energy conservation and symplecticity can be expressed in terms of these coefficients. Series expansions of the form (4.4), called B-series, have their origin in the paper of Butcher (1972) and were introduced by Hairer & Wanner (1974).

Such B-series integrators are comprised of Runge–Kutta methods (RK), Taylor series methods, the underlying one-step method (in the sense of Kirchgraber (1986), see also chapter XIV of HLW02[1]) of linear multistep methods (lmm), and all their extensions such as general linear methods (glm) and multistep-multistage-multiderivative methods (mmm); see the left cube of Figure 4.1.

For the numerical treatment of Hamiltonian systems, partitioned methods that allow one to treat the p and q components in a different manner are even more important. The basic method is the symplectic Euler discretisation, which combines the explicit and implicit Euler methods. Taylor series methods are replaced by the generating function methods of Feng Kang,

---

[1] The monograph "Geometric Numerical Integration" of Hairer, Lubich & Wanner (2002) will be cited frequently. Reference to it will be abbreviated by HLW02.

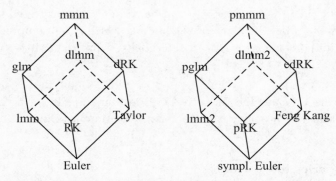

Fig. 4.1. The cube of B-series methods (left) and its extension to partitioned methods (right).

multistep methods by their variants for second-order differential equations (lmm2), etc.; see the right cube of Figure 4.1.

4.1.3 Exact energy conservation

For a numerical solution obtained by $y_{n+1} = \Phi_h(y_n)$ we would like to have energy conservation (i.e., $H(y_n) = const$ for all n) and symplecticity (i.e., $\Phi_h'(y)^T J \Phi_h'(y) = J$) at the same time. Unfortunately, this is not possible. Ge & Marsden (1988) proved that for Hamiltonian systems without further conserved quantities such a method has to be a re-parametrisation of the exact flow. An algebraic proof for the impossibility of having an energy conserving symplectic B-series integrator (which is different from the exact flow) is given by Chartier, Faou & Murua (2005).

Let us study what happens when we force energy conservation and thus give up symplecticity. We consider the three-body problem (Sun–Jupiter–Saturn) which is a Hamiltonian system with

$$H(p, q) = \frac{1}{2} \sum_{i=0}^{2} \frac{1}{m_i} p_i^T p_i - G \sum_{i=1}^{2} \sum_{j=0}^{i-1} \frac{m_i m_j}{\|q_i - q_j\|}.$$

The initial values $q_i(0), p_i(0) \in \mathbf{R}^3$ and the parameters G and m_i are taken from HLW02, page 11. To this problem we apply two kinds of integrators that exactly conserve the Hamiltonian along the numerical solution. Notice, however, that neither of these methods is symmetric so that the considerations of Section 4.4 do not apply.

Projection method. The most obvious approach for achieving exact energy conservation is by projection. Assume that y_n is an approximation

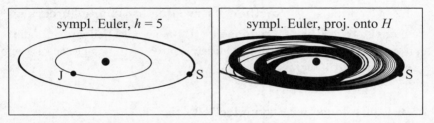

Fig. 4.2. Numerical solution of the Sun–Jupiter–Saturn system.

to $y(t_n)$ satisfying $H(y_n) = H(y_0)$. We compute $\widehat{y}_{n+1} = \Phi_h(y_n)$ with some basic method, and then project $\widehat{y}_{n+1}$ orthogonally onto the energy surface $\{y \mid H(y) = H(y_0)\}$ yielding y_{n+1}. If we take as basic method the symplectic Euler method (see method (4.8) below), an integration with step size $h = 5$ over an interval of $1.3 \cdot 10^6$ Earth days gives the result of Figure 4.2. The left picture shows the numerical solution obtained without any projection. Although the energy is not exactly conserved, we observe a qualitatively correct behaviour (for an explanation see Section 4.2). The picture on the right in Figure 4.2 shows the result obtained by the same method, but with an additional projection onto the energy surface after every step. Clearly, this does not improve the result; in fact, it even destroys the good long-time behaviour.

Energy conserving B-series method. In the previous experiment one can criticise the fact that the projection step throws out the method from the class of B-series integrators. This is correct. Motivated by the results of Faou, Hairer & Pham (2005) we therefore consider the method $y_{n+1} = \Phi_h(y_n)$, where $\Phi_h(y)$ is the exact flow at time $t = h$ of

$$\dot{y} = f(y) + h^3\big(f'f''(f,f) + f''(f, f'f)\big)(y). \tag{4.5}$$

It is not difficult to check that this is a B-series method, and that for $f(y) = J^{-1}\nabla H(y)$ the energy $H(y)$ is a first integral of (4.5), so that $H(y_n) = const$. Since the perturbation in (4.5) is not Hamiltonian, the method is not symplectic. We do not claim that this method can be realised by a Runge–Kutta or multistep method. Application to the Sun–Jupiter–Saturn system gives a result that is very similar to that of the projection method (right picture of Figure 4.2).

From these experiments we conclude that exact energy conservation alone is certainly not sufficient for a qualitatively correct long-time integration of Hamiltonian systems.

4.2 Methods that exactly conserve quadratic first integrals

We now turn our attention to property (P2) – symplecticity – of Hamiltonian systems. This turns out to be closely related to the exact conservation of quadratic first integrals.

4.2.1 Equivalence with symplecticity

Consider the Hamiltonian system together with its variational equation,

$$\dot{y} = J^{-1}\nabla H(y), \qquad \dot{\Psi} = J^{-1}\nabla^2 H(y)\Psi, \tag{4.6}$$

where $\Psi(t)$ is the derivative of $y(t)$ with respect to its initial value. The symplecticity condition (4.2) just expresses the fact that $\Psi^T J \Psi$ is a first integral of the system (4.6); i.e., it is constant along solutions of (4.6).

Theorem 4.1 (criterion for symplecticity) *A B-series integrator is symplectic (i.e., it satisfies $\Phi_h'(y)^T J \Phi_h'(y) = J$) if and only if it exactly conserves all quadratic first integrals of a system $\dot{y} = f(y)$.*

Conservation of quadratic first integrals implies symplecticity. Bochev & Scovel (1994) have shown for Runge–Kutta and general linear methods that the derivative of the numerical solution with respect to the initial value, $\Psi_n := \partial y_n / \partial y_0$, is the result of the same numerical method applied to the augmented system (4.6). This implies the statement, because $\Psi^T J \Psi$ is a quadratic first integral of the system. The extension to B-series methods is straight-forward.

Symplecticity implies conservation of quadratic first integrals. Calvo & Sanz-Serna (1994) have given a characterisation of the symplecticity of B-series methods in terms of the coefficients $a(\tau)$ of (4.4); see also HLW02, page 201. Chartier, Faou & Murua (2005) show that exactly the same conditions on the coefficients imply that the method conserves exactly all quadratic first integrals of the differential equation.

Implicit midpoint rule. Let us illustrate the above characterisation of symplectic B-series methods for the implicit midpoint rule

$$y_{n+1} - y_n = hf\big((y_{n+1} + y_n)/2\big). \tag{4.7}$$

For this we assume that $Q(y) = y^T C y$ (with a symmetric matrix C) is a first integral of $\dot{y} = f(y)$, i.e., $y^T C f(y) = 0$ for all y. Left-multiplication of (4.7) with $(y_{n+1} + y_n)^T C$ yields a vanishing right-hand side and

$$0 = (y_{n+1} + y_n)^T C(y_{n+1} - y_n) = y_{n+1}^T C y_{n+1} - y_n^T C y_n.$$

This proves that $Q(y_n)$ is exactly conserved, implying that the implicit midpoint rule is a symplectic integrator.

4.2.2 Partitioned methods

As mentioned in the introduction, partitioned methods play an important role for solving Hamiltonian systems. They allow one to treat the variables p and q in the Hamiltonian system (4.1) in a different way. For partitioned methods the above characterisation remains valid only if one restricts the statements to quadratic first integrals of the special form $Q(p,q) = p^T E q$ with an arbitrary matrix E.

Symplectic Euler method. Consider the combination of the implicit and explicit Euler methods. This yields the discretisation

$$p_{n+1} = p_n - h\nabla_q H(p_{n+1}, q_n)$$
$$q_{n+1} = q_n + h\nabla_p H(p_{n+1}, q_n). \tag{4.8}$$

If $p^T E q$ is a first integral of (4.1), a multiplication of the first relation of (4.8) by $(Eq_n)^T$, of the second relation by $p_{n+1}^T E$, and addition of the two proves the exact conservation of this first integral. Consequently, the method is symplectic.

Störmer–Verlet scheme. Composing a half-step of method (4.8) with its adjoint (explicit in p and implicit in q) gives

$$p_{n+1/2} = p_n - \frac{h}{2}\nabla_q H(p_{n+1/2}, q_n)$$
$$q_{n+1} = q_n + \frac{h}{2}\Big(\nabla_p H(p_{n+1/2}, q_n) + \nabla_p H(p_{n+1/2}, q_{n+1})\Big) \tag{4.9}$$
$$p_{n+1} = p_{n+1/2} - \frac{h}{2}\nabla_q H(p_{n+1/2}, q_{n+1}).$$

As the composition of symplectic mappings it is a symplectic integrator.

For a separable Hamiltonian $H(p,q) = p^T p/2 + U(q)$, this scheme implies

$$q_{n+1} - 2q_n + q_{n-1} = -h^2 \nabla_q U(q_n), \tag{4.10}$$

which is a natural discretisation of $\ddot{q} = -\nabla_q U(q)$ that can already be found in the Principia of Newton; c.f., Hairer, Lubich & Wanner (2003).

4.2.3 Near energy conservation

The aim of this article is to study energy conservation of numerical integrators. In general, symplectic B-series methods cannot conserve the Hamiltonian exactly (see Section 4.1.3). However, we have the following central result, which was intuitively clear since the use of symplectic methods and was rigorously proved by Benettin & Giorgilli (1994); see also Section IX.8 of HLW02.

Theorem 4.2 *Consider*

- *a Hamiltonian system with analytic* $H : U \to \mathbf{R}$, *and*
- *a symplectic B-series method* $\Phi_h(y)$ *of order r.*

As long as $\{y_n\}$ *stays in a compact set, we have for* $t_n = nh$ *and* $h \to 0$,

$$H(y_n) = H(y_0) + \mathcal{O}(h^r) + \mathcal{O}(t_n e^{-\gamma/\omega h}), \tag{4.11}$$

where $\gamma > 0$ *only depends on the method, and* ω *is related to the Lipschitz-constant (or highest frequency) of the differential equation.*

If h is small enough, the second error term in (4.11) is exponentially small on exponentially long time intervals. Thus we have error conservation up to a bounded $\mathcal{O}(h^r)$ term on such long intervals.

Let us illustrate this behaviour with the symplectic Euler method (4.8) applied to the Sun–Jupiter–Saturn system. Figure 4.3 shows the relative error in the Hamiltonian on an interval of 500 000 Earth days. The energy oscillates around the correct constant value and does not show any drift. The non-symplectic explicit Euler method, however, has a linear drift in the energy that makes the method useless even when applied with a much smaller step size.

Idea of the proof (backward error analysis). Let us indicate the proof of the previous theorem. This is a welcome opportunity for mentioning backward error analysis which is one of the most important tools in the analysis of geometric integrators.

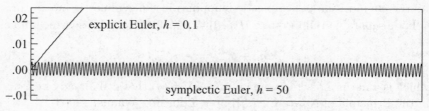

Fig. 4.3. Energy conservation of numerical methods.

Formal analysis. The numerical solution of a B-series method can be interpreted as the exact solution of a modified differential equation which, very similar to (4.4), is of the form

$$\dot{y} = f(y) + hb(\,\mathbf{\cdot}\!\!-\!\!\mathbf{\cdot}\,)f'(y)f(y) \tag{4.12}$$

$$+ h^2\left(\frac{b(\mathbf{V})}{2}f''(y)\big(f(y), f(y)\big) + b(\,\mathbf{\cdot}\,)f'(y)f'(y)f(y)\right) + \dots$$

The coefficients $b(\tau)$ of this modified equation are obtained recursively by comparing the series (4.4) with the Taylor series expansion of the solution of (4.12) at $t = h$. Consequently, we have $y_n = \widetilde{\varphi}_{nh}(y_0)$, where $\widetilde{\varphi}_t(y)$ is the exact flow of (4.12).

It turns out that for symplectic B-series integrators (4.4) and for $f(y) = J^{-1}\nabla H(y)$, (4.12) is Hamiltonian with modified Hamiltonian

$$\widetilde{H}(y) = H(y) + h^r H_{r+1}(y) + h^{r+1} H_{r+2}(y) + \dots \tag{4.13}$$

Since the exact flow $\widetilde{\varphi}_t(y)$ conserves the Hamiltonian $\widetilde{H}(y)$, it follows that $\widetilde{H}(y_n) = const$, and thus $H(y_n) = H(y_0) + \mathcal{O}(h^r)$. Unfortunately, the above series are asymptotic series and usually diverge. This is why we call this part of the proof a *formal analysis*.

Rigorous analysis. Whereas the formal analysis is relatively simple and gives already much insight into long-time integration, the rigorous analysis is rather technical. One has to truncate the series so that the resulting error in $y_n - \widetilde{\varphi}_{nh}(y_0)$ is as small as possible. This induces the linearly increasing exponentially small error term in the statement of Theorem 4.2.

Illustration with the Lennard–Jones potential.

To illustrate the result of the previous theorem, consider the Hamiltonian

$$H(p, q) = \frac{1}{2}p^2 + q^{-12} - q^{-6},$$

which models the motion of a particle against a strongly repelling wall. The variable q represents the distance to the wall, and p the velocity of the particle. With initial values $q_0 = 10$ and $p_0 < 0$ such that $H(p_0, q_0) = 1$, the particle moves against the wall until $t \approx 6.5$, and then it bounces off. The solid lines in Figure 4.4 show the error in the Hamiltonian $H(p, q)$ for the Störmer–Verlet method applied with 3 different step sizes. In the beginning this error is very small, then it increases and is maximal when the particle approaches the wall. At $t \approx 6.5$ the value of q is very small, so that the Lipschitz constant of the system (i.e., ω in (4.11)) is large. For relatively large step sizes h (upper pictures of Figure 4.4) when $\omega h \approx 1$, the exponential term in (4.11) becomes dominant, and energy

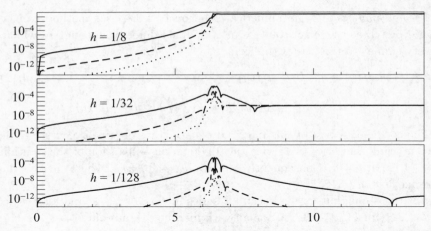

Fig. 4.4. Error in the truncated modified Hamiltonian for the Lennard–Jones potential and the Störmer–Verlet method.

conservation breaks down. In such situations, the value of the energy can change dramatically. We thus see that in spite of the use of a symplectic integrator, the energy drifts off the correct value if the step size is too large.

For this simple problem, we have computed the perturbations $H_3(y)$ and $H_5(y)$ in the modified Hamiltonian (the functions $H_{2k}(y)$ vanish identically, because the Störmer–Verlet method is symmetric). In Figure 4.4 we have included the functions $H(y_n) + h^2 H_3(y_n) - const$ (dashed) and $H(y_n) + h^2 H_3(y_n) + h^4 H_5(y_n) - const$ (dotted), where the constants are chosen so that the expressions vanish at the initial value. We see that $H(y_n) + h^2 H_3(y_n) + h^4 H_5(y_n) = const$ up to round-off on a large part of the interval considered. This nicely illustrates that the modified Hamiltonian is much better conserved than the energy $H(y)$.

4.3 Methods that nearly conserve quadratic first integrals

We next study the question whether a larger class of numerical integrators can have the same good energy conservation as symplectic methods. Let us begin with an instructive example.

4.3.1 Trapezoidal rule

Consider the trapezoidal rule

$$y_{n+1} = y_n + \frac{h}{2}\big(f(y_n) + f(y_{n+1})\big), \tag{4.14}$$

and apply it to the Sun–Jupiter–Saturn system as in the previous experiments. Figure 4.5 shows the error in the Hamiltonian (upper picture) and in the angular momentum (lower picture), which is a quadratic first integral of the system. We have used the same initial data and the same integration interval as in Figure 4.3, and constant step size $h = 50$.

We first notice that the trapezoidal rule cannot be symplectic. Otherwise, by Theorem 4.1, the angular momentum would be exactly conserved along the numerical solution. Nevertheless, we observe an excellent conservation of the total energy, very similar to that for the symplectic Euler method of Figure 4.3. We shall give two explanations of this good long-time behaviour.

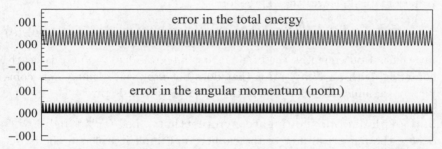

Fig. 4.5. Error in the total energy and in the angular momentum for the trapezoidal rule.

1st explanation: near-conservation of quadratic first integrals.
Let $u(t)$ be a solution of the modified differential equation in the sense of backward error analysis, cf. (4.12). For the moment we only need that $u(t)$ is smooth and $u(t_n) = y_n$. This function thus satisfies

$$u(t + h) = u(t) + \frac{h}{2}\Big(f\big(u(t + h)\big) + f\big(u(t)\big)\Big).$$

Writing the Taylor series expansion as $u(t+h) = e^{hD}u(t)$, where D denotes differentiation with respect to time t, this relation becomes

$$\Big(e^{hD} - 1\Big)u(t) = \frac{h}{2}\Big(e^{hD} + 1\Big)f\big(u(t)\big) \tag{4.15}$$

or, equivalently,

$$\Big(1 - \frac{h^2}{12}D^2 + \frac{h^4}{120}D^4 + \ldots\Big)\dot{u}(t) = f\big(u(t)\big). \tag{4.16}$$

Assume now that $Q(y) = y^T C y$ is a first integral of the differential equation (like the angular momentum in the Sun–Jupiter–Saturn system). This is equivalent to the condition $y^T C f(y) = 0$ for all y. Multiplying (4.16) from the left by $u(t)^T C$ gives a vanishing right-hand side and (omitting the obvious argument t)

$$u^T C \left(\dot{u} - \frac{h^2}{12} u^{(3)} + \frac{h^4}{120} u^{(5)} + \dots \right) = u^T C f(u) = 0. \qquad (4.17)$$

Miraculously, this expression can be written as a total differential and, after multiplication by 2, becomes

$$\frac{d}{dt} \left(u^T C u - \frac{h^2}{12} \left(2u^T C \ddot{u} - \dot{u}^T C \dot{u} \right) + \dots \right) = 0. \qquad (4.18)$$

This means that the function $u(t)$ and hence the numerical solution of the trapezoidal rule leaves the expression

$$\tilde{Q}(y) = y^T C y - \frac{h^2}{12} \left(2y^T C f'(y) f(y) - f(y)^T C f(y) \right) + \dots \qquad (4.19)$$

invariant. Consequently, the original first integral $Q(y) = y^T C y$ is nearly conserved with an $\mathcal{O}(h^2)$ error that does not grow with time. This completely explains the behaviour of the lower picture of Figure 4.5.

2nd explanation: conjugate symplecticity. Let Φ_h^{Eulex} and Φ_h^{Eulim} denote the numerical flow of the explicit and implicit Euler methods, respectively. Those of the trapezoidal rule (4.14) and of the midpoint rule (4.7) are then given by

$$\Phi_h^{\text{trap}} = \Phi_{h/2}^{\text{Eulim}} \circ \Phi_{h/2}^{\text{Eulex}}, \qquad \Phi_h^{\text{midp}} = \Phi_{h/2}^{\text{Eulex}} \circ \Phi_{h/2}^{\text{Eulim}}.$$

This implies that the trapezoidal rule and the implicit midpoint rule are connected by the conjugacy relation $\Phi_{h/2}^{\text{Eulex}} \circ \Phi_h^{\text{trap}} = \Phi_h^{\text{midp}} \circ \Phi_{h/2}^{\text{Eulex}}$. With $\chi_h := (\Phi_h^{\text{midp}})^{-1/2} \circ \Phi_{h/2}^{\text{Eulex}}$, which is $\mathcal{O}(h^2)$ close to the identity,

$$\chi_h \circ \Phi_h^{\text{trap}} = \Phi_h^{\text{midp}} \circ \chi_h \quad \text{and} \quad \chi_h \circ \left(\Phi_h^{\text{trap}} \right)^n = \left(\Phi_h^{\text{midp}} \right)^n \circ \chi_h, \qquad (4.20)$$

so that a numerical solution $\{y_n\}_{n\geq 0}$ of the trapezoidal rule is connected via $\chi_h(y_n) = z_n$ to the numerical solution $\{z_n\}_{n\geq 0}$ of the midpoint rule obtained with starting value $z_0 = \chi_h(y_0)$. This explains why the non-symplectic trapezoidal rule has the same good long-time behaviour as the symplectic midpoint rule (upper picture of Figure 4.5).

4.3.2 Symmetric linear multistep methods

Symmetric linear multistep methods form an important class of numerical integrators that have properties similar to those of the trapezoidal rule. However, one has to take care of the stability of parasitic solutions.

Multistep methods for first-order differential equations. Since the numerical solution of a multistep method

$$\alpha_k y_{n+k} + \ldots + \alpha_0 y_n = h\big(\beta_k f(y_{n+k}) + \ldots + \beta_0 f(y_n)\big) \qquad (4.21)$$

depends on k starting approximations $y_0, \ldots, y_{k-1}$, it is not at all obvious how symplecticity or the conservation of first integrals should be interpreted. The key idea is to consider the so-called *underlying one-step method* $\Phi_h(y)$, which is formally a series like (4.4), whose coefficients $a(\bullet)$, $a(\int), \ldots$ are determined by

$$\alpha_k \Phi_h^k(y) + \ldots + \alpha_1 \Phi_h(y) + \alpha_0 y = h\big(\beta_k f(\Phi_h^k(y)) + \ldots + \beta_0 f(y)\big).$$

This means that for starting approximations given by $y_j = \Phi_h^j(y_0)$ for $j = 0, \ldots, k-1$, the numerical solutions of the multistep method and of its underlying one-step method are identical. We say that the linear multistep method (4.21) is symplectic (conserves energy, conserves quadratic first integrals, ...), if its underlying one-step method is symplectic (conserves energy, conserves quadratic first integrals, ...).

It has been shown by Tang (1993) that linear multistep methods cannot be symplectic. However, they have an interesting property that will be explained next. If we let $u(t)$ be the solution of the modified differential equation of $\Phi_h(y)$, we have

$$\rho(e^{hD})u(t) = h\sigma(e^{hD})f\big(u(t)\big),$$

where $\rho(\zeta)$ and $\sigma(\zeta)$ are the generating polynomials of the coefficients α_j and β_j, respectively. This equation reduces to (4.15) for the trapezoidal rule. If the method is symmetric ($\alpha_{k-j} = -\alpha_j$ and $\beta_{k-j} = \beta_j$), the analysis of Section 4.3.1 extends straight-forwardly to the present situation, and shows that for problems having $Q(y) = y^T C y$ as first integral there exists $\widetilde{Q}(y)$ of the form (4.19) which is exactly conserved by the method (see Hairer & Lubich (2004)). Moreover, Chartier, Faou & Murua (2005) have shown that a method with this property is conjugate to a symplectic integrator.

Attention! In spite of these nice properties of symmetric linear multistep methods (4.21), they are not recommended for the long-time integration of Hamiltonian systems. The difficulty is that for nonlinear problems no results on stable propagation of the parasitic solution components are known, and numerical experiments reveal that they are unstable in most cases.

Multistep methods for second-order differential equations. An important class of Hamiltonian systems have $H(p, q) = \frac{1}{2}p^T M^{-1}p + U(q)$ with constant mass matrix M (for convenience we assume in the following that M is the identity). Such problems are equivalent to the second-order differential equation

$$\ddot{q} = -\nabla U(q), \qquad (4.22)$$

and it is natural to consider multistep methods adapted to this form:

$$\alpha_k q_{n+k} + \ldots + \alpha_0 q_n = -h^2 \big(\beta_k \nabla U(q_{n+k}) + \ldots + \beta_0 \nabla U(q_n)\big). \quad (4.23)$$

Notice that the Störmer–Verlet method (4.10) is a special case of this formulation. The statements for methods (4.21) can all be extended to the situation of second-order differential equations (near conservation of quadratic first integrals, conjugate symplecticity of the underlying one-step method).

To also obtain bounds on the parasitic solution components for arbitrary starting approximations, we extend the idea of backward error analysis (Hairer 1999). We write the numerical solution as

$$q_n = v(nh) + \sum \zeta_j^n w_j(nh) \qquad (4.24)$$

where the ζ_j's stand for zeros (different from 1) of the characteristic polynomial $\rho(\zeta) = \sum_{j=0}^k \alpha_j \zeta^j$ and products thereof. Inserting (4.24) into the linear multistep formula (4.23), expanding into a Taylor series, and comparing the expressions multiplying ζ_j^n and powers of h, we get the modified equations – a second-order differential equation for $v(t)$, first-order differential equations for $w_j(t)$ if ζ_j is a simple zero of $\rho(\zeta)$, and algebraic relations, if ζ_j is a product of zeros of $\rho(\zeta)$.

If we are interested in energy conservation we have to complement (4.23) with an approximation of the derivative, which in general is given by a formula of the form

$$\dot{q}_n = \frac{1}{h} \sum_{j=-\ell}^{\ell} \delta_j q_{n+j}.$$

Exploiting the Hamiltonian structure in the modified differential equations, Hairer & Lubich (2004) prove the following result.

Theorem 4.3 *Consider the Hamiltonian system (4.22) with analytic potential function $U(q)$, and assume that the linear multistep method (4.23) has order r and is*

- *symmetric, i.e., $\alpha_{k-j} = \alpha_j$ and $\beta_{k-j} = \beta_j$;*

- *without weak instability, i.e., $\rho(\zeta) = (\zeta - 1)^2 \widehat{\rho}(\zeta)$ and the zeros of $\widehat{\rho}(\zeta)$ lie on the unit circle, are simple, and different from 1.*

If the starting approximations $q_0, \ldots, q_{k-1}$ are $\mathcal{O}(h^{r+1})$ close to the exact solution, and the numerical solution stays in a compact set, then we have on intervals of length $T = \mathcal{O}(h^{-r-2})$

- $\|w_j(t)\| \leq C h^{r+1}$ *with C independent of t;*
- $\{q_n, \dot{q}_n\}$ *nearly conserves the total energy (without drift);*
- $\{q_n, \dot{q}_n\}$ *nearly conserves quadratic first integrals of the form $q^T E \dot{q}$ (without drift).*

We do not have energy conservation on exponentially long time intervals (as for symplectic integrators), but the intervals are sufficiently long for practical computations. This result justifies the use of high-order symmetric linear multistep methods (4.23) for long-time integrations in celestial mechanics.

4.4 Energy conservation with symmetric methods

There is still another class of numerical integrators – symmetric methods – which, for special kinds of Hamiltonian systems, give good energy conservation. Since in certain situations (such as variable step size integration, multiple time stepping, reversible averaging) it is much easier to design symmetric discretisations than symplectic ones, it is of interest to characterise the problems for which symmetric methods have a good long-time behaviour.

4.4.1 Symmetric non-symplectic methods

Let us start with a numerical experiment. We consider the 3-stage Lobatto IIIB method which is an implicit Runge–Kutta method (its coefficients can be found in Section II.1.4 of HLW02). It is neither symplectic nor conjugate to a symplectic method (see Section VI.7.4 of HLW02), but it is a symmetric integrator. This means that its numerical flow satisfies $\Phi_{-h}^{-1}(y) = \Phi_h(y)$. We apply this integrator with step size $h = 400$ in the usual way to the three-body Sun–Jupiter–Saturn system. The result can be seen in Figure 4.6. There is apparently no difference compared with the results obtained by the trapezoidal rule. What is the reason?

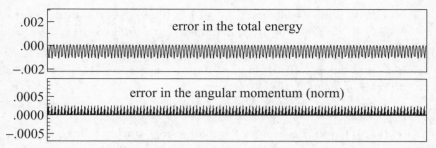

Fig. 4.6. Error in the total energy and in the angular momentum for the 3-stage Lobatto IIIB method.

4.4.2 Integrable reversible systems

Unfortunately, very little is known about the energy conservation of symmetric B-series methods. The only exceptions are integrable reversible systems. In this survey article we present some results without giving all the technical assumptions, and we refer the interested reader to Chapter XI of HLW02.

We assume that the differential equation (not necessarily Hamiltonian) can be written in the form

$$\dot{u} = f(u, v), \qquad \dot{v} = g(u, v), \tag{4.25}$$

and that it is *reversible* with respect to the involution $(u, v) \mapsto (u, -v)$. This means that

$$f(u, -v) = -f(u, v), \qquad g(u, -v) = g(u, v). \tag{4.26}$$

Hamiltonian systems, for which the Hamiltonian is quadratic in p, satisfy these relations with q in the role of u, and p in the role of v.

Such a system is called an *integrable reversible system* if there exists a reversibility-preserving change of coordinates $(a, \theta) \mapsto (u, v)$ such that in the new coordinates the system is of the form

$$\dot{a} = 0, \qquad \dot{\theta} = \omega(a). \tag{4.27}$$

This system can be solved exactly. The *action* variables a are constant (i.e., first integrals), and the *angle* variables θ grow linearly with time. The Kepler problem with $H(p_1, p_2, q_1, q_2) = \frac{1}{2}(p_1^2 + p_2^2) - (q_1^2 + q_2^2)^{-1/2}$ satisfies all these conditions if we put $u = (q_1, p_2)$ and $v = (-p_1, q_2)$. The Sun–Jupiter–Saturn system is a small perturbation of an integrable reversible system.

Under certain technical assumptions (analyticity of the vector field, strong non-resonance condition, etc.) it is proved in HLW02 that all action variables are nearly conserved over long times for symmetric B-series methods. Moreover, the global error grows at most linearly with time. Since these results hold also for small reversible perturbations of integrable reversible systems, the behaviour shown in Figure 4.6 is explained.

4.4.3 An example: the perturbed pendulum

Let us illustrate with a simple example the difficulties that can be encountered by a symmetric method. Consider the one-degree-of-freedom Hamiltonian system with (see Figure 4.7)

$$H(p,q) = \frac{1}{2}\,p^2 - \cos q + 0.2\sin(2q). \qquad (4.28)$$

With $u = q$ and $v = p$ it is of the form (4.25) and satisfies the condition (4.26). Considered as a Hamiltonian system, it is also integrable.

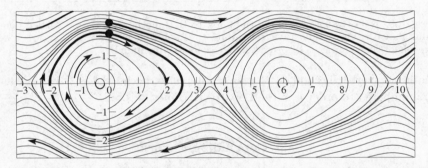

Fig. 4.7. Phase portrait of the perturbed pendulum problem.

We consider two different initial values (thick points in Figure 4.7). The values $q_0 = 0$ and $p_0 = 1.8$ produce a periodic solution whose orbit is invariant with respect to the reflection $p \mapsto -p$. For $q_0 = 0$ and $p_0 = 2.2$ the solution is still periodic (on the cylinder), but it does not contain any symmetry.

As in the previous experiment, we apply the 3-stage Lobatto IIIB method. We use the step size $h = 0.2$ and consider an interval of length 200. For the initial values with symmetric solution (Figure 4.8) the energy is well conserved without any drift. For the second set of initial values, however, there is a clear drift in the energy along the numerical solution (Figure 4.9).

Symplectic methods and methods that are conjugate to a symplectic method will have bounded energy error for this problem. We have included in Figure 4.9 the numerical result obtained with the symplectic

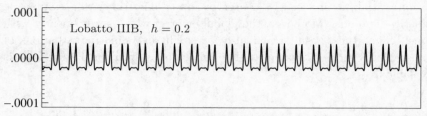

Fig. 4.8. Energy conservation of the Lobatto IIIB method for the perturbed pendulum with initial values corresponding to a symmetric solution.

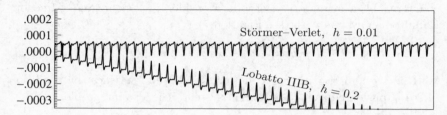

Fig. 4.9. Energy conservation of the Lobatto IIIB method for the perturbed pendulum with initial values corresponding to a solution without symmetry. The result of the symplectic Störmer–Verlet scheme is included.

Störmer–Verlet method (with smaller step size, because it is only of order 2, compared to order 4 of the Lobatto IIIB method).

4.5 Concluding remarks

In many applications, and in particular in long-time integrations of mechanical systems, it is important that the energy along the numerical solution does not drift from the correct value. Within the class of B-series methods we have studied the following properties:

- *symplecticity (Section 4.2):* the energy is nearly conserved for all Hamiltonian systems (integrable or chaotic) provided the step size is sufficiently small;
- *conjugate symplecticity (Section 4.3):* methods with this property have the same long-time behaviour as symplectic methods and are well suited for the integration of Hamiltonian systems;
- *symmetry (Section 4.4):* for reversible Hamiltonian systems and a solution with a certain symmetry, symmetric methods usually give excellent results; a complete explanation is missing in many situations.

Figure 4.10 shows the connections between these properties. Symplecticity and exact energy conservation are not compatible. However, it is

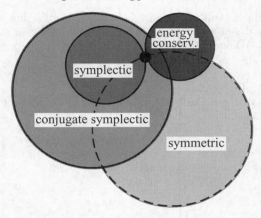

Fig. 4.10. Survey of geometric integrators for Hamiltonian systems.

possible to have symmetric methods that exactly conserve energy. Examples are energy-momentum methods (see Section V.5 of HLW02) which, however, do not fall into the class of B-series methods.

References

G. Benettin and A. Giorgilli (1994), On the Hamiltonian interpolation of near to the identity symplectic mappings with application to symplectic integration algorithms, *J. Statist. Phys.* **74**, 1117–1143.

P. B. Bochev and C. Scovel (1994), On quadratic invariants and symplectic structure, *BIT* **34**, 337–345.

J. C. Butcher (1972), An algebraic theory of integration methods, *Math. Comput.* **26**, 79–106.

M. P. Calvo and J. M. Sanz-Serna (1994), Canonical B-series, *Numer. Math.* **67**, 161–175.

P. Chartier, E. Faou, and A. Murua (2005), An algebraic approach to invariant preserving integrators: the case of quadratic and Hamiltonian invariants, *Report, February 28*.

E. Faou, E. Hairer, and T.-L. Pham (2004), Energy conservation with non-symplectic methods: examples and counter-examples, *BIT* **44**, 699–709.

Z. Ge and J. E. Marsden (1988), Lie-Poisson Hamilton-Jacobi theory and Lie-Poisson integrators, *Phys. Lett. A* **133**, 134–139.

E. Hairer (1999), Backward error analysis for multistep methods, *Numer. Math.* **84**, 199–232.

E. Hairer and C. Lubich (2004), Symmetric multistep methods over long times, *Numer. Math.* **97**, 699–723.

E. Hairer, C. Lubich, and G. Wanner (2002), *Geometric Numerical Integration. Structure-Preserving Algorithms for Ordinary Differential Equations*, Springer Series in Computational Mathematics 31, Springer, Berlin.

E. Hairer, C. Lubich, and G. Wanner (2003), Geometric numerical integration illustrated by the Störmer–Verlet method, *Acta Numerica* **12**, 399–450.

E. Hairer and G. Wanner (1974), On the Butcher group and general multi-value methods, *Computing* **13**, 1–15.

U. Kirchgraber (1986), Multi-step methods are essentially one-step methods, *Numer. Math.* **48**, 85–90.

Y.-F. Tang (1993), The symplecticity of multi-step methods, *Computers Math. Applic.* **25**, 83–90.

5

Dispersive Properties of Numerical Schemes for Nonlinear Schrödinger Equations

Liviu I. Ignat

Departamento de Matemáticas
Universidad Autónoma de Madrid
Madrid, Spain
e-mail: liviu.ignat@uam.es

Enrique Zuazua

Departamento de Matemáticas
Universidad Autónoma de Madrid
Madrid, Spain
e-mail: enrique.zuazua@uam.es

Abstract

In this article we report on recent work on building numerical approximation schemes for nonlinear Schrödinger equations. We first consider finite-difference space semi-discretizations and show that the standard conservative scheme does not reproduce at the discrete level the dispersion properties of the continuous Schrödinger equation. This is due to high frequency numerical spurious solutions. In order to damp out or filter these high-frequencies and to reflect the properties of the continuous problem we propose two remedies. First, adding a suitable extra numerical viscosity term at a convenient scale, and, second, a two-grid filter of the initial datum with meshes of ratio $1/4$. We prove that these alternate schemes preserve the dispersion properties of the continuous model. We also present some applications to the numerical approximation of nonlinear Schrödinger equations with initial data in L^2. Despite the fact that classical energy methods fail, using these dispersion properties, the numerical solutions of the semi-discrete nonlinear problems are proved to converge to the solution of the nonlinear Schrödinger equation. We also discuss some open problems and some possible directions of future research.

5.1 Introduction

Let us consider the $1 - d$ linear Schrödinger Equation (LSE) on the whole line

$$\begin{cases} iu_t + u_{xx} = 0, \ x \in \mathbf{R}, \ t \neq 0, \\ u(0, x) = \varphi(x), \ x \in \mathbf{R}. \end{cases} \tag{5.1}$$

The solution is given by $u(t) = S(t)\varphi$, where $S(t) = e^{it\Delta}$ is the free Schrödinger operator which defines a unitary transformation group in $L^2(\mathbf{R})$. The linear semigroup has two important properties, the conservation of the L^2-norm

$$\|u(t)\|_{L^2(\mathbf{R})} = \|\varphi\|_{L^2(\mathbf{R})} \tag{5.2}$$

and a dispersive estimate :

$$|u(t, x)| \leq \frac{1}{\sqrt{4\pi|t|}} \|\varphi\|_{L^1(\mathbf{R})}. \tag{5.3}$$

By classical arguments in the theory of dispersive equations the above estimates imply more general space-time estimates for the linear semigroup which allow proving the well-posedness of a wide class of nonlinear Schrödinger equations (cf. Strichartz (1977), Tsutsumi (1987), Cazenave (2003)).

In this paper we present some recent results on the qualitative properties of some numerical approximation schemes for the linear Schrödinger equation and its consequences in the context of nonlinear problems.

More precisely, we analyze whether these numerical approximation schemes have the same dispersive properties, uniformly with respect to the mesh-size h, as in the case of the continuous Schrödinger equation (5.1). In particular we analyze whether the decay rate (5.3) holds for the solutions of the numerical scheme, uniformly in h. The study of these dispersion properties of the numerical scheme in the linear framework is relevant also for proving their convergence in the nonlinear context. Indeed, since the proof of the well-posedness of the nonlinear Schrödinger equations in the continuous framework requires a delicate use of the dispersion properties, the proof of the convergence of the numerical scheme in the nonlinear context is hopeless if these dispersion properties are not verified at the numerical level.

To better illustrate the problems we shall address, let us first consider the conservative semi-discrete numerical scheme

$$\begin{cases} i\dfrac{du^h}{dt} + \Delta_h u^h = 0, \ t \neq 0, \\ u^h(0) = \varphi^h. \end{cases} \tag{5.4}$$

Here u^h stands for the infinite unknown vector $\{u_j^h\}_{j \in \mathbf{Z}}$, $u_j^h(t)$ being the approximation of the solution at the node $x_j = jh$, and Δ_h the classical second-order finite difference approximation of ∂_x^2:

$$(\Delta_h u^h)_j = \frac{u_{j+1}^h - 2u_j^h + u_{j-1}^h}{h^2}.$$

This scheme satisfies the classical properties of consistency and stability which imply L^2-convergence. In fact stability holds because of the conservation of the discrete L^2-norm under the flow (5.4):

$$\frac{d}{dt}\left(h \sum_{j \in \mathbf{Z}} |u_j^h(t)|^2 \right) = 0. \tag{5.5}$$

The same convergence results hold for semilinear equations (NSE):

$$iu_t + u_{xx} = f(u) \tag{5.6}$$

provided that the nonlinearity f is globally Lipschitz continuous. But, it is by now well known (cf. Tsutsumi (1987), Cazenave (2003)) that the NSE is also well-posed for some nonlinearities that superlinearly grow at infinity. This well-posedness result cannot be proved simply as a consequence of the L^2 conservation property and the dispersive properties of the LSE play a key role.

Accordingly, one may not expect to prove convergence of the numerical scheme in this class of nonlinearities without similar dispersive estimates that should be uniform in the mesh-size parameter $h \to 0$. In particular, a discrete version of (5.3) is required to hold, uniformly in h. This difficulty may be avoided considering more smooth initial data φ, say, in $H^1(\mathbf{R})$, a space in which the Schrödinger equation generates a group of isometries and the nonlinearity is locally Lipschitz. But here, in order to compare the dynamics of the continuous and semi-discrete systems we focus on the $L^2(\mathbf{R})$-case, which is a natural class in which to solve the nonlinear Schrödinger equation.

In this article we first prove that the conservative scheme (5.4) fails to have uniform dispersive properties. We then introduce two numerical schemes for which the estimates are uniform. The first one uses an artificial numerical viscosity term and the second one involves a two-grid algorithm to precondition the initial data. Both approximation schemes of the linear semigroup converge and have uniform dispersion properties. This allows us to build two convergent numerical schemes for the NSE in the class of $L^2(\mathbf{R})$ initial data.

5.2 Notation and Preliminaries

In this section we introduce some notation that will be used in what follows: discrete l^p spaces, semidiscrete Fourier transform, discrete fractional differentiation, as well as the standard Strichartz estimates for the continuous equations.

The spaces $l^p(h\mathbf{Z})$, $1 \leq p < \infty$, consist of all complex-valued sequences $\{c_k\}_{k \in \mathbf{Z}}$ with

$$\|\{c_k\}\|_{l^p(h\mathbf{Z})} = \left(h \sum_{k \in \mathbf{Z}} |c_k|^p \right)^{1/p} < \infty.$$

In contrast to the continuous case, these spaces are nested:

$$l^1(h\mathbf{Z}) \subseteq l^2(h\mathbf{Z}) \subseteq l^\infty(h\mathbf{Z}).$$

The semidiscrete Fourier transform is a natural tool for the analysis of numerical methods for partial differential equations, where we are always concerned with functions defined on discrete grids. For any $u \in l^1(h\mathbf{Z})$, the semidiscrete Fourier transform of u at the scale h is the function $\hat{u}$ defined by

$$\hat{u}(\xi) = (\mathcal{F}_h v)(\xi) = h \sum_{j \in \mathbf{Z}} e^{-ijh\xi} u_j.$$

A priori, this sum defines a function $\hat{u}$ for all $\xi \in \mathbf{R}$. We remark that any wave number ξ is indistinguishable on the grid from all other wave numbers $\xi + 2\pi m/h$, where m is an integer, a phenomenon called aliasing. Thus, it is sufficient to consider the restriction of $\hat{u}$ to wave numbers in the range $[-\pi/h, \pi/h]$. Also u can be recovered from $\hat{u}$ by the inverse semidiscrete Fourier transform

$$v_j = (\mathcal{F}_h^{-1}\hat{v})_j = \int_{-\pi/h}^{\pi/h} e^{ijh\xi} \hat{u}(\xi) d\xi, \quad j \in \mathbf{Z}.$$

We will also make use of a discrete version of fractional differentiation. For $\varphi \in l^2(h\mathbf{Z})$ and $0 \leq s < 1$ we define

$$(D^s\varphi)_j = \int_{-\pi/h}^{\pi/h} |\xi|^s \hat{\varphi}(\xi) e^{ijh\xi} d\xi.$$

Now, we make precise the classical dispersive estimates for the linear continuous Schrödinger semigroup $S(t)$. The energy and decay estimates (5.2) and (5.3) lead, by interpolation (cf. Bergh & Löfström (1976)), to the following $L^{p'} - L^p$ decay estimate:

$$\|S(t)\varphi\|_{L^p(\mathbf{R})} \lesssim t^{-(\frac{1}{2} - \frac{1}{p})} \|\varphi\|_{L^{p'}(\mathbf{R})},$$

for all $p \geq 2$ and $t \neq 0$. More refined space-time estimates known as the *Strichartz inequalities* (cf. Strichartz (1977), Ginibre & Velo (1992), Kell & Tao (1998)) show that, in addition to the decay of the solution as $t \to \infty$, a gain of spatial integrability occurs for $t > 0$. Namely

$$\|S(\cdot)\varphi\|_{L^q(\mathbf{R},L^r(\mathbf{R}))} \leq C\|\varphi\|_{L^2(\mathbf{R})}$$

for suitable values of q and r, the so-called 1/2-admissible pairs. We recall that an α-admissible pair (q,r) satisfies

$$\frac{1}{q} = \alpha\left(\frac{1}{2} - \frac{1}{r}\right).$$

Also a local gain of 1/2 space derivative occurs in $L^2_{x,t}$ (cf. Constantin & Saut (1988), Kenig, Ponce & Vega (1991)):

$$\sup_{x_0,R} \frac{1}{R} \int_{B(x_0,R)} \int_{-\infty}^{\infty} |D_x^{1/2} e^{it\Delta} u_0|^2 dt dx \leq C\|u_0\|_{L^2(\mathbf{R})}^2.$$

5.3 Lack of Dispersion of the Conservative Semi-Discrete Scheme

Using the discrete Fourier transform, we remark that there are *slight* (see Fig. 5.1) but important differences between the symbols of the operators $-\Delta$ and $-\Delta_h$: $p(\xi) = \xi^2$, $\xi \in \mathbf{R}$ for $-\Delta$ and $p_h(\xi) = 4/h^2 \sin^2(\xi h/2)$, $\xi \in [\pi/h, \pi/h]$ for $-\Delta_h$. The symbol $p_h(\xi)$ changes convexity at the points $\xi = \pm\pi/2h$ and has critical points also at $\xi = \pm\pi/h$, two properties that the continuous symbol does not fulfil. As we will see, these pathologies affect the dispersive properties of the semi-discrete scheme.

Firstly we remark that $e^{it\Delta_h} = e^{it\Delta_1/h^2}$. Thus, by scaling, it is sufficient to consider the case $h = 1$ and the large time behavior of solutions for this mesh-size.

A useful tool to study the decay properties of solutions to dispersive equations is the classical Van der Corput lemma. Essentially it says that

$$\left| \int_a^b e^{it\psi(\xi)} d\xi \right| \lesssim t^{-1/k}$$

provided that ψ is real valued and smooth in (a,b) satisfying $|\partial^k \psi(x)| \geq 1$ for all $x \in (a,b)$. In the continuous case, i.e., with $\psi(\xi) = \xi^2$, using that the second derivative of the symbol is identically two ($\psi''(\xi) = 2$), one easily obtains (5.3). However, in the semi-discrete case the symbol of the semidiscrete approximation $p_1(\xi)$ satisfies

$$|\partial^2 p_1(\xi)| + |\partial^3 p_1(\xi)| \geq c$$

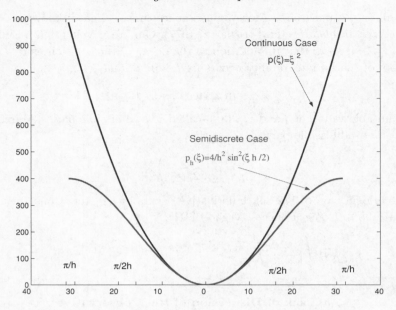

Fig. 5.1. The two symbols

for some positive constant c, a property that the second derivative does not satisfy. This implies that for any t

$$\|u^1(t)\|_{l^\infty(\mathbf{Z})} \lesssim \left(\frac{1}{t^{1/2}} + \frac{1}{t^{1/3}} \right) \|u^1(0)\|_{l^1(\mathbf{Z})}. \tag{5.7}$$

This estimates was obtained in Stefanov & Kevrekidis (2005) for the semi-discrete Schrödinger equation in the lattice $\mathbf{Z}$. But here, we are interested on the behavior of the system as the mesh-size h tends to zero.

The decay estimate (5.7) contains two terms. The first one $t^{-1/2}$, is of the order of that of the continuous Schrödinger equation. The second term $t^{-1/3}$ is due to the discretization scheme and, more precisely, to the behavior of the semi-discrete symbol at the frequencies $\pm \pi/2$.

A scaling argument implies that

$$\frac{\|u^h(t)\|_{l^\infty(h\mathbf{Z})}}{\|u^h(0)\|_{l^1(h\mathbf{Z})}} \lesssim \frac{1}{t^{1/2}} + \frac{1}{(th)^{1/3}},$$

an estimate which fails to be uniform with respect to the mesh size h.

As we have seen, the $l^\infty(\mathbf{Z})$ norm of the discrete solution $u^1(t)$ behaves as $t^{-1/3}$ as $t \to \infty$. This is illustrated in Fig. 5.2 by choosing the discrete Dirac delta δ_0 as initial datum such that $u(0)_j = \delta_{0j}$ where δ is the Kronecker symbol. More generally one can prove that there is no gain of

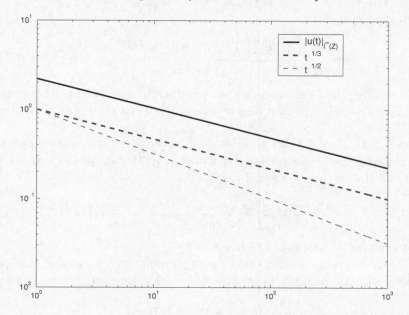

Fig. 5.2. Log-log plot of the time evolution of the l^∞ norm of u^1
with initial datum δ_0.

integrability, uniformly with respect to the mesh size h. The same occurs
in what concerns the gain of the local regularity. The last pathology is due
to the fact that, in contrast with the continuous case, the symbol $p_h(\xi)$ has
critical points also at $\pm\pi/h$. These negative results are summarized in the
following two theorems.

Theorem 5.1 *Let* $T > 0$, $q_0 \geq 1$ *and* $q > q_0$. *Then,*

$$\sup_{h>0,\varphi^h \in l^{q_0}(h\mathbf{Z})} \frac{\|S^h(T)\varphi^h\|_{l^q(h\mathbf{Z})}}{\|\varphi^h\|_{l^{q_0}(h\mathbf{Z})}} = \infty \qquad (5.8)$$

and

$$\sup_{h>0,\varphi^h \in l^{q_0}(h\mathbf{Z})} \frac{\|S^h(\cdot)\varphi^h\|_{L^1((0,T),l^q(h\mathbf{Z}))}}{\|\varphi^h\|_{l^{q_0}(h\mathbf{Z})}} = \infty. \qquad (5.9)$$

Theorem 5.2 *Let* $T > 0$, $q \in [1, 2]$ *and* $s > 0$. *Then,*

$$\sup_{h>0,\varphi^h \in l^q(h\mathbf{Z})} \frac{\left(h\sum_{j=0}^{1/h} |(D^s S^h(T)\varphi^h)_j|^2\right)^{1/2}}{\|\varphi^h\|_{l^q(h\mathbf{Z})}} = \infty \qquad (5.10)$$

and

$$\sup_{h>0,\varphi^h\in l^q(h\mathbf{Z})} \frac{\left(\int_0^T h\sum_{j=0}^{1/h}|(D^s S^h(t)\varphi^h)_j|^2 dt\right)^{1/2}}{\|\varphi^h\|_{l^q(h\mathbf{Z})}} = \infty. \tag{5.11}$$

According to these theorems the semi-discrete conservative scheme fails to have uniform dispersive properties with respect to the mesh-size h.

Proof of Theorem 5.1. As we mentioned before, this pathological behavior of the semi-discrete scheme is due to the contributions of the frequencies $\pm\pi/2h$. To see this we argue by scaling:

$$\frac{\|S^h(T)\varphi^h\|_{l^q(h\mathbf{Z})}}{\|\varphi^h\|_{l^{q_0}(h\mathbf{Z})}} = \frac{h^{\frac{1}{q}}}{h^{\frac{1}{q_0}}} \frac{\|S^1(T/h^2)\varphi^h\|_{l^q(\mathbf{Z})}}{\|\varphi^h\|_{l^{q_0}(\mathbf{Z})}}, \tag{5.12}$$

reducing the estimates to the case $h = 1$.

Using that $p_1(\xi)$ changes convexity at the point $\pi/2$, we choose as initial data a wave packet with its semidiscrete Fourier transform concentrated at $\pi/2$.

We introduce the operator $S_1 : \mathcal{S}(\mathbf{R}) \to \mathcal{S}(\mathbf{R})$ as

$$(S_1(t)\varphi)(x) = \int_{-\pi}^{\pi} e^{-4it\sin^2\frac{\xi}{2}} e^{ix\xi}\hat{\varphi}(\xi). \tag{5.13}$$

Using the results of Plancherel & Pólya (1937) and Magyar, Stein & Wainger (2002) concerning band-limited functions, i.e., with compactly supported Fourier transform, it is convenient to replace the discrete norms by continuous ones:

$$\sup_{\varphi\in l^{q_0}(\mathbf{Z})} \frac{\|S^1(t)\varphi\|_{l^q(\mathbf{Z})}}{\|\varphi\|_{l^{q_0}(\mathbf{Z})}} \gtrsim \sup_{\text{supp}\,\hat{\varphi}\subset[-\pi,\pi]} \frac{\|S_1(t)\varphi\|_{L^q(\mathbf{R})}}{\|\varphi\|_{L^{q_0}(\mathbf{R})}}. \tag{5.14}$$

According to this we may consider that x varies continuously in $\mathbf{R}$. To simplify the presentation we set $\psi(\xi) = -4t\sin^2\frac{\xi}{2} + x\xi$. For any interval $[a, b] \subset [-\pi, \pi]$, applying the Mean Value Theorem to $e^{it\psi(\xi)}$, we have

$$\left|\int_a^b e^{i\psi(\xi)}\hat{\varphi}(\xi)d\xi\right| \geq (1 - |b-a| \sup_{\xi\in[a,b]}|\psi'(\xi)|) \int_a^b \hat{\varphi}(\xi)d\xi$$

provided that $\hat{\varphi}$ is nonnegative. Observe that

$$\psi'(\xi) = -2t\sin\xi + x \sim -2t\left[1 + O((\xi - \frac{\pi}{2})^2)\right] + x$$

for $\xi \sim \pi/2$. Let ϵ be a small positive number that we shall fix below and $\hat{\varphi}_\epsilon$ supported on the set $\{\xi : \xi - \pi/2 = O(\epsilon)\}$. Then, $|\psi'(\xi)| = O(\epsilon^{-1})$ as long as $x - 2t = O(\epsilon^{-1})$ and $t = O(\epsilon^{-3})$. This implies that

$$\left| \int_{-\pi}^{\pi} e^{-4it \sin^2 \frac{\xi}{2} + ix\xi} \hat{\varphi}_\epsilon d\xi \right| \gtrsim \int_{\frac{\pi}{2}-\epsilon}^{\frac{\pi}{2}+\epsilon} \hat{\varphi}_\epsilon(\xi) d\xi.$$

Integrating over $x - 2t = O(\epsilon^{-1})$ we get, for all $t = O(\epsilon^{-3})$,

$$\|S_1(t)\varphi_\epsilon\|_{L^q(\mathbf{R})} \gtrsim \epsilon^{-\frac{1}{q}} \int_{\frac{\pi}{2}-\epsilon}^{\frac{\pi}{2}+\epsilon} \hat{\varphi}_\epsilon(\xi) d\xi. \tag{5.15}$$

Then,

$$\frac{\|S_1(t)\varphi_\epsilon\|_{L^q(\mathbf{R})}}{\|\varphi_\epsilon\|_{L^{q_0}(\mathbf{R})}} \gtrsim \epsilon^{-\frac{1}{q}} \frac{\int_{\frac{\pi}{2}-\epsilon}^{\frac{\pi}{2}+\epsilon} \hat{\varphi}_\epsilon(\xi) d\xi}{\|\varphi_\epsilon\|_{L^{q_0}(\mathbf{R})}}. \tag{5.16}$$

We now choose a function φ such that its Fourier transform $\hat{\varphi}$ has compact support and satisfies $\hat{\varphi}(0) > 0$. Then we choose φ_ϵ in the following manner

$$\hat{\varphi}_\epsilon(\xi) = \epsilon^{-1} \hat{\varphi}\left(\epsilon^{-1}\left(\xi - \frac{\pi}{2}\right)\right).$$

For such φ_ϵ, using the properties of the Fourier transform, we obtain that $\|\varphi_\epsilon\|_{L^{q_0}(\mathbf{R})}$ behaves as ϵ^{-1/q_0} and

$$\frac{\|S_1(t)\varphi_\epsilon\|_{L^q(\mathbf{R})}}{\|\varphi_\epsilon\|_{L^{q_0}(\mathbf{R})}} \gtrsim \epsilon^{-\frac{1}{q}+\frac{1}{q_0}}$$

as long as $t = O(\epsilon^{-3})$.

Finally we choose ϵ such that $\epsilon^{-3} \simeq h^{-2}$. Then, $T/h^2 \sim \epsilon^{-3}$ and the above results imply

$$h^{\frac{1}{q}-\frac{1}{q_0}} \sup_{\text{supp}\,\hat{\varphi}\subset[-\pi,\pi]} \frac{\|S_1(T/h^2)\varphi^h\|_{L^q(\mathbf{R})} dt}{\|\varphi^h\|_{L^{q_0}(\mathbf{R})}} \gtrsim h^{\frac{1}{q}-\frac{1}{q_0}} h^{\frac{2}{3}(-\frac{1}{q}+\frac{1}{q_0})}$$

$$\gtrsim h^{\frac{1}{3}(\frac{1}{q}-\frac{1}{q_0})}. \tag{5.17}$$

This, together with (5.12) and (5.14), finishes the proof. $\square$

Proof of Theorem 5.2. The proof uses the same ideas as in the case of Theorem 5.1 with the difference that we choose wave packets concentrated at π. $\square$

5.4 The Viscous Semi-discretization Scheme

As we have seen in the previous section a simple conservative approximation with finite differences does not reflect the dispersive properties of the LSE. In general, a numerical scheme introduces artificial numerical dispersion, which is an intrinsic property of the scheme and not of the original PDE. A possible remedy is to introduce a dissipative term to compensate the artificial numerical dispersion.

We propose the following viscous semi-discretization of (5.1)

$$
\begin{cases}
i\dfrac{du^h}{dt} + \Delta_h u^h = ia(h)\mathrm{sgn}(t)\Delta_h u^h, \ t \neq 0, \\
u^h(0) = \varphi^h,
\end{cases}
\tag{5.18}
$$

where $a(h)$ is a positive function which tends to 0 as h tends to 0. We remark that the proposed scheme is a combination of the conservative approximation of the Schrödinger equation and a semidiscretization of the heat equation in a suitable time-scale. More precisely, the scheme

$$
\frac{du^h}{dt} = a(h)\Delta_h u^h
$$

which underlines in (5.18) may be viewed as a discretization of

$$
u_t = a(h)\Delta u,
$$

which is, indeed, a heat equation in the appropriate time-scale. The scheme (5.18) generates a semigroup $S_+^h(t)$, for $t > 0$. Similarly one may define $S_-^h(t)$, for $t < 0$. The solution u^h satisfies the following energy estimate

$$
\frac{d}{dt}\left[\frac{1}{2}\|u^h(t)\|_{l^2(h\mathbf{Z})}^2\right] = -a(h)\mathrm{sgn}(t)\left[h\sum_{j\in\mathbf{Z}}\left|\frac{u_{j+1}^h(t) - u_j^h(t)}{h}\right|^2\right]. \tag{5.19}
$$

In this energy identity the role that the numerical viscosity term plays is clearly reflected. In particular it follows that

$$
a(h)\int_{\mathbf{R}}\|D^1 u^h(t)\|_{l^2(h\mathbf{Z})}^2 dt \leq \frac{1}{2}\|\varphi^h\|_{l^2(h\mathbf{Z})}^2. \tag{5.20}
$$

Therefore in addition to the L^2-stability property we get some partial information on $D^1 u^h(t)$ in $l^2(h\mathbf{Z})$ that, despite the vanishing multiplicative factor $a(h)$, gives some extra control on the high frequencies.

The following result holds.

Theorem 5.3 *Let us fix $p \in [2,\infty]$ and $\alpha \in (1/2,1]$. Then, for $a(h) = h^{2-1/\alpha}$, $S_\pm^h(t)$ maps continuously $l^{p'}(h\mathbf{Z})$ into $l^p(h\mathbf{Z})$ and there exists some positive constant $c(p)$ such that*

$$
\|S_\pm^h(t)\varphi^h\|_{l^p(h\mathbf{Z})} \leq c(p)(|t|^{-\alpha(1-\frac{2}{p})} + |t|^{-\frac{1}{2}(1-\frac{2}{p})})\|\varphi^h\|_{l^{p'}(h\mathbf{Z})} \tag{5.21}
$$

holds for all $|t| \neq 0$, $\varphi \in l^{p'}(h\mathbf{Z})$ and $h > 0$.

As Theorem 5.3 indicates, when $\alpha > 1/2$, roughly speaking, (5.18) reproduces the decay properties of LSE as $t \to \infty$.

Proof of Theorem 5.3. We consider the case of $S_+^h(t)$, the other one being similar. We point out that $S_+^h(t)\varphi^h = \exp((i + a(h)\mathrm{sgn}(t))t\Delta_h)\varphi^h$. The term $\exp(a(h)\mathrm{sgn}(t)t\Delta_h)\varphi^h$ represents the solution of the semi-discrete heat equation

$$v_t^h - \Delta_h v^h = 0 \tag{5.22}$$

at time $|t|a(h)$. This shows that, as we mentioned above, the viscous scheme is a combination of the conservative one and the semi-discrete heat equation.

Concerning the semidiscrete approximation (5.22) we have, as in the continuous case, the following uniform (with respect to h) norm decay :

$$\|v^h(t)\|_{l^p(h\mathbf{Z})} \lesssim |t|^{-1/2(1/q-1/p)}\|v_0^h\|_{l^q(h\mathbf{Z})} \tag{5.23}$$

for all $1 \le q \le p \le \infty$. This is a simple consequence of the following estimate (that is obtained by multiplying (5.22) by the test function $|v_j^h|^{p-1}v_j^h$)

$$\frac{d}{dt}\left(\|v^h(t)\|_{l^p(h\mathbf{Z})}^p\right) \le -c(p)\|\nabla^+|v^h|^{p/2}\|_{l^2(h\mathbf{Z})}$$

and discrete Sobolev inequalities (see Escobedo & Zuazua (1991) for its continuous counterpart).

In order to obtain (5.21) it suffices to consider the case $p' = 1$ and $p' = 2$, since the others follow by interpolation. The case $p' = 2$ is a simple consequence of the energy estimate (5.19). The terms $t^{-\alpha(1-2/p)}$ and $t^{-1/2(1-2/p)}$ are obtained when estimating the high and low frequencies, respectively. The numerical viscosity term contributes to the estimates of the high frequencies. The low frequencies are estimated by applying the Van der Corput Lemma (cf. Stein (1993), Proposition 2, Ch. VIII. Sect. 1, p. 332).

We consider the projection operator P^h on the low frequencies $[-\pi/4h, \pi/4h]$ defined by $\widehat{P^h\varphi^h} = \widehat{\varphi^h}\chi_{(-\pi/4h,\pi/4h)}$. Using that

$$S_+^h\varphi^h = e^{it\Delta_h}e^{ta(h)\Delta_h}[P^h\varphi^h + (I - P^h)\varphi^h]$$

it is sufficient to prove that

$$\|e^{it\Delta_h}e^{ta(h)\Delta_h}P^h\varphi^h\|_{l^\infty(\mathbf{Z})} \lesssim \frac{1}{t^{1/2}}\|\varphi^h\|_{l^1(h\mathbf{Z})} \tag{5.24}$$

and

$$\|e^{it\Delta_h}e^{ta(h)\Delta_h}(I - P^h)\varphi^h\|_{l^\infty(\mathbf{Z})} \lesssim \frac{1}{t^\alpha}\|\varphi^h\|_{l^1(h\mathbf{Z})} \tag{5.25}$$

for all $t > 0$, uniformly in $h > 0$. By Young's Inequality it is sufficient to obtain upper bounds for the kernels of the operators involved:

$$K_1^h(t) = \chi_{(-\pi/4h, \pi/4h)} e^{-4it \sin^2(\frac{\xi h}{2}) - 4ta(h) \sin^2(\frac{\xi h}{2})}$$

and

$$K_2^h(t) = \chi_{(-\pi/h, \pi/h) \setminus (-\pi/4h, \pi/4h)} e^{-4it \sin^2(\frac{\xi h}{2}) - 4ta(h) \sin^2(\frac{\xi h}{2})}.$$

The second estimate comes from the following

$$\|K_2^h(t)\|_{l^\infty(h\mathbf{Z})} \leq \int_{\frac{\pi}{4h}}^{\frac{\pi}{h}} e^{-4t \frac{a(h)}{h^2} \sin^2(\frac{\xi h}{2})} d\xi \lesssim \frac{1}{h} \left(\frac{ta(h)}{h^2} \right)^{-\alpha} \lesssim \frac{1}{t^\alpha}.$$

The first kernel is rewritten as $K_1^h(t) = K_3^h(t) * H^h(ta(h))$, where $K_3^h(t)$ is the kernel of the operator $P^h e^{it\Delta_h}$ and H^h is the kernel of the semidiscrete heat equation (5.22). Using the Van der Corput lemma we obtain $\|K_3^h(t)\|_{l^\infty(h\mathbf{Z})} \lesssim 1/\sqrt{t}$. Also by (5.23) we get $\|H^h(ta(h))\|_{l^1(h\mathbf{Z})} \lesssim 1$. Finally by Young's inequality we obtain the desired estimate for $K_1^h(t)$. $\qquad\square$

As a consequence of the above theorem, the following TT^* estimate is satisfied.

Lemma 5.1 *For $r \geq 2$ and $\alpha \in (1/2, 1]$, there exists a constant $c(r)$ such that*

$$\|(S_{\mathrm{sgn}(t)}^h(t))^* S_{\mathrm{sgn}(s)}^h(s) f^h\|_{l^r(h\mathbf{Z})} \leq$$
$$\leq c(r)(|t - s|^{-\alpha(1 - \frac{2}{r})} + |t - s|^{-1/2(1 - \frac{2}{r})}) \|f^h\|_{l^{r'}(h\mathbf{Z})}$$

holds for all $t \neq s$.

As a consequence of this, we have the following result.

Theorem 5.4 *The following properties hold :*

(i) For every $\varphi^h \in l^2(h\mathbf{Z})$ and finite $T > 0$, the function $S_{\mathrm{sgn}(t)}^h(t)\varphi^h$ belongs to $L^q([-T, T], l^r(h\mathbf{Z})) \cap C([-T, T], l^2(h\mathbf{Z}))$ for every α-admissible pair (q, r). Furthermore, there exists a constant $c(T, r, q)$ depending on $T > 0$ such that

$$\|S_{\mathrm{sgn}(\cdot)}^h(\cdot)\varphi^h\|_{L^q([-T,T], l^r(h\mathbf{Z}))} \leq c(T, r, q)\|\varphi^h\|_{l^2(h\mathbf{Z})}, \tag{5.26}$$

for all $\varphi^h \in l^2(h\mathbf{Z})$ and $h > 0$.

(ii) If (γ, ρ) is an α-admissible pair and $f \in L^{\gamma'}([-T, T], l^{\rho'}(h\mathbf{Z}))$, then for every α-admissible pair (q, r), the function

$$t \mapsto \Phi_f(t) = \int_{\mathbf{R}} S_{\mathrm{sgn}(t-s)}^h(t - s) f(s) ds, \ t \in [-T, T] \tag{5.27}$$

belongs to $L^q([-T,T], l^r(h\mathbf{Z})) \cap C([-T,T], l^2(h\mathbf{Z}))$. *Furthermore, there exists a constant* $c(T,q,r,\gamma,\rho)$ *such that*

$$\|\Phi f\|_{L^q([-T,T], l^r(h\mathbf{Z}))} \leq c(T,q,r,\gamma,\rho)\|f\|_{L^{\gamma'}([-T,T], l^{\rho'}(h\mathbf{Z}))}, \quad (5.28)$$

for all $f \in L^{\gamma'}([-T,T], l^{\rho'}(h\mathbf{Z}))$ *and* $h > 0$.

Proof All the above estimates follow from Lemma 5.1 as a simple consequence of the classical TT^* argument (cf. Cazenave (2003), Ch. 2, Section 3, p. 33). □

We remark that all the estimates are local in time. This is a consequence of the different behavior of the operators $S_\pm^h$ at $t \sim 0$ and $t \sim \pm\infty$. Despite their local (in time) character, these estimates are sufficient to prove well-posedness and convergence for approximations of the nonlinear Schrödinger equation. Global estimates can be obtained by replacing the artificial viscosity term $a(h)\Delta_h$ in (5.18) by a higher order one : $\tilde{a}(h)\Delta_h^2$ with a convenient $\tilde{a}(h)$. The same arguments as before ensure the same decay as in (5.21) as $t \sim 0$ and $t \sim \infty$, namely $t^{-\frac{1}{2}(1-\frac{2}{p})}$.

Remark 5.1 *Using similar arguments one can also show that a uniform (with respect to h) gain of s space derivatives locally in $L_{x,t}^2$ holds for $0 < s < 1/2\alpha - 1/2$. In fact one can prove the following stronger result.*

Theorem 5.5 *For all* $\varphi^h \in l^2(h\mathbf{Z})$ *and* $0 < s < 1/2\alpha - 1/2$

$$\sup_{j \in \mathbf{Z}} \int_{-\infty}^{\infty} |(D^s S_{\text{sgn}(t)}^h(t)\varphi^h)_j|^2 dt \lesssim \|\varphi\|_{l^2(h\mathbf{Z})}^2 \quad (5.29)$$

holds uniformly in $h > 0$.

This is a consequence of the energy estimate (5.20) for the high frequencies and of dispersive arguments for the low ones (cf. Constantin & Saut (1988) and Kenig, Ponce & Vega (1991)).

5.5 A Viscous Approximation of the NSE

We concentrate on the semilinear NSE equation in $\mathbf{R}$:

$$\begin{cases} iu_t + \Delta u = |u|^p u, \, x \in \mathbf{R}, \, t > 0, \\ u(0,x) = \varphi(x), \, x \in \mathbf{R}. \end{cases} \quad (5.30)$$

It is convenient to rewrite the problem (5.30) in the integral form

$$u(t) = S(t)\varphi - i \int_0^t S(t-s)|u(s)|^p u(s)ds, \quad (5.31)$$

where the Schrödinger operator $S(t) = e^{it\Delta}$ is a one-parameter unitary group in $L^2(\mathbf{R})$ associated with the linear continuous Schrödinger equation. The first result, due to Tsutsumi (1987), on the global existence for L^2-initial data, is the following theorem.

Theorem 5.6 *(Global existence in L^2, Tsutsumi (1987)). For $0 \le p < 4$ and $\varphi \in L^2(\mathbf{R})$, there exists a unique solution u of (5.30) in $C(\mathbf{R}, L^2(\mathbf{R})) \cap L^q_{loc}(\mathbf{R}, L^{p+2}(\mathbf{R}))$ with $q = 4(p+1)/p$ that satisfies the L^2-norm conservation property*

$$\|u(t)\|_{L^2(\mathbf{R})} = \|\varphi\|_{L^2(\mathbf{R})}.$$

This solution depends continuously on the initial condition φ in $L^2(\mathbf{R})$.

Local existence is proved by applying a fixed point argument in the integral formulation (5.31). Global existence holds because of the $L^2(\mathbf{R})$-conservation property which allows excluding finite-time blow-up.

We now consider the following viscous semi-discretization of (5.30):

$$\begin{cases} i\dfrac{du^h}{dt} + \Delta_h u^h = i\,\mathrm{sgn}(t)a(h)\Delta_h u^h + |u^h|^p u^h, t \ne 0, \\ u^h(0) = \varphi^h, \end{cases} \tag{5.32}$$

with $0 \le p < 4$ and $a(h) = h^{2 - \frac{1}{\alpha(h)}}$ such that $\alpha(h) \downarrow 1/2$ and $a(h) \to 0$ as $h \downarrow 0$. The following $l^2(h\mathbf{Z})$-norm dissipation law holds:

$$\frac{d}{dt}\left(\frac{1}{2}\|u^h(t)\|^2_{l^2(h\mathbf{Z})}\right) = -a(h)\mathrm{sgn}(t)\left[h\sum_{j\in\mathbf{Z}}\left|\frac{u^h_{j+1} - u^h_j}{h}\right|^2\right]. \tag{5.33}$$

Concerning the well posedness of (5.32) the following holds:

Theorem 5.7 *(Ignat and Zuazua (2005a)). Let $p \in (0,4)$ and $\alpha(h) \in (1/2, 2/p]$. Set*

$$\frac{1}{q(h)} = \alpha(h)\left(\frac{1}{2} - \frac{1}{p+2}\right)$$

so that $(q(h), p+2)$ is an $\alpha(h)$-admissible pair. Then, for every $\varphi^h \in l^2(h\mathbf{Z})$, there exists a unique global solution

$$u^h \in C(\mathbf{R}, l^2(h\mathbf{Z})) \cap L^{q(h)}_{loc}(\mathbf{R}; l^{p+2}(h\mathbf{Z}))$$

of (5.32) which satisfies the following estimates, independently of h:

$$\|u^h\|_{L^\infty(\mathbf{R}, l^2(h\mathbf{Z}))} \le \|\varphi^h\|_{l^2(h\mathbf{Z})} \tag{5.34}$$

and, for all finite $T > 0$,

$$\|u^h\|_{L^{q(h)}([-T,T],l^{p+2}(h\mathbf{Z}))} \leq c(T)\|\varphi^h\|_{l^2(h\mathbf{Z})}. \tag{5.35}$$

Sketch of the Proof. The proof uses Theorem 5.4 and a standard fixed point argument as in Tsutsumi (1987) and Cazenave (2003) in order to obtain local solutions. Using the a priori estimate (5.33) we obtain a global in time solution.

Let us now address the problem of convergence as $h \to 0$. Given $\varphi \in L^2(\mathbf{R})$, for the semi-discrete problem (5.32) we consider a family of initial data $(\varphi_j^h)_{j \in \mathbf{Z}}$ such that

$$E_h\varphi^h \to \varphi$$

weakly in $L^2(\mathbf{R})$ as $h \to 0$. Here and in the sequel E_h denote the piecewise constant interpolator $E_h : l^2(h\mathbf{Z}) \to L^2(\mathbf{R})$.

The main convergence result is contained in the following theorem.

Theorem 5.8 *The sequence $E_h u^h$ satisfies*

$$E_h u^h \overset{\star}{\rightharpoonup} u \ \ in \ \ L^\infty(\mathbf{R}, L^2(\mathbf{R})), \tag{5.36}$$

$$E_h u^h \rightharpoonup u \ \ in \ \ L_{loc}^s(\mathbf{R}, L^{p+2}(\mathbf{R})) \ \forall \, s < q, \tag{5.37}$$

$$E_h u^h \to u \ \ in \ \ L_{loc}^2(\mathbf{R} \times \mathbf{R}), \tag{5.38}$$

$$|E_h u^h|^p E_h u^h \rightharpoonup |u|^p u \ \ in \ \ L_{loc}^{q'}(\mathbf{R}, L^{(p+2)'}(\mathbf{R})) \tag{5.39}$$

where u is the unique solution of NSE and $2/q = 1/2 - 1/(p+2)$.

Remark 5.2 *Our method works similarly in the critical case $p = 4$ for small initial data. It suffices to modify the approximation scheme by taking a nonlinear term of the form $|u^h|^{2/\alpha(h)}u^h$ in the semi-discrete equation (5.32) with $a(h) = h^{2-1/\alpha(h)}$ and $\alpha(h) \downarrow 1/2$, $a(h) \downarrow 0$, so that, asymptotically, it approximates the critical nonlinearity of the continuous Schrödinger equation. In this way the critical continuous exponent $p = 4$ is approximated by semi-discrete critical problems.*

The critical semi-discrete problem presents the same difficulties as the continuous one. Thus, the initial datum needs to be assumed to be small. But the smallness condition is independent of the mesh-size $h > 0$. More precisely, the following holds.

Theorem 5.9 *Let $\alpha(h) > 1/2$ and $p(h) = 2/\alpha(h)$. There exists a constant ϵ, independent of h, such that for all $\|\varphi^h\|_{l^2(h\mathbf{Z})} < \epsilon$, the semi-discrete critical equation has a unique global solution*

$$u^h \in C(\mathbf{R}, l^2(h\mathbf{Z})) \cap L_{loc}^{p(h)+2}(\mathbf{R}, l^{p(h)+2}(h\mathbf{Z})). \tag{5.40}$$

Moreover $u^h \in L_{loc}^q(\mathbf{R}, l^r(h\mathbf{Z}))$ for all $\alpha(h)$- admissible pairs (q, r) and

$$\|u^h\|_{L^q((-T,T), l^r(h\mathbf{Z}))} \le C(q, T)\|\varphi^h\|_{l^2(h\mathbf{Z})}. \tag{5.41}$$

Observe that, in particular, $(3/\alpha(h), 6)$ is an $\alpha(h)$-admissible pair. This allows us to bound the solutions u^h in a space $L_{loc}^s(\mathbf{R}, L^6(\mathbf{R}))$ with $s < 6$. With the same notation as in the subcritical case the following convergence result holds.

Theorem 5.10 *When $p = 4$ and under the smallness assumption of Theorem 5.9, the sequence Eu^h satisfies*

$$Eu^h \overset{\star}{\rightharpoonup} u \ in \ \ L^\infty(\mathbf{R}, L^2(\mathbf{R})), \tag{5.42}$$

$$Eu^h \to u \ in \ \ L_{loc}^s(\mathbf{R}, L^6(\mathbf{R})) \,\forall\, s < 6, \tag{5.43}$$

$$Eu^h \to u \ in \ \ L_{loc}^2(\mathbf{R} \times \mathbf{R}), \tag{5.44}$$

$$|Eu^h|^{p(h)}|Eu^h| \rightharpoonup |u|^4 u \ in \ \ L_{loc}^{6'}(\mathbf{R}, L^{6'}(\mathbf{R})) \tag{5.45}$$

where u is the unique weak solution of critical (NSE).

5.6 A Two-Grid Scheme

As an alternative to the previous scheme based on numerical viscosity, we propose a two-grid algorithm introduced in Ignat & Zuazua (2005b), which allows constructing conservative and convergent numerical schemes for the nonlinear Schrödinger equation. As we shall see, the two-grid method acts as a preconditioner or filter that eliminates the unwanted high-frequency components from the initial data and nonlinearity. This method is inspired by that used in Glowinski (1992) and Negreanu & Zuazua (2004) in the context of the propagation and control of the wave equation. We emphasize that, by this alternative approach, the purely conservative nature of the scheme is preserved. But, for that to be the case, the nonlinearity needs to be approximated in a careful way.

The method is roughly as follows. We consider two meshes: the coarse one $4h\mathbf{Z}$ of size $4h$, $h > 0$, and the fine one $h\mathbf{Z}$, of size $h > 0$. The computational mesh is the fine one, of size h. The method relies basically on solving the finite-difference semi-discretization (5.4) on the fine mesh $h\mathbf{Z}$, but only for slowly oscillating data and nonlinearity, interpolated from the coarse grid $4h\mathbf{Z}$. As we shall see, the $1/4$ ratio between the two meshes is important to guarantee the convergence of the method. This choice of the mesh-ratio guarantees a particular structure of the data that cancels the two pathologies of the discrete symbol mentioned above. Indeed, a careful Fourier analysis of those initial data shows that their discrete Fourier transforms vanish quadratically at the points $\xi = \pm\pi/2h$ and $\xi = \pm\pi/h$. As we shall see, this suffices to recover the dispersive properties of the continuous model.

To make the analysis rigorous we introduce the space of slowly oscillating sequences (SOS). The SOS on the fine grid $h\mathbf{Z}$ are those which are obtained from the coarse grid $4h\mathbf{Z}$ by an interpolation process. Obviously there is a one to one correspondence between the coarse grid sequences and the space

$$\mathbf{C}_4^{h\mathbf{Z}} = \{\psi \in \mathbf{C}^{h\mathbf{Z}} : \text{supp}\,\psi \subset 4h\mathbf{Z}\}.$$

We introduce the extension operator E:

$$(E\psi)((4j + r)h) = \frac{4 - r}{4}\psi(4jh) + \frac{r}{4}\psi((4j + 4)h), \qquad (5.46)$$

for all $j \in \mathbf{Z}$, $r = \overline{0,3}$ and $\psi \in \mathbf{C}_4^{h\mathbf{Z}}$. This associates to each element of $\mathbf{C}_4^{h\mathbf{Z}}$ an SOS on the fine grid. The space of slowly oscillating sequences on the fine grid is as follows

$$V_4^h = \{E\psi : \psi \in C_4^{h\mathbf{Z}}\}.$$

We also consider the projection operator $\Pi : \mathbf{C}^{h\mathbf{Z}} \to \mathbf{C}_4^{h\mathbf{Z}}$:

$$(\Pi\phi)((4j + r)h) = \phi((4j + r)h)\delta_{4r} \,\forall j \in \mathbf{Z}, r = \overline{0,3}, \phi \in \mathbf{C}^{h\mathbf{Z}} \qquad (5.47)$$

where δ is Kronecker's symbol. We remark that $E : C_4^{h\mathbf{Z}} \to V_4^h$ and $\Pi : V_4^h \to \mathbf{C}_4^{h\mathbf{Z}}$ are bijective linear maps satisfying $\Pi E = I_{\mathbf{C}_4^h\mathbf{Z}}$ and $E\Pi = I_{V_4^h}$, where I_X denotes the identity operator on X. We now define $\tilde{\Pi} = E\Pi : \mathbf{C}^{h\mathbf{Z}} \to V_4^h$, which acts as a smoothing or filtering operator and associates to each sequence on the fine grid a slowly oscillating one. As we said above, the restriction of this operator to V_4^h is the identity.

Concerning the discrete Fourier transform of SOS, by means of explicit computations, one can prove that:

Lemma 5.2 *Let $\phi \in l^2(h\mathbf{Z})$. Then,*

$$\widehat{\tilde{\Pi}\phi}(\xi) = 4\cos^2(\xi h)\cos^2\left(\frac{\xi h}{2}\right)\widehat{\Pi\phi}(\xi). \qquad (5.48)$$

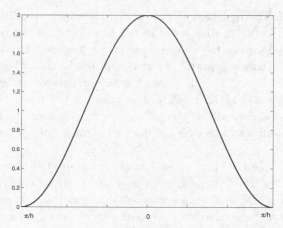

Fig. 5.3. The multiplicative factor $2\cos^2(\xi h/2)$
for the two-grid method with mesh ratio $1/2$

Remark 5.3 *One could think on a simpler two-grid construction, using mesh-ratio $1/2$ and, consequently, considering meshes of size h and $2h$. We then get $\widehat{\tilde{\Pi}\varphi}(\xi) = 2\cos^2(\xi h/2)\widehat{\Pi\varphi}(\xi)$. This cancels the spurious numerical solutions at the frequencies $\pm\pi/h$ (see Fig. 5.3), but not at $\pm\pi/2h$. In this case, as we proved in Section 5.3, the Strichartz estimates fail to be uniform on h. Thus instead we choose the ratio between grids to be $1/4$. As the Figure 5.4 shows, the multiplicative factor occurring in (5.48) will cancel the spurious numerical solutions at $\pm\pi/h$ and $\pm\pi/2h$.*

As we have proved in Section 5.3, there is no gain (uniformly in h) of integrability of the linear semigroup $e^{it\Delta_h}$. However the linear semigroup has appropriate decay properties when restricted to V_4^h uniformly in $h > 0$. The main results we get are the following.

Theorem 5.11 *Let $p \geq 2$. The following properties hold:*

i) $\|e^{it\Delta_h}\tilde{\Pi}\varphi\|_{l^p(h\mathbf{Z})} \lesssim |t|^{-1/2(1/p'-1/p)}\|\tilde{\Pi}\varphi\|_{l^{p'}(h\mathbf{Z})}$ *for all* $\varphi \in l^{p'}(h\mathbf{Z})$, $h > 0$ *and* $t \neq 0$.

ii) For every sequence $\varphi \in l^2(h\mathbf{Z})$, the function $t \to e^{it\Delta_h}\tilde{\Pi}\varphi$ belongs to $L^q(\mathbf{R}, l^r(h\mathbf{Z})) \cap C(\mathbf{R}, l^2(h\mathbf{Z}))$ for every admissible pair (q, r). Furthermore,

$$\|e^{it\Delta_h}\tilde{\Pi}\varphi\|_{L^q(\mathbf{R}, l^r(h\mathbf{Z}))} \lesssim \|\tilde{\Pi}\varphi\|_{l^2(h\mathbf{Z})},$$

uniformly in $h > 0$.

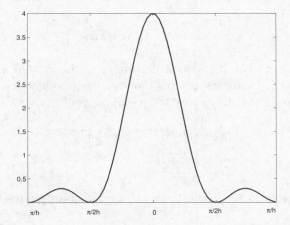

Fig. 5.4. The multiplicative factor $4\cos^2(\xi h)\cos^2\left(\frac{\xi h}{2}\right)$
for the two-grid method with mesh ratio $1/4$

iii) Let (q, r), $(\tilde{q}, \tilde{r})$ be two admissible pairs. Then,

$$\left\|\int_{s<t} e^{i(t-s)\Delta_h}\tilde{\Pi}F(s)ds\right\|_{L^q(\mathbf{R}, l^r(h\mathbf{Z}))} \lesssim \|\tilde{\Pi}F\|_{L^{\tilde{q}}(\mathbf{R}, l^{\tilde{r}}(h\mathbf{Z}))}$$

for all $F \in L^{\tilde{q}}(\mathbf{R}, l^{\tilde{r}}(h\mathbf{Z}))$, uniformly in $h > 0$.

Concerning the local smoothing properties we can prove the following result.

Theorem 5.12 *The following estimate*

$$\sup_{j\in\mathbf{Z}}\int_{-\infty}^{\infty}\left|(D^{1/2}e^{it\Delta_h}\tilde{\Pi}f)_j\right|^2 dt \lesssim \|\tilde{\Pi}f\|_{l^2(h\mathbf{Z})}^2 \tag{5.49}$$

holds for all $f \in l^2(h\mathbf{Z})$, uniformly in $h > 0$.

Proof of Theorem 5.11. The estimates ii) and iii) easily follow by the classical TT^* argument (cf. Keel & Tao (1998), Cazenave (2003)) once one proves i) with $p' = 1$ and $p' = 2$. The case $p' = 2$ is a consequence of the conservation of energy property. For $p' = 1$, by a scaling argument, we may assume that $h = 1$. The same arguments as in Section 5.4, reduce the proof to the following upper bound for the kernel

$$\|K^1(t)\|_{l^\infty(h\mathbf{Z})} \lesssim \frac{1}{t^{1/2}},$$

where

$$\widehat{K^1(t)} = 4e^{-4it\sin^2\frac{\xi}{2}}\cos^2(\xi)\cos^2\left(\frac{\xi}{2}\right).$$

Using the fact that the second derivative of the symbol $4\sin^2(\xi/2)$ is given by $2\cos\xi$, by means of oscillatory integral techniques (cf. Kenig, Ponce & Vega (1991), Corollary 2.9, p. 46) we get

$$\|K^t\|_{l^\infty(\mathbf{Z})} \lesssim \frac{1}{t^{1/2}} \left\| 2|\cos(\xi)|^{3/2}\cos^2\frac{\xi}{2} \right\|_{L^\infty([-\pi,\pi])} \lesssim \frac{1}{t^{1/2}}.$$

$\square$

Proof of Theorem 5.12. The estimate (5.49) is equivalent to the following one

$$\sup_{j\in\mathbf{Z}} \int_{-\infty}^{\infty} \left| (e^{it\Delta_h}\tilde{\Pi}f)_j \right|^2 dt \lesssim \|D^{-1/2}\tilde{\Pi}f\|^2_{l^2(h\mathbf{Z})}. \qquad (5.50)$$

By scaling we consider the case $h = 1$. Applying the results of Kenig, Ponce & Vega (1991) (Theorem 4.1, p. 54) we get

$$\sup_{j\in\mathbf{Z}} \int_{-\infty}^{\infty} \left| (e^{it\Delta_h}\tilde{\Pi}f)_j \right|^2 dt \lesssim \int_{-\pi}^{\pi} \frac{|\widehat{\Pi f}(\xi)|^2\cos^4\xi\cos^4\frac{\xi}{2}}{|\sin\xi|} d\xi$$

$$\lesssim \int_{-\pi}^{\pi} \frac{|\widehat{\Pi f}(\xi)|^2}{|\xi|} d\xi \lesssim \|D^{-1/2}f\|^2_{l^2(\mathbf{Z})}.$$

Observe that the key point in the above proof is that the factor $\cos(\xi/2)$ in the amplitude of the Fourier representation of the initial datum compensates the effects of the critical points of the symbol $\sin^2(\xi/2)$ near the points $\pm\pi$. $\square$

The results proved in Theorem 5.11 i) are plotted in Fig. 5.7. We choose an initial datum as in Fig. 5.6, obtained by interpolation of the Dirac delta: $\Pi u(0) = \delta_0$ (see Fig. 5.5). Figure 5.7 shows the different behavior of the solutions of the conservative and the two-grid schemes. The $l^\infty(\mathbf{Z})$-norm of the solution $u^1(t)$ for the two-grid algorithm behaves like $t^{-1/2}$ as $t \to \infty$, with the decay rate predicted above, while the solutions of the conservative scheme, without the two-grid filtering, decay like $t^{-1/3}$.

5.7 A Conservative Approximation of the NSE

We consider the following semi-discretization of the NSE :

$$\begin{cases} i\dfrac{du^h}{dt} + \Delta_h u^h = \tilde{\Pi}f(u^h), \ t \neq 0, \\ u^h(0) = \tilde{\Pi}\varphi^h, \end{cases} \qquad (5.51)$$

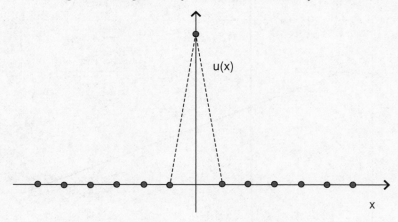

Fig. 5.5. $u^1(0) = \delta_0$

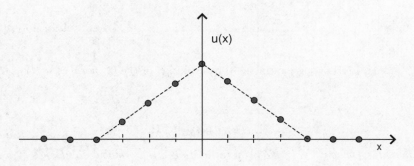

Fig. 5.6. $u^1(0) = E\delta_0$

where $f(u^h)$ is a suitable approximation of $|u|^p u$ with $0 < p < 4$. In order to prove the global well-posedness of (5.51), we need to guarantee the conservation of the $l^2(h\mathbf{Z})$ norm of solutions, a property that the solutions of NSE satisfy. For that the nonlinear term $f(u^h)$ has to be chosen so that $(\tilde{\Pi} f(u^h), u^h)_{l^2(h\mathbf{Z})} \in \mathbf{R}$. For that to be the case, it is not sufficient to discretize the nonlinearity as for the viscous scheme, by simply sampling it on the discrete mesh. A more careful choice is needed. The following result holds.

Theorem 5.13 *Let $p \in (0, 4)$, $q = 4(p + 2)/p$ and $f : \mathbf{C}^{h\mathbf{Z}} \to \mathbf{C}^{h\mathbf{Z}}$ be such that*

$$\|\tilde{\Pi} f(u)\|_{l^{(p+2)'}(h\mathbf{Z})} \lesssim \||u|^p u\|_{l^{(p+2)'}(h\mathbf{Z})} \tag{5.52}$$

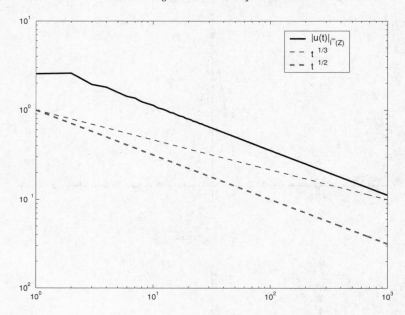

Fig. 5.7. Log-log plot of the time evolution of the l^∞ norm of $u^1(t)$

and

$$(\tilde{\Pi}f(u), u)_{l^2(h\mathbf{Z})} \in \mathbf{R}.$$

Then, for every $\varphi^h \in l^2(h\mathbf{Z})$, there exists a unique global solution

$$u^h \in C(\mathbf{R}, l^2(h\mathbf{Z})) \cap L^q_{loc}(\mathbf{R}; l^{p+2}(h\mathbf{Z})) \tag{5.53}$$

of (5.51) which satisfies the estimates

$$\|u^h\|_{L^\infty(\mathbf{R}, l^2(h\mathbf{Z}))} \le \|\tilde{\Pi}\varphi\|_{l^2(h\mathbf{Z})} \tag{5.54}$$

and

$$\|u^h\|_{L^q(I, l^{p+2}(h\mathbf{Z}))} \le c(I)\|\tilde{\Pi}\varphi\|_{l^2(h\mathbf{Z})} \tag{5.55}$$

for all finite intervals I, where the above constants are independent of h.

Remark 5.4 *The conditions above on the nonlinearity are satisfied if one chooses*

$$(f(u^h))_{4j} = g\left((u^h_{4j} + \sum_{r=1}^{3} \frac{4-r}{4}(u^h_{4j+r} + u^h_{4j-r}))/4\right); \quad g(s) = |s|^p s. \tag{5.56}$$

With this choice it is easy to check that (5.52) holds with $C > 0$ independent of $h > 0$. Furthermore $(\tilde{\Pi} f(u^h), u^h)_{l^2(h\mathbf{Z})} \in \mathbf{R}$ since

$$(\tilde{\Pi} f(u^h), u^h)_{l^2(h\mathbf{Z})} =$$

$$= h \sum_{r=0}^{3} \sum_{j \in \mathbf{Z}} \left(\frac{4-r}{4} (f(u^h))_{4j} + \frac{r}{4} (f(u^h))_{4j+4} \right) \overline{u}^h_{4j+r}$$

$$= h \sum_{j \in \mathbf{Z}} (f(u^h))_{4j} \left(\sum_{r=0}^{3} \frac{4-r}{4} \overline{u}^h_{4j+r} + \sum_{r=0}^{3} \frac{r}{4} \overline{u}^h_{4j+r-4} \right)$$

$$= h \sum_{j \in \mathbf{Z}} g \left((u^h_{4j} + \sum_{r=1}^{3} \frac{4-r}{4} (u^h_{4j+r} + u^h_{4j-r}))/4 \right)$$

$$\times (\overline{u}^h_{4j} + \sum_{r=1}^{3} \frac{4-r}{4} (\overline{u}^h_{4j+r} + \overline{u}^h_{4j-r})).$$

Proof of Theorem 5.13. Local existence and uniqueness are a consequence of the Strichartz estimates (Theorem 5.11) and of a fixed point argument. The fact that $(\tilde{\Pi} f(u^h), u^h)_{l^2(h\mathbf{Z})}$ is real guarantees the conservation of the discrete energy $h \sum_{j \in \mathbf{Z}} |u_j(t)|^2$. This allows excluding finite-time blow-up.

□

The main convergence result is the following

Theorem 5.14 *Let u^h be the unique solution of (5.51) with discrete initial data φ^h such that $E_h \varphi^h \rightharpoonup \varphi$ weakly in $L^2(\mathbf{R})$. Then, the sequence $E_h u^h$ satisfies*

$$E_h u^h \overset{\star}{\rightharpoonup} u \text{ in } L^\infty(\mathbb{R}, L^2(\mathbf{R})), \tag{5.57}$$

$$E_h u^h \rightharpoonup u \text{ in } L^q_{loc}(\mathbb{R}, L^{p+2}(\mathbf{R})), \tag{5.58}$$

$$E_h u^h \to u \text{ in } L^2_{loc}(\mathbb{R} \times \mathbf{R}), \tag{5.59}$$

$$E_h \tilde{\Pi} f(u^h) \rightharpoonup |u|^p u \text{ in } L^{q'}_{loc}(\mathbb{R}, L^{(p+2)'}(\mathbf{R})) \tag{5.60}$$

where u is the unique solution of NSE and $2/q = 1/2 - 1/(p+2)$.

The critical nonlinearity $p = 4$ may also be handled by the two-grid algorithm. In this case one can take directly $p = 4$ in the semi-discrete scheme since the two-grid algorithm guarantees the dispersive estimates to be true for all 1/2-admissible pairs.

5.8 Open Problems

- **Time Splitting Methods.** In Besse, Bidégaray & Descombes (2002), (see also Sanz-Serna & Calvo (1994), Descombres & Schatzman (2002)) the authors consider the NSE with initial data in $H^2(\mathbf{R}^2)$ and the nonlinear term $|u|^2 u$. A time splitting method is used in order to approximate the solution. More precisely, the nonlinear Schrödinger equation is split into the flow X^t generated by the linear Schrödinger equation

$$\begin{cases} v_t - i\Delta v = 0,\ x \in \mathbf{R}^2,\ t > 0, \\ v(0,x) = v_0(x),\ x \in \mathbf{R}^2. \end{cases} \tag{5.61}$$

and the flow Y^t for the differential equation

$$\begin{cases} w_t - i|w|^2 w = 0,\ x \in \mathbf{R}^2,\ t > 0, \\ w(0,x) = w_0(x),\ x \in \mathbf{R}^2. \end{cases} \tag{5.62}$$

One can then approximate the flow of NSE by combining the two flows X^t and Y^t using some of the classical splitting methods: the Lie formula $Z_L^t = X^t Y^t$ or the Strang formula $Z_S^t = X^{t/2} Y^t X^{t/2}$. In Besse, Bidégaray & Descombes (2002) the convergence of these methods is proved for initial data in $H^2(\mathbf{R}^2)$. Note however that the nonlinearity $|u|^2 u$ is locally Lipschitz in $H^2(\mathbf{R}^2)$. Consequently this nonlinearity in this functional setting can be dealt with by means of classical energy methods, without using the Strichartz type estimate.

   A possible problem for future research is to replace the above equations (5.61), (5.62), which are continuous in the variable x, by discrete ones and to analyze the convergence of the splitting method for the initial data in $L^2(\mathbf{R})$. As we saw in Section 5.3 the simpler approximation of (5.61) by finite differences does not have the dispersive properties of the continuous model. It is then natural to consider one of the two remedies we have designed: to add numerical viscosity or to regularize the initial data by a two grid algorithm. The convergence of the splitting algorithm is open because of the lack of dispersion of the ODE (5.62) and its semi-discretizations.

- **Discrete Transparent Boundary Conditions.** In Arnold, Ehrhardt & Sofronov (2003) the authors introduce a discrete transparent boundary condition for a Crank–Nicolson finite difference discretization of the Schrödinger equation. The same ideas allow constructing similar DTBC for various numerical approximations of the LSE. It would be interesting to study the dispersive properties of these approximations by means of the techniques of Markowich & Poupaud (1999) based on microlocal analysis. Supposing that the approximation fails to have the appropriate dispersive properties, one could apply the methods presented here in order to recover the dispersive properties of the continuous model.

- **Fully Discrete Schemes.** It would be interesting to develop a similar analysis for fully discrete approximation schemes. We present two schemes, one which is implicit and the other one which is explicit in time. The first one:

$$i\frac{u_j^{n+1} - u_j^n}{\Delta t} + \frac{u_{j+1}^{n+1} - 2u_j^{n+1} + u_{j-1}^{n+1}}{(\Delta x)^2} = 0,\ n \geq 0,\ j \in \mathbf{Z}, \qquad (5.63)$$

introduces time viscosity and consequently has the right dispersive properties. The second one is conservative and probably will present some pathologies. As an example we choose the following approximation scheme:

$$i\frac{u_j^{n+1} - u_j^{n-1}}{2\Delta t} + \frac{u_{j+1}^n - 2u_j^n + u_{j-1}^n}{(\Delta x)^2} = 0,\ n \geq 1,\ j \in \mathbf{Z}. \qquad (5.64)$$

In this case it is expected that the dispersive properties will not hold for any Courant number $\lambda = \Delta t/(\Delta x)^2$ which satisfies the stability condition. Giving a complete characterization of the fully discrete schemes satisfying the dispersive properties of the continuous Schrödinger equation is an open problem.

- **Bounded Domains.** In Bourgain (1993) the LSE is studied on the torus $\mathbf{R}/\mathbf{Z}$ and the following estimates are proved :

$$\|e^{it\Delta}\varphi\|_{L^4(\mathbf{T}^2)} \lesssim \|\varphi\|_{L^2(\mathbf{T})}. \qquad (5.65)$$

This estimate allows one to show the well posedness of a NSE on $\mathbf{T}^2$. As we prove in Ignat (2006), in the case of the semidiscrete approximations, similar l_x^2-$L_t^4 l_x^4$ estimates fail to be uniform with respect to the mesh size Δx. It is an open problem to establish what is the complete range of (q, r) (if any) for which the estimates l_x^2-$L_t^q l_x^r$ are uniform with respect to the mesh size. It is then natural to consider schemes with numerical viscosity or with a two-grid algorithm.

More recently, the results by Burq, Gérard and Tzvetkov (2004) show Strichartz estimates with loss of derivatives on compact manifolds without boundary. The corresponding results on the discrete level remain to be studied.

- **Variable Coefficients.** In Banica (2003) the global dispersion and the Strichartz inequalities are proved for a class of one-dimensional Schrödinger equations with step-function coefficients having a finite number of discontinuities. Staffilani & Tataru (2002) proved the Strichartz estimates for C^2 coefficients. As we proved in Section 5.3, even in the case of the approximations of the constant coefficients, the Strichartz estimates fail to be uniform with respect to the mesh size h. It would be interesting to study if the two remedies we have presented in this article are also efficient for a variable-coefficient problem.

Acknowledgements

This work has been supported by Grant BFM2002-03345 of the Spanish MCYT and the Network "Smart Systems" of the EU. Liviu I. Ignat was also supported by a doctoral fellowship of MEC (Spain) and by Grant 80/2005 of CNCSIS (Romania).

References

Arnold, A., Ehrhardt M. and Sofronov I., (2003) Discrete transparent boundary conditions for the Schrödinger equation: fast calculation, approximation and stability, *Commun. Math. Sci.* **1**, 501–556.

Banica, V., (2003) Dispersion and Strichartz inequalities for Schrödinger equations with singular coefficients, *Siam J. Math. Anal.* **35**, 868–883.

Bergh, J. and Löfström, J., (1976) *Interpolation Spaces, An Introduction*, Springer Verlag, Berlin-New York.

Besse, C., Bidégaray, B. and Descombes, S., (2002) Order estimates in time of splitting methods for the nonlinear Schrödinger equation, *SIAM J. Numer. Anal.* **40**, 26–40.

Bourgain, J., (1993) Fourier transform restriction phenomena for certain lattice subsets and applications to nonlinear evolution equations. *Geometric and Functional Analysis* **3**, 107–156.

Burq, N., Gérard, P. and Tzvetkov, N., (2004) Strichartz Inequalities and the nonlinear Schrödinger equation on compact manifolds. *Amer. J. Math.* **126**, 569–605.

Cazenave, T., (2003) *Semilinear Schrödinger Equations*, Courant Lecture Notes in Mathematics, 10, New York.

Constantin, P. and Saut, J., (1988) Local smoothing properties of dispersive equations, *J. Am. Math. Soc.* **1**, 413–439.

Descombes, S. and Schatzman, M., (2002) Strangs formula for holomorphic semigroups, *J. Math. Pures Appl.* **81**, 93-114.

Escobedo M. and Zuazua E., (1991) Large time behavior for convection-diffusion equations in $\mathbf{R}^N$, *J. Funct. Anal.* **100**, 119–161.

Ginibre, J. and Velo, G., (1992) Smoothing properties and retarded estimates for some dispersive evolution equations, *Commun. Math. Phys.* **144**, 163–188.

Glowinski, R., (1992) Ensuring well-posedness by analogy: Stokes problem and boundary control for the wave equation, *J. Comput. Phys.* **103**, 189–221.

Ignat, I. L., (2006) Ph.D. thesis. Universidad Autónoma de Madrid. In preparation.

Ignat, I. L. and Zuazua, E., (2005a) Dispersive properties of a viscous numerical scheme for the Schrödinger equation, *C. R. Math. Acad. Sci. Paris* **340**, 529–534.

Ignat, I. L. and Zuazua, E., (2005b) A two-grid approximation scheme for nonlinear Schrödinger equations: Dispersive properties and convergence, *C. R. Math. Acad. Sci. Paris* **341**, 381-386.

Keel, M. and Tao, T., (1998) Endpoint Strichartz estimates, *Am. J. Math.* **120**, 955–980.

Kenig, C. E. , Ponce, G. and Vega, L., (1991) Oscillatory integrals and regularity of dispersive equations, *Indiana Univ. Math. J.* **40**, 33–69.

Magyar, A., Stein, E. M. and Wainger, S., (2002) Discrete analogues in harmonic analysis: Spherical averages, *Annals of Mathematics* **155**, 189-208.

Markowich, P. A. and Poupaud, F., (1999) The pseudo-differential approach to finite differences revisited, *Calcolo* **36**, 161–186.

Negreanu, M. and Zuazua, E., (2004) Convergence of a multigrid method for the controllability of a 1-d wave equation, *C. R. Math. Acad. Sci. Paris, Ser. I* **338**, 413–418.

Plancherel, M. and Pólya, G., (1937) Fonctions entières et intégrales de Fourier multiples, *Commentarii mathematici Helvetici* **10**, 110–163.

Sanz-Serna, J. M. and Calvo, M. P., (1994) *Numerical Hamiltonian Problems*, Chapman and Hall, London.

Simon, J., (1987) Compact sets in the space $L^p(0,T;B)$, *Ann. Mat. Pura Appl.* **146**, 65–96.

Staffilani, G. and Tataru, D., (2002) Strichartz estimates for a Schrödinger operator with non-smooth coefficients, *Comm. Partial Differential Equations* **27**, 1337–1372.

Stefanov, A. and Kevrekidis, P. G., (2005) Asymptotic Behaviour of Small Solutions for the Discrete Nonlinear Schrödinger and Klein-Gordon Equations, *Nonlinearity* **18**, 1841–1857.

Stein, E., (1993) *Harmonic analysis: real-variable methods, orthogonality, and oscillatory integrals*, Monographs in Harmonic Analysis, III, Princeton University Press.

Strichartz, R. S., (1977) Restrictions of Fourier transforms to quadratic surfaces and decay of solutions of wave equations, *Duke Math. J.* **44**, 705–714.

Trefethen, L. N., (1996) *Finite Difference and Spectral Methods for Ordinary and Partial Differential Equations*, http://web.comlab.ox.ac.uk /oucl/work/nick.trefethen/pdetext.html.

Tsutsumi Y., (1987) L^2-solutions for nonlinear Schrödinger equations and nonlinear groups, *Funkc. Ekvacioj, Ser. Int.* **30**, 115–125.

Zuazua, E., (2005) Propagation, observation, and control of waves approximated by finite difference methods, *Siam Review* **47**, 197–243.

6

Eigenvalues and Nonsmooth Optimization

Adrian Lewis

School of Operations Research and Industrial Engineering
Cornell University
Ithaca, NY, USA
e-mail: aslewis@orie.cornell.edu

Abstract

Variational analysis concerns the geometry and calculus of nonsmooth sets and functions, often viewed from an optimization perspective. Over several decades, variational analysis has matured into a powerful and elegant theory. One rich source of concrete examples involves the eigenvalues of symmetric and nonsymmetric matrices, sometimes deriving from dynamical or feedback control questions. This essay presents some central ideas of variational analysis, developed from first principles, including convexity and duality, generalized gradients, sensitivity, Clarke regularity, and numerical nonsmooth optimization. Illustrative examples from eigenvalue optimization, many from joint work with J.V. Burke and M.L. Overton, include semidefinite programming, asymptotic stability, simultaneous plant stabilization, and the distance to instability.

6.1 Introduction

The eigenvalues of a matrix vary nonsmoothly as we perturb the matrix. For example, as the real parameter τ decreases through zero, the eigenvalues of the matrix

$$\begin{bmatrix} 0 & \tau \\ 1 & 0 \end{bmatrix}$$

coalesce at zero from opposite sides of the real axis and then split along the imaginary axis. This inherent nonsmoothness constrains standard developments of eigenvalue perturbation theory, such as Kato (1982), Bhatia (1997). The traditional theory, albeit a powerful tool in many applications, primarily focuses either on precise sensitivity results with respect to a single parameter, or on broader bounds for more general perturbations.

The modern theory of variational analysis offers an elegant attack on this dilemma. Growing originally out of the calculus of variations, and driven by the broad successes of the systematic approach to convexity popularized by Rockafellar's *Convex Analysis* (1970), the nonconvex theory pioneered by Clarke (1973, 1983) has now blossomed into a comprehensive and powerful framework for optimization and variational problems beyond the realm of classical calculus. The monographs Clarke (1998) and Rockafellar and Wets (1998) give excellent overviews of variational analysis; Borwein and Lewis (2000) is a broad introduction.

This essay sketches the symbiotic relationship between variational analysis and eigenvalue perturbation theory. I illustrate the main themes with examples chosen heavily from my own recent work on symmetric matrices and my collaboration with Jim Burke and Michael Overton on nonsymmetric matrices. On the one hand, the language and tools of variational analysis and nonsmooth optimization crystallize spectral properties of matrices beyond the usual reach of eigenvalue perturbation theory. On the other hand, classical mathematical knowledge about matrix spectra, and their broad applicability, ensure that nonsmooth spectral analysis serves as a significant testing ground for nonsmooth optimization theory.

6.2 Convexity, hyperbolic polynomials, and Lidskii's theorem

Modern variational analysis grew originally from a systematic study of convexity, so it is with convexity that we begin. Eigenvalues of real symmetric matrices exhibit remarkable convexity properties, underlying an explosion of interest throughout the optimization community over the last decade in a far-reaching generalization of linear programming known as *semidefinite programming*: see Ben-Tal and Nemirovski (2001) and Todd (2001).

Denote by $\mathbf{S}^n$ the Euclidean space of n-by-n real symmetric matrices, equipped with the inner product $\langle X, Y \rangle = \text{trace}(XY)$. Within this space, the positive semidefinite matrices $\mathbf{S}^n_+$ constitute a closed convex cone. Semidefinite programming is the study of linear optimization over intersections of $\mathbf{S}^n_+$ with affine subspaces.

An illuminating and strikingly general framework in which to consider the most basic convexity properties of symmetric matrix eigenvalues is that of *hyperbolic polynomials*, a notion originally associated with partial differential equations— see Gårding (1951). The determinant is a hyperbolic polynomial on $\mathbf{S}^n$ relative to the identity matrix I: in other words, it is homogeneous (of degree n), and for any $X \in \mathbf{S}^n$, the polynomial $\lambda \mapsto \det(X - \lambda I)$ has all real roots, namely the eigenvalues $\lambda_1(X) \geq \cdots \geq \lambda_n(X)$. With this notation, we can consider the *characteristic map* $\lambda : \mathbf{S}^n \to \mathbf{R}^n$.

A *spectral set* in $\mathbf{S}^n$ is an inverse image of the form

$$\lambda^{-1}(S) = \{X \in \mathbf{S}^n : \lambda(X) \in S\}$$

for any set $S \subset \mathbf{R}^n$.

The core perturbation property of the eigenvalues of symmetric matrices is the following result (which forms a central theme of Bhatia (1997), for example). We denote the group of n-by-n permutation matrices by $\mathbf{P}^n$. For a vector $x \in \mathbf{R}^n$, we denote by $\mathbf{P}^n x$ the set $\{Px : P \in \mathbf{P}^n\}$. Analogously, for a set $S \subset \mathbf{R}^n$, we denote by $\mathbf{P}^n S$ the set $\cup_{x \in S} \mathbf{P}^n x$, and we call S *symmetric* if $\mathbf{P}^n S = S$. We denote the convex hull operation by conv, the standard Euclidean norm on $\mathbf{R}^n$ by $\| \cdot \|$, and the positive orthant and its interior by $\mathbf{R}^n_+$ and $\mathbf{R}^n_{++}$ respectively.

Theorem 6.1 (Lidskii, 1950) *Any matrices $X, Y \in \mathbf{S}^n$ satisfy*

$$\lambda(X) - \lambda(Y) \in \mathrm{conv}\,(\mathbf{P}^n \lambda(X - Y)).$$

Immediate corollaries include many important classical properties of eigenvalues of symmetric matrices, some of which are collected below: see Horn and Johnson (1985), and Stewart and Sun (1990). Lidskii's theorem is not the easiest avenue to any of these results, but it does provide a unifying perspective: see Bhatia (1997).

Corollary 6.1 (characteristic map behavior) *The characteristic map* $\lambda : \mathbf{S}^n \to \mathbf{R}^n$ *has the following properties.*

Monotonicity *The map λ is monotone relative to the orderings induced by the cones $\mathbf{S}^n_+$ and $\mathbf{R}^n_+$: any matrices $X, Y \in \mathbf{S}^n$ satisfy*

$$X - Y \in \mathbf{S}^n_+ \quad \Rightarrow \quad \lambda(X) - \lambda(Y) \in \mathbf{R}^n_+.$$

Convexity *If the set $C \subset \mathbf{R}^n$ is symmetric and convex, then the spectral set $\lambda^{-1}(C)$ is convex. In particular, the* **hyperbolicity cone** $\lambda^{-1}(\mathbf{R}^n_{++})$ *is convex.*

Nonexpansivity *The map λ is nonexpansive: $\|\lambda(X) - \lambda(Y)\| \leq \|X - Y\|$. The same inequality also holds using the infinity norm on $\mathbf{R}^n$ and the spectral norm on $\mathbf{S}^n$.*

Hyperbolic polynomials are strikingly simple to define, and form a broad, rich class: see Bauschke et al. (2001). Nonetheless, hyperbolic polynomials in three or fewer variables have a very specific structure. In one or two variables, this observation is easy and uninteresting; in three variables, it is neither. The following result, conjectured in Lax (1958), was observed in Lewis et al. (2005) to be equivalent to a recent result of Helton and Vinnikov (2002).

Theorem 6.2 ("Lax conjecture") *A polynomial p on $\mathbb{R}^3$ is homogeneous of degree 3, hyperbolic relative to the direction $e = (1, 0, 0)$, and satisfies $p(e) = 1$, if and only if it has the form*

$$p(x) = \det(x_1 I + x_2 A + x_3 B)$$

for some matrices $A, B \in \mathbf{S}^n$.

At first sight, the Lax conjecture looks rather narrow in it applicability. However, as the next corollary due to Gurvits (2004) exemplifies, it is a much more general tool than first appearances suggest.

Corollary 6.2 *Lidskii's theorem holds for any hyperbolic polynomial.*

Proof Suppose the degree-n polynomial p is hyperbolic on $\mathbb{R}^k$ relative to the direction d. By normalizing, we can suppose $p(d) = 1$. For any vectors x, y, we want to prove $\lambda(x) - \lambda(y) \in \operatorname{conv}(\mathbf{P}^n \lambda(x-y))$. Apply the Lax conjecture to the polynomial on $\mathbb{R}^k$ defined by $w \in \mathbb{R}^3 \mapsto p(w_1 d + w_2 x + w_3 y)$, which is itself hyperbolic relative to e. The result now follows by appealing to Lidskii's theorem on $\mathbf{S}^n$. □

As an immediate consequence of this result, or alternatively, by directly applying the same proof technique, each part of Corollary 6.1 also holds for any hyperbolic polynomial. Each of these results has a more direct proof. The monotonicity result appeared in Gårding (1959), which also contains a short proof of the central fact that the hyperbolicity cone is convex. The more general convexity result appears in Bauschke et al. (2001), along with the nonexpansive property, for which we need to make the nondegeneracy assumption $\lambda(x) = 0 \Rightarrow x = 0$ and define $\|x\| = \|\lambda(x)\|$.

6.3 Duality and normal cones

A characteristic feature of convex analysis and optimization is the heavy use of duality arguments, featuring separating hyperplanes in various guises: see Rockafellar (1970). The most basic form of this idea is duality for cones. The *dual cone* of a set $S \subset \mathbf{R}^n$ is the closed convex cone

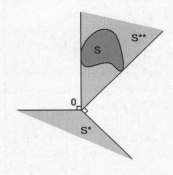

$$S^* = \bigcap_{x \in S} \{y : \langle x, y \rangle \leq 0\}$$

(interpreting $\emptyset^* = \mathbf{R}^n$). The set S is itself a closed convex cone if and only $S = S^{**}$.

In addition to the "primal" properties of the characteristic map $\lambda : \mathbf{S}^n \to \mathbf{R}^n$ listed in Corollary 6.1, λ also behaves well under duality operations. The following basic theorem is one of several analogous results concerning polar sets and Fenchel conjugate functions in Lewis (1996c).

Theorem 6.3 (dual spectral cones) *For symmetric sets* $S \subset \mathbf{R}^n$,

$$(\lambda^{-1}(S))^* = \lambda^{-1}(S^*).$$

This result is reminiscent of von Neumann's 1937 characterization of unitarily invariant matrix norms on the Euclidean space of n-by-n complex matrices $\mathbf{M}^n$ (equipped with the Frobenius norm). Part of von Neumann's development is the formula

$$(\sigma^{-1}(G))^D = \sigma^{-1}(G^D),$$

where $\sigma : \mathbf{M}^n \to \mathbf{R}^n$ maps any matrix to a vector with components its singular values (in decreasing order), G is any symmetric norm-unit-ball satisfying $x \in G \Leftrightarrow |x| \in G$ (the absolute value applied componentwise), and G^D denotes the dual unit ball. Semisimple Lie theory provides one algebraic framework for exploring the parallels between von Neumann's duality formula and Theorem 6.3 (dual spectral cones): see Lewis (2000). Other authors have investigated results like Theorem 6.3 for Euclidean Jordan algebras, a popular setting in which to study interior-point optimization algorithms: see Baes (2004) and Sun and Sun (2004).

A principal application of the dual cone idea is in the development of optimality conditions for constrained optimization problems. Given a convex set $C \subset \mathbf{R}^n$, the *normal cone* to C at a point $\bar{x} \in C$ is

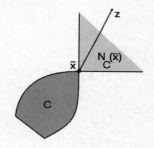

$$N_C(\bar{x}) = (C - \bar{x})^*.$$

Using this notation, we have the *best approximation condition*:

$$\bar{x} \in \operatorname{argmin}\{\|z - x\| : x \in C\} \Rightarrow z - \bar{x} \in N_C(\bar{x}), \tag{6.1}$$

(and in fact the converse also holds). Theorem 6.3 (dual spectral cones) is a special case of the following characterization of normal cones to spectral sets: see Lewis (1996a).

Theorem 6.4 (spectral normal cones) *If the set* $C \subset \mathbf{R}^n$ *is symmetric and convex, then the spectral set* $\lambda^{-1}(C)$ *is convex, and matrices* $X, Y \in \mathbf{S}^n$

satisfy $Y \in N_{\lambda^{-1}(C)}(X)$ *if and only if there exists vectors* $x, y \in \mathbf{R}^n$ *and a real n-by-n matrix* U *satisfying*

$$X = U^T(\operatorname{Diag} x)U, \quad U^T U = I \tag{6.2}$$

$$Y = U^T(\operatorname{Diag} y)U, \quad y \in N_C(x). \tag{6.3}$$

In other words, if we can recognize normals to the symmetric convex set C, then we can recognize normals to the convex spectral set $\lambda^{-1}(C)$ via simultaneous spectral decompositions.

6.4 Normals to nonconvex sets and Clarke regularity

The normal cone to a convex set $C \subset \mathbf{R}^n$ has the following key elementary properties, which may be found in Rockafellar (1970), for example.

- (i) $N_C(x)$ is a **convex cone** for any point $x \in C$.
- (ii) The **best approximation** condition (6.1) holds.
- (iii) The set-value mapping $x \in C \mapsto N_C(x)$ has **closed graph**: if $(x_r, y_r) \to (x, y)$ in $\mathbf{R}^n \times \mathbf{R}^n$ and $y_r \in N_C(x_r)$, then $y \in N_C(x)$.

This latter property guarantees some robustness for the normal cone, in theory and algorithmic practice.

To broaden the context of variational analysis to nonconvex closed sets $S \subset \mathbf{R}^n$ (such as smooth manifolds), we define the *Clarke normal cone mapping* $N_S : S \to \mathbf{R}^n$ to be the set-valued mapping satisfying properties (i), (ii), (iii) with minimal graph: see Clarke (1973) and Clarke et al. (1998). Thus the normal cone at a point $\bar{x} \in S$ consists of all convex combinations of limits of directions from points near $\bar{x}$ to their projections on S.

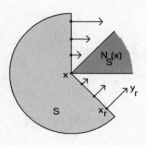

Variational analysis can also be developed in a parallel fashion without the assumption of convexity in property (i): see Mordukhovich (1976) and Rockafellar and Wets (1998). However, the Clarke cone suffices for this essay.

The Clarke normal cone is a useful tool for describing necessary optimality conditions for variational problems. For example, the best approximation condition (6.1) generalizes as follows: see Clarke (1983).

Theorem 6.5 (necessary optimality condition) *If the point* $\bar{x}$ *minimizes the smooth function* $f : \mathbf{R}^n \to \mathbb{R}$ *on the closed set* $S \subset \mathbf{R}^n$, *then* $-\nabla f(\bar{x}) \in N_S(\bar{x})$.

We call a closed set S *Clarke regular* at a
point $x \in S$ if any tangent direction to S at x
lies in $N_S(x)^*$: see Clarke (1983). Geometrically, for any sequences of points $w_r \in S$ and
$z_r \in \mathbf{R}^n$ approaching x, if z_r has a nearest
point x_r in S, and the angle between $z_r - x_r$
and $w_r - x$ converges to θ, then θ is ob-
tuse. Clarke regularity is in fact independent
of the inner product. Convex sets and manifolds are regular at every point x, in fact having the stronger property of *prox-regularity*:
every point near x has a unique nearest point
in S (see Poliquin et al. (2000)).

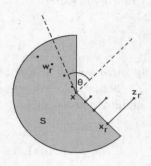

Clarke regularity is a recurrent theme in this essay, and plays a central
role both in theoretical variational analysis and in algorithmic matters. The
following result, from Clarke (1983), is an example of the kind of calculus
that Clarke regularity expedites.

Theorem 6.6 (chain rule) *Suppose that the map* $\Phi : \mathbf{R}^m \to \mathbf{R}^n$ *is smooth
around the point* $y \in \mathbf{R}^m$*, and that the closed set* $S \subset \mathbf{R}^n$ *is Clarke regular
at the point* $\Phi(y)$*. If*

$$N_S(\Phi(y)) \cap \ker(\nabla\Phi(y))^* = \{0\},$$

then the inverse image $\Phi^{-1}(S)$ *is Clarke regular at* y*, with Clarke normal
cone given by*

$$N_{\Phi^{-1}(S)}(y) = (\nabla\Phi(y))^* N_S(\Phi(y)).$$

We return to the implications of Clarke regularity for nonsmooth optimization algorithms towards the end of this essay.

Remarkably, the characteristic map behaves just as well with respect to
the Clarke normal cone as it does for normal cones to convex sets: see
Lewis (1996b). Furthermore, Clarke regularity "lifts" from $\mathbf{R}^n$ to $\mathbf{S}^n$.

Theorem 6.7 (spectral Clarke normal cones) *If the set* $S \subset \mathbf{R}^n$ *is
symmetric and closed, then matrices* $X, Y \in \mathbf{S}^n$ *satisfy* $Y \in N_{\lambda^{-1}(S)}(X)$ *if
and only if equations (6.2) and (6.3) hold. Furthermore, the spectral set*
$\lambda^{-1}(S)$ *is Clarke regular at* X *if and only if* S *is Clarke regular at the point*
$\lambda(X)$*.*

This result even remains unchanged for the nonconvex normal cone: see
Lewis (1999b).

As an example, consider the optimization problem

$$\sup\{\langle X, Y \rangle : X \in \mathbf{S}^n,\ \lambda(X) = x\},$$

for a given vector $x \in \mathbf{R}^n$ with nonincreasing components and a given matrix $Y \in \mathbf{S}^n$. The characteristic map λ is nonexpansive, by Corollary 6.1 (characteristic map behavior), so in particular continuous, and $\|\lambda(X)\| = \|X\|$ for all $X \in \mathbf{S}^n$. Hence continuity and compactness ensure this problem has an optimal solution X_0. Applying Theorem 6.5 (necessary optimality condition) shows $Y \in N_\Omega(X_0)$, where Ω is the spectral set $\lambda^{-1}(x) = \lambda^{-1}(\mathbf{P}^n x)$, so Theorem 6.7 (spectral Clarke normal cones) shows that the matrices X_0 and Y have a simultaneous spectral decomposition. An elementary argument then shows $\langle X_0, Y \rangle = x^T \lambda(Y)$, so we deduce the well-known inequality (essentially due to von Neumann (1937)).

$$\langle X, Y \rangle \leq \lambda(X)^T \lambda(Y), \quad \text{for any } X, Y \in \mathbf{S}^n. \tag{6.4}$$

6.5 Stability and the Belgian chocolate problem

We turn next to eigenvalues of nonsymmetric matrices. Our primary focus is on the set of *stable* matrices $\mathbf{M}^n_{\text{st}}$, which consists of those matrices in $\mathbf{M}^n$ having all eigenvalues in the closed left halfplane. The stability of a matrix $A \in \mathbf{M}^n$ is closely related to the asymptotic behavior of the dynamical system $\dot{x} = Ax$: specifically, as time t increases, e^{At} decays like $e^{\alpha t}$ if and only if $A - \alpha I$ is stable.

Analogously, a polynomial $p(z)$ is *stable* if all its roots lie in the closed left halfplane: if in fact they lie in the *open* halfplane, we call the polynomial *strictly* stable. Thus a matrix is stable exactly when its characteristic polynomial is stable. The set of stable monic polynomials

$$\Delta_n = \left\{ w \in \mathbf{C}^n : z^n + \sum_{j=0}^{n-1} w_j z^j \text{ stable} \right\}$$

has the following beautiful variational property: see Burke and Overton (2001b).

Theorem 6.8 (regularity of stable polynomials) *The set of stable monic polynomials Δ_n is everywhere Clarke regular.*

The corresponding property for the stable matrices $\mathbf{M}^n_{\text{st}}$ elegantly illustrates the power of nonsmooth calculus. We consider the map $\Phi : \mathbf{M}^n \to$

$\mathbf{C}^n$ taking a matrix $X \in \mathbf{M}^n$ to its characteristic polynomial:

$$\det(X - zI) = z^n + \sum_{j=0}^{n-1} \Phi(X)_j z^j.$$

With this notation we have $\mathbf{M}_{\mathrm{st}}^n = \Phi^{-1}(\Delta_n)$. Even if X has a multiple eigenvalue (as a root of its characteristic polynomial), the *nonderogatory* case where each eigenspace is one-dimensional is "typical" (from the perspective of Arnold's stratification of $\mathbf{M}^n$ into manifolds with fixed Jordan structure—see Arnold (1971)). In this case, the derivative $\nabla\Phi(X)$ is onto, so we can calculate the Clarke normal cone to $\mathbf{M}_{\mathrm{st}}^n$ at X easily using the chain rule (Theorem 6.6), thereby recapturing the central result of Burke and Overton (2001a).

Corollary 6.3 (regularity of stable matrices) *The set of stable matrices $\mathbf{M}_{\mathrm{st}}^n$ is Clarke regular at any stable nonderogatory matrix $X \in \mathbf{M}^n$, with Clarke normal cone*

$$N_{\mathbf{M}_{\mathrm{st}}^n}(X) = \nabla\Phi(X)^* N_{\Delta_n}(\Phi(X)).$$

An instructive two-part problem involving sets of stable polynomials was proposed by Blondel (1994), as a challenge to illustrate the difficulty of simultaneous plant stabilization in control. This problem illustrates the interplay between modelling (in this case, control-theoretic), computational experiments, and nonsmooth optimization theory.

Problem Given a real parameter δ, consider the problem of finding real stable polynomials p, q, r satisfying

$$r(z) = (z^2 - 2\delta z + 1)p(z) + (z^2 - 1)q(z). \qquad (6.5)$$

(Notice the problem admits no solution if $\delta = 1$.) Solve this problem when $\delta = 0.9$, and calculate how far δ can increase before the problem is unsolvable.

Blondel offered a prize of one kilogram of Belgian chocolate for each part of this problem. The first part was solved by a randomized search in Patel et al. (2002). The second part remains open, although, following work surveyed in Patel et al. (2002), the answer is known to be strictly less than one.

Consider the following variational approach. We vary polynomials p and q of fixed degree in order to move the roots of p, q, and r as far to the left in the complex plane as possible. After normalizing so that the product pqr is monic, we arrive at the following numerical optimization problem.

$$(\mathrm{Bl}_\delta) \quad \begin{cases} \text{minimize} \ \ \alpha \\ \text{subject to } p(z + \alpha)q(z + \alpha)r(z + \alpha) \text{ stable monic} \\ \qquad\qquad p, q \text{ cubic}, \ r \text{ given by equation (6.5)} \end{cases}$$

In Section 6.8 we describe a simple, general-purpose, "gradient sampling" method for numerical nonsmooth optimization. Computational experiments with this technique suggest that, for all values of the parameter δ near 0.9, the optimal solution $\bar{p}, \bar{q}, \bar{r}, \bar{\alpha}$ of the problem (Bl_δ) has a persistent structure:

- the polynomial $\bar{q}$ is scalar;
- the polynomial $z \mapsto \bar{r}(z + \bar{\alpha})$ is a multiple of z^5;
- the objective value satisfies $\bar{\alpha} < 0$ (**solving Blondel's problem**);
- the polynomial $z \mapsto \bar{p}(z + \bar{\alpha})$ is strictly stable.

The figure below (from Burke et al. (2005a)) shows the roots of optimal polynomials $\bar{p}$ $(+)$ and $\bar{r}$ $(\times)$ for various values of δ.

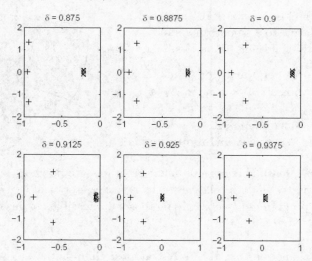

Having observed this structure computationally, some simple algebra shows that for any value of δ near

$$\bar{\delta} = \frac{1}{2}\sqrt{2 + \sqrt{2}} = 0.92\ldots,$$

the problem (Bl_δ) has a unique feasible solution with this structure, solving Blondel's problem for $\delta \le \bar{\delta}$. Furthermore, a little nonsmooth calculus using Theorem 6.8 (regularity of stable polynomials) shows, at least with the extra restriction that q is scalar, that this solution is a strict local minimizer for (Bl_δ): see Burke et al. (2005a).

6.6 Partly smooth sets and sensitivity

The persistent structure of optimal solutions for Blondel's problem in the previous section exemplifies a widespread phenomenon in optimization. Assuming appropriate nondegeneracy conditions, optimal solutions for linear, nonlinear, semidefinite and semi-infinite programs all have structures that persist under small perturbations to the problem: in linear programs, the optimal basis is fixed, in nonlinear programs, the active set stays unchanged, and the rank of the optimal matrix in a semidefinite program is constant. Variational analysis offers a unifying perspective on this phenomenon.

Nonsmoothness abounds in optimization, but is usually **structured**. The following definition from Lewis (2003) captures a key structural idea for the sensitivity analysis of smooth and nonsmooth optimization problems.

We call a closed set $S \subset \mathbf{R}^n$ *partly smooth* relative to a smooth manifold $M \subset S$ if the following properties hold.

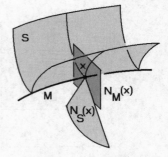

- S is **Clarke regular** throughout M.
- M is a **"ridge"** of S: that is, $N_S(x)$ spans $N_M(x)$ for all points $x \in M$.
- The set-valued mapping $x \in M \mapsto N_S(x)$ is **continuous**.

For example, feasible regions of linear programs are polyhedral: any polyhedron is partly smooth relative to the relative interior of any of its faces. Nonlinear programming considers more general feasible regions of the form

$$S \ = \ \{x \in \mathbf{R}^n : g_i(x) \le 0 \text{ for } i = 1, 2, \ldots, m\},$$

for smooth functions $g_i : \mathbf{R}^n \to \mathbf{R}$. Suppose the point $\bar{x} \in S$ satisfies the *linear independence constraint qualification*:

$$\{\nabla g_i(\bar{x}) : i \in I\} \text{ is linearly independent, where } I = \{i : g_i(\bar{x}) = 0\}.$$

In this case the set defined by the active constraints

$$M \ = \ \left\{ x : g_i(x) = 0 \text{ for } i \in I, \ \|x - \bar{x}\| < \epsilon \right\}$$

is a manifold for small $\epsilon > 0$, relative to which the set S is partly smooth.

As a final example, consider the semidefinite cone $\mathbf{S}_+^n$. In the space $\mathbf{S}^n$, for any integer $r = 0, 1, \ldots, n$, the set of matrices in $\mathbf{S}^n$ of rank r constitute a manifold, relative to which $\mathbf{S}_+^n$ is partly smooth. Feasible regions of semidefinite programs are inverse images of $\mathbf{S}_+^n$ under affine maps. We can see that such sets are also partly smooth, using a chain rule analogous to Theorem 6.6.

The notion of partial smoothness unifies a variety of **active set** ideas in optimization. Typical sensitivity analysis for variational problems shows that smooth perturbation of the parameters defining a problem leads to a solution that varies smoothly while retaining a constant underlying structure, often reflecting a persistent set of binding or "active" constraints. Partial smoothness abstracts this general observation, generalizing earlier work on convex optimization in Burke and Moré (1988) and Wright (1993).

Consider for example a feasible region $S \subset \mathbf{R}^n$ and an optimization problem

$$(\mathrm{P}_y) \qquad \inf\{\langle y, x \rangle : x \in S\},$$

depending on the parameter $y \in \mathbf{R}^n$. By Theorem 6.5 (necessary optimality condition), any optimal solution x for (P_y) must satisfy

$$(\mathrm{OC}_y) \qquad -y \in N_S(x).$$

Suppose the instance $(\mathrm{P}_{\bar{y}})$ (for some particular vector $\bar{y} \in \mathbf{R}^n$) has an optimal solution $\bar{x}$ lying on a manifold $M \subset S$ relative to which S is partly smooth. Let us make two further assumptions, typical in sensitivity analysis:

(i) the Clarke normal cone $N_S(\bar{x})$ contains the vector $-\bar{y}$ in its relative interior (that is, relative to its span);

(ii) perturbing the point $\bar{x}$ on M leads to quadratic growth of the linear function $\langle \bar{y}, \cdot \rangle$.

Condition (i) is a strengthening of condition $(\mathrm{OC}_{\bar{y}})$ typically known as a *strict complementarity condition*. Condition (ii) is a *second-order sufficient condition*. With these assumptions, for any y near $\bar{y}$, the optimality condition (OC_y) has a unique solution $x(y) \in M$ near $\bar{x}$, depending smoothly on y. If we assume that S is in fact prox-regular (rather than simply Clarke regular) throughout M, then $x(y)$ must be a local minimizer for the instance (P_y). Furthermore, in this case, a variety of common conceptual algorithms applied to (P_y) "identify" the manifold M finitely: the algorithm generates iterates eventually lying in this manifold— see Hare and Lewis (2004).

Partial smoothness offers a simple unifying language to illuminate the persistent structure of the optimal solutions of perturbed linear, nonlinear, and semidefinite programs. We next apply this idea to the Belgian chocolate problem.

If a polynomial lies on the boundary of the set of stable monics, then it has some purely imaginary roots $iy_1, iy_2, \ldots, iy_r$ (where we assume $y_1 > y_2 > \ldots > y_r$). If each such root iy_j has multiplicity m_j, we call the sequence $m_1, m_2, \ldots, m_r$ the *imaginary multiplicity list*. In the example to the right, the multiplicity list is $3, 1, 2$.

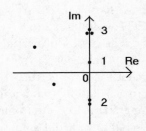

The root cause for the persistent structure in the solutions to the Belgian chocolate problem is then the following result.

Theorem 6.9 (partial smoothness of stable polynomials) *Consider a polynomial lying in the set of stable monics Δ_n. The set of nearby monics with the same imaginary multiplicity list constitute a manifold, with respect to which Δ_n is partly smooth.*

Applying a suitable chain rule using the characteristic polynomial map, just as we derived Corollary 6.3 (regularity of stable matrices) from Theorem 6.8 (regularity of stable polynomials), we deduce the analogous matrix version below: see Lewis (2003).

Corollary 6.4 (partial smoothness of stable matrices) *Consider a nonderogatory matrix lying in the stable set $\mathbf{M}_{\mathrm{st}}^n$. The set of nearby matrices with the same imaginary eigenvalue multiplicity list constitute a manifold, with respect to which $\mathbf{M}_{\mathrm{st}}^n$ is partly smooth.*

In practice, varying a parametrized matrix in order to move its eigenvalues as far as possible into the left halfplane typically leads to nonderogatory optimal solutions with multiple eigenvalues: see Burke et al. (2002b, 2005b). The above result crystallizes the underlying theoretical cause of this phenomenon: see Burke et al. (2000, 2001).

6.7 Nonsmooth analysis and the distance to instability

So far in this essay we have taken a geometric approach to variational analysis and nonsmooth optimization, emphasizing the role of the Clarke normal cone. Conceptually, however, the theory is much broader, encompassing powerful generalizations of the derivative and of classical calculus:

see Clarke et al. (1998) and Rockafellar et al. (1998). We next briefly sketch the beginnings of this development, building on the geometric ideas we have already introduced.

Consider a function $f : \mathbf{R}^n \to [-\infty, \infty]$ with closed *epigraph*

$$\mathrm{epi}(f) = \{(x, r) \in \mathbf{R}^n \times \mathbf{R} : r \geq f(x)\}.$$

By analogy with the smooth case, we define the *Clarke generalized derivative* by

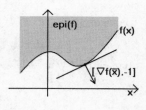

$$\partial f(\bar{x}) = \{y : (y, -1) \in N_{\mathrm{epi}(f)}(\bar{x}, f(\bar{x}))\}.$$

Theorem 6.5 (necessary optimality condition) implies the following central role for the generalized derivative in optimization:

$$\bar{x} \text{ minimizes } f \quad \Rightarrow \quad 0 \in \partial f(\bar{x}). \tag{6.6}$$

We call f *Clarke regular* at $\bar{x}$ if $\mathrm{epi}(f)$ is Clarke regular at $(\bar{x}, f(\bar{x}))$, and make the analogous definition for prox-regularity: see Poliquin and Rockafellar (1996). For example, any smooth function f is Clarke regular, with generalized derivative $\partial f(\bar{x}) = \{\nabla f(\bar{x})\}$. Any convex function is also Clarke regular, with generalized derivative agreeing with the classical convex subdifferential: see Rockafellar (1970).

Our approach to the generalized derivative above is appealing in its theoretical economy, but is conceptually opaque. The definition makes little obvious connection with classical differentiation. The following result from Clarke (1973) relates the generalized derivative of a Lipschitz function to the local behavior of its derivative, which is defined almost everywhere by virtue of Rademacher's theorem.

Theorem 6.10 (generalized derivatives of Lipschitz functions) *The Clarke generalized derivative of a Lipschitz function $f : \mathbf{R}^n \to \mathbf{R}$ at a point $\bar{x} \in \mathbf{R}^n$ is given by*

$$\partial f(\bar{x}) = \mathrm{conv} \{\lim \nabla f(x_r) : x_r \to \bar{x}\}.$$

The function f is Clarke regular at $\bar{x}$ if and only if its directional derivative satisfies

$$f'(\bar{x}; d) = \limsup_{x \to \bar{x}} \langle \nabla f(x), d \rangle$$

for every direction $d \in \mathbf{R}^n$

Without Clarke regularity, the optimality condition (6.6) may be weak. For example, zero *maximizes* minus the absolute value function, yet $0 \in \partial(-|\cdot|)(0)$. With regularity however, (6.6) strengthens to the more intuitive condition $f'(\bar{x}; d) \geq 0$ for all directions d.

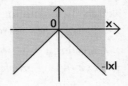

A class of functions very common in applications are those $f : \mathbf{R}^n \to \mathbf{R}$ that can be written locally in the form

$$f(x) = \max_{t \in T} f_t(x),$$

where the parameter set T is compact, and the map $(x, t) \mapsto \nabla^k f_t(x)$ is continuous for $k = 0, 1, 2$. Such functions are called *lower-$C^{(2)}$*: they are prox-regular, so in particular, Clarke regular—see Rockafellar et al. (1998).

A typical example of a lower-$C^{(2)}$ function, arising in robust control systems design, is called the *distance to instability* in Byers (1988), and is also known as the *complex stability radius*—see Hinrichson and Pritchard (1986). It is the distance from a matrix in $\mathbf{M}^n$ to the set of unstable matrices. An easy argument shows that, for any matrix $X_0 \in \mathbf{M}^n$, this function $\beta : \mathbf{M}^n \to \mathbf{R}$ can be written in the form

$$\beta(X) \quad = \quad \min\{\|Xu - zu\| : \|u\| = 1, \ \mathrm{Re}\, z \geq 0, \ |z| \leq k\}$$

for all matrices $X \in \mathbf{M}^n$ near X_0, where the constant k depends on X_0. If X_0 is strictly stable, then the quantity $\|Xu - zu\|$ is bounded away from zero for all X near X_0, unit vectors $u \in \mathbf{C}^n$, and complex z with $\mathrm{Re}\, z \geq 0$. Consequently, the function $-\beta$ is lower-$C^{(2)}$ on the strictly stable matrices. For the $\mathbf{H}^\infty$-norm in robust control (see Zhou et al. (1996)), a similar analysis applies.

The figure below, from Burke et al. (2005a), shows the results of maximizing the minimum of the two distances to instability of the companion matrices corresponding to the polynomials p and r in the chocolate problem. We restrict p to be a monic cubic, plotting its roots as $\diamondsuit$, and q to be a scalar; we plot the roots of r as $\circ$. To compare, we leave the old optimally stable roots in the plot. Notice how maximizing the stability radius causes the root of order five to split, moving the roots closer to the imaginary axis but nonetheless increasing the distance to instability.

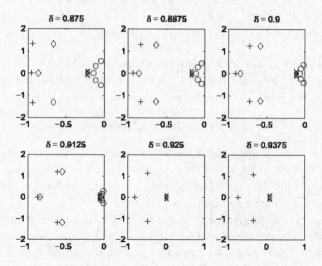

6.8 The gradient sampling method

Despite half a century of advances in computational optimization, and several decades of development in the theory of nonsmooth optimization, numerical minimization of nonsmooth nonconvex functions remains challenging: dependable publicly-available code is scarce. The results described above for the Belgian chocolate problem were obtained in Burke et al. (2005a) using a simple intuitive numerical method based on *gradient sampling*: see Burke et al. (2002b, 2005b).

To motivate this method, consider a Lipschitz function $f : \mathbf{R}^n \to \mathbf{R}$ that is Clarke regular at the point $x \in \mathbf{R}^n$. The *direction of steepest descent* is the unit vector $u \in \mathbf{R}^n$ minimizing the directional derivative $f'(x; u)$. Theorem 6.10 (generalized derivatives of Lipschitz functions) shows that this vector lies in the direction of the vector

$$\overline{d} = -\lim_{\epsilon \downarrow 0} \operatorname{argmin}\{\|d\| : d \in \operatorname{cl} \operatorname{conv} \nabla f(x + \epsilon B)\}$$

(where B is the unit ball in $\mathbf{R}^n$). For example, if f is smooth at x, then $\overline{d} = -\nabla f(x)$.

To approximate the direction $\overline{d}$, we fix some small radius $\epsilon > 0$, and sample some number $m > n$ random, independent, uniformly distributed points $Y_j \in x + \epsilon B$. Almost surely, f is differentiable at each Y_j: we assume, as is often the case in practice, that the gradients $\nabla f(Y_j)$ are readily available. We then define an approximate steepest descent direction by

$$\widehat{d} \ = \ -\operatorname{argmin}\{\|d\| : d \in \operatorname{conv}\{\nabla f(Y_j) : j = 1, \ldots, m\}\}.$$

In practice, we choose $Y_1 = x$, to guarantee $f'(x; \widehat{d}) < 0$.

Finally, we imitate the classical steepest descent method for smooth minimization. We perform a simple *linesearch* to choose a *stepsize*

$$\bar{t} \ \approx \ \operatorname{argmin}_{t \geq 0} f(x + t\widehat{d}),$$

(in practice often simply requiring the descent condition $f(x + \bar{t}\widehat{d}) < f(x)$). We then update $x \leftarrow x + \bar{t}\widehat{d}$, and repeat the whole process. The loop terminates when the vector $\widehat{d}$ is small, at which point we may stop, or restart with a smaller radius ϵ.

For nonsmooth nonconvex minimization problems with cheap function and gradient evaluations and involving not too many variables, computational experience suggests that the gradient sampling method is a robust and reliable tool: see Burke et al. (2005b). The random sampling approach to approximating the Clarke generalized derivative is motivated theoretically in Burke et al. (2002a). Under reasonable conditions and suitably implemented, the minimization method converges almost surely to a point whose generalized derivative contains zero (see Burke et al. (2005b)): as discussed above, assuming Clarke regularity, this condition guarantees that there are no descent directions. Random sampling helps the method avoid a common difficulty in designing nonsmooth optimization algorithms: the expected value of the random search direction $\widehat{d}$ depends continuously on the current point x (see Lewis (2005)), so the algorithm does not "jam".

6.9   Lidskii's theorem again

The Clarke normal cone and generalized derivative are powerful and versatile tools. Our discussion in the previous section indicates their use in understanding algorithms for nonsmooth optimization. Our analysis of the Belgian chocolate problem and subsequent sketch of the idea of partial smoothness suggests the utility of nonsmooth language for optimality conditions and sensitivity analysis. To bring this essay full circle, we end with a purely analytic application of nonsmooth optimization ideas, using variational analysis to prove Lidskii's eigenvalue perturbation result (Theorem 6.1). The argument follows Lewis (1999a).

Given two matrices $X, Y \in \mathbf{S}^n$, we wish to show

$$\lambda(X) - \lambda(Y) \in \operatorname{conv}(\mathbf{P}^n \lambda(X - Y)).$$

If this inclusion fails, the separating hyperplane theorem implies the existence of a vector $w \in \mathbf{R}^n$ satisfying the inequality

$$w^T(\lambda(X) - \lambda(Y)) > \max_{P \in \mathbf{P}^n} w^T P \lambda(X - Y).$$

An elementary argument identifies the right hand side as $[w]^T \lambda(X - Y)$, where the vector $[w] \in \mathbf{R}^n$ has the same components as w rearranged into nonincreasing order.

Turning our attention to the left hand side of the above inequality, we consider the (nonconvex) spectral function $F = w^T \lambda$. A suitable nonsmooth version of the mean value theorem (see Clarke (1983)) applied to this function shows that, for some matrix V on the line segment between X and Y and some matrix $Z \in \partial F(V)$, we have

$$F(X) - F(Y) = \langle Z, X - Y \rangle \le \lambda(Z)^T \lambda(X - Y),$$

using von Neumann's inequality (6.4). The analogous result to Theorem 6.7 (spectral Clarke normal cones) for generalized derivatives shows there exists vectors $v, z \in \mathbf{R}^n$ and a real n-by-n matrix U satisfying

$$V = U^T(\operatorname{Diag} v)U, \quad U^T U = I$$
$$Z = U^T(\operatorname{Diag} z)U, \quad z \in \partial f(v),$$

where $f : \mathbf{R}^n \to \mathbf{R}$ is the function defined by $f(v) = w^T[v]$. Theorem 6.10 (generalized derivatives of Lipschitz functions) implies $\partial f(v) \subset \operatorname{conv}(\mathbf{P}^n w)$, so

$$\lambda(Z) = [z] \in \operatorname{conv}(\mathbf{P}^n[w]).$$

We quickly deduce the contradiction $\lambda(Z)^T \lambda(X - Y) \le [w]^T \lambda(X - Y)$, completing the proof.

6.10 Final thoughts

Modern variational analysis deserves a wider mathematical audience than it has so far reached. This essay aims to make converts, by illustrating the elegant interplay between eigenvalues and nonsmooth optimization.

Convexity is a ubiquitous mathematical idea, and its significance for spectral properties of symmetric matrices is well known: Lidskii's theorem is a central example. Hyperbolic polynomials provide a starkly simple setting in which to view many of these classical properties. Conversely, the truth of the Lax conjecture extends a wealth of symmetric matrix results and methods, including Lidskii's theorem, to hyperbolic polynomials.

Over several decades, convex analysis has developed into a widely-used language for diverse areas of mathematics beyond just optimization, neatly unifying the geometry, duality, and calculus of convex sets and functions, and their normal cones and generalized derivatives. Building on this success, the nonconvex theory pioneered by Clarke has matured into a versatile toolkit. The spectral behavior of symmetric matrices provides a powerful illustration of this theory in action.

Far from being pathological or rare, nonsmoothness is fundamental to our understanding of optimization. Its occurrence in concrete problems is typically structured. In particular, Clarke regularity is often a reasonable assumption, with far-reaching implications both in theory and computational practice. Many nonsmooth optimization problems exhibit partial smoothness, an advantageous mix of smooth and nonsmooth behavior helping our understanding of sensitivity analysis and algorithm convergence.

Optimizing the stability of polynomials or matrices exemplifies partially smooth behavior: active roots or eigenvalues cluster at optimal solutions. The Belgian chocolate problem is a typical illustration, both of the theory and of the robust effectiveness of Gradient Sampling, a simple and intuitive computational approach to nonsmooth optimization.

Variational analysis, in its full generality, is less familiar and more challenging than in the convex case. However, the nonsmooth nature of eigenvalues makes it a natural ingredient for perturbation theory in particular, and matrix analysis in general. It seems likely that the interplay between eigenvalues and nonsmooth optimization, briefly sketched in this essay, will continue to flourish.

References

Arnold, V.I. (1971), On matrices depending on parameters, *Russian Mathematical Surveys* **26**, 29–43.

Baes, M. (2004), Spectral functions on Jordan algebras: differentiability and convexity properties, Technical report, CORE, Louvain, Belgium.

Bauschke, H.H., Güler, O., Lewis, A.S., and Sendov, H.S. (2001), Hyperbolic polynomials and convex analysis, *Canadian Journal of Mathematics* **53**, 470–488.

Ben-Tal, A. and Nemirovski, A. (2001), *Lectures on Modern Convex Optimization: Analysis, Algorithms, and Engineering Applications*, SIAM, Philadelphia.

Bhatia, R. (1997), *Matrix Analysis*, Springer, New York.

Blondel, V.D. (1994), *Simultaneous Stabilization of Linear Systems.*, Springer, Berlin.

Borwein, J.M. and Lewis, A.S. (2000), *Convex Analysis and Nonlinear Optimization*, Springer, New York.

Burke, J.V., Henrion, D., Lewis, A.S., and Overton, M.L. (2005a), Analysis of a Belgian chocolate stabilization problem, *IEEE Transactions on Automatic Control*, to appear.

Burke, J.V., Lewis, A.S., and Overton, M.L. (2000), Optimizing matrix stability, *Proceedings of the American Mathematical Society* **129**, 1635–1642.

Burke, J.V., Lewis, A.S., and Overton, M.L. (2001), Optimal stability and eigenvalue multiplicity, *Foundations of Computational Mathematics* **1**, 205–225.

Burke, J.V., Lewis, A.S., and Overton, M.L. (2002a), Approximating subdifferentials by random sampling of gradients, *Mathematics of Operations Research* **27**, 567–584.

Burke, J.V., Lewis, A.S., and Overton, M.L. (2002b), Two numerical methods for optimizing matrix stability, *Linear Algebra and its Applications* **351/2**, 117–145.

Burke, J.V., Lewis, A.S., and Overton, M.L. (2003), A nonsmooth, nonconvex optimization approach to robust stabilization by static output feedback and low-order controllers, technical report, Courant Institute, to appear, Proceedings of the 4th IFAC Symposium on Robust Control Design, ROCOND'03.

Burke, J.V., Lewis, A.S., and Overton, M.L. (2005b), A robust gradient sampling algorithm for nonsmooth, nonconvex optimization, *SIAM Journal on Optimization* **15**, 751–779.

Burke, J.V. and Moré, J.J. (1988), On the identification of active constraints, *SIAM Journal on Numerical Analysis* **25**, 1197–1211.

Burke, J.V. and Overton, M.L. (2001a), Variational analysis of non-Lipschitz spectral functions, *Mathematical Programming* **90**, 317–352.

Burke, J.V. and Overton, M.L. (2001b), Variational analysis of the abscissa mapping for polynomials, *SIAM Journal on Control and Optimization* **39**, 1651–1676.

Byers, R. (1988), A bisection method for computing the distance of a stable matrix to the unstable matrices, *SIAM Journal on Scientific and Statistical Computing* **9**, 875–881.

Clarke, F.H. (1973), *Necessary Conditions for Nonsmooth Problems in Optimal Control and the Calculus of Variations*, PhD thesis, University of Washington, Seattle.

Clarke, F.H. (1983), *Optimization and Nonsmooth Analysis*, Wiley, New York. Republished as Vol. 5, Classics in Applied Mathematics, SIAM, 1990.

Clarke, F.H., Ledyaev, Yu.S., Stern, R.J., and Wolenski, P.R. (1998), *Nonsmooth Analysis and Control Theory*, Springer-Verlag, New York.

Gårding, L. (1951), Linear hyperbolic differential equations with constant coefficients, *Acta Mathematica* **85**, 2–62.

Gårding, L. (1959), An inequality for hyperbolic polynomials, *Journal of Mathematics and Mechanics* **8**, 957–965.

Gurvits, L. (2004), Combinatorics hidden in hyperbolic polynomials and related topics, technical report, Los Alamos National Laboratory.

Hare, W.L. and Lewis, A.S. (2004), Identifying active constraints via partial smoothness and prox-regularity, *Journal of Convex Analysis* **11**, 251–266.

Helton, J.W. and Vinnikov, V. (2002), Linear matrix inequality representation of sets, technical report, Department of Mathematics, University of California San Diego. Available as `arXiv:math.OC/0306180v1 11 Jun 2003`.

Hinrichsen, D. and Pritchard, A.J. (1986), Stability radii of linear systems, *Systems and Control Letters* **7**, 1–10.

Horn, R.A. and Johnson, C. (1985), *Matrix Analysis*, Cambridge University Press, Cambridge, U.K.

Kato, T. (1982), *A Short Introduction to Perturbation Theory for Linear Operators*, Springer-Verlag, New York.

Lax, P.D. (1958), Differential equations, difference equations and matrix theory, *Communications on Pure and Applied Mathematics* **6**, 175–194.

Lewis, A.S. (1996a), Convex analysis on the Hermitian matrices, *SIAM Journal on Optimization* **6**, 164–177.

Lewis, A.S. (1996b), Derivatives of spectral functions, *Mathematics of Operations Research* **6**, 576–588.

Lewis, A.S. (1996c), Group invariance and convex matrix analysis, *SIAM Journal on Matrix Analysis and Applications* **17**, 927–949.

Lewis, A.S. (1999a), Lidskii's theorem via nonsmooth analysis, *SIAM Journal on Matrix Analysis and Applications* **21**, 379–381.

Lewis, A.S. (1999b), Nonsmooth analysis of eigenvalues, *Mathematical Programming* **84**, 1–24.

Lewis, A.S. (2000), Convex analysis on Cartan subspaces, *Nonlinear Analysis, Theory, Methods and Applications* **42**, 813–820.

Lewis, A.S. (2003), Active sets, nonsmoothness and sensitivity, *SIAM Journal on Optimization* **13**, 702–725.

Lewis, A.S. (2005), Local structure and algorithms in nonsmooth optimization, in Jarre, F., Lemaréchal, C., and Zowe, J., editors, *Oberwolfach Proceedings: Workshop on Optimization*.

Lewis, A.S., Parrilo, P.A., and Ramana, M.V. (2005), The Lax conjecture is true, *Proceedings of the American Mathematical Society* **133**, 2495–2499.

Mordukhovich, B.S. (1976), Maximum principle in the problem of time optimal response with nonsmooth constraints, *Journal of Applied Mathematics and Mechanics* **40**, 960–969.

von Neumann, J. (1937), Some matrix inequalities and metrization of matricspace, *Tomsk University Review* **1**, 286–300, in: *Collected Works*, Pergamon, Oxford, 1962, Volume IV, 205-218.

Patel, V.V., Doehare, G., and Viswanath, T. (2002), Some applications of a randomized algorithms for control system design, *Automatica* **28**, 2085–2092.

Poliquin, R.A. and Rockafellar, R.T. (1996), Prox-regular functions in variational analysis, *Transactions of the American Mathematical Society* **348**, 1805–1838.

Poliquin, R.A., Rockafellar, R.T., and Thibault, L. (2000), Local differentiability of distance functions, *Transactions of the American Mathematical Society* **352**, 5231–5249.

Rockafellar, R.T. (1970), *Convex Analysis*, Princeton University Press, Princeton, N.J.

Rockafellar, R.T. and Wets, R.J.-B. (1998), *Variational Analysis*, Springer, Berlin.

Stewart, G.W. and Sun, J.G. (1990), *Matrix Perturbation Theory*, Academic Press, Boston.

Sun, D. and Sun, J. (2004), Löwner's operator and spectral functions in Euclidean Jordan algebras, technical report, National University of Singapore.

Todd, M.J. (2001), Semidefinite optimization, *Acta Numerica* **10**, 515–560.

Wright, S.J. (1993), Identifiable surfaces in constrained optimization, *SIAM Journal on Control and Optimization* **31**, 1063–1079.

Zhou, K., Doyle, J.C., and Glover, K. (1996), *Robust and Optimal Control*, Prentice-Hall, Upper Saddle River, NJ.

Noether's Theorem for Smooth, Difference and Finite Element Systems

Elizabeth L. Mansfield

Institute of Mathematics, Statistics and Actuarial Science
University of Kent
Canterbury, United Kingdom
e-mail: E.L.Mansfield@kent.ac.uk

Dedicated to the memory of Karin Gatermann *1961–2005*

7.1 Introduction

The question, "Is the long term qualitative behaviour of numerical solutions accurate?" is increasingly being asked. One way of gauging this is to examine the success or otherwise of the numerical code to maintain certain conserved quantities such as energy or potential vorticity. For example, numerical solutions of a conservative system are usually presented together with plots of energy dissipation. But what if the conserved quantity is a less well studied quantity than energy or is not easily measured in the approximate function space? What if there is more than one conserved quantity? Is it possible to construct an integrator that maintains, *a priori*, several laws at once?

Arguably, the most physically important conserved quantities arise via Noether's theorem; the system has an underlying variational principle and a Lie group symmetry leaves the Lagrangian invariant. A Lie group is a group whose elements depend in a smooth way on real or complex parameters. Energy, momentum and potential vorticity, used to track the development of certain weather fronts, are conserved quantities arising from translation in time and space, and fluid particle relabelling respectively. The Lie groups for all three examples act on the base space which is discretised. It is not obvious how to build their automatic conservation into a discretisation, and expressions for the conserved quantities must be known exactly in order to track them.

The study of Lie group symmetries of differential equations is one of the success stories of symbolic computation, (Hereman (1997)). Not only symmetries but integration techniques based on them are now commercially available, (Cheb-Terrab, Duarte and da Mota (1998)). Moreover, these can usually be obtained without understanding the underlying theory: no human interaction with the software is required. This success is built on the fact that explicit, exact, analytic formulae are known for all the requisite quantities (see for example Olver (1993)), and the algorithms which are required for the intermediate processing are well understood (Hubert (2000), Hubert (2003), Mansfield and Clarkson (1997), Reid (1991), Reid, Wittkopf and Boulton (1996)).

One possibility is to use symbolic methods to study symmetries and conservation laws of discrete systems. One might then calculate intrinsically conserved quantities of existing schemes, but so far this line of research has been less successful for a variety of reasons. The philosophical points of view that are possible for such a theory are still debated, and the computations involved are less tractable than those for smooth systems (Hydon (2000), Levi, Tremblay and Winternitz (2005), Quispel, Capel and Sahadevan (1992)).

The key objective of the present article is to examine the idea of making a conservation law an *intrinsic* property of a scheme by building in a symmetry and a discrete variational principle. There are several challenges to this approach. The first is to show how a group action that takes place in a base space that gets discretised is nevertheless still present in some sense. The second is to present a mathematical structure that allows a discrete conservation law to be proven rigorously from the existence of the symmetry.

At the simplest level, the proof of Noether's theorem for smooth systems involves symbolic manipulation of the formulae involved. It is necessary to dig a little deeper to see what might transfer to a discrete setting. The algebraic foundation and the mathematical structures using which Noether's theorem can be proved and elucidated involve the construction of a *variational complex* (see Olver (1993) and references therein). A complex is an *exact* sequence of maps, that is, the kernel of one map equals the image of the previous map in the sequence. The familiar grad – curl – div sequence is *locally exact*, that is, is exact provided the domain of the functions involved is diffeomorphic to a disc. The variational complex involves the extended sequence of operators, grad – curl – div – Euler-Lagrange – Helmholtz. This extension makes sense if the coefficient functions involve arbitrary dependent variables and their derivatives, as indeed a Lagrangian does. Exactness means, for example, an expression is a divergence if and only if it maps to zero under the Euler–Lagrange operator.

Variational methods for difference systems have been available in the literature for some time (Kupershmidt (1985)). The complete set of proofs showing the difference variational complex is exact were given more recently (Hydon and Mansfield (2004)). There is no tangent structure on a discrete lattice, and so no "top down" construction for the variational complex for difference systems can exist. Yet the formulae involved in the difference version of Noether's theorem are amazingly similar, to the point where a "syntax translation" tells you how to convert one to its counterpart. An independent view and derivation of Noether's theorem for difference systems has been given by Dorodnitsyn (2001).

In the third part of this paper, we discuss how the algebraic arguments transfer to moment-based approximations on an arbitrary triangulation. Classical constructions from algebraic topology, such as simplicial spaces, chains and cochains, boundary and coboundary operators, are needed for this. These ideas are of increasing interest to both physicists and numerical analysts (Chard and Shapiro (2000), Hiptmair (2002), Mattiussi (1997), Schwalm, Moritz, Giona and Schwalm (1999), Tonti (1975)). The interplay of such notions with physical quantities and systems is being explored as a way to ensure that the correct geometry of a problem is encoded in the discretisation.

Our arguments require that the set of moments used fits into an exact scheme as described by Arnold (2002). This means that the various projections to finite dimensional function spaces need to maintain the exactness of the grad – curl – div sequence. Exactness guarantees the conditions for numerical stability given by Brezzi's theorem (Brezzi (1974)), so these ideas have innate meaning for the finite element method quite apart from those presented here. The variational complex for such schemes is detailed in Mansfield and Quispel (2005). Here, we develop those ideas further to investigate Noether's Theorem for finite element approximations.

In Section 7.1, a brief look at Noether's theorem for smooth systems tells the story in a way that the analogies for finite difference and finite element can be easily seen. This is followed by a discussion of the variational complex for difference systems. We define the difference Euler-Lagrange operator, explain how group actions are inherited, and give some examples. Also included is a discussion of how the theory of moving frames can be used to find difference invariants of given Lie group actions. These are used to construct a Lagrangian which *a priori* will have a conservation law corresponding to the given group.

The main result of this paper can be summarised as follows: instead of proving approximate conservation of an exact quantity, we demonstrate the possibility of exact conservation of an associated approximate quantity. The examples are deliberately small and straightforward.

7.2 Brief review of Noether's theorem for smooth systems

Definition 7.1 *A conservation law* for a system of differential equations is a divergence expression which is zero on solutions of the system.

For example, the heat equation $u_t + (-u_x)_x = 0$ is its own conservation law. To move from the divergence form to the more usual integral form, integrate over an arbitrary domain, assume ∂_t and $\int$ commute, and apply Stokes' Theorem, to obtain,

$$\frac{\partial}{\partial t} \int_\Omega u + (-u_x)]_{\partial\Omega} = 0.$$

In words, this equation reads, "the rate of change of total heat in Ω equals the net of comings and goings of heat across the boundary." The conserved quantity is (usually) that behind the time derivative in the divergence expression.

Noether's theorem provides a conservation law for an Euler–Lagrange system where the Lagrangian is invariant under a Lie group action. The Lagrangian here includes the volume form in the action integral, so we speak of the *Lagrangian form*. Table 7.1 gives the standard names of the conserved quantity for the most common group actions arising in physical applications.

Symmetry	Conserved Quantity
$\begin{cases} t^* = t + c \\ \text{translation in time} \end{cases}$	Energy
$\begin{cases} x_i^* = x_i + c \\ \text{translation in space} \end{cases}$	Linear Momenta vector
$\begin{cases} \mathbf{x}^* = \mathcal{R}\mathbf{x} \\ \text{rotation in space} \end{cases}$	Angular Momenta vector
$\begin{cases} a^* = \phi(a,b), b^* = \psi(a,b) \\ \phi_a\psi_b - \phi_b\psi_a \equiv 1 \\ \text{Particle relabelling} \end{cases}$	Potential vorticity

Table 7.1. *The usual examples*

7.2.1 The Euler–Lagrange Equations

The Euler–Lagrange equations are the result of applying a "zero derivative" condition when the dependent variable in a Lagrangian form is varied.

Example 6 *If the Lagrangian is $L[u] = \frac{1}{2}\left(u_x^2 + u_{xx}^2\right)\mathrm{d}x$ then the variation of $L[u]$ in the direction v is, by definition,*

$$
\begin{aligned}
\widehat{\mathrm{d}}L[u](v) &= \frac{\mathrm{d}}{\mathrm{d}\epsilon}\Big|_{\epsilon=0} L[u + \epsilon v] \\
&= (u_x v_x + u_{xx} v_{xx})\mathrm{d}x \\
&= (-u_{xx}v + u_{xxxx}v)\mathrm{d}x + \frac{\mathrm{D}}{\mathrm{D}x}\left(u_x \mathrm{d}u - 2u_{xx}v_x + \frac{\mathrm{D}}{\mathrm{D}x}\left(u_{xx}v\right)\right) \\
&= E(L)v\mathrm{d}x + \frac{\mathrm{D}}{\mathrm{D}x}\eta(L, v).
\end{aligned}
$$

The Euler–Lagrange equation for this Lagrangian is $u_{xxxx} - u_{xx} = 0$.

For the purposes of this article, the way to think of the Euler–Lagrange operator is as $E = \pi \circ \widehat{\mathrm{d}}$, where π projects out the total derivative (total divergence) term. In the case of more than one dependent variable, where each one varies separately, we obtain an equation for each dependent variable. For example,

$$
\begin{aligned}
\widehat{\mathrm{d}}L[u_1, u_2](v_1, v_2) &= \frac{\mathrm{d}}{\mathrm{d}\epsilon}\Big|_{\epsilon=0} L[u^1 + \epsilon v^1, u^2 + \epsilon v^2] \\
&= E^1(L)v^1\mathrm{d}x + E^2(L)v^2\mathrm{d}x + \frac{\mathrm{D}}{\mathrm{D}x}\eta(L, v).
\end{aligned}
$$

The Euler–Lagrange system is then $E^i(L) = 0$, $i = 1, 2$. General formulae, explicit, exact, symbolic, for E^i and $\eta(L, v)$ are known, (Olver (1993)).

7.2.2 Variational Symmetries

Symmetries of differential structures are studied in terms of Lie group actions. A Lie group is one whose elements can be parametrised smoothly by real (or complex) numbers. (More technically, a Lie group is a differentiable manifold with a group product, such that the multiplication and the inverse maps are smooth functions.) It turns out it is sufficient to study actions of one-parameter subgroups of Lie groups.

Definition 7.2 *A subgroup of the Lie group G is called a* one-parameter subgroup *if it is parametrised by $\mathbb{R}$, so that $g(\epsilon) \in G$ for all $\epsilon \in \mathbb{R}$, and*

$$
g(\epsilon) \cdot g(\delta) = g(\epsilon + \delta).
$$

For example, the set

$$
\left\{ \begin{pmatrix} \exp(\epsilon) & 0 \\ 0 & \exp(-\epsilon) \end{pmatrix} \mid \epsilon \in \mathbb{R} \right\}
$$

is a one-parameter subgroup of $SL(2, \mathbb{R})$, the special linear group of 2×2 matrices with determinant equal to one.

Definition 7.3 *A (right) action of a group G on a space M is a smooth map*

$$G \times M \to M, \qquad (g, z) \mapsto g \cdot z$$

such that

$$g_1 \cdot (g_2 \cdot z) = (g_2 g_1) \cdot z.$$

For a one-parameter subgroup this becomes

$$g(\delta) \cdot (g(\epsilon) \cdot z) = g(\delta + \epsilon) \cdot z.$$

Example 7 *For the group $G = (\mathbb{R}, +)$, that is, the real numbers under addition, the* projective action *on the plane is given by*

$$\epsilon \cdot x = x^* = \frac{x}{1 - \epsilon x}, \qquad \epsilon \cdot u = u^*(x^*) = \frac{u(x)}{1 - \epsilon x}. \tag{7.1}$$

This is actually only a *local* action since ϵ is restricted to a neighbourhood of $0 \in \mathbb{R}$, where the neighbourhood depends on x. We demonstrate the group action property:

$$\delta \cdot (\epsilon \cdot x) = \delta \cdot \left(\frac{x}{1 - \epsilon x} \right) = \frac{\dfrac{x}{1 - \delta x}}{1 - \epsilon \dfrac{x}{1 - \delta x}} = \frac{x}{1 - (\epsilon + \delta)x} = (\epsilon + \delta) \cdot x.$$

For actions on $\mathcal{X} \times \mathcal{U}$ where $\mathcal{X}$ is the space of independent variables and $\mathcal{U}$ the space of dependent variables, then an action is induced on the associated jet bundle. This is called the *prolongation* action and is obtained using the chain rule of undergraduate calculus. Thus, continuing Example 7,

$$\epsilon \cdot u_x = u^*_{x^*} = \frac{\partial u^*(x^*)}{\partial x} \Big/ \frac{\partial x^*}{\partial x} = \frac{u_x}{(1 - \epsilon x)^2}$$

and checking this indeed gives a group action,

$$\delta \cdot (\epsilon \cdot u_x) = \frac{\delta \cdot u_x}{(1 - \epsilon(\delta \cdot x))^2} = \frac{u_x}{(1 - (\delta + \epsilon)x)^2}.$$

Given a prolongation action, we then have an induced action on the integral of the Lagrangian form, given by

$$\epsilon \cdot \int_\Omega L(x, u, u_x, \dots)\, \mathrm{d}x := \int_{\epsilon \cdot \Omega} L(\epsilon \cdot x, \epsilon \cdot u, \epsilon \cdot u_x, \cdots)\, \mathrm{d}\epsilon \cdot x$$

$$= \int_\Omega L(\epsilon \cdot x, \epsilon \cdot u, \epsilon \cdot u_x, \cdots) \frac{\mathrm{d}\epsilon \cdot x}{\mathrm{d}x}\, \mathrm{d}x \tag{7.2}$$

where the first line is the definition of a group action on an integral, and the second follows by regarding the group action as a change of variable, back to the original domain. If the Lagrangian is invariant under this group action for arbitrary Ω, we call the group action a *variational symmetry*. By standard arguments (involving the Hilbert space L^2),

$$L(x, u, u_x, \dots) = L(\epsilon \cdot x, \epsilon \cdot u, \epsilon \cdot u_x, \cdots)\frac{\mathrm{d}\epsilon \cdot x}{\mathrm{d}x}.$$

for all ϵ.

Definition 7.4 *The infinitesimal action corresponding to that of a one-parameter group with parameter ϵ is obtained by applying $\dfrac{\mathrm{d}}{\mathrm{d}\epsilon}\Big|_{\epsilon=0}$ to the transformed variables.*

Continuing Example 7, we have

$$\frac{\mathrm{d}}{\mathrm{d}\epsilon}\Big|_{\epsilon=0}\epsilon \cdot x = x^2, \qquad \frac{\mathrm{d}}{\mathrm{d}\epsilon}\Big|_{\epsilon=0}\epsilon \cdot u = xu, \qquad \frac{\mathrm{d}}{\mathrm{d}\epsilon}\Big|_{\epsilon=0}\epsilon \cdot u_x = 2xu_x.$$

If i indexes the independent variables and α indexes the dependent variables, we denote the infinitesimal action on these by

$$\phi^\alpha = \frac{\mathrm{d}}{\mathrm{d}\epsilon}\Big|_{\epsilon=0}\epsilon \cdot u^\alpha, \quad \xi_i = \frac{\mathrm{d}}{\mathrm{d}\epsilon}\Big|_{\epsilon=0}\epsilon \cdot x_i. \tag{7.3}$$

Definition 7.5 *With ϕ^α and ξ_i as defined in (7.3), the characteristic of the group action is the vector $Q = (Q^\alpha)$, with*

$$Q^\alpha = \phi^\alpha - \sum_i \xi_i u_{x_i}^\alpha.$$

We can now state **Noether's Theorem**.

Theorem 7.1 *If Q^α are the characteristics of a variational symmetry of a Lagrangian form, then*

$$Q \cdot E(L) = \sum_\alpha Q^\alpha E^\alpha(L) = \mathrm{Div}(\mathcal{A}(Q, L))$$

where precise expressions (symbolic, exact, analytic) for $\mathcal{A}(L, Q)$ are known (Olver (1993), Proposition 5.74).

In words, given a symmetry of a Lagrangian, there is a divergence expression, $\mathrm{Div}(\mathcal{A}(L, Q))$ which is zero on solutions of the Euler–Lagrange system, $E^\alpha(L) = 0$.

On the simplest level, the proof involves a manipulation of the expressions involved. In order to translate the theorem to a discrete setting, we need to look at the *algebraic* underpinning of the proof. This consists of the variational complex which we now briefly describe. Full details may be found in (Olver (1993)).

7.2.3 The variational complex

The variational complex based on a p-dimensional space is constructed from the commutative diagram,

$$\xrightarrow{\textbf{D}} \Lambda^{p-1} \xrightarrow{\textbf{D}} \Lambda^p \xrightarrow{\widehat{\mathrm{d}}} \widehat{\Lambda}_1 \xrightarrow{\widehat{\mathrm{d}}} \widehat{\Lambda}_2 \xrightarrow{\widehat{\mathrm{d}}}$$
$$\qquad\qquad\qquad \downarrow \pi \qquad \downarrow \pi \qquad\qquad (7.4)$$
$$\qquad\qquad\qquad \Lambda^1{}_* \xrightarrow{\delta} \Lambda^2{}_* \xrightarrow{\delta}$$

Brief description of the components of (7.4):
The spaces Λ^k on the left of the diagram (7.4) are k-forms in the independent variables, but where the coefficients may depend, in a smooth way, on a finite number of dependent variables and their derivatives. The map $\textbf{D}$ is the *total* exterior derivative. For example, in two dimensions, $\textbf{D}(u_x \mathrm{d}y) = u_{xx}\mathrm{d}x\mathrm{d}y$. The spaces $\widehat{\Lambda}^j$ are the so-called *vertical k-forms*, that is, forms in the dependent variables and their derivatives, multiplied by the volume form on the base space. For example, in a two dimensional space, $x^2 u_x u_y \mathrm{d}u\mathrm{d}u_x\mathrm{d}x\mathrm{d}y \in \widehat{\Lambda}^2$. The map $\widehat{\mathrm{d}}$ is the exterior derivatives in the vertical direction. Thus, $\widehat{\mathrm{d}}(xyu_x^2\mathrm{d}x\mathrm{d}y) = 2xyu_x\mathrm{d}u_x\mathrm{d}x\mathrm{d}y$.

The first step in the calculation of the Euler–Lagrange operator in Example 6 is indeed the map $\widehat{\mathrm{d}}$. Using the exterior form notation, the calculation becomes

$$\begin{aligned}
\widehat{\mathrm{d}}(L\mathrm{d}x) &= \widehat{\mathrm{d}}\left(\tfrac{1}{2}\left(u_x^2 + u_{xx}^2\right)\mathrm{d}x\right) \\
&= (u_x\mathrm{d}u_x + u_{xx}\mathrm{d}u_{xx})\mathrm{d}x \\
&= (-u_{xx}\mathrm{d}u + u_{xxxx}\mathrm{d}u)\mathrm{d}x \\
&\quad + \tfrac{\mathrm{D}}{\mathrm{D}x}\left(u_x\mathrm{d}u - 2u_{xx}\mathrm{d}u_x + \tfrac{\mathrm{D}}{\mathrm{D}x}\left(u_{xx}\mathrm{d}u\right)\right) \\
&= E(L)\mathrm{d}u\mathrm{d}x + \tfrac{\mathrm{D}}{\mathrm{D}x}\eta(L).
\end{aligned}$$

As is seen in this example, the "integration by parts" step uses an action of the operator $\mathrm{D}/\mathrm{D}x$ on the forms, for example

$$\frac{\mathrm{D}}{\mathrm{D}x}(u^2\mathrm{d}u_x) = 2u\mathrm{d}u_x + u^2\mathrm{d}u_{xx}$$

and so forth. This generalises to higher dimensions, so that there is an action of the total divergence operator on the $\widehat{\Lambda}^k$.

The spaces $\Lambda^k{}_*$ are defined as equivalence classes of vertical forms; two forms are equivalent if they differ by a total divergence. The map d_* is then the maps $\widehat{\mathrm{d}}$ as induced on these classes, while the maps π are the projection maps. The Euler–Lagrange operator is then $\widehat{\mathrm{d}} \circ \pi$.

Definition 7.6 *The variational complex, given here for a p-dimensional base space, is*

$$\cdots \xrightarrow{\textbf{D}} \Lambda^{p-1} \xrightarrow{\textbf{D}} \Lambda^p \xrightarrow{E} \Lambda^1{}_* \xrightarrow{\mathrm{d}_*} \Lambda^2{}_* \xrightarrow{\mathrm{d}_*} \cdots \qquad (7.5)$$

Note that the map d_* is denoted by δ in (Olver (1993)). We reserve the notation δ for the simplicial coboundary map needed in section 11.3.

Theorem 7.2 *(Olver (1993)) The complex (7.5) is exact. That is, the image of one map equals the kernel of the next.*

Thus, if $E(L) = 0$ then L is necessarily in the image of **D**. Since **D** on Λ^{p-1} is essentially the total divergence operator, this means that E is zero, and only zero on, total divergences. The proof of this result is constructive, that is, formulae for the pre-images are known. These formulae are given in terms of homotopy operators which can be used, at least in principle, in ansatz methods for finding conservation laws not necessarily arising from Noether's Theorem. In practice, more direct methods are often used (Wolf (2000), Wolf (2003), Wolf, Brand and Mohammadzadeh (1999)).

The infinitesimal form of a Lie group action induces an action on forms. To describe this, we make the following definitions.

Definition 7.7 *Given the characteristic of an action $Q = (Q^\alpha)$ given in Definition 7.5, we define the* characteristic vector

$$\mathbf{v}_Q = \sum_{\alpha, K} \mathrm{D}^K (Q^\alpha) \frac{\partial}{\partial u^\alpha_K}$$

where α indexes the dependent variables and K is a multi-index of differentiation.

The inner product of a vector with a form is given by

$$\frac{\partial}{\partial u^\alpha_K} \lrcorner \, \mathrm{d}u^\beta_J = \delta^\alpha_\beta \delta^K_J, \qquad \frac{\partial}{\partial u^\alpha_K} \lrcorner \, \mathrm{d}x_i = 0$$

where δ is the Kronecker delta, and acts on products as a signed derivation. Thus, for example,

$$\mathbf{v} \lrcorner \, f(u) \mathrm{d}u_x \mathrm{d}u_{xx} = f(u) \frac{\mathrm{D}Q}{\mathrm{D}x} \mathrm{d}u_{xx} - f(u) \frac{\mathrm{D}^2 Q}{\mathrm{D}x^2} \mathrm{d}u_x.$$

Noether's theorem is obtained by considering the map $\mathbf{v}_Q \lrcorner \circ \widehat{\mathrm{d}}$ and hence $\mathbf{v}_Q \lrcorner \circ E$;

$$\xrightarrow{\ \mathbf{D}\ } \Lambda^p \underset{\mathbf{v}_Q \lrcorner}{\overset{E}{\underset{\longleftarrow}{\longrightarrow}}} \Lambda^1_* \xrightarrow{\ \mathrm{d}_*\ } \tag{7.6}$$

It is straightforward to show that the induced infinitesimal action of a Lie group on a Lagrangian form has the formula,

$$\frac{\mathrm{d}}{\mathrm{d}\epsilon}\Big|_{\epsilon=0} \epsilon \cdot L[u] = \mathbf{v}_Q \lrcorner \, \widehat{\mathrm{d}}L[u] + \mathrm{Div}(L\boldsymbol{\xi}) \tag{7.7}$$

where $\boldsymbol{\xi} = (\xi_1, \ldots, \xi_p)$.

If the Lagrangian is invariant, the left-hand side of (7.7) will be zero. Since $\widehat{\mathrm{d}}L[u] = \sum E^\alpha(L)\mathrm{d}u^\alpha \mathrm{d}\mathbf{x} + \mathrm{Div}(\eta(L))$ and $\mathbf{v}_Q$ has no $\partial/\partial x_i$ terms, so that Div and $\mathbf{v}_Q \lrcorner$ commute, Noether's theorem follows.

Example 8 *Consider the Lagrangian, $L[u] = \frac{1}{2}\left(\frac{u_x}{u}\right)^2 \mathrm{d}x$, which is invariant under both translation in x and scaling in u. The associated Euler–Lagrange equation is*

$$E(L) = \frac{u_x^2}{u^3} - \frac{u_{xx}}{u^2}.$$

For the translation action, $Q = -u_x$ since $\phi = 0$ and $\xi = 1$. And indeed,

$$-u_x E(L) = \frac{1}{2}\frac{\mathrm{d}}{\mathrm{d}x}\left(\frac{u_x^2}{u^2}\right).$$

For the scaling action, $Q = u$, as $\phi = u$ and $\xi = 0$, and so

$$uE(L) = -\frac{\mathrm{d}}{\mathrm{d}x}\left(\frac{u_x}{u}\right).$$

A more significant example can be found in (Bila, Mansfield and Clarkson (2005)) where conservation laws arising from symmetries of a meteorological model are classified.

In summary, the algebraic part of the proof of Noether's theorem involves a variational complex and an infinitesimal group action. Emulating the algebraic pattern, rather than the analysis, is the key to success for the construction and proof of the discrete Noether's Theorems.

We next look at the translation of these concepts for difference systems.

7.3 Difference Systems

We will consider a difference Lagrangian $L[u_n^\alpha]$ to be a smooth function of a finite number of difference variables and their shifts. Such difference Lagrangians may result from a discretisation of a smooth Lagrangian, but not necessarily. Since there exist inherently discrete systems with perhaps no continuum limit, we limit the types of calculations we perform here strictly to those operations pertinent to such systems.

We regard the lattice coordinates $\mathbf{n} = (n^1, \ldots, n^p) \in \mathbb{Z}^p$ as the independent variables. The dependent variables $\mathbf{u_n} = (u_\mathbf{n}^1, \ldots, u_\mathbf{n}^q)$ are assumed to vary continuously and to take values in $\mathbb{R}$. Let $\mathbf{1}_k$ be the p-tuple whose only nonzero entry is in the kth place; this entry is 1. Then the kth shift map acts as

$$S_k : \mathbf{n} \mapsto \mathbf{n} + \mathbf{1}_k, \qquad S_k : f(\mathbf{n}) \mapsto f(\mathbf{n} + \mathbf{1}_k) \qquad S_k : u_{\mathbf{n}}^{\beta} \mapsto u_{\mathbf{n}+\mathbf{1}_k}^{\beta},$$
$$S_k : f(\mathbf{n}, \ldots, u_{\mathbf{n}+\mathbf{m}}^{\beta}, \cdots) \mapsto f(\mathbf{n} + \mathbf{1}_k, \ldots, u_{\mathbf{n}+\mathbf{m}+\mathbf{1}_k}^{\beta}, \cdots)$$

where f is a smooth function of its arguments. Note that the shift maps commute (i.e., $S_k S_j = S_j S_k$), We write the composite of shifts using multi-index notation as

$$S^{\mathbf{m}} = S_1^{m_1} \ldots S_p^{m_p} \tag{7.8}$$

so that, for example, $u_{\mathbf{n}+\mathbf{m}}^{\beta} = S^{\mathbf{m}} u_{\mathbf{n}}^{\beta}$.

Definition 7.8 *A function* $F[u_n^{\alpha}]$ *is said to be a* total difference *if there is a vector* $(A_1[u_n^{\alpha}], \cdots A_p[u_n^{\alpha}])$ *such that*

$$F = (S_1 - \text{id})A_1 + \cdots + (S_p - \text{id})A_p.$$

Definition 7.9 *A* difference conservation law *for a difference system is a total difference which is zero on solutions.*

Example 9 *The standard discretisation of the heat equation,*

$$u_{n,m+1} - u_{n,m} = u_{n+1,m} - 2u_{n,m} + u_{n-1,m}$$

is a difference conservation law for itself, since it can be written

$$(S_1 - \text{id})[(S_1 - \text{id})(-u_{n-1,m})] + (S_2 - \text{id})u_{n,m} = 0.$$

Just as an integral of a total divergence depends only on the boundary data, so does the sum over a lattice domain of a total difference.

7.3.1 The difference Euler–Lagrange operator

As with smooth systems, the *difference* Euler–Lagrange equations result from a "zero derivative" condition when a difference Lagrangian is varied with respect to its variables. The "integration by parts" step of the calculation is replaced by, in one dimension,

$$\sum (Sf)_n g_n = \sum f_n (S^{-1}g)_n + (S - \text{id}) \sum (f_n (S^{-1}g)_n).$$

Example 10

$$\widehat{d}(L_n) = \widehat{d}\left(\tfrac{1}{2}u_n^2 + u_n u_{n+1}\right)$$

$$= (u_n du_n + u_{n+1} du_n + u_n du_{n+1})$$

$$= (u_n + u_{n+1} + u_{n-1})du_n + (S - \mathrm{id})(u_n du_{n+1})$$

$$= E(L_n)du_n + (S - \mathrm{id})(\eta(L_n)).$$

General formulae, (explicit, exact, symbolic), for E and $\eta(L_n)$ are known (Hydon and Mansfield (2004)).

As for the smooth case, we define the difference Euler–Lagrange operator to be $E = \pi \circ \widehat{d}$, where π projects out the total difference term. If there is more than one dependent variable, we obtain one equation for each dependent variable, for example in one dimension,

$$\widehat{d}(L_n[u, v]) = E^u(L_n)du_n + E^v(L_n)dv_n + (S - \mathrm{id})(\eta(L_n)).$$

7.3.2 Difference variational symmetries

If the difference equation arises as a discretisation of a smooth system, where there is a group action on the base space, then we can treat the mesh variables $x_\mathbf{n}$ as *dependent* variables (recall the independent variables are now the integer lattice co-ordinates), see Example 11 below. The induced group action will satisfy the property that the group action commutes with the shift map:

$$\epsilon \cdot S^j(u_n) = \epsilon \cdot u_{n+j} = S^j \epsilon \cdot u_n$$

for all j. For example,

$$\epsilon \cdot u_n = \frac{u_n}{1 - \epsilon x_n} \qquad \text{implies} \qquad \epsilon \cdot u_{n+j} = \frac{u_{n+j}}{1 - \epsilon x_{n+j}}.$$

We will assume this property for any group action on a difference system, not just those arising from discretisations.

The symmetry condition is that $L[u_n^\alpha]$ is an invariant function,

$$L_\mathbf{n}\big(u_\mathbf{n}^{\alpha_1}, \cdots, u_{\mathbf{n+k}}^{\alpha_\ell}\big) = L_\mathbf{n}\big(\epsilon \cdot u_\mathbf{n}^{\alpha_1}, \cdots, \epsilon \cdot u_{\mathbf{n+k}}^{\alpha_\ell}\big). \tag{7.9}$$

Defining the characteristics of the symmetry to be

$$Q_\mathbf{n}^\alpha = \left.\frac{\mathrm{d}}{\mathrm{d}\epsilon}\right|_{\epsilon=0} \epsilon \cdot u_\mathbf{n}^\alpha, \tag{7.10}$$

and applying

$$\left.\frac{\mathrm{d}}{\mathrm{d}\epsilon}\right|_{\epsilon=0}$$

to both sides of (7.9) yields

$$0 = \sum_k \frac{\partial L_{\mathbf{n}}}{\partial u_{\mathbf{n+k}}^\alpha} Q_{\mathbf{n+k}}^\alpha \qquad (7.11)$$

Since by our assumption,

$$Q_{\mathbf{n+k}}^\alpha = S^k(Q_{\mathbf{n}}^\alpha),$$

equation (7.11) can be written as

$$0 = X_Q \lrcorner \widehat{\mathrm{d}} L_{\mathbf{n}}$$

where

$$X_Q = \sum_{\alpha, \mathbf{j}} S^{\mathbf{j}}(Q_{\mathbf{n}}^\alpha) \frac{\partial}{\partial u_{\mathbf{n+j}}^\alpha}.$$

Theorem 7.3 The difference Noether's theorem. *If the symmetry condition (7.9) holds, then with the characteristics of the symmetry defined in (7.10),*

$$Q \cdot E(L_{\mathbf{n}}) = \sum_j (S_j - \mathrm{id})\big(\mathcal{A}_{\mathbf{n}}^j(Q_{\mathbf{n}}, L_{\mathbf{n}})\big).$$

Thus a symmetry yields a total difference expression which is zero on solutions of the difference Euler–Lagrange system. Explicit formulae for $\mathcal{A}_{\mathbf{n}}(Q_{\mathbf{n}}, L_{\mathbf{n}})$ are known (Hereman, Colagrosso, Sayers, Ringler, Deconinck, Nivala and Hickman (2005), Hereman, Sanders, Sayers and Wang (2005), Hydon and Mansfield (2004)). As for the smooth system, these quantities are defined in terms of homotopy operators which may be used to obtain conservation laws, not necessarily arising from Noether's theorem, in ansatz-based methods.

The similarity of the formulae to those of the smooth case is striking, particularly when the formulae for $\mathcal{A}_{\mathbf{n}}(Q_{\mathbf{n}}, L_{\mathbf{n}}))$ and $\mathcal{A}(Q, L))$ are compared. In fact, the algebraic underpinning of the difference Noether's theorem matches that of the smooth. One can build a diagram in complete analogy to (7.4), and the locally exact variational complex for difference systems is

$$\xrightarrow{\Delta} \mathbf{Ex}^{p-1} \xrightarrow{\Delta} \mathbf{Ex}^p \xrightarrow{E} \Lambda_*^1 \xrightarrow{\mathrm{d}_*} \Lambda_*^2 \xrightarrow{\mathrm{d}_*}$$

where $\mathbf{Ex}^n$ is a difference analogue of Λ^n and Λ_*^j are j-forms in the difference dependent variables and their shifts, modulo total differences. The diagram corresponding to (7.6) is

$$\xrightarrow{\Delta} \mathbf{Ex}^p \underset{\mathbf{X}_Q \,\lrcorner}{\overset{E}{\underset{\longleftarrow}{\longrightarrow}}} \Lambda^1 {}_* \xrightarrow{\mathrm{d}_*} \tag{7.12}$$

Note that the map d_* is denoted by δ in (Hydon and Mansfield (2004)). We reserve the notation δ for the simplicial coboundary map needed in section 11.3.

Remark 7.1 *The difference Noether's theorem is independent of any continuum limit. This is important since there are difference systems with multiple limits, or even no continuum limits at all. In cases where the difference system does have a continuum limit, it is interesting to note that in the examples studied, the Euler–Lagrange system and the conservation law also have continuum limits, and indeed limit to their corresponding quantities, but no proof of a general result is known.*

Example 11 *This elementary example is taken from the Introduction of Lee (1987), and concerns a difference model for the Lagrangian, $\int(\frac{1}{2}\dot{x}^2 - V(x))\,\mathrm{d}t$. Define*

$$\bar{V}(n) = \frac{1}{x_n - x_{n-1}} \int_{x_{n-1}}^{x_n} V(x)\,\mathrm{d}x$$

and take

$$L_n = \left[\frac{1}{2}\left(\frac{x_n - x_{n-1}}{t_n - t_{n-1}}\right)^2 - \bar{V}(n)\right](t_n - t_{n-1}).$$

The group action is translation in time, $t_n^* = t_n + \epsilon$, with x_n invariant. The conserved quantity is thus "energy". Now, $Q_n^t = 1$ for all n, and $Q_n^x = 0$. The Euler–Lagrange equation for the t_n, viewed as a dependent variable, is

$$0 = E^t(L_n) = \frac{\partial}{\partial t_n} L_n + S\left(\frac{\partial}{\partial t_{n-1}} L_n\right)$$

and since L_n is a function of $(t_n - t_{n-1})$,

$$0 = E^t(L_n) = (S - \mathrm{id})\left(\frac{\partial}{\partial t_n} L_n\right)$$

verifying the difference Noether Theorem in this case. The first integral (conservation law) is thus

$$\frac{1}{2}\left(\frac{x_n - x_{n-1}}{t_n - t_{n-1}}\right)^2 + \bar{V}(n) = c.$$

Note that the energy in the smooth case is

$$\frac{1}{2}\dot{x}^2 + V$$

showing the continuum limit of the energy for the difference system is the energy for the smooth system.

Remark 7.2 *The Euler–Lagrange equations for the mesh variables could well be regarded as an equation for a moving mesh. It may be appropriate to add terms to the difference Lagrangian that keep the mesh from collapsing or folding.*

7.3.3 Building in a conservation law to a difference variational system

If we know the group action for a particular conservation law, we can "design in" that conservation law into a discretisation by taking a Lagrangian composed of invariants. The Fels and Olver formulation of moving frames (Fels and Olver (1998), Fels and Olver (1999)) is particularly helpful here. A sample theorem concerning difference rotation invariants on $\mathbb{Z}^2$ follows. Consider the action,

$$\epsilon \cdot \begin{pmatrix} x_n \\ y_n \end{pmatrix} = \begin{pmatrix} \cos\epsilon & -\sin\epsilon \\ \sin\epsilon & \cos\epsilon \end{pmatrix} \begin{pmatrix} x_n \\ y_n \end{pmatrix}. \tag{7.13}$$

Theorem 7.4 *Let (x_n, y_n), (x_m, y_m) be two points in the plane. Then*

$$I_{n,m} = x_n y_n + x_m y_m, \quad J_{n,m} = x_n y_m - x_m y_n$$

generate the invariants under the action (7.13); any planar rotation difference invariant is a function of these.

Example 12 *We consider a difference Lagrangian which is invariant under the action (7.13). Suppose*

$$L_n = \frac{1}{2} J_{n,n+1}^2 = \frac{1}{2}(x_n y_{n+1} - x_{n+1} y_n)^2.$$

Then the Euler–Lagrange equations are

$$E_n^x = J_{n,n+1} y_{n+1} - J_{n-1,n} y_{n-1},$$
$$E_n^y = -J_{n,n+1} x_{n+1} + J_{n-1,n} x_{n-1}.$$

Now, $Q_n = (Q_n^x, Q_x^y) = (-y_n, x_n) = \frac{\mathrm{d}}{\mathrm{d}\epsilon}\big|_{\epsilon=0}(x_n^*, y_n^*)$ *and thus*

$$\begin{aligned}
Q_n \cdot E_n &= J_{n,n+1}(-y_n y_{n+1} - x_n x_{n+1}) \\
&\quad + J_{n-1,n}(y_n y_{n-1} + x_n x_{n-1}) \\
&= -J_{n,n+1}I_{n,n+1} + J_{n-1,n}I_{n-1,n} \\
&= -(S - \mathrm{id})(J_{n-1,n}I_{n-1,n})
\end{aligned}$$

gives the conserved quantity. Since the group action is a rotation, the conserved quantity is "angular momentum". Note that $I_{n,m} = I_{m,n}$ and $J_{n,m} = -J_{m,n}$.

Knowing the invariants is actually only half the battle, if you also require that the difference Lagrangian has a particular continuum limit. For one-dimensional systems, the theory of multispace can be used to obtain invariance under a given group action and a given limit simultaneously, see Olver (2001), Mansfield and Hydon (2001).

7.4 Finite Element systems

In obtaining a Noether's theorem for finite element approximations, we base our discussion on the variational complex developed in Mansfield and Quispel (2005). This, in turn, is based on the discussion of numerically stable finite element approximations given in Arnold (2002). We first look at a simple one-dimensional example. The analogies with the finite difference case here are sufficiently strong that we can obtain immediate results. We then discuss the higher-dimensional case.

7.4.1 The one dimensional case

We give an example of a system of moments that fit a commutative diagram and show how the Euler–Lagrange equations are derived. Let the "triangulation" of $\mathbb{R}$ be given by $\ldots x_{n-1}, x_n, x_{n+1}, \ldots$. We choose moment-based approximations for 0-forms (functions), and 1-forms so that the following diagram is commutative in each (x_n, x_{n+1});

$$\begin{array}{ccccccc}
0 \longrightarrow \mathbb{R} \longrightarrow & \Lambda^0 & \xrightarrow{\mathrm{d}} & \Lambda^1 & \longrightarrow 0 \\
& \Pi_0 \downarrow & & \Pi_1 \downarrow & \\
0 \longrightarrow \mathbb{R} \longrightarrow & \mathcal{F}_0 & \xrightarrow{\mathrm{d}} & \mathcal{F}_1 & \longrightarrow 0
\end{array} \qquad (7.14)$$

The maps Π_i are projections to piecewise defined forms.

Example 13 *In this example, the piecewise projection of 1-forms is*

$$f(x)\mathrm{d}x|_{(x_n,x_{n+1})} \mapsto \left(\int_{x_n}^{x_{n+1}} f(x)\psi_n(x)\mathrm{d}x \right) \mathrm{d}x$$

where ψ_n is given diagrammatically as

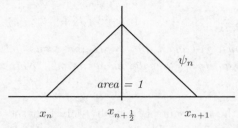

Here $x_{n+\frac{1}{2}}$ is any intermediate point, and $\psi_n(x_{n+\frac{1}{2}})$ is chosen so that the integral $\int_{x_n}^{x_{n+1}} \psi_n = 1$. The moments used to approximate functions are

$$\alpha_n = \frac{1}{x_{n+\frac{1}{2}} - x_n} \int_{x_n}^{x_{n+\frac{1}{2}}} u(x)\,\mathrm{d}x, \quad \beta_n = \frac{1}{x_{n+1} - x_{n+\frac{1}{2}}} \int_{x_{n+\frac{1}{2}}}^{x_{n+1}} u(x)\,\mathrm{d}x.$$

Commutativity means that

$$\Pi_1(u_x\mathrm{d}x) = \left(\frac{\mathrm{d}}{\mathrm{d}x}\Pi_0(u) \right) \mathrm{d}x, \tag{7.15}$$

while the projection property is that

$$\Pi_i \circ \Pi_i = \Pi_i.$$

So, we take the projection of $u|_{(x_n,x_{n+1})}$ using α_n, β_n to be

$$u \mapsto 2\frac{\beta_n - \alpha_n}{x_{n+1} - x_n}x + \left(\frac{x_{n+\frac{1}{2}} + x_{n+1}}{x_{n+1} - x_n} \right)\alpha_n$$

$$- \left(\frac{x_{n+\frac{1}{2}} + x_n}{x_{n+1} - x_n} \right)\beta_n$$

The moments α_n and β_n are not unrelated, however. The formulae are the same, only the domains differ. In effect, $\beta_n = \alpha_{n+\frac{1}{2}}$. So, we can define a shift map S so that

$$S(n) = n + \tfrac{1}{2},$$
$$S(\alpha_n) = \beta_n,$$
$$S(\beta_n) = \alpha_{n+1}.$$

We consider the simplest Lagrangian $\mathcal{L} = \int \frac{1}{2}u_x^2\,\mathrm{d}x$ which projects to

$$\Pi(\mathcal{L}) = \sum_n \int_{x_n}^{x_{n+1}} \frac{1}{2}\Pi(u)_x^2\,\mathrm{d}x = \sum_n 2\frac{(\beta_n - \alpha_n)^2}{x_{n+1} - x_n} = \sum L_n.$$

Then

$$\hat{\mathrm{d}}L_n = 4\frac{\beta_n - \alpha_n}{x_{n+1} - x_n}(\mathrm{d}\beta_n - \mathrm{d}\alpha_n)$$

$$= 4\frac{S(\alpha_n) - \alpha_n}{x_{n+1} - x_n}(\mathrm{d}S(\alpha_n) - \mathrm{d}\alpha_n)$$

$$= 4\left(S^{-1}\left(\frac{S(\alpha_n) - \alpha_n}{x_{n+1} - x_n}\right) - \frac{S(\alpha_n) - \alpha_n}{x_{n+1} - x_n}\right)\mathrm{d}\alpha_n + (S - \mathrm{id})(something).$$

The discrete Euler–Lagrange equation is the coefficient of $\mathrm{d}\alpha_n$. *After "integration", and setting* $\beta_n = \alpha_{n+\frac{1}{2}}$,

$$\frac{\alpha_{n+\frac{1}{2}} - \alpha_n}{x_{n+1} - x_n} \equiv c$$

which has the correct continuum limit.

We note that usually the approximation of functions is chosen so that the result is still continuous. This is an additional requirement that our calculations don't seem to need.

The main conjecture is that provided the system of moments used to project the forms fits the commutative diagram (7.14), then an Euler–Lagrange system, in the form of a recurrence system and having the correct continuum limit, can be derived (Mansfield and Quispel (2005)). As earlier, this will be a zero derivative condition obtained when the projected Lagrangian is varied with respect to the independent moments, modulo the analogue of a total difference.

We will show in Section 7.4.3 how a group action acting on dependent and independent variables induces an action on moments given as integrals.

7.4.2 The higher-dimensional case

We give the three-dimensional case; there are no significant changes for higher (or lower!) dimensions.

Given a system of moments and sundry other data, also known as degrees of freedom, we require that these yield projection operators such that the diagram (written here for three-dimensional space) commutes:

$$0 \longrightarrow \mathbb{R} \longrightarrow \Lambda^0 \longrightarrow \Lambda^1 \longrightarrow \Lambda^2 \longrightarrow \Lambda^3 \longrightarrow 0$$
$$\qquad \Pi_0 \downarrow \quad \Pi_1 \downarrow \quad \Pi_2 \downarrow \quad \Pi_3 \downarrow \qquad (7.16)$$
$$0 \longrightarrow \mathbb{R} \longrightarrow \mathcal{F}^0 \longrightarrow \mathcal{F}^1 \longrightarrow \mathcal{F}^2 \longrightarrow \mathcal{F}^3 \longrightarrow 0$$

all relative to some triangulation.

In general, a Lagrangian is composed of wedge products of 1-, 2- , ... p-forms. In Arnold (2002) it is argued that if the approximation of an n-form is taken to be its projection in $\mathcal{F}^n$, then commutativity implies conditions for Brezzi's theorem (Brezzi (1974)), guaranteeing numerical stability, will hold.

Thus a *finite element Lagrangian* is built up of wedge products of forms in $\mathcal{F}_0$, $\mathcal{F}_1$, ... $\mathcal{F}_{p-1}$, $\mathcal{F}_p$, with *unevaluated degrees of freedom*. Call the space of such products, $\tilde{\mathcal{F}}_p$. In each top-dimensional (p-dimensional) simplex, denoted τ, integrate to get

$$L = \sum_\tau L_\tau(\alpha_\tau^1, \cdots \alpha_\tau^p)$$

where α_τ^j is the jth degree of freedom in τ. Note that L can also depend on mesh data $x_{\mathbf{n}}$. We can now take the finite element vertical exterior derivative, $\hat{\mathrm{d}}$, to be the variation with respect to the α_τ^j.

There will be analogues of the shift maps that take moments defined on one simplex to moments defined on nearby simplexes.

The analogue of total divergence or total difference is the *coboundary* concept from simplicial algebraic topology which we define next. A coboundary has the key property that for topologically trivial domains, its integral depends only on data defined on the boundary of the domain of integration. It is the generalisation, to an arbitrary mesh, of a telescoping sum.

Definition 7.10 *Let X be a simplicial (triangulated) space. Denote by $\bar{C}_n(\mathcal{R})$ the vector space formed by all formal, finite sums of the n-simplexes of X with coefficients in $\mathcal{R}$. There is a boundary map*

$$\partial : \bar{C}_n \longrightarrow \bar{C}_{n-1}$$

obtained by mapping each simplex to the sum of its boundary edges, signed according to whether the orientation of the edge is that induced by the orientation of the simplex or its opposite, and extended linearly.

Example 14 *In Figure 7.1 we show an oriented simplex τ together with its oriented edges e_i. The boundary $\partial\tau = e_1 - e_2 + e_3$, where the signs are determined by whether the orientation on τ induces the given orientation on the edge, or not.*

See Frankel (1997), Chapter 13 (in particular 13.2b), for a readable account of oriented chains and the boundary map.

Definition 7.11 *For the simplicial space X, an n-cochain with coefficients in $\mathcal{R}$ is a map $\phi : \bar{C}_n \longrightarrow \mathcal{R}$. The set of simplicial n-cochains is denoted $\bar{C}^n$. The coboundary map*

$$\delta : \bar{C}^n \longrightarrow \bar{C}^{n+1}$$

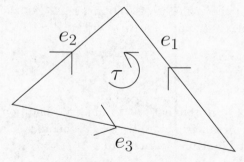

Fig. 7.1. $\partial(\tau) = e_1 - e_2 + e_3$

is defined by

$$(\delta\phi)(\sigma) = \phi(\partial\sigma).$$

For the simplex in Example 14, if $\phi(e_i) = c_i$, then $(\delta\phi)(\tau) = c_1 - c_2 + c_3$.

If a cochain ψ is of the form $\delta\phi$ for some cochain ϕ, we say simply that ψ *is* a coboundary. If the simplicial space is in fact a regular triangulation, it will be possible to write coboundaries in the form of a total difference.

For the variational calculations that we consider, the coefficients $\mathcal{R}$ are vertical forms in the $d\alpha^\tau$ which themselves have coefficient functions of the moments, the mesh variables, and so forth.

We can finally define the Finite Element Euler–Lagrange operator to be

$$E = \pi \circ \widehat{d} \circ \int$$

where $\int$ is the integration over each p-dimensional simplex that is used to obtain the projected Lagrangian, and π is the projection map to equivalence classes, where two forms are equivalent if they differ by a p-dimensional coboundary. (Recall p is the dimension of the base space.)

The variational complex for the Finite Element Method (Mansfield and Quispel (2005)), is then

$$\xrightarrow{\text{d}} \widetilde{\mathcal{F}}^{p-1} \xrightarrow{\text{d}} \widetilde{\mathcal{F}}^p \xrightarrow{E} \mathcal{F}^1_* \xrightarrow{\text{d}_*} \mathcal{F}^2_* \xrightarrow{\text{d}_*}$$

where:

- $\widetilde{\mathcal{F}}_*$ is the algebra generated by the $\mathcal{F}_i$ with unevaluated degrees of freedom
- $\text{d}_* = \pi \circ \widehat{d}$ is the vertical exterior derivative, that is, with respect to the degrees of freedom, modulo coboundaries,
- $\mathcal{F}^*_*$ is the algebra of vertical forms modulo p-dimensional coboundaries.

Looking now at the analogue of (7.6) and (7.12) for the Finite Element variational complex, we can tentatively write

$$\longrightarrow \widetilde{\mathcal{F}}^p \xrightarrow{\ E\ } \widetilde{\mathcal{F}}^1_* \longrightarrow \atop \overleftarrow{\ \ \ } \atop v_Q \lrcorner \qquad\qquad (7.17)$$

Taking $L_\tau \in \widetilde{\mathcal{F}}^p$, if the natural symmetry condition holds, that is $v_Q \lrcorner \hat{\mathrm{d}}(L_\tau) = 0$ (or, more generally, is a coboundary), we will have the **Finite Element Noether's Theorem**,

$$0 = \sum_\tau Q_\tau \cdot E(L_\tau) + \delta(\eta(L, Q)). \qquad (7.18)$$

There are two problems. One is to find the general formula for both E and $\eta(L, Q)$ for an arbitrary mesh. If the mesh is regular, then E and $\eta(L, Q)$ will be easily derivable from arguments analogous to those for the difference case. There are increasingly many computational arguments in favour of considering *cubical* simplicial spaces; see Kaczynski, Mischaikow and Mrozek (2003), for an exposition. In this case, coboundaries are essentially total differences.

The second problem is to define v_Q, which requires determining the infinitesimal action that is induced on the moments and other degrees of freedom. In the next section, we address this second problem.

7.4.3 Group actions on moments

For degrees of freedom that are values of a function at a particular point, the induced action is the same as for the function itself, and the discussion in Section 7.3.2 applies. For degrees of freedom that are moments defined by integrals, we can use results for the variational symmetry group action on Lagrangians derived earlier. Thus, given a group action on the independent and dependent variables, we take as a definition of the induced group action on the moment with weight function ψ,

$$\epsilon \cdot \int_\tau f(x, u, \cdots) \psi(x) \, \mathrm{d}x = \int_\tau f(\epsilon \cdot x, \epsilon \cdot u, \cdots) \psi(\epsilon \cdot x) \frac{\mathrm{D}\epsilon \cdot x}{\mathrm{D}x} \, \mathrm{d}x. \quad (7.19)$$

Example 15 *Suppose the group action is translation in x, so that $\epsilon \cdot x = x + \epsilon$, while the dependent variables are invariants, $\epsilon \cdot u = u$. Then the induced action on the moments*

$$\alpha^j = \int_\tau x^j u \, \mathrm{d}x \qquad (7.20)$$

is

$$\epsilon \cdot \alpha_\tau^0 = \alpha_\tau^0,$$
$$\epsilon \cdot \alpha_\tau^1 = \alpha_\tau^1 + \epsilon\alpha_\tau^0,$$
$$\epsilon \cdot \alpha_\tau^2 = \alpha_\tau^2 + 2\epsilon\alpha_\tau^1 + \epsilon^2\alpha_\tau^0,$$

and so forth. Thus,

$$Q_\tau^j = \frac{\mathrm{d}}{\mathrm{d}\epsilon}\bigg|_{\epsilon=0} (\epsilon \cdot \alpha_\tau^j) = j\alpha_\tau^{j-1},$$

whence

$$\mathbf{v}_Q = \sum_{\tau,j} j\alpha_\tau^{j-1} \frac{\partial}{\partial\alpha_\tau^j}.$$

If the mesh variables x_n are also regarded as dependent (i.e., movable) then one would add

$$\sum_n \frac{\partial}{\partial x_n}$$

to this vector. The zeroth and first-order moment invariants are generated by

$$\alpha_\tau^0, \quad \alpha_\tau^0\alpha_\sigma^1 - \alpha_\tau^1\alpha_\sigma^0$$

while the second-order invariants are generated by

$$\alpha_\tau^2(\alpha_\sigma^0)^2 - 2\alpha_\sigma^1\alpha_\tau^1\alpha_\tau^0 + (\alpha_\tau^1)^2\alpha_\sigma^0,$$

where τ and σ are not necessarily distinct simplexes. The method of moving frames shows that any moment invariant (to order two) is a function of these. Allowing movable mesh variables, we have that $x_n - x_m$ is an invariant, as is $\alpha_\tau^0 x_n - \alpha_\tau^1$.

7.4.4 Building in a conservation law

The algebra underpinning Noether's theorem shows that designing a conservation law into a numerically stable scheme requires a number of conditions to be met on the choice of moments and how the Lagrangian is approximated.

- The approximation of forms is required to fit into a commutative diagram, (7.16) not only for stability but for the variational complex to be applicable.

- The induced group action probably should involve essentially a finite number of moments, so it may be necessary to use symmetry-adapted moments. For example, if the group action is the projective action, (7.1), then the index j in the moments (7.20) needs to be in the range $-N, \cdots, -3$.
- The projected Lagrangian form needs to be invariant under the induced action and have the correct continuum limit.

7.5 Conclusions

Instead of the approximate conservation of an exact law, the algebraic arguments offered here yield exact conservation of an approximate law! Clearly much remains to be done to bring these ideas into the practical arena, in particular the analytic problem of achieving everything listed in Section 7.4.4 for some interesting applications. Another problem is, how does the order of approximation of the conservation law compare to the order of approximation of the scheme? Whether such schemes prove efficient and useful is for the future to decide. Nevertheless, schemes with guaranteed conservation laws appear to be possible.

Acknowledgments

It is a pleasure to acknowledge the support and encouragement of Arieh Iserles and Peter Olver, discussions with Peter Hydon and Reinout Quispel, financial assistance from the Leverhulme Trust, UK and the hospitality of the Institute of Advanced Studies, LaTrobe University, Australia.

References

D. Arnold (2002), *Differential Complexes and Numerical Stability*, Proceedings of the International Congress of Mathematicians, Beijing 2002, Volume I: Plenary Lectures. http://www.ima.umn.edu/~arnold/papers/icm2002.pdf

N. Bila, E. L. Mansfield and P. A. Clarkson (2006), Symmetry group analysis of the shallow water and semi-geostrophic systems, *Q. J. Mech. Appl. Math.*, **59**, 95–123.

F. Brezzi (1974), On the existence, uniqueness and approximation of saddle point problems arising from Lagrange multipliers, *Rev. Française Automat. Informat. Recherche Opérationnelle Sér. Rouge Anal. Numér.* **8**, 129–151.

J. A. Chard and V. Shapiro (2000), A multivector data structure for differential forms and equations, *Math. Comp. Sim.* **54**, 33–64.

E. S. Cheb-Terrab, L. G. S. Duarte and L. A. da Mota (1998), Computer algebra solving of second order ODEs using symmetry methods, *Computer Physics Communications* **108**, 90–114.

V. Dorodnitsyn (2001), Noether-type theorems for difference equations, *App. Num. Math.* **39**, 307–321.

M. Fels and P. J. Olver (1998), Moving coframes I, *Acta Appl. Math.* **51**, 161–213.

M. Fels M. and P. J. Olver (1999), Moving coframes II, *Acta Appl. Math.* **55**, 127–208.

T. Frankel (1997), *The Geometry of Physics*, Cambridge University Press, Cambridge.

W. Hereman (1997), Review of symbolic software for Lie symmetry analysis *Math. Comp. Model.* **25**, 115–132.

W. Hereman, M. Colagrosso, R. Sayers, A. Ringler, B. Deconinck, M. Nivala and M. S. Hickman (2005), *Continuous and discrete homotopy operators with applications in integrability testing*, in: *Differential Equations with Symbolic Computation*, Eds.: D. Wang and Z. Zheng, World Scientific, Singapore.

W. Hereman, J. A. Sanders, J. Sayers and J. P. Wang (2005), Symbolic computation of polynomial conserved densities, generalized symmetries, and recursion operators for nonlinear differential-difference equations, in: *Group Theory and Numerical Analysis*, Eds.: P. Winternitz, D. Gomez-Ullate, A. Iserles, D. Levi, P. J. Olver, R. Quispel and P. Tempesta, CRM Proceedings and Lecture Notes **39**, AMS.

R. Hiptmair (2002), Finite elements in computational electromagnetism, *Acta Numerica* **11**, 237–340.

E. Hubert (2000), Factorization free decomposition algorithms in differential algebra, *J. Symb. Comp.* **29**, 641–662.

E. Hubert (2003), *Notes on triangular sets and triangulation-decomposition algorithms II: Differential Systems*, in: *Symbolic and Numerical Scientific Computations*, Eds.: U. Langer and F. Winkler, Lect. Notes Comp. Sci., **2630**, Springer-Verlag, Heidelberg, 40–87.

P. E. Hydon (2000), Symmetries and first integrals of ordinary difference equations, *Phil. Trans. R. Soc. Lond. A* **456**, 2835–2855.

P. E. Hydon and E. L. Mansfield (2004), A variational complex for difference equations, *Foundations of Computational Mathematics* **4**, 187–217.

T. Kaczynski, K. Mischaikow and M. Mrozek (2003), Computing homology, *Homology, Homotopy and Applications*, **5** 233–256.

B. Kupershmidt (1985), Discrete Lax equations and differential-difference calculus, *Astérisque* **123**, SMF, Paris.

T. D. Lee (1987), Difference equations and conservation laws, *J. Stat. Phys.* **46**, 843–860.

D. Levi, S. Tremblay and P. Winternitz (2005), Lie symmetries of multidimensional difference equations, *J. Phys. A: Math. Gen.* **34**, 9507–9524.

E. L. Mansfield and P. A. Clarkson (1997), Applications of the differential algebra package `diffgrob2` to classical symmetries of differential equations, *J. of Symb. Comp.* **23**, 517–533.

E. L. Mansfield and P. E. Hydon (2001), Towards approximations which preserve integrals, in: *ISSAC 2001*, Eds.: B. Mourrain, ACM Publishing, New York, 217–222.

E. L. Mansfield and G. R. W. Quispel (2005), Towards a variational complex for the Finite Element Method, in: *Group Theory and Numerical Analysis*, Eds.:

P. Winternitz, D. Gomez-Ullate, A. Iserles, D. Levi, P. J. Olver, R. Quispel and P. Tempesta, CRM Proceedings and Lecture Notes **39**, AMS.

C. Mattiussi (1997), An analysis of finite volume, finite element, and finite difference methods using some concepts from algebraic topology, *J. Comp. Phys* **133**, 289–309.

P. J. Olver (1993), *Applications of Lie Groups to Differential Equations*, Graduate Texts Math., **107**, Second Edition, Springer Verlag, New York.

P. J. Olver (2001), Geometric foundations of numerical algorithms and symmetry, *Appl. Alg. Engin. Comp. Commun.* **11**, 417–436.

G. R. W. Quispel, H. W. Capel and R. Sahadevan (1992), Continuous symmetries of differential-difference equations – the Kac-van Moerbeke equation and Painlevé reduction, *Phys. Lett.* **170A**, 379–383.

G. J. Reid (1991), Algorithms for reducing a system of PDEs to standard form, determining the dimension of its solution space and calculating its Taylor series solution, *Europ. J. Appl. Math.* **2**, 293–318.

G. J. Reid, A. D. Wittkopf and A. Boulton (1996), Reduction of systems of nonlinear partial differential equations to simplified involutive forms, *Europ. J. Appl. Math.* **7**, 604–635.

W. Schwalm, B. Moritz, M. Giona and M. Schwalm (1999), Vector difference calculus for physical lattice models, *Phys. Rev. E* **59**, 1217-1233.

E. Tonti (1975) *On the formal structure of physical theories*, Istituto di Matematica del Politecnico di Milano, Milan.
`http://www.dic.univ.trieste.it/perspage/tonti/papers.htm`

T. Wolf (2000), The symbolic integration of exact PDEs, *J. Symb. Comp.* **30**, 619–629.

T. Wolf (2002), A comparison of four approaches to the calculation of conservation laws, *Europ. J. Appl. Math.* **13**, 129–152.

T. Wolf, A. Brand and M. Mohammadzadeh (1999), Computer algebra algorithms and routines for the computation of conservation laws and fixing of gauge in differential expressions, *J. Symb. Comp.* **27**, 221–238.

8

Hyperbolic 3-Manifolds and Their Computational Aspect

G. Robert Meyerhoff

Department of Mathematics
Boston College
Boston, MA, USA
e-mail: robert.meyerhoff@bc.edu

Abstract

In the late 1970's, W. Thurston proved that the set of volumes of hyperbolic 3-manifolds has a rich and intriguing structure. In particular, the set of volumes is well-ordered and of order type ω^ω:

$$v_1 < v_2 < v_3 < \ldots \to v_\omega < v_{\omega+1} < v_{\omega+2} < \ldots$$

The question of which hyperbolic 3-manifolds realize these low volumes has resisted a variety of assaults for almost 30 years. Now, it appears that new ideas have put us on the verge of answering these questions in a satisfying way. I will present one of these new ideas and show how the implementation of the idea makes natural use of the computer in two different ways. This is joint work with D. Gabai and P. Milley. I will start from the beginning and slowly work my way up to the main ideas.

8.1 Historical Ramblings

We begin with selected topics in the history of Geometry. These remarks are meant to be taken casually. In particular, I do not claim that all I say is entirely accurate. But at least it should present some general themes that will provide context for hyperbolic 3-manifolds.

Our story begins with the Pythagorean mathematicians of 2500 to 2600 years ago. They held an atomistic view of the Universe. As such, it is not surprising that they believed that all lengths are *commensurable*. That is, given two lengths, they believed that these lengths were both integral multiples of some common unit. (As an aside, it's interesting to speculate as to how the very practical Pythagoreans would have physically found the common unit if presented with two lengths.)

The Pythagoreans also knew the Pythagorean Theorem: in a right triangle the square of the hypotenuse is equal to the sum of the squares of the other two sides. When confronted with an equal-legged right triangle, the Pythagoreans would have assumed the hypotenuse was commensurable with the sides. Of course, this isn't true because, in modern terms, $\sqrt{2}$ is irrational.

The Pythagoreans were profoundly shaken by this discovery. It showed them that their intuitive method of doing mathematics could lead to catastrophe. To avoid this, they invented the axiomatic method: start with self-evident truths, use incontrovertible rules of reasoning, and produce mathematical truths.

After many years of trial and error, Greek mathematicians subsequent to the Pythagoreans came up with the following five postulates for geometry:

P1) Two points determine a (unique) line.

P2) Lines can be extended indefinitely

P3) A circle with any center and radius can be constructed.

P4) All right angles are equivalent.

P5) Given two lines and a third line that intersects both of them in interior angles less than a right angle (on the same side), then the two lines (if extended indefinitely) meet at some point on the same side of the third line as the two less-than-right angles.

Much later commentators complained that Postulate 5 is not self-evident. These commentators wished to prove that P5 followed from P1 through P4, thereby reducing Euclidean Geometry to self-evident postulates. The Euclidean mathematicians, who spent countless years fine-tuning the postulates, would have laughed at this complaint. They knew full well that Postulate 5 was absolutely necessary to produce a sufficiently rich theory of Geometry.

Nonetheless, the problem of proving that P5 followed from P1 $\rightarrow$ P4 occupied numerous mathematicians for over a thousand years. One noteworthy example is provided by the Jesuit mathematician Saccheri ($\sim$1700), a masterful teacher who delighted in the method of proof by contradiction. Saccheri decided to assume P1 $\rightarrow$ P4 and the negation of P5, and show that this leads to a contradiction.

Given this starting point, Saccheri proved many theorems without reaching a contradiction. These theorems are theorems in non-Euclidean Geometry, although Saccheri didn't look at it this way. He thirsted to find a contradiction, and finally, in desperation, claimed a (lame) contradiction ("this is repugnant to the nature of a straight line!"). In his mind, this "contradiction" meant that he had succeeded in his task of vindicating Euclid.

I recently heard the following interpretation of Saccheri's behavior. The claim is that Saccheri knew full well that he was unable to produce a legitimate contradiction, and hence that he had likely constructed a new type of geometry. However, as a Jesuit mathematician well aware of the persecution of Galileo and other scientists with heretical views, Saccheri realized that he needed to state that the accepted wisdom (Euclidean Geometry is perfect) was true. So he put forth a lame contradiction, and thereby kept the regressive elements at bay.

Despite this perhaps fanciful interpretation, three mathematicians subsequent to Saccheri are given credit for discovering Non-Euclidean Geometry: Bolyai, Lobachevsky, and Gauss. The work of Bolyai and Lobachevsky was done, independently, in the 1820's. Lobachevsky and Bolyai said that P1 → P4 together with the negation of P5 produces a new, legitimate Geometry, and then they vigorously developed the theory of this geometry. Gauss probably made the discovery prior to Bolyai and Lobachevsky but declined to announce his results because he feared the "stings and clamors of the Boeotians."

Bolyai and Lobachevsky proved *lots* of theorems about their new geometry. But all of their theorems were synthetically derived (directly from the axioms) hence mysterious. It is not surprising that mathematicians were slow to embrace this new geometry when surprising theorems (e.g., there are no rectangles in this geometry) are presented with no intuitive explanation to make them palatable.

As an aside, we note that in 1816 Wachter showed that if the negation of P5 is assumed, then a sphere of infinite radius (sphere with center at infinity) is Euclidean (that is, its induced geometry is Euclidean). Bolyai and Lobachevsky rediscovered this.

After Bolyai and Lobachevsky, Non-Euclidean Geometry (also called Hyperbolic Geometry) languished for 30 years. Then Beltrami produced models for Hyperbolic Geometry, thereby providing a concrete and intuitive basis for the geometry. Beltrami also proved that Hyperbolic Geometry is consistent if and only if Euclidean Geometry is consistent.

8.2 Models for Hyperbolic Geometry

By studying a model of Hyperbolic Geometry we can get a more intuitive sense of the geometry. The first model we will study is the *Poincaré Disk Model* **PD**. This consists of the open unit disk $\{(x, y) : x^2 + y^2 < 1\}$ together with the hyperbolic metric ds_H determined by $ds_H = \frac{ds_E}{(1/2)(1-(x^2+y^2))}$

where ds_E is the usual Euclidean metric for the plane. Thus, the Poincaré Disk Model can be thought of as a distorted version of the Euclidean disk and we can attempt to understand **PD** by exploiting its underlying Euclidean aspects. Of course, this can lead to confusion as to what is hyperbolic and what is Euclidean.

Casually looking at the metric ds_H we see that a small curve near the origin would have hyperbolic and Euclidean lengths that are quite close (here we are ignoring the factor of $1/2$ in the denominator which is needed to make certain calculations work out nicely), but that as we get close to the boundary circle, a short Euclidean curve would have a long hyperbolic length. Of course, to the inhabitants of hyperbolic space, this analysis is nonsensical: all points have the same geometric properties. The trick is that the interplay between the hyperbolic and the Euclidean makes points appear different to our extrinsic eyes.

The first step towards understanding a geometry is to analyze the various geometric objects that live there. First, the geodesics in **PD**, that is, the curves that locally minimize the metric. It turns out that the set of geodesics consists of Euclidean lines through the origin (hence perpendicular to the boundary at both ends) and Euclidean circles perpendicular to the boundary. This description of the geodesics gives a comforting feeling because it is a description in terms of familiar geometric objects. Further, given this collection of geodesics, it is easy to see that the Euclidean fifth postulate is not satisfied; although perhaps it's best to use Playfair's version of the Euclidean Parallel Postulate: Given a line l and a point P not on l, there is exactly one line through P that is parallel to l.

Next we describe the set of hyperbolic circles in **PD**. Given the rotational symmetry of ds_H around the origin, it is not surprising that, as sets of points, the hyperbolic circles centered at the origin are precisely the Euclidean circles of radius less than one centered at the origin. But it might be surprising to find out that the hyperbolic circles centered at points other than the origin are also Euclidean circles, although now the hyperbolic center is not the same as the Euclidean center. The explanation for this surprising fact is that the set of orientation-preserving isometries of **PD** are precisely the set of fractional linear transformations that take the unit circle onto itself. Because fractional linear transformations preserve angles, and take the set of Euclidean circles and lines to the set of Euclidean circles and lines, it follows that the Euclidean/hyperbolic circles centered at the origin are transformed by isometries to Euclidean/hyperbolic circles.

Different models for hyperbolic geometry are useful in different contexts. For example, the Poincaré Disk Model is well-suited for analyzing problems involving geodesics through a point: by the homogeneous nature of hyperbolic space, the point can be assumed to be the origin and the hyperbolic geodesics through this point are Euclidean straight lines. Our next model is particularly good at analyzing situations involving a distinguished point at infinity

The *Upper Half-Plane Model* consists of the set $\{(x, y) : y > 0\}$ together with the metric $ds_H = \frac{ds_E}{y}$. Although all points look the same to the inhabitants, to our extrinsic eyes we see that Euclidean curves high above the bounding $x-$axis are much shorter than their Euclidean counterparts with y close to 0. Hence if one were trying to find the hyperbolically shortest path between two points with the same y coordinates, one would be inclined to head up (higher y) and then down rather than follow the Euclidean straight line between the two points. In fact, the geodesics in this model are Euclidean circles perpendicular at both ends to the bounding $x-$axis, and vertical Euclidean lines. The explanation again uses fractional linear transformations: the orientation-preserving isometries of **UHP** are precisely the fractional linear transformations of the form $z \rightarrow (az+b)/(cz+d)$ where a, b, c, d are real numbers. In fact, $Isom_+(\textbf{UHP}) = PSL(2, \textbf{R})$, the set of 2×2 matrices with real entries and determinant one, modulo plus-or-minus the identity.

Another explanation is that there is a fractional linear transformation taking **PD** to **UHP** which transforms the metric appropriately. This also makes it clear that hyperbolic circles are Euclidean circles (as point sets) in **UHP** although now the hyperbolic centers *always* differ from the Euclidean centers.

The generalizations of these models to 3 dimensions are straightforward. For example, the *Upper-Half-Space Model* consists of the set $\{(x, y, t) : t > 0\}$ together with the metric $ds_H = \frac{ds_E}{t}$. The geodesics are Euclidean circles perpendicular at both ends to the bounding $xy-$plane, and vertical Euclidean lines. And hyperbolic spheres are Euclidean spheres.

What is a sphere of infinite radius like? Start with a point in **UHS** and construct a sphere with that point as its maximum (as a set in $\textbf{R}^3$). Now, hold that maximum point but let the sphere expand. The radius is expanding and the center is drifting downwards. Eventually (Euclideanly) the bottom of the sphere bumps into the bounding $xy-$plane and the process stops. At this point, this Euclidean sphere is not a true hyperbolic sphere, it's a hyperbolic sphere of infinite radius and it's called a *horosphere*; the point of tangency on the $xy-$plane is called the *center* of the horosphere. Wachter's claim (see above) is that horospheres have induced Euclidean geometries. Why is this so?

It's easy to see that this claim is true if we use a hyperbolic isometry to send the center of a horosphere to the point at infinity for **UHS**. The horosphere centered at this point must be a plane parallel to the xy−plane. Hence it has constant t coordinate and hence its induced metric is a constant times the Euclidean metric, hence Euclidean. Here we use the fact, discovered by Poincaré in the 1880's, that $Isom_+(\mathbf{UHS}) = PSL(2, \mathbf{C})$, the set of 2×2 matrices with complex entries and determinant one, modulo plus-or-minus the identity. These fractional linear transformations act on the bounding extended complex plane and extend naturally to $\mathbf{UHS}$: a point in $\mathbf{UHS}$ can be located as the point of intersection of a Euclidean hemisphere and two vertical planes; then these three objects are sent to vertical planes and/or hemispheres by the natural extension of the fractional linear transformation and they locate the image of the original point as their intersection.

As an example, we note that the matrix $[2\ \ 0\ \ 0\ \ (1/2)]$ takes $z = x + iy$ to $\frac{2z+0}{0z+(1/2)}$. That is, $z \to 4z$. Hence hemispheres centered at the origin are blown up by a factor of 4 as point sets. If the matrix is $[2e^{i\theta}\ \ 0\ \ 0\ \ e^{-i\theta}/2]$, then $z \to 4e^{2i\theta}z$ and the hemispheres are rotated through angle 2θ around the t−axis, as well as being blown up by a factor of 4.

8.3 Surfaces

Definition (naive): A 2-manifold is a space that locally looks like the xy−plane.

All compact 2-manifolds are well-known. They are the sphere, the torus, the 2-holed torus, the 3-holed torus, the 4-holed torus, and so on.

Note that here we are using the topological notion of equivalence. Two 2-manifolds are topologically equivalent or *homeomorphic* if there exists an invertible map from one onto the other which is continuous and whose inverse is continuous.

The geometric notion of equivalence would produce a different list because, for example, spheres of various radii in Euclidean 3-space would be considered geometrically different because the means of measuring distances (induced from Euclidean 3-space) would produce different calculations in each sphere.

We now give a definition of *geometric 2-manifold*.

Definition (naive): A geometric 2-manifold is a 2-manifold with a metric that locally produces exactly the same calculations as the geometric model space.

For our purposes we will restrict to 2-dimensional geometric model spaces that are simply connected (all loops shrink to points) and whose metrics are homogeneous (have the same metric properties at each point). It is well-known that the only such 2-dimensional model spaces are the sphere **S** (normalized to have radius 1), the Euclidean plane **E**, and the hyperbolic plane **H**(normalized as well).

These three model geometries are easy to tell apart via some simple local measurements. For example, a circle of radius r could be constructed in each space ($r < \pi$ in **S**) and the circumference would be $2\pi \sin(r)$ in **S**, 2π in **E**, and $2\pi \sinh(r)$ in **H**. Note that as r approaches zero, the spherical and hyperbolic circumferences approach the Euclidean circumference.

Another simple local measurement would be the sum of the angles in a triangle. In **S** the sum is always greater than π, in **E** it is always equal to π, and in **H** it is always less than π. Again, as the triangles get smaller in **S** and **H** they become more Euclidean. Note also that in **S** the area of a triangle is $(\alpha + \beta + \gamma) - \pi$ where α, β, γ are the angles in the triangle. Similarly, in **H** the area of a triangle is $\pi - (\alpha + \beta + \gamma)$. In **E** the angles do not determine the triangle and the area is not determined by the angles.

Example: A sphere of radius 1 is a spherical 2-manifold.

Q: Does a sphere admit a metric which is Euclidean?

The most obvious attempt to make a sphere Euclidean fails: Consider the surface of a cube. This is topologically a sphere, and has an unusual metric induced from the ambient Euclidean 3-space $\mathbf{E}^3$. It comes pretty close to being Euclidean. Points on the cubical faces have nice Euclidean neighborhoods, and points on the edges have nice Euclidean neighborhoods (flatten out a neighborhood to see this), but the vertex points don't work because there is only $(3/2)\pi$ angle. But perhaps if we tried harder we might succeed in putting a Euclidean structure on a sphere? The answer is no, as we will see shortly.

However, there is an immediate example of a 2-manifold which admits a Euclidean structure:

Example: A torus admits a metric which makes it a Euclidean 2-manifold. The metric for the torus must be chosen with a little care. The usual torus (the surface of a doughnut) sitting in $\mathbf{E}^3$ with the induced metric is not Euclidean: one can show that it's a bit too curved. However if we take a square in $\mathbf{E}^2$ and abstractly identify opposite sides we get a Euclidean torus (don't try to glue it up in $\mathbf{E}^3$: the first gluing of opposite edges preserves the Euclidean structure, but the second gluing breaks it). If we analyze the three types of points (points in the square, points on an edge, vertex points) we see that all have neighborhoods that are exactly Euclidean. The vertices work because all four are glued to give one point, and that point has 2π angle.

We now make a 2-paragraph digression to re-think the Euclidean torus:

If you were a 2-dimensional creature living in a Euclidean torus gotten by identifying opposite edges of a square, what would you see? If you looked straight ahead through the right edge, your line of sight would come around in back of you and you would see the back of your head. If you looked straight up through the top edge you would see the bottom of your feet. Similarly, with some work one could analyze various lines of sight. But a better way to think of this is as follows. When you look ahead through the right edge, construct another copy of the square (with a model of you inside it) and glue its left edge to your original copy's right edge, and continue this process to the right infinitely often and then to the left infinitely often. Then do it through the top edge infinitely often in both directions, and then move up to various of the new squares and continue the process until the entire plane is covered by copies of the original square. This object so constructed is the universal cover of the Euclidean torus; it turns out to be our friend the Euclidean plane $\mathbf{E}$, and it contains infinitely many copies of you. It is now easy to determine what you would see along any sightline.

In reverse, we can think of this Euclidean torus as the orbit space $\mathbf{E}/G$ where G is the group of (Euclidean) isometries generated by $z \to z + 1$ and $z \to z + i$. Hence, we can think of our Euclidean torus as a geometric object or an algebraic object G.

Q: Which compact 2-manifolds admit which geometric structures?

At this point we know that the sphere admits a spherical structure, the torus admits a Euclidean structure, and that the most obvious attempt to put a Euclidean structure on a sphere fails. It might seem rather daunting to show that a particular 2-manifold does not admit a Euclidean structure, because naively one would have to try an infinite number of possibilities and eliminate them all.

However, we are rescued by the Gauss-Bonnet Theorem (a good exposition of this material is in Weeks (1985)).

Theorem 8.1 (Gauss-Bonnet) *If (X, ρ) is a geometric 2-manifold consisting of a compact 2-manifold X and a geometric metric ρ then*

$$2\pi\chi(X) = \kappa(\rho)\mathrm{Area}(X, \rho)$$

where $\chi(X)$ is the Euler characteristic of the topological manifold X which is computed by decomposing X into triangles then computing $V - E + F$; $\kappa(\rho)$ is the curvature of the geometric model space ($\kappa(\mathbf{S}) = 1$, $\kappa(\mathbf{E}) = 0$, $\kappa(\mathbf{H}) = -1$), and $\mathrm{Area}(X, \rho)$ is the area of the geometric manifold.

Because the sphere has Euler characteristic 2 and the area of any geometric 2-manifold is positive, we see that the sphere can admit neither a Euclidean nor a hyperbolic structure. Similarly, because the torus has Euler characteristic 0, we see that the torus can admit neither a spherical nor a hyperbolic structure. Finally, because the $n-$holed tori for $n > 1$ have Euler characteristic less than 0, we see that they can admit neither spherical nor Euclidean structures.

Our knowledge of which compact 2-manifolds admit which geometric structures will be satisfyingly completed by showing that all $n-$holed tori for $n > 1$ admit hyperbolic structures. The construction utilizes the remarkable properties of the hyperbolic plane. To start, it is easy to show that a 2-holed torus can be obtained by identifying opposite edges of an $8-$gon, and that given this identification all 8 vertices get identified to one point. If we took a regular Euclidean $8-$gon with this identification scheme and analyzed all the points, we would see that the one vertex would have much too much angle and Euclideaness would break down.

But if we could find a regular $8-$gon in $\mathbf{H}$ with all angles $\pi/4$ then this identification scheme (via hyperbolic isometries) would produce a hyperbolic 2-holed torus. This can be done as follows, where we use the Poincaré Disk model because of its symmetry about the origin. Regularly space 8 rays coming out of the origin. Take a circle centered at the origin and mark its intersection with these 8 rays. Connect these 8 intersection points by geodesic segments in the obvious way to produce a regular 8-gon in $\mathbf{PD}$. If the circle is very small then the 8-gon is nearly Euclidean and the angles at the vertices are much too big. If the circle is very large and the points are near the boundary then the geodesic segments are nearly perpendicular to the boundary circle and the 8 angles are close to 0. For some circle in between we get a regular 8-gon with all angles $\pi/4$ and with the above identifications this produces a hyperbolic 2-holed torus.

The same process works for a 3-holed torus via a 12-gon, and so on.

8.4 Hyperbolic 3-Manifolds

In the 2-dimensional case, all compact manifolds admit spherical, Euclidean, or hyperbolic geometric structures. The biggest category being the hyperbolic category: exactly one 2-manifold is spherical, exactly one 2-manifold is Euclidean, and all the rest are hyperbolic. Further, the hyperbolic 2-manifolds are the most complicated topologically. The situation in 3 dimensions turns out to be similar, but it was a surprise when it was discovered. The trick is that, despite their abundance, the complicated topological nature of hyperbolic 3-manifolds made them difficult to find.

In contrast, examples of spherical and Euclidean 3-manifolds were easy to come by. For example, the 3-sphere is spherical, and the 3-torus (which is gotten by identifying opposite faces of a cube) is Euclidean.

But it wasn't until the 1930's that the first examples of compact hyperbolic 3-manifolds were discovered by Lobell (by gluing together 8 copies of a 14-sided right-angled hyperbolic polyhedron; see the historical note in Ratcliffe (1994) on page 500). We comment that all of our manifolds are assumed to be orientable.

A simpler example was found by Seifert and Weber in a 1933 paper.

Example (Seifert-Weber): Take a regular dodecahedron and identify opposite faces with a $3\pi/5$ twist. This results in the edges being identified five to one. Hence, to get a geometric manifold we need dihedral angles of $2\pi/5$. This can be accomplished in hyperbolic 3-space. The vertex points work as well with these angles, and we have a hyperbolic 3-manifold. The fact that a regular dodecahedron with all angles $2\pi/5$ exists in $\mathbf{H}^3$ is not as obvious as the analogous 2-dimensional fact. In 2 dimensions, the limiting case has all angles 0 and there's a lot of breathing room in our attempts to find small dihedral angles. In 3 dimensions the limiting case has vertices at infinity and a small neighborhood of a vertex point cut off by a horosphere must be a Euclidean regular triangle, hence have all angles $2\pi/6$, just barely enough to allow for a smaller regular dodecahedron to have all dihedral angles $2\pi/5$.

It wasn't until the 1980's that W. Thurston, inspired by work of T. Jorgensen, showed that hyperbolic 3-manifolds were abundant within the category of closed 3-manifolds, and that it was likely that "most" topological 3-manifolds admit hyperbolic structures. In fact, Thurston made a broader *Geometrization Conjecture*: that all compact 3-manifolds have a natural decomposition into pieces that admit geometric structures from 8 natural 3-dimensional geometries. G. Perelman has put forth a proof of this conjecture.

The situation in higher dimensions is quite different. Attempts to put geometric structures on higher dimensional manifolds must contend with a greater number of combinatorial constraints and this is sufficiently difficult that the geometric n-manifolds are sparse within the category of n-manifolds for $n \geq 4$.

But, because we are in 3 dimensions we get to follow Thurston's lead and focus on hyperbolic 3-manifolds. The question that arises is how to go about understanding this fundamentally important class of 3-manifolds? One great tool in studying manifolds is the use of *invariants*; they distill the essence of the manifold down to simple mathematical objects. In 2 dimensions, the Euler characteristic is a simple-to-compute invariant (just triangulate the manifold and count $V - E + F$) which turns out to completely classify compact 2-manifolds. By the Gauss-Bonnet Theorem, the area of a hyperbolic 2-manifold is determined by its Euler Characteristic, hence the area can be used to tell us what topological 2-manifold a given hyperbolic 2-manifold is.

One has to be a little careful here, because a given 2-manifold that admits a hyperbolic structure actually admits many non-equivalent hyperbolic structures. For example, one can take the shortest geodesic in a hyperbolic 2-manifold and start pinching it (making it shorter) and with some care still have a hyperbolic 2-manifold. One aspect of the needed care is to ensure that the shorter the geodesic, the bigger the (embedded) collar around it. Although the area can tell us what topological 2-manifold a given hyperbolic 2-manifold is, by the Gauss-Bonnet Theorem, it doesn't distinguish between the various hyperbolic structures on the 2-manifold.

In 3 dimensions the situation is markedly different. Mostow's Rigidity Theorem tells us that if a compact 3-manifold admits a hyperbolic structure, then this hyperbolic structure is unique. Mostow's Theorem was generalized to complete hyperbolic 3-manifolds of finite volume by A. Marden and G. Prasad. Here the restriction to finite volume is necessary, and there is in fact a rich theory of hyperbolic structures on infinite-volume hyperbolic 3-manifolds.

8.5 Volumes of Hyperbolic 3-Manifolds

The most natural geometric invariant is the volume, which simply uses the metric to measure the hyperbolic 3-manifold.

Example: Consider $C = \{(x, y, t) : 0 \leq x \leq 1,\ 0 \leq y \leq 1,\ t \geq 1\}$ in the Upper-Half-Space Model of hyperbolic 3-space, and glue opposite vertical faces by the hyperbolic isometries $z \to z + 1$ and $z \to z + i$. This is a slightly odd example in that it is not quite a hyperbolic 3-manifold according to our criteria because it has a boundary, which is a Euclidean torus T (of course, the boundary could be thrown out, but then it wouldn't be complete). Nonetheless, we can compute the volume of this hyperbolic object (note that topologically this object is $T^2 \times [1, \infty]$) by integrating the hyperbolic volume form $\frac{dxdydt}{t^3}$ (which is simply the size of an infinitesimal cube in **UHS**) over C. Interestingly, despite the infinite nature of C, the volume turns out to be finite:

$$\int \int \int \frac{dxdydt}{t^3} = \text{Area}(T) \int_1^\infty \frac{dt}{t^3} = \text{Area}(T) \left(\frac{-1}{2t^2} \right]_1^\infty = \frac{1}{2}\text{Area}(T).$$

In the late 1970's W. Thurston, utilizing work of Jorgensen and M. Gromov, proved a remarkable structure theorem for the set of volumes of hyperbolic 3-manifolds. Here is a statement of a part of his theorem.

Theorem 8.2 (Thurston) *The set of volumes of hyperbolic 3-manifolds is a well-ordered subset of* $\mathbf{R}_+$ *of order type* ω^ω. *The set of manifolds with any given volume is finite.*

So, there is a lowest volume of a hyperbolic 3-manifold, then a next lowest volume, then a third lowest, and so on. Further, Thurston showed that these volumes limit on the volume v_ω of a complete, non-compact hyperbolic 3-manifold (a *cusped* hyperbolic 3-manifold), and that then the process continued with a next lowest volume and so on.

$$v_1 < v_2 < v_3 < \ldots \to v_\omega < v_{\omega+1} < v_{\omega+2} < \ldots$$

This raises several questions. What hyperbolic 3-manifold has minimum volume? What properties do the limiting manifolds have? What hyperbolic 3-manifold realizes this least limiting volume?

We begin by investigating the middle question, in one lower dimension. As mentioned, in 2 dimensions one can alter the geometry of a hyperbolic 2-manifold by, for example, pinching a short geodesic. The more you pinch the geodesic, the bigger the collar around it must grow. In the limit, the geodesic disappears (goes off to infinity) and there remains an infinite collar neighborhood, called a *cusp neighborhood*. Actually, there are 2 such cusp neighborhoods in the 2-dimensional case and in fact the limit of the pinching process is 2 complete punctured hyperbolic 2-manifolds if the pinching geodesic is separating, or 1 complete punctured hyperbolic 2-manifold with 2 punctures if the geodesic is not separating.

In 3 dimensions the limiting behavior of volumes is generally produced by taking a short geodesic with a maximal embedded solid tube neighborhood and shrinking the geodesic. As the geodesic shrinks, the maximal embedded solid tube around it grows and the geodesic drifts away from the "thick" part of the manifold. The geodesic gets shorter and shorter and the embedded solid tube gets bigger and bigger, and in the limit the geodesic disappears and the embedded solid tube neighborhood becomes a cusp neighborhood. Such a limiting manifold is a cusped hyperbolic 3-manifold. There is a big difference from the 2-dimensional case though: by Mostow's Theorem, the topology of the manifolds in the sequence must change. Basically, the embedded solid tube must be "reseated" on its boundary torus; that is, Dehn surgery is being performed.

Limits are good, so we include cusped hyperbolic 3-manifolds in the class of hyperbolic 3-manifolds. Consider the short geodesic in one of the approximating hyperbolic 3-manifolds $M = \mathbf{H}^3/\Gamma$ where $\Gamma \subset \mathrm{Isom}_+(\mathbf{H}^3)$. When we lift the short geodesic from M to its universal cover $\mathbf{H}^3$ we get a collection of infinite length geodesics (compare with our previous discussion about the universal cover of a Euclidean torus). If we focus on one such infinite geodesic we see that there is an isometry $\gamma \in \Gamma$ which takes the geodesic to itself by translating along it. In a sense, it winds the infinite geodesic on itself to produce the original closed (short) geodesic. We note that the embedded solid tube around the short geodesic lifts to a collection of infinite embedded solid tubes whose cores comprise the collection of infinite geodesics.

Having analyzed the lifts of a short geodesic, we turn to the limit of the short geodesics, which is the cusp. What does a cusp neighborhood in a cusped manifold M look like in the universal cover, $\mathbf{H}^3$? The answer is that it consists of a collection of horoballs, all of which are equivalent under the action of the group Γ where $M = \mathbf{H}^3/\Gamma$.

Given a collection of horoballs it is generally a good idea to view them in the Upper-Half-Space model. It is also a good idea to normalize so that one of the horoballs is centered at infinity. Now, expand the horoballs equivariantly until the first time that two horoballs bump into each other. Note that if the cusp neighborhood had been the maximal (embedded) cusp neighborhood, then no additional expansion would have been needed. In any case, we can normalize so that one of the bumping points is the point $(0, 0, 1)$. If we focus on the cusp at infinity, we see that there is a subgroup Γ_∞ of Γ which fixes the point at infinity and takes the horoball at infinity to itself, thereby producing the original cusp neighborhood in M.

8.6 The Maximal Cusp Diagram

What do we see when we look down from infinity at all these horoballs? From infinity each of these horoballs looks like a disk, and the biggest disks have radius $\frac{1}{2}$ because of our normalization. This radius of $\frac{1}{2}$ can be thought of as being measured in the induced Euclidean metric on the bounding horosphere $\{(x, y, t) : t = 1\}$. The transformations in Γ_∞ are simply translations of this bounding horosphere and, by construction, must move points a distance greater than or equal to 1. Thus we see that the picture from infinity contains infinite repetition, and that all the information is contained in a fundamental domain determined by Γ_∞. In the finite-volume case Γ_∞ is generated by two elements $z \to z + a$ and $z \to z + b$ where a, b are non-trivial complex numbers that are not multiples of each other.

This view from infinity of horoballs together with the generators of Γ_∞ produces the *Maximal Cusp Diagram*, which contains within it a description of the torus boundary of the maximal cusp neighborhood. We now exploit this maximal cusp torus T_c to get some control over the volume of the cusped manifold. This control arises from our previous calculation of the hyperbolic volume of the region $A \times [1, \infty]$ where A was a square (more generally, a parallelogram) with opposite edges identified (forming a torus) in the horosphere $\{t = 1\}$. The answer was that the volume is one-half the area of the boundary torus, and this relationship holds for the maximal cusp and its maximal torus boundary.

By our set-up there is an embedded disk of radius $\frac{1}{2}$ in the maximal cusp torus T_c. Hence $\pi(1/2)^2 = \pi/4$ is a lower bound for the area of T_c. Of course, we've missed area outside this disk. It is a well-known result of Thue that the densest packing of equal-radius disks in the plane is the hexagonal packing, which has density $\frac{\pi}{2\sqrt{3}}$. In our set-up this means that $\frac{\pi/4}{\text{area}(T_c)} \leq \pi/(2\sqrt{3})$ and this tells us that area$(T_c) \geq \sqrt{3}/2$. This produces a lower bound for the volume of a cusped hyperbolic 3-manifold: $\sqrt{3}/4 \leq v_\omega$.

If one can use packing of disks to improve the area bound for the boundary of the maximal cusp torus, then one should be able to use packing of horoballs to improve the volume bound for the maximal cusp. The maximal density of horoballs packed in $\mathbf{H}^3$ (ignoring technical problems associated with packings in hyperbolic space) was discovered by K. Boroczky and it is $\sqrt{3}/(2V_I)$ where V_I is the volume of the ideal regular tetrahedron in $\mathbf{H}^3$. So we have an improved lower bound $V_I/2 \leq v_\omega$.

C. Adams showed that in the maximal cusp diagram there is another disk of radius $1/2$ which is not a translate under Γ_∞ of the original disk of radius $1/2$. This produces a bound of $V_I \leq v_\omega$ (see Adams (1987)). Subsequently,

C. Cao and R. Meyerhoff proved that, in fact, $2V_I = v_\omega$ (this volume is realized by the figure-eight knot complement in the 3-sphere and by its sibling); see Cao-Meyerhoff (2001). However, we plan to ignore the details of the proof of Cao-Meyerhoff and begin to present our new method for studying volume bounds.

We will consider the *orthodistance spectrum* for the horoball lifts of a maximal cusp in a cusped hyperbolic 3-manifold. That is we look at the distances between various horoball lifts in $\mathbf{H}^3$. Let $o(1)$ be the distance between the nearest horoballs. Of course, in our set-up $o(1) = 0$ because we expanded our cusp neighborhood out to the first "hit" which occurred (after normalizing) at $(0, 0, 1)$, among other points. In the maximal cusp diagram we see that the disks associated with horoballs a distance $o(1)$ from the horoball B_∞ at infinity are the disk of radius $1/2$ centered at $(0, 0)$ and its translates under Γ_∞ as well as the disk of radius $1/2$ guaranteed by Adams, and its translates. We call this collection of disks $D(1)$. Now let $o(2)$ be the second nearest distance between horoball lifts other than the distances subsumed under $o(1)$ and described in the previous sentence. (Note that $o(2)$ could be 0.)

If we analyze the disks in the maximal cusp diagram that are $o(2)$ away from B_∞ we see that they all have the same radius and that if $o(2)$ is close to zero then this radius is close to $1/2$ and that if $o(2)$ is far from zero then this radius is considerably less than $1/2$, in fact the formula for the radius is $(\frac{1}{2})e(2)^{-2}$, where $e(2)$ is the Euclidean distance between the centers of horoballs abutting B_∞ that are separated by distance $o(2)$. We can compute that $e(n) = e^{o(n)/2}$.

Consider the collection of radius $1/2$ disks $D(1)$ described two paragraphs above, and expand these disks to radius $(\frac{1}{2})e(2)$. By construction (and a bit of work) it can be shown that these expanded disks must be disjoint (or perhaps tangent). Hence we can improve our bound on area by using $o(2)$. In particular, if $o(2)$ is big and the horoballs, hence disk shadows, are widely separated then we get plenty of area from the expanded disks. But if $o(2)$ is small, and the expansion is hence small, then we still do well because we know that a horoball a distance $o(2)$ away from B_∞ must have radius close to $\frac{1}{2}$ and this means we have disks of radius close to $\frac{1}{2}$ in our maximal cusp diagram; this collection of disks is denoted $D(2)$. This sort of trade-off argument was introduced in Gabai-Meyerhoff-Milley (2001).

8.7 The Mom Technology

Now, we will be greedy and push this analysis out to $o(4)$ the fourth nearest distance between horoball lifts (after taking into account the notion of equivalence described above) and look at the collections of disks $D(1)$, $D(2)$, $D(3)$, $D(4)$. In particular, we will expand these disks according to $o(4)$; for example, the $D(1)$ disks will be expanded to radius $(\frac{1}{2})e(4)$.

The problem arises that the various expanded disks may overlap, for example if there are disks in $D(1)$ whose associated horoballs are separated by $o(2)$ (and $o(2) < o(4)$). If there are few overlaps, then we're happy because there won't be much punishment for double-counting of area. In fact, we can set up a parameter space argument on the computer where the parameters are the orthodistances $o(2)$, $o(3)$, $o(4)$, and then carry out the various area calculations including the accounting for overlaps. This is one use of the computer in our work, and it produces a bound of 2.7 for the volume of a cusped hyperbolic 3-manifold in the case where there are few overlaps arising from $o(4)-$related disk expansion.

What if there are lots of overlaps arising from $o(4)-$related expansion? The computations become unpleasant, and worse, we find ourselves losing area rapidly as we have to adjust for double counting of area associated with the overlaps. This sounds bad. But in this case, the bad situation turns out to be good. The fact that there are lots of overlaps turns out to provide powerful topological information about the cusped manifold.

In particular, the overlaps correspond to 2-handles. So we now take a moment to talk about building a manifold by joining handles. We work through the classical example: building a 2-dimensional torus by starting with a disk and adding handles appropriately. Take a disk and deform it to produce a (lower) hemisphere, and think of its boundary as the equator. Now add a line connecting diametrically opposed points on the equator; this line is a 1-handle which we can thicken to produce something that looks like an actual handle for our "basket". After some mild deformation of the basket, we see that it is topologically equivalent to a sphere with two disks cut out. This, in turn, is topologically equivalent to a cylinder, and we can bend this cylinder to get half of a classical torus. Now, the two circle boundary components of this cylinder, which is in the shape of half a torus, can be connected by a 1-handle. The 1-handle can be thickened and we note that this new manifold has one boundary component (a circle). In fact, we can deform this new manifold to get a torus with a disk removed. Adding in a disk (a 2-handle) completes the handlebody construction of the torus.

Handles arise naturally in our situation. For example, if we have two horoball lifts that are separated by $o(2)$ then the shortest geodesic connecting them (note that it is orthogonal to both horoballs) is a 1-handle. In fact, associated with $o(1)$, $o(2)$, $o(3)$ we have a collection of 1-handles. Note that the 1-handles associated with $o(1)$ have length 0, but that's not a problem.

Overlaps in the maximal cusp diagram generally correspond to 2-handles. For example, if a disk D_1 in $D(1)$ and a disk D_3 in $D(3)$ are separated by distance $o(2)$ then we have an overlap from our $o(4)$−expanded disks, and we also have a 2-handle which is a totally geodesic surface with a boundary made up of six pieces: the 1-handle from D_1 to B_∞, then the Euclidean line on B_∞ running from the base of the 1-handle connecting D_1 and B_∞ to the base of the 1-handle connecting B_∞ and D_3, then the 1-handle connecting B_∞ and D_3, then a curve running on the horosphere associated with D_3, then the 1-handle connecting D_3 and D_1, and finally a curve on the horosphere associated with D_1. This is a 2-handle of type $(1, 3, 2)$.

The idea is that if we have sufficient overlaps, then we can build up the manifold from the cusp neighborhood and the handles (if we fill in the torus boundary components that remain). That is, our handle structure produces a cusped hyperbolic 3-manifold (with more than one cusp) and we can obtain our original manifold by filling in the cusps appropriately (via hyperbolic Dehn surgery).

So, *the dichotomy that enables us to answer low-volume questions works as follows (in the cusped category).* Take a cusped hyperbolic 3-manifold. On the one hand, if the maximal cusp diagram has little overlap between the expanded disks (the expansion is associated with $o(4)$) associated with the disks in $D(1)$, $D(2)$, $D(3)$ then a computer analysis shows that the volume of the cusped hyperbolic 3-manifold must be at least 2.7. On the other hand, if there is sufficient overlap among the expanded disks then we can use the handle information residing in the overlaps to build a 2-cusped (or more) hyperbolic 3-manifold from which the original hyperbolic 3-manifold is obtained by Dehn surgery. So, the plan in the sufficient-overlap case is to find all the relevant 2-or-more-cusped hyperbolic 3-manifolds that arise as potential parents for low-volume 1-cusped hyperbolic 3-manifolds.

There are two sufficient-overlap subcases that we consider for this work. First, we look at the subcase where we have two 1-handles arising from $o(1)$ and $o(2)$ and two 2-handles arising from overlaps of expanded disks coming from $D(1)$ and $D(2)$ (similarly we could have two 1-handles arising from $o(1)$ and $o(3)$ and two 2-handles arising from overlaps of expanded disks coming from $D(1)$ and $D(3)$; and once more with 2 and 3). Second, we consider the subcase where we have three 1-handles arising from $o(1)$, $o(2)$, $o(3)$ and three 2-handles arising from overlaps of expanded disks coming from $D(1)$, $D(2)$, $D(3)$.

Given a certain amount of topological and geometric preparation, we find that the handle structures can be used to construct 2-complexes (in the given 3-manifold) with good dual triangulations. In particular, in the two 1-handles and two 2-handles subcase, we find that the dual triangulations are made up of either four 3-sided pyramids, or two 4-sided pyramids; and in the three 1-handles and three 2-handles subcase, we find that the dual triangulations are made up of either six 3-sided pyramids; two 3-sided pyramids and an octahedron; or two 5-sided pyramids.

We wish to find the cusped hyperbolic 3-manifolds (with two or more cusps) that can arise from these triangulations. These cusped manifolds will be the parents (via Dehn surgery) of the low-volume cusped hyperbolic 3-manifolds. We call these parent cusped hyperbolic 3-manifolds *Mom(2)*'s in the two 1-handles and two 2-handles case, and *Mom(3)*'s in the three 1-handles and three 2-handles case.

Thus we come to our second use of the computer. In each of the five situations (as listed two paragraphs up), we have the computer enumerate all ways to identify faces (in pairs) to get a 3-manifold, and then we use J. Weeks' SnapPea program to test whether the 3-manifold (with boundary) is hyperbolic. Of course, we significantly reduce the amount of work the computer needs to do by eliminating natural redundancies, and by throwing out those 3-manifolds that have boundary components that are not tori (note, they automatically can't be hyperbolic).

We arrive at a list of Mom(2)'s: $m125$, $m129$, $m202$ where they are described using the notation in J. Weeks' census of hyperbolic 3-manifolds obtained by gluing seven or fewer tetrahedra. Note that the manifold $m129$ is the Whitehead Link complement in S^3. The list of Mom(3)'s could also be given, but perhaps it wouldn't be too illuminating. Suffice it to say that there are 18 of them.

Thus we arrive at our theorem.

Theorem 8.3 (Gabai, Meyerhoff, Milley) *A cusped orientable hyperbolic 3-manifold has volume greater than 2.7, or is obtained by Dehn surgery on a Mom(2) or a Mom(3).*

In fact, Weeks has analyzed the Dehn surgery spaces for these 21 Mom's and, using Weeks' work, we have that the first six lowest volume cusped orientable hyperbolic 3-manifolds are $m003$, $m004$, $m006$, $m007$, $m009$, $m010$ which are the first 6 orientable cusped hyperbolic 3-manifolds in the SnapPea census.

References

C. Adams (1987), The noncompact hyperbolic 3-manifold of minimum volume, *Proc. Amer. Math. Soc.* **100**, 601–606.

C. Cao and G. R. Meyerhoff (2001), The orientable cusped hyperbolic 3-manifolds of minimum volume, *Invent. Math.* **146**, 451–478.

D. Gabai, G. R. Meyerhoff, and P. Milley (2001), Volumes of tubes in hyperbolic 3-manifolds, *J. Diff. Geometry* **57**, 23–46.

J. Ratcliffe (1994), *Foundations of Hyperbolic Manifolds* Springer-Verlag, New York.

J. Weeks (1985), *The Shape of Space*, Marcel Dekker, New York.

J. Weeks, *SnapPea*, computer program available at geometrygames.org

9

Smoothed Analysis of Algorithms and Heuristics: Progress and Open Questions

Daniel A. Spielman

Applied Mathematics and Computer Science, Yale University
New Heaven, Connecticut, USA
e-mail: spielman@cs.yale.edu

Shang-Hua Teng

Computer Science, Boston University
and Akamai Technologies Inc
Boston, Massachusetts, USA
e-mail: steng@cs.bu.edu

Abstract

In this paper, we survey some recent progress in the smoothed analysis of algorithms and heuristics in mathematical programming, combinatorial optimization, computational geometry, and scientific computing. Our focus will be more on problems and results rather than on proofs. We discuss several perturbation models used in smoothed analysis for both continuous and discrete inputs. Perhaps more importantly, we present a collection of emerging open questions as food for thought in this field.

9.1 Prelinminaries

The quality of an algorithm is often measured by its time complexity (Aho, Hopcroft & Ullman (1983) and Cormen, Leiserson, Rivest & Stein (2001)). There are other performance parameters that might be important as well, such as the amount of space used in computation, the number of bits needed to achieve a given precision (Wilkinson (1961)), the number of cache misses in a system with a memory hierarchy (Aggarwal et al. (1987), Frigo et al. (1999), and Sen et al. (2002)), the error probability of a decision algorithm (Spielman & Teng (2003a)), the number of random bits needed in a randomized algorithm (Motwani & Raghavan (1995)), the number of calls to a particular "oracle" program, and the number of iterations of an iterative algorithm (Wright (1997), Ye (1997), Nesterov & Nemirovskii (1994), and Golub & Van Loan (1989)). The quality of an approximation algorithm could be its approximation ratio (Vazirani (2001)) and the quality of an online algorithm could be its competitive ratio (Sleator & Tarjan (1985) and Borodin & El-Yaniv (1998)).

Once we fix a quality parameter Q, there might still be more than one way to measure an algorithm A. If our universe of inputs happens to have only one instance $\mathbf{x}$ then the most natural measure is the *instance-based complexity*, given by $Q(A, \mathbf{x})$. In such a case, if we have a few algorithms $A_1, \ldots, A_k$ in our repertoire, we can easily decide which one is better. If our universe of inputs has two instances $\mathbf{x}$ and $\mathbf{y}$, then the instance-based measure of an algorithm A defines a two dimensional vector $(Q(A, \mathbf{x}), Q(A, \mathbf{y}))$. For two algorithms A_1 and A_2, if $Q(A_1, \mathbf{x}) < Q(A_2, \mathbf{x})$ but $Q(A_1, \mathbf{y}) > Q(A_2, \mathbf{y})$, then strictly speaking, they are not comparable.

The universe D of inputs is much more complex, both in theory and in practice. The instance-based measure defines a high-dimensional vector when D is finite. Otherwise, it can be viewed as a function from D to $\mathbb{R}$. How should one measure the quality of an algorithm? How should one compare two algorithms?

Traditionally, one partitions an input domain D into a collection of subdomains $\{D_1, \ldots, D_n, \ldots\}$ according to the *input size*. The set D_n represents all instances in D whose input size is n. Given an algorithm A, for each D_n, one comes up with a scalar $t_{Q,A}(n)$ that "summarizes" the performance of A over D_n, as given by the restriction $Q_n(A)$ of $Q(A, \cdot)$ to D_n. Then $t_{Q,A}(n)$ is a function of n. With the help of big-O or big-Θ notations, one often characterizes the behavior of A by evaluating $t_{Q,A}(n)$ asymptomatically.

The definition of input sizes could be a source of discussion, for example,

- in optimization, scientific computing, and computational geometry, the input size could be the number of real scalars in the input;
- in number-theoretical algorithms, it could be the total number of bits in the input;
- in comparison-based sorting, it could the number of elements, while in some other sorting algorithms, it could be the total number of letters in the input;
- in the knapsack problem, it could be the total magnitude (or the size of the unary representation) of the input.

Whatever the definition of the input size is, we need to find a way to measure and to summarize the performance of an algorithm over an input subdomain D_n.

The most commonly used measure is the *worst-case measure*. It is given by

$$\mathrm{W}\left[Q_n(A)\right] = \max_{\mathbf{x} \in D_n} Q(A, \mathbf{x}).$$

When the worst-case measure of an algorithm A is small[1], we have an absolute guarantee on the performance of algorithm A no matter which input it is given. Algorithms with good worst-case performance have been developed for a great number of problems including some seemingly difficult ones such as primality testing (Solovay & Strassen (1977), Miller (1975), Adleman & Huang (1987), and Agrawal, Kayal & Saxena (2004)) and convex programming (Nesterov & Nemirovskii (1994)). These algorithms have time complexity upper-bounded by a (low-degree) polynomial function in n.

However, with an even greater number of problems, ranging from network design to industrial optimizations, we have been less lucky. Scientists and engineers often use heuristic algorithms for these problems. Most of these algorithms, after years of improvements, work well in practice. But, their worst-case complexities might be still be very poor. For example, they could be exponential in their input sizes. For theorists who are also concerned about the practical performance of algorithms, it has long been observed that the worst-case instances of such an algorithm might not be "typical" and might never occur in practice. Thus, worst-case analysis can pessimistically suggest that the performance of the algorithm is poor. Trying to rigorously understand and model the practical performance of heuristic algorithms has been a major challenge in Theoretical Computer Science[2] (cf. the report of Condon et al. (1999)).

Average-case analysis was introduced to overcome this difficulty. In it, one first determines a distribution of inputs and then measures the expected performance of an algorithm assuming inputs are drawn from this distribution. If we suppose that $\mathcal{S}$ is a distribution over D_n, the average-case measure according to $\mathcal{S}$ is

$$\mathrm{AVG}_{\mathcal{S}}\left[Q_n(A)\right] = \mathrm{E}_{\mathbf{x} \in_{\mathcal{S}} D_n}\left[Q(A, \mathbf{x})\right],$$

where we use $\mathbf{x} \in_{\mathcal{S}} D_n$ to denote that $\mathbf{x}$ is randomly chosen from D_n according to distribution $\mathcal{S}$.

Ideally, one should use a mathematically analyzable distribution that is also the same as or close to the "practical distribution." But finding such

---

[1] For example, the number of comparisons needed by the merge-sort algorithm to sort any sequence of n elements is bounded above by $n \log n$.

[2] The theory-practice gap is not limited to heuristics with exponential complexities. Many polynomial time algorithms, such as the interior-point method for linear programming (Karmarkar (1984)) and the conjugate gradient method for linear systems (Hestenes & Stiefel (1952)), are often much faster than their worst-case bounds. In addition, various heuristics are used to speed up the practical performance of codes that are based on worst-case polynomial time algorithms. These heuristics might in fact worsen the worst-case performance, or make the worst-case complexity hard to analyze.

a distribution and analyzing it could be a difficult or even impossible task. As most average-case analyses are conducted on simpler distributions than what might occur in practice, the inputs encountered in applications may bear little resemblance to the random inputs that dominate the analysis. For example, a randomly chosen graph with average degree around six is rarely similar to a finite-element graph in two dimensions, even though the latter also has average degree around six. Random objects such as random graphs or random matrices might have some special properties with all but exponentially low probability, and these special properties might dominate the average-case analysis.

Smoothed analysis (Spielman & Teng (2004)) is a recently developed framework for analyzing algorithms and heuristics. It is partially motivated by the observation that input parameters in practice are often subject to a small degree of random noise: In industrial optimization and market predictions, the input parameters could be obtained by physical measurements, and measurements usually have some random uncertainties of low magnitudes. In computer aided design, the input parameters could be the output of another computer program, e.g., a geometric modeling program, that might have numerical imprecision due to rounding and approximation errors. Even in applications where inputs are discrete, there might be randomness in the formation of inputs. For example, the network structure of the Internet may very well be governed by some "blueprints" of the government and industrial giants, but it is still "perturbed" by the involvements of smaller Internet service providers. Thus it may be neither completely random nor arbitrary.

In smoothed analysis, we assume that an input to an algorithm is subject to a slight random perturbation. The *smoothed measure* of an algorithm on an input instance is its expected performance over the perturbations of that instance. We define the *smoothed complexity* of an algorithm as the maximum smoothed measure over its inputs.

In this paper, we survey the progress made in smoothed analysis in recent years. We discuss several perturbation models considered for both continuous and discrete problems. We then present some open questions in this field.

9.2 Basic Perturbation Models and Polynomial Smoothed Complexity

To conduct smoothed analysis, we need a perturbation model that can capture the randomness and imprecision in the formation of inputs. To be concrete in the discussion below, we first consider the case when our sub-universe is $D_n = \mathbb{R}^n$, as often considered in optimization, scientific computing, and computational geometry. For these continuous inputs, for example, the family of Gaussian distributions (cf. Feller (1968, 1970)) provide a perturbation model for noise.

Recall that a univariate Gaussian distribution with mean 0 and standard deviation σ has density

$$\frac{1}{\sqrt{2\pi}\sigma}e^{-x^2/2\sigma^2}.$$

A *Gaussian random vector* of variance σ^2 centered at the origin in $\mathbb{R}^n$ is a vector where each entry is a Gaussian random variable of standard deviation σ and mean 0. It has density

$$\frac{1}{\left(\sqrt{2\pi}\sigma\right)^d}e^{-\|\mathbf{x}\|^2/2\sigma^2}.$$

Definition 9.1 (Gaussian Perturbations) *Let $\bar{\mathbf{x}} \in \mathbb{R}^n$. A σ-Gaussian perturbation of $\bar{\mathbf{x}}$ is a random vector $\mathbf{x} = \bar{\mathbf{x}} + \mathbf{g}$, where $\mathbf{g}$ is a Gaussian random vector of variance σ^2.*

Definition 9.2 (Smoothed Complexity with Gaussian Perturbations) *Suppose $Q_n : D_n = \mathbb{R}^n \to \mathbb{R}^+$ is a quality function. Then the smoothed complexity of Q_n under σ-Gaussian perturbations is given as*

$$\max_{\bar{\mathbf{x}}\in\mathbb{R}^n} \mathbf{E}_{\mathbf{g}}\left[Q_n\left(\bar{\mathbf{x}} + \|\bar{\mathbf{x}}\|_2\, \mathbf{g}\right)\right],$$

where $\mathbf{g}$ is a Gaussian random vector of variance σ^2.

Each instance $\bar{\mathbf{x}}$ of a computational problem has a neighborhood which, intuitively, contains the set of instances that are close to and similar to $\bar{\mathbf{x}}$. A perturbation model defines a distribution over the neighborhood of $\bar{\mathbf{x}}$. The closer $\mathbf{x}$ is to $\bar{\mathbf{x}}$, the higher $\mathbf{x}$ and $\bar{\mathbf{x}}$ might be correlated due to the randomness in the formation of input instances. In Gaussian perturbations, the closeness and similarity among inputs are measured by their Euclidean distance. As the density function decreases exponentially in distance, the variance parameter σ defines the magnitude of perturbations and also captures the radius of the most likely neighborhood of an instance. The smoothed complexity is measured in terms of the input length n as well as σ, the magnitude of the perturbations. As σ increases continuously starting from 0, the smoothed complexity interpolates between the worst-case and average-case complexities (Spielman & Teng (2004)).

Of course, not all computational problems deal with continuous inputs. A commonly used communication model with a noisy channel assumes inputs are subject to Boolean perturbations of probability σ:

Definition 9.3 (Boolean Perturbations) *Let $\bar{\mathbf{x}} = (\bar{x}_1, \ldots, \bar{x}_n) \in \{0,1\}^n$ or $\{-1,1\}^n$. A σ-Boolean perturbation of $\bar{\mathbf{x}}$ is a random string $\mathbf{x} = (x_1, \ldots, x_n) \in \{0,1\}^n$ or $\{-1,1\}^n$, where $x_i = \bar{x}_i$ with probability $1 - \sigma$.*

In Boolean perturbations, the closeness and similarity of instances are measured by their Hamming distances. Again, the parameter σ defines the magnitude of perturbations as well as the radius of the most likely neighborhood of an instance.

In scheduling, packing, and sorting, the inputs are often integers of certain magnitudes. Banderier, Beier, and Mehlhorn (2003) propose to use the partial bit randomization model:

Definition 9.4 (Partial Bit Randomization) *Let $\bar{z}$ be an integer and k be a positive integer indicating the magnitude of the perturbation. A k-partial bit randomization of $\bar{z}$ is an integer z obtained from $\bar{z}$ by replacing its k least significant bits by a random number in $[0 : 2^{k-1}]$ according to some distribution over $[0 : 2^{k-1}]$.*

In comparison-based sorting and online problems, each input consists of a sequence of elements. Banderier, Beier, and Mehlhorn (2003) introduce the following *partial permutation model*:

Definition 9.5 (Partial Permutation Perturbations) *Let $\bar{\mathbf{s}}$ be a sequence of n elements. Let $0 \leq \sigma \leq 1$ be the magnitude of perturbations. A σ-partial permutation of $\bar{\mathbf{s}}$ is a random sequence $\mathbf{s}$ obtained from $\bar{\mathbf{s}}$ by first building a subset S by independently selecting each index number from $\{1, 2, \ldots, n\}$ with probability σ, and then randomly permuting elements of $\bar{\mathbf{s}}$ in position S while retaining the positions of all other elements.*

The perturbation model that most naturally captures the imprecision in the formation of inputs can vary from application to application. For instance, it might be more suitable to use uniform random perturbations within a properly-centered ball to analyze some computational geometry algorithms.

Definition 9.6 (Uniform Ball Perturbations) *Let $\bar{\mathbf{x}} \in \mathbb{R}^n$. A uniform ball perturbation of radius σ of $\bar{\mathbf{x}}$ is a random vector $\mathbf{x}$ chosen uniformly from the ball of radius σ centered at $\bar{\mathbf{x}}$.*

For any of the basic perturbation models we have discussed, there might be some refinements and variants worthy of considerations.

For example, Eppstein[1] proposed the following refinement of the partial permutation model: Let $\bar{\mathbf{s}}$ be a sequence of n elements that have a total ordering. Let $\|\mathbf{s}\|$ denote the number of elements of the input that must be moved to make the input sorted or reverse-sorted. To obtain a perturbed element, one randomly chooses a set S of $(\sigma \cdot \|\mathbf{s}\|)$ elements, and randomly permutes them. In this model, one does not perturb the already-sorted input or the reverse-sorted input at all, and the perturbations of other inputs depend on their distance to these inputs. This definition is inspired by the definition of smoothed analysis for problems that take inputs from $\mathbb{R}^n$: we do not perturb the zero vector, and perturb other vectors in proportion to their norm. For sorting, one may view the already-sorted input as a zero, and distance-to-sorted as a norm.

In analyzing scientific computing algorithms that take advantage of the sparsity in the problem instances, one may find relative Gaussian perturbations or zero-preserving Gaussian perturbations better models of imprecision:

Definition 9.7 (Relative Gaussian Perturbations) *Let $\bar{\mathbf{x}}$ be a vector $(\bar{x}_1, \ldots, \bar{x}_n) \in \mathbb{R}^n$. A relative σ-Gaussian perturbation of $\bar{\mathbf{x}}$ is a random vector $\mathbf{x} = (x_1, \ldots, x_n)$ where $x_i = \bar{x}_i(1 + g_i)$, where g_i is a Gaussian random variable with standard deviation σ.*

Definition 9.8 (Zero-Preserving Gaussian Perturbations) *For any $\bar{\mathbf{x}} = (\bar{x}_1, \ldots, \bar{x}_n) \in \mathbb{R}^n$, a zero-preserving σ-Gaussian perturbation of $\bar{\mathbf{x}}$ is a vector $\mathbf{x} = (x_1, \ldots, x_n)$ where $x_i = \bar{x}_i + (1 - \mathbf{IsZero}(\bar{x}_i)) g_i$, where g_i is a Gaussian random variable with standard deviation σ and $\mathbf{IsZero}(x) = 1$ if $x = 0$, and $\mathbf{IsZero}(x) = 0$, otherwise.*

When time complexity is the main concern, the central questions in smoothed analysis naturally are:

Does an algorithm have polynomial smoothed complexity? Is a decision or search/ optimization problem in smoothed polynomial time?

In addition to the notion of input size, one needs a model of perturbations and a notion of magnitudes of perturbations to define polynomial smoothed complexity. Given a model and notion of magnitudes of perturbations, there might still be several possible definitions of polynomial smoothed complexity.

---

[1] Personal Communication

Spielman and Teng (2004) define polynomial smoothed complexity as:

Definition 9.9 (Polynomial Smoothed Complexity) *Given a problem P with input domain $D = \cup_n D_n$ where D_n represents all instances whose input size is n. Let $\mathcal{R} = \cup_{n,\sigma} R_{n,\sigma}$ be a family of perturbations where $R_{n,\sigma}$ defines for each $\bar{\mathbf{x}} \in D_n$ a perturbation distribution of $\bar{\mathbf{x}}$ with magnitude σ. Let A be an algorithm for solving P and $T_A(\mathbf{x})$ be the time complexity for solving an instance $\mathbf{x} \in D$. Then algorithm A has polynomial smoothed complexity if there exist constants n_0, σ_0, c, k_1 and k_2 such that for all $n \geq n_0$ and $0 \leq \sigma \leq \sigma_0$,*

$$\max_{\bar{\mathbf{x}} \in D_n} \left(E_{\mathbf{x} \leftarrow R_{n,\sigma}(\bar{\mathbf{x}})} [T_A(\mathbf{x})] \right) \leq c \cdot \sigma^{-k_2} \cdot n^{k_1}, \tag{9.1}$$

where $\mathbf{x} \leftarrow R_{n,\sigma}(\bar{\mathbf{x}})$ means $\mathbf{x}$ is chosen according to distribution $R_{n,\sigma}(\bar{\mathbf{x}})$.

The problem P is in smoothed polynomial time with perturbation model $\mathcal{R}$ if it has an algorithm with polynomial smoothed complexity.

For example, Spielman and Teng show that the simplex method with the shadow-vertex pivoting rule (Gass & Saaty (1955)) has polynomial smoothed complexity under Gaussian perturbations. We can relax or strengthen the dependency on σ in the definition of the polynomial smoothed complexity.

Definition 9.10 (Polynomial Smoothed Complexity: II) *Let P, A, D, D_n, $\mathcal{R}$, $R_{n,\sigma}$ be the same as in Definition 9.9. Then algorithm A has polynomial smoothed complexity if there exist constants n_0, σ_0, c, k, and a function $g : \mathbb{R}^+ \to \mathbb{R}^+$ such that for all $n \geq n_0$ and $0 \leq \sigma \leq \sigma_0$,*

$$\max_{\bar{\mathbf{x}} \in D_n} \left(E_{\mathbf{x} \leftarrow R_{n,\sigma}(\bar{\mathbf{x}})} [T_A(\mathbf{x})] \right) \leq c \cdot g(\sigma) \cdot n^k.$$

In particularly, when $g(\sigma)$ is a poly-logarithmic function of $1/\sigma$, we say the algorithm has polynomial smoothed complexity with poly-logarithmic dependency on $1/\sigma$.

By Markov's inequality (cf. Alon & Spencer (1992) and Feller (1968)), if an algorithm A has smoothed complexity $T(n,\sigma)$, then

$$\min_{\bar{\mathbf{x}} \in D_n} \Pr_{\mathbf{x} \leftarrow R_{n,\sigma}(\bar{\mathbf{x}})} \left[T_A(\mathbf{x}) \leq \delta^{-1} T(n,\sigma) \right] \geq 1 - \delta. \tag{9.2}$$

In other words, if A has polynomial smoothed complexity, then for any $\bar{\mathbf{x}}$, with high probability, say with $(1 - \delta)$, A can solve a random perturbation of $\bar{\mathbf{x}}$ in time polynomial in n, $1/\sigma$, and $1/\delta$

However, the probabilistic upper bound given in (9.2) does not usually imply that the smoothed complexity of A is $O(T(n, \sigma))$. In fact Eqn (9.2) may not even imply that

$$\max_{\bar{\mathbf{x}} \in D_n} \left(\mathbf{E}_{\mathbf{x} \leftarrow R_{n,\sigma}(\bar{\mathbf{x}})} \left[T_A(\mathbf{x}) \right] \right) \text{ is finite.}$$

Eqn (9.2) suggests a relaxed extension of polynomial smoothed complexity.

Definition 9.11 (Probably Polynomial Smoothed Complexity) *Let P, A, D, D_n, $\mathcal{R}$, $R_{n,\sigma}$ be the same as in Definition 9.9. Then algorithm A has probably polynomial smoothed complexity if there exist constants n_0, σ_0, c, k_1, k_2, k_3, such that for all $n \geq n_0$ and $0 \leq \sigma \leq \sigma_0$,*

$$\max_{\bar{\mathbf{x}} \in D_n} \left(\mathbf{Pr}_{\mathbf{x} \leftarrow R_{n,\sigma}(\bar{\mathbf{x}})} \left[T_A(\mathbf{x}) > c \cdot \sigma^{-k_1} \cdot \delta^{-k_2} \cdot n^{k_3} \right] \right) \leq \delta. \qquad (9.3)$$

Equivalently, there exist constants n_0, σ_0, c, and α, such that for all $n \geq n_0$ and $0 \leq \sigma \leq \sigma_0$,

$$\max_{\bar{\mathbf{x}} \in D_n} \left(\mathbf{E}_{\mathbf{x} \leftarrow R_{n,\sigma}(\bar{\mathbf{x}})} \left[(T_A(\mathbf{x}))^{\alpha} \right] \right) \leq c \cdot \sigma^{-1} \cdot n \qquad (9.4)$$

The relaxation of polynomial smoothed complexity given in Eqn (9.3) is introduced by Blum and Dunagan (2002) in their analysis of the perceptron algorithm. They show that the perceptron algorithm has probably polynomial smoothed complexity, in spite of the fact that its smoothed complexity according to Definition 9.9 is unbounded. Beier and Vöcking (2004), in their study of the binary optimization problem, introduce the alternative form given in Eqn (9.4).

9.3 Progress in Smoothed Analysis

We cluster the materials in this section into four subsections.

- Linear Programming.
- Combinatorial Optimization.
- Scientific Computing.
- Discrete and Geometric Structures.

Although these topics appear to be diverse, the approaches developed for conducting smoothed analysis in these areas are quite similar. In fact, most approaches consist of two basic steps:

- **Geometric/Combinatorial Conditions of Nice Instances:** Establish a set of analyzable geometric or combinatorial conditions under which the algorithm performs well on an instance.

- **Probabilistic Analysis:** Prove that for every input, these geometric/combinatorial conditions hold with high probability over its perturbations.

The challenge in the first step is to establish *manageable* conditions. The instance-based complexity itself provides the most accurate characterization of nice and bad input instances of an algorithm, but this characterization is hardly useful in analysis. What we often look for are conditions that are accurate enough for predicting the performance and simple enough for probabilistic analysis. For example, the number of iterations of the Conjugate Gradient Method (CG) (Hestenes & Stiefel (1952)) for solving a symmetric positive definite linear system $\mathbf{A}\mathbf{x} = \mathbf{b}$ can be bounded above by $O\left(\sqrt{\kappa(\mathbf{A})}\right)$ (Golub & Van Loan (1989)), where $\kappa(\mathbf{A})$ is the condition number of $\mathbf{A}$ – the ratio of the largest eigenvalue of $\mathbf{A}$ to the smallest eigenvalue of $\mathbf{A}$. Thus, if $\mathbf{A}$ is from a distribution where $\kappa(\mathbf{A})$ is small with high probability, then we can use $\kappa(\mathbf{A})$ as our condition of nice inputs, even though there might exist $\mathbf{A}$ with very large $\kappa(\mathbf{A})$ and $\mathbf{b}$ for which the CG converges rapidly. But if $\mathbf{A}$ is from a distribution where the condition numbers are mostly very large, and the CG has been observed and believed to perform well, then we need to find some other conditions for its good performance.

To establish a lower bound on the worst-case complexity of an algorithm we rely a lot on our intuition of the properties for bad instances. In contrast, to prove a smoothed upper bound, we need to work with our imagination to find properties of nice instances. However, these two studies are not completely unrelated, and if all of our worst-case instances are unstable in a perturbation model, then there might be reasons to believe that the smoothed measure is good.

9.3.1 Linear Programming

In a linear program, one is asked to optimize a linear objective function subject to a set of linear constraints. Mathematically, according to one standard form of linear programming, one is solving

$$\max \quad \mathbf{c}^T\mathbf{x} \quad \text{subject to} \quad \mathbf{A}\mathbf{x} \leq \mathbf{b},$$

where $\mathbf{A}$ is an $m \times n$ matrix, $\mathbf{b}$ is an m-dimensional vector, and $\mathbf{c}$ is an n-dimensional vector.

If the constraints are feasible, then they define a convex polyhedron $\{\mathbf{x} : \mathbf{A}\mathbf{x} \leq \mathbf{b}\}$. This polyhedron could be unbounded in the direction of $\mathbf{c}$

in which case the optimal value of the linear program is infinite. Otherwise, the optimal value is finite and the solution point $\mathbf{x}$ that achieves this optimal value must be a vertex of the polyhedron $\{\mathbf{x} : \mathbf{Ax} \leq \mathbf{b}\}$. Note that a vertex is determined by a subset of equations from $\mathbf{Ax} = \mathbf{b}$.

Linear programming is perhaps the most fundamental optimization problem (Dantzig (1991)). Several methods for solving linear programs have been developed since its introduction (Dantzig (1951), Khachiyan (1979), and Karmarkar (1984)). The most commonly used approaches for solving a linear program are the simplex method (Dantzig (1951)) and the interior-point method (Karmarkar (1984)).

We start our discussion with results in the smoothed analysis of the simplex method. We then continue with three other methods for solving linear programs: the perceptron method, its variant with scaling, and the interior-point method.

Smoothed Analysis of the Simplex Method

The simplex method provides a family of linear programming algorithms. Most of them are two-phase algorithms: In Phase I, they determine whether a given linear program is infeasible and, if the program is feasible, they also compute an initial vertex $\mathbf{v}_0$ of the feasible region and enter Phase II, where they iterate: in the ith iteration, they find a neighboring vertex $\mathbf{v}_i$ of $\mathbf{v}_{i-1}$ with better objective value, or terminate with an extreme ray from $\mathbf{v}_{i-1}$ on which the objective function is unbounded above, or terminate with an optimal solution $\mathbf{v}_{i-1}$. Some two-phase simplex methods can determine whether a feasible linear program is unbounded in the objective direction in Phase I.

Spielman and Teng (2004) consider the smoothed complexity of the simplex method under Gaussian perturbations: For any $\bar{\mathbf{A}}, \bar{\mathbf{b}}, \bar{\mathbf{c}}$, the perturbations of the linear program defined by $(\bar{\mathbf{A}}, \bar{\mathbf{b}}, \bar{\mathbf{c}})$ is

$$\max \quad \mathbf{c}^T \mathbf{x} \qquad \text{subject to} \quad \mathbf{Ax} \leq \mathbf{b},$$

where $\mathbf{A}$, $\mathbf{b}$, and $\mathbf{c}$, respectively, are obtained from $\bar{\mathbf{A}}, \bar{\mathbf{b}}, \bar{\mathbf{c}}$ by a Gaussian perturbations of variance

$$\left(\left\| \bar{\mathbf{A}}, \bar{\mathbf{b}}, \bar{\mathbf{c}} \right\|_F \sigma \right)^2,$$

where $\|(\mathbf{A}, \mathbf{b}, \mathbf{c})\|_F$ is the square root of the sum of squares of the entries in $\mathbf{A}, \mathbf{b}$, and $\mathbf{c}$.

In this smoothed setting, with probability 1, every vertex of the feasible region is determined by exactly n equations. Two vertices $\mathbf{v}$ and $\mathbf{u}$ of the feasible region are neighbors if their associated sets of equations differ by only one equation. So with probability 1, apart from the *extreme* vertices from which there is a feasible ray, each vertex of a perturbed linear program has n neighbors.

Spielman and Teng prove the following theorem.

Theorem 9.1 (Smoothed Complexity of the Simplex Method)
There exists a two-phase simplex algorithm with polynomial smoothed complexity under Gaussian perturbations.

Let $\mathbf{A}$ be a σ-Gaussian perturbation of an $m \times n$ matrix $\bar{\mathbf{A}}$ with $\|\bar{\mathbf{A}}\|_F \leq 1$ and $\mathbf{1}$ be the m-vector all of whose entries are equal to 1. Then the polyhedron $\{\mathbf{x} : \mathbf{Ax} \leq \mathbf{1}\}$ is always feasible with $\mathbf{0}$ as a feasible point. For any two n-vectors $\mathbf{c}$ and $\mathbf{t}$ the projection of the polyhedron $\{\mathbf{x} : \mathbf{Ax} \leq \mathbf{1}\}$ on the two-dimensional plane spanned by $\mathbf{c}$ and $\mathbf{t}$ is called the *shadow* of the polyhedron onto the plane spanned by $\mathbf{c}$ and $\mathbf{t}$. We denote this shadow by $\mathbf{Shadow_{t,c}}\,(\mathbf{A})$, and its size, the number of its vertices, by $|\mathbf{Shadow_{t,c}}\,(\mathbf{A})|$. Theorem 9.1 is built upon the smoothed analysis of $|\mathbf{Shadow_{t,c}}\,(\mathbf{A})|$.

Theorem 9.2 (Smoothed Shadow Size) *For any $m \times n$ matrix $\bar{\mathbf{A}}$ with $\|\bar{\mathbf{A}}\|_F \leq 1$, let $\mathbf{A}$ be a σ-Gaussian perturbation of $\bar{\mathbf{A}}$. For any two n-dimensional vectors $\mathbf{c}$ and $\mathbf{t}$*

$$\mathrm{E}_{\mathbf{A}}\left[|\mathbf{Shadow_{t,c}}\,(\mathbf{A})|\right] = O\left(\frac{mn^3}{\min(\sigma, 1/\sqrt{n \log m})^6}\right).$$

This probabilistic geometric theorem provides a smoothed upper bound on the Phase II complexity of the simplex method algorithm with the shadow-vertex pivot rule (Gass & Saaty (1955)). Theorem 9.1 was established by a reduction of Phase I computation to Phase II computation in $n + 1$ dimensions.

Recently, Deshpande and Spielman (2005) improve Theorem 9.2 with a greatly simplified proof.

Theorem 9.3 (Deshpande-Spielman) *For any $m \times n$ matrix $\bar{\mathbf{A}}$ with $\|\bar{\mathbf{A}}\|_F \leq 1$, let $\mathbf{A}$ be a σ-Gaussian perturbation of $\bar{\mathbf{A}}$. For any two n-dimensional vectors $\mathbf{c}$ and $\mathbf{t}$*

$$\mathrm{E}_{\mathbf{A}}\left[|\mathbf{Shadow_{t,c}}\,(\mathbf{A})|\right] = O\left(\frac{m^3 n^{1.5}}{\min(\sigma, 1/\sqrt{n \log m})^3}\right).$$

Perceptron Algorithms

The perceptron algorithm (Agmon (1954) and Rosenblatt (1962)) is an iterative algorithm for finding a feasible solution to linear programs in the form

$$\mathbf{Ax} \geq \mathbf{0}, \quad \mathbf{x} \neq \mathbf{0}. \tag{9.5}$$

It is commonly used in Machine Learning (Minsky & Papert (1988)) for finding a linear separator of two sets of points $R = \{\mathbf{p}_1, \ldots, \mathbf{p}_{m_1}\}$ and $B = \{\mathbf{q}_{m_1+1}, \ldots, \mathbf{q}_{m_2}\}$ in $\mathbb{R}^n$. If R and B are separable by a hyperplane through $\mathbf{0}$ with normal vector $\mathbf{x}$, then

$$\mathbf{p}_i^T \mathbf{x} \geq 0 \text{ for each } i \in \{1 : m_1\}, \quad \text{and}$$
$$\mathbf{q}_i^T \mathbf{x} \leq 0 \text{ for each } i \in \{m_1 + 1 : m_1 + m_2\}.$$

Letting $\mathbf{a}_i = \mathbf{p}_i$ for $i \in \{1 : m_1\}$ and $\mathbf{a}_i = -\mathbf{q}_i$ for $\{m_1 + 1 : m_1 + m_2\}$, the problem becomes the linear program given in (9.5).

The perceptron algorithm starts with an initial unit vector $\mathbf{x}_0$. During the kth iteration, if there exists a row $\mathbf{a}_i^T$ of $\mathbf{A}$ with $\mathbf{a}_i^T \mathbf{x}_{k-1} < 0$ then it sets $\mathbf{x}_k = \mathbf{x}_{k-1} + \mathbf{a}_i / \|\mathbf{a}_i\|_2$. The complexity question is: how many iterations does the perceptron algorithm take when given a feasible linear program of form (9.5)? The following theorem of Block (1962) and Novikoff (1962) gives an upper bound in term of a geometric quantity.

Theorem 9.4 (Block-Novikoff) *For a linear program of form (9.5), let*

$$\rho(\mathbf{A}) = \max_{\mathbf{x}: \|\mathbf{x}\|_2 = 1, \mathbf{Ax} \geq 0} \min_i \left(\frac{\mathbf{a}_i^T}{\|\mathbf{a}_i\|_2} \mathbf{x} \right).$$

The perceptron algorithm terminates in $O(1/\rho^2(\mathbf{A}))$ iterations.

The parameter $\rho(\mathbf{A})$ is known as the *wiggle room* of the perceptron problem. By establishing a probabilistic lower bound on the wiggle room, Blum and Dunagan (2002) obtain the following result.

Theorem 9.5 (Blum-Dunagan) *For any $\bar{\mathbf{A}} \in \mathbb{R}^{m \times n}$ with $\|\bar{\mathbf{A}}\|_F \leq 1$, let $\mathbf{A}$ be a σ-Gaussian perturbation of $\bar{\mathbf{A}}$ for $\sigma \leq \sqrt{1/2n}$. Then, for any δ, with probability at least $1 - \delta$, in*

$$O \left(\frac{n^3 m^2 \log^2(m/\delta)}{\sigma^2 \delta^2} \right)$$

iterations, the perceptron algorithm finds a feasible solution or correctly concludes that the linear program defined by $\mathbf{A}$ is infeasible.

Blum and Dunagan's result does not imply that the smoothed complexity of the perceptron algorithm is polynomial in m, n, and $1/\sigma$. It only states that with high probability, the perceptron algorithm with a polynomial number of iterations would correctly solve a perturbed linear program. See Definition 9.11. But this discrepancy in the definitions of polynomial smoothed complexities in the results of the perceptron method (Theorem 9.5) and the simplex method (Theorem 9.1) might provide some insights on why the simplex method usually performs better in practice than the perceptron algorithm.

Recently, Dunagan and Vempala (2004) improve the performance of a randomized version of the perceptron algorithm by applying periodic rescalings. They show that, with high probability, their algorithm finds a feasible solution to feasible linear programs of form (9.5) in time $O(mn^4 \log n \log(1/\rho(\mathbf{A})))$. It is not hard to combine the analysis of Blum-Dunagan with the the result of Dunagan-Vempala to prove the following result.

Theorem 9.6 (Smoothed complexity perceptron algorithms with rescaling) *For any $\bar{\mathbf{A}} \in \mathbb{R}^{m \times n}$ with $\|\bar{\mathbf{A}}\|_F \leq 1$, let $\mathbf{A}$ be a σ-Gaussian perturbation of $\bar{\mathbf{A}}$, for $\sigma \leq \sqrt{1/2n}$. Then, for any δ, with probability at least $1 - \delta$, in random*

$$O\left(mn^4 \log\left(\frac{nm}{\sigma\delta}\right)\right)$$

time, the Dunagan-Vempala perceptron algorithm finds a feasible solution or correctly concludes that the linear program defined by $\mathbf{A}$ is infeasible.

Condition Number of Linear Programs and the Smoothed Complexity of Interior-Point Methods

The parameter $\rho(\mathbf{A})$ aforementioned is a special case of the condition number of a linear program introduced by Renegar (1994), (1995a). Consider the following four common canonical forms of linear programs and their dual forms:

$$\max\ \mathbf{c}^T\mathbf{x} \ \text{s.t.}\ \mathbf{A}\mathbf{x} \leq \mathbf{b}$$
$$\min\ \mathbf{b}^T\mathbf{y} \ \text{s.t.}\ \mathbf{A}^T\mathbf{y} = \mathbf{c}, \ \ \mathbf{y} \geq \mathbf{0}, \tag{1}$$

$$\max\ \mathbf{c}^T\mathbf{x} \ \text{s.t.}\ \mathbf{A}\mathbf{x} \leq \mathbf{b}, \ \mathbf{x} \geq \mathbf{0}$$
$$\min\ \mathbf{b}^T\mathbf{y} \ \text{s.t.}\ \mathbf{A}^T\mathbf{y} \geq \mathbf{c}, \ \mathbf{y} \geq \mathbf{0}, \tag{2}$$

$$\max\ \mathbf{c}^T\mathbf{x} \ \text{s.t.}\ \mathbf{A}\mathbf{x} = \mathbf{b}, \ \mathbf{x} \geq \mathbf{0}$$
$$\min\ \mathbf{b}^T\mathbf{y} \ \text{s.t.}\ \mathbf{A}^T\mathbf{y} \geq \mathbf{c}, \tag{3}$$

$$\text{find } \mathbf{x} \neq \mathbf{0} \ \text{ s.t. } \ \mathbf{A}\mathbf{x} \leq \mathbf{0}$$
$$\text{find } \mathbf{y} \neq \mathbf{0} \ \text{ s.t. } \ \mathbf{A}^T\mathbf{y} = \mathbf{0}, \ \mathbf{y} \geq \mathbf{0}. \tag{4}$$

A key concept in defining the condition number is the set of ill-posed linear programs. A linear program is *ill-posed* if the program can be made both feasible and infeasible by arbitrarily small changes to its data.

In his pioneering work (Renegar (1994), (1995a), (1995b)), Renegar defines the condition number of a linear program as the scale-invariant reciprocal of the distance of that program to "ill-posedness". Any linear program may be expressed in each of the first three canonical forms. However, transformations among formulations do not in general preserve their condition numbers (Renegar (1995a)). Therefore, the condition number is defined for each normal form.

Let F be the property that a linear program is feasible. For each $(\mathbf{A}, \mathbf{b}, \mathbf{c})$ and $i \in \{1, 2, 3\}$, let $\mathbf{PLP}_i(\mathbf{A}, \mathbf{b}, \mathbf{c})$ and $\mathbf{DLP}_i(\mathbf{A}, \mathbf{b}, \mathbf{c})$ be the primal and dual linear programs, respectively, in normal form (i) defined by data $(\mathbf{A}, \mathbf{b}, \mathbf{c})$. Let

$$\rho_i(\mathbf{A}, \mathbf{b}, \mathbf{c}) = \sup \{\delta : \|\Delta\mathbf{A}, \Delta\mathbf{b}, \Delta\mathbf{c}\|_F \leq \delta \text{ implies}$$
$$F(\mathbf{PLP}_i(\mathbf{A}, \mathbf{b}, \mathbf{c})) = F(\mathbf{PLP}_i(\mathbf{A} + \Delta\mathbf{A}, \mathbf{b} + \Delta\mathbf{b}, \mathbf{c} + \Delta\mathbf{c})) \ \&$$
$$F(\mathbf{DLP}_i(\mathbf{A}, \mathbf{b}, \mathbf{c})) = F(\mathbf{DLP}_i(\mathbf{A} + \Delta\mathbf{A}, \mathbf{b} + \Delta\mathbf{b}, \mathbf{c} + \Delta\mathbf{c}))\}.$$

The condition number of the linear program defined by data $(\mathbf{A}, \mathbf{b}, \mathbf{c})$ in normal form (i) is

$$C_i(\mathbf{A}, \mathbf{b}, \mathbf{c}) = \frac{\|\mathbf{A}, \mathbf{b}, \mathbf{c}\|_F}{\rho_i(\mathbf{A}, \mathbf{b}, \mathbf{c})}.$$

We can similarly define the distance $\rho_4(\mathbf{A})$ to ill-posedness and the condition number $C_4(\mathbf{A})$ of a linear program in normal form (4).

Dunagan, Spielman and Teng (2002) prove the following theorem:

Theorem 9.7 (Smoothed condition number) *For any* $(\bar{\mathbf{A}}, \bar{\mathbf{b}}, \bar{\mathbf{c}})$ *with* $\|\bar{\mathbf{A}}, \bar{\mathbf{b}}, \bar{\mathbf{c}}\|_F \leq 1$ *and* $\sigma \leq 1$, *let* $\mathbf{A}$, $\mathbf{b}$ *and* $\mathbf{c}$ *be* σ-*Gaussian perturbations of* $\bar{\mathbf{A}}$, $\bar{\mathbf{b}}$ *and* $\bar{\mathbf{c}}$, *respectively. Then,*

$$\mathrm{E}_{\mathbf{A}, \mathbf{b}, \mathbf{c}}\left[\log C_i(\mathbf{A}, \mathbf{b}, \mathbf{c})\right] = O(\log(mn/\sigma)).$$

For any linear program of form (i), $i \in \{1, 2, 3\}$, specified by $(\mathbf{A}, \mathbf{b}, \mathbf{c})$ and parameter $\epsilon \leq 1$, there is an interior-point algorithm that determines whether the program is infeasible or unbounded or finds a feasible solution $\mathbf{x}$ with duality gap $\epsilon \|\mathbf{A}, \mathbf{b}, \mathbf{c}\|_F$ in $O(N^3 \log(N \cdot C_i(\mathbf{A}, \mathbf{b}, \mathbf{c})/\epsilon))$ operations (Renegar (1994), (1995a), (1995b), Vera (1996), Freund & Vera (1999)). where $N = \max(m, n)$. Let $T_i((\mathbf{A}, \mathbf{b}, \mathbf{c}), \epsilon)$ be the time complexity of these interior-point algorithms. For a linear program of form (4) given by $\mathbf{A}$, there is an algorithm that finds a feasible solution $\mathbf{x}$ or determines that the program is infeasible in $O(N^3 \log(N \cdot C_4(\mathbf{A})))$ operations (Cucker & Peña (2001) and Freund & Vera (1999)). Let $T_i((\mathbf{A}, \mathbf{b}, \mathbf{c}), \epsilon)$ be the time complexity of these interior-point algorithms. We have,

Theorem 9.8 (Smoothed Complexity of IPM: Approximation) *For any* $\sigma \leq 1$, *for any* $\bar{\mathbf{A}} \in \mathbb{R}^{m \times n}$, $\bar{\mathbf{b}} \in \mathbb{R}^m$ *and* $\bar{\mathbf{c}} \in \mathbb{R}^n$ *such that* $\left\| \bar{\mathbf{A}}, \bar{\mathbf{b}}, \bar{\mathbf{c}} \right\|_F \leq 1$, *let* $(\mathbf{A}, \mathbf{b}, \mathbf{c})$ *be a* σ-*Gaussian perturbation of* $(\bar{\mathbf{A}}, \bar{\mathbf{b}}, \bar{\mathbf{c}})$. *Then,*

$$\mathrm{E}_{(\mathbf{A}, \mathbf{b}, \mathbf{c})} \left[T_i((\mathbf{A}, \mathbf{b}, \mathbf{c}), \epsilon) \right] = O \left(\max(m, n)^3 \log \left(\frac{mn}{\sigma \epsilon} \right) \right).$$

When an exact solution of a linear program is desired, one can find an optimal solution in two steps: First, apply the interior-point method to find a feasible point that is close enough to optimal solution. Then run a termination algorithm that "jumps" from the close-to-optimal solution to the optimal solution. For a feasible program defined by $(\mathbf{A}, \mathbf{b}, \mathbf{c})$, there is a precision quantity $\delta(\mathbf{A}, \mathbf{b}, \mathbf{c})$ such that for all $\epsilon \leq \delta(\mathbf{A}, \mathbf{b}, \mathbf{c})$, one could jump from any ϵ-accurate solution to an exact solution. Spielman and Teng (2003b) show that under σ-Gaussian perturbations, the smoothed value of $\max(1, \log(1/\delta(\mathbf{A}, \mathbf{b}, \mathbf{c})))$ is $O(\log nm/\sigma)$. Putting everything together, we obtain the following theorem.

Theorem 9.9 (Smoothed Complexity of IPM: Exact Solution) *For any* $\bar{\mathbf{A}} \in \mathbb{R}^{m \times n}$, $\bar{\mathbf{b}} \in \mathbb{R}^m$ *and* $\bar{\mathbf{c}} \in \mathbb{R}^n$ *such that* $\left\| \bar{\mathbf{A}}, \bar{\mathbf{b}}, \bar{\mathbf{c}} \right\|_F \leq 1$ *and* $\sigma \leq 1$, *let* $(\mathbf{A}, \mathbf{b}, \mathbf{c})$ *be a* σ-*Gaussian perturbation of* $(\bar{\mathbf{A}}, \bar{\mathbf{b}}, \bar{\mathbf{c}})$. *One can apply the interior-point algorithm with periodic applications of a termination procedure to exactly solve a linear program in normal form 1, 2, or 3 in an expected*

$$O \left(\max(m, n)^3 \log \left(\frac{mn}{\sigma} \right) \right)$$

arithmetic operations.

Smoothed Complexity of Linear Programming in Low Dimensions

A linear program is often referred to as a low-dimensional linear program if $m \gg n$. Clarkson (1995) introduces a remarkable reduction algorithm and proves the following lemma.

Lemma 9.1 (Clarkson's Reduction) *For any linear program with* m *constraints in* n *variables, with random sampling, one can reduce the* *problem of solving this program to the problems of solving $O(n^2 \log m)$ programs with $9n^2$ constraints in n variables.*

One can solve a low-dimensional linear program in two steps:

1. Apply Clarkson's reduction to the input program.
2. Use an interior-point algorithm to solve these smaller linear programs.

We can use Theorem 9.9 to prove the following theorem.

Theorem 9.10 (Smoothed complexity of low-dimensional linear programming) *There is a linear programming algorithm with smoothed complexity*

$$O\left(n^2 m + n^8 \log m \log(mn/\sigma)\right).$$

So far, this is the best smoothed bound for low-dimensional linear programming. In contrast, the best worst-case upper bound for low-dimensional linear programming is obtained by the combination of Clarkson's reduction with the randomized simplex algorithm of Kalai (1992) and Matoušek, Sharir, and Welzl (1992). The complexity of this combination is

$$n^2 m + n^{O\left(\sqrt{n/\log n} + \log \log m\right)}.$$

9.3.2 Combinatorial Optimization

In combinatorial optimization, the solutions are discrete. However, not all input parameters of combinatorial optimization problems are necessarily discrete. For example, in optimization problems defined on weighted graphs such as the Traveling Salesman Problem, the Minimum Steiner Tree Problem, and the Multi-Commodity Flow Problem, the weights could be continuous while the graph structure remains discrete (Papadimitriou & Steiglitz (1982)). In integer linear programming (Schrijver (1986)), all input parameters can be continuous, though the solutions must be integer vectors. For these problems, we can still consider the effects of Gaussian perturbations. In this subsection, we discuss results on the smoothed analysis of the binary optimization problem, integer programming, and some problems in online optimization.

Binary Optimization Problems

Beier and Vöcking (2004) consider the following Binary Optimization Problem:

$$\max \quad \mathbf{c}^T \mathbf{x} \quad \text{subject to} \quad \mathbf{A}\mathbf{x} \le \mathbf{b} \text{ and } \mathbf{x} \in \{0, 1\}^n$$

Several classical discrete optimization problems can be expressed as binary optimization problems. One example is the 0/1-Knapsack Problem: Given a set of n items $\{(w_1, v_1), \ldots, (w_n, v_n)\}$ where item i has weight $w_i \geq 0$ and value $v_i \geq 0$, and a knapsack of capacity c, find a subset $S \subseteq \{1, \ldots, n\}$ with $\sum_{i \in S} w_i \leq c$ that maximizes $\sum_{i \in S} v_i$. By setting $x_i = 1$ if $i \in S$ and $x_i = 0$ if $i \notin S$, one can express the knapsack problem as a binary optimization problem:

$$\max \sum_i v_i x_i, \text{ subject to } \sum_i w_i x_i \leq c \text{ and } x_i \in \{0, 1\} \ \forall \ i \in \{1, \ldots, n\}.$$

Another example is the Constrained Shortest Path Problem (Ziegelmann (2001)): Given a graph $G = (V, E)$ where each edge $e \in E$ has distance $d_e > 0$ and latency $l_e \geq 0$, a source vertex s, a destination vertex t, and a latency tolerance parameter L, find a path P from s to t with $\sum_{e \in P} l_e \leq L$ that minimizes $\sum_{e \in P} d_e$.

Let $\mathcal{P}$ be the set of all simple paths from s to t. In any feasible solution, there must be a path in $\mathcal{P}$ with all of its edges chosen. Let $\mathcal{C}$ be all subsets of the edges whose removal disconnects s from t. By the duality relation between cuts and paths, in any feasible solution and for any cut $C \in \mathcal{C}$, at least one of its edges is chosen. For each $e \in E$, let x_e be a binary variable with $x_e = 1$ if e is chosen. One can then reformulate the constrained shortest path problem as:

$$\max \quad -\sum_{e \in E} d_e x_e \quad \text{subject to} \sum_{e \in E} l_e x_e \leq L$$
$$\sum_{e \in C} x_e \geq 1 \quad \text{for all } C \in \mathcal{C} \quad \text{and}$$
$$x_e \in \{0, 1\} \quad \text{for all } e \in E.$$

In their smoothed analysis of the binary optimization problem, Beier and Vöcking distinguish two types of expressions: *deterministic expressions* and *stochastic expressions*. Unlike the stochastic constraints, the deterministic constraints are not subject to perturbations. For instance, in the smoothed analysis of the Constrained Shortest Path Problem, one could assume the distances and latencies are subject to some perturbations while the set of combinatorial structure of the graph is not subject to any perturbation, making the constraints $\sum_{e \in C} x_e \geq 1$ for all $C \in \mathcal{C}$ deterministic.

One way to capture the deterministic constraints in the binary optimization problem is the following reformulation:

$$\min \quad \mathbf{c}^T \mathbf{x} \quad \text{subject to} \quad \mathbf{A}\mathbf{x} \leq \mathbf{b} \text{ and } \mathbf{x} \in S \cap \{0, 1\}^n, \quad (9.6)$$

where S is the intersection of the feasible sets of the deterministic linear constraints. Then, in smoothed analysis, one assumes entries of $(\mathbf{A}, \mathbf{b})$ are always subject to perturbations and only needs to consider the possibility of whether the objective function is also subject to perturbations.

Beier and Vöcking (2004) also introduce a quite general way to extend Gaussian perturbations: Let f be a piece-wise continuous univariate density function of a probability distribution (i.e., $f(x) \geq 0$ and $\int_{\mathbb{R}} f(x)dx = 1$), with finite mean $\int_{\mathbb{R}} |x| f(x)dx$ and $\sup_x f(x) = 1$. For $\sigma \leq 1$, let f_σ be a scaling of f such that $f_\sigma(x) = f(x/\sigma)/\sigma$. It is easy to check that the mean of f_σ satisfies

$$\int_{\mathbb{R}} |x| \cdot f_\sigma(x)dx = \sigma \left(\int_{\mathbb{R}} |x| \cdot f(x)dx \right),$$

and

$$\int_{\mathbb{R}} f_\sigma(x)dx = \int_{\mathbb{R}} f(x/\sigma)/\sigma dx = \int_{\mathbb{R}} f(y)dy = 1.$$

For any $\bar{x}$, an f-perturbation with magnitude $\sigma < 1$ is a random variable $x = \bar{x}+r$, where r is randomly chosen according to a density f_σ. To perturb a vector, one independently perturbs each of its entries. For example, the σ-Gaussian perturbation is an f-perturbation with magnitude σ when $f(x) = e^{-x^2/2}/\sqrt{2\pi}$. By setting f to be $1/2$ in $[-1, 1]$ and 0 outside $[-1, 1]$, one obtains a family of uniform perturbations within a box of side-length 2σ centered at a vector. By setting f to be 1 in $[0, 1]$ and 0 outside $[0, 1]$, one obtains a family of uniform perturbations with a box of side-length σ in the positive quadrant of a vector.

Before stating the main result of Beier and Vöcking, let us first review a few concepts from complexity theory (Papadimitriou (1994) and Sipser (1996)). Let RP denote the class of decision problems solvable by a randomized polynomial time algorithm such that for every "yes"-instance, the algorithm accepts with probability at least $1/2$, and for every "no"-instance, the algorithm always rejects. Let coRP be the complement of RP. Let ZPP be the intersection of RP and coRP. In other words, ZPP is the class of decision problems solvable by a randomized algorithm that always returns the correct answer, and whose expected running time (on every input) is polynomial.

For a binary optimization problem Π, let Π_u be the "unary" representation of Π – in Π_u, all parameters in the stochastic expressions are assumed to be integers in unary representation.

Theorem 9.11 (Beier and Vöcking's Characterization) *For every density function f with finite mean and $\sup_x f(x) \leq 1$, in the perturbation model defined by f-perturbations, a binary optimization problem Π is in* *smoothed polynomial time probabilistically (see (9.4)) if and only if $\Pi_u \in$* ZPP.[1] *Moreover, a binary optimization problem Π is in smoothed polynomial time as in (9.1), if Π_u can be solved in linear time.*

For instance, because the unary version of the 0/1-knapsack problem can be solved in linear time using dynamic programming, as a corollary of Theorem 9.11, the 0/1-knapsack problem is in smoothed polynomial time. Like the 0/1-knapsack problem, the Constrained Shortest Path Problem is NP-complete while its unary version is in P. Thus, the Constrained Shortest Path Problem is in smoothed polynomial time probabilistically. One can similarly prove the Constrained Minimum Spanning Tree Problem (Ravi & Goemans (1996)), the Constrained Minimum Weighted Matching Problem, and several other instances of packing problems are in smoothed polynomial time in the sense of Definition 9.11.

In contrast, even though 0/1-integer programming with a fixed number of constraints is in smoothed polynomial time, general 0/1-integer program is not (unless NP = RP).

In the proof of Theorem 9.11, Beier and Vöcking examine the distribution of three quantities of a binary optimization problem Π:

- **Winner Gap:** If Π is feasible and has more than one feasible solution, then the *winner gap* is the difference between the objective value of the optimal solution and the objective value of the second best feasible solution.
- **Loser Gap:** Let $I^+(\Pi) \subseteq \{0,1\}^n$ be the set of infeasible binary vectors with better objective values than the optimal solution. Suppose the feasible region is given by $\mathbf{a}_i^T \mathbf{x} \leq b_i$, for $i \in [1, m]$. The *loser gap* is then equal to

$$\min_{\mathbf{x} \in I^+(\Pi)} \max_i (\mathbf{a}_i^T \mathbf{x} - b_i),$$

  that is, the minimum amount of violation of binary vectors in $I^+(\Pi)$.
- **Feasibility Gap:** Suppose $\mathbf{x}^*$ is the optimal solution. Then the *feasibility gap* is equal to

$$\min_{i:\ \mathbf{a}_i^T \mathbf{x} \leq b_i \text{ is stochastic}} (b_i - \mathbf{a}_i^T \mathbf{x}^*),$$

  that is, the minimum slack of the optimal solution with respect to stochastic constraints.

---

[1] Usually by saying Π has a *pseudo-polynomial time* algorithm, one means $\Pi_u \in$ P. So $\Pi_u \in$ ZPP means that Π are solvable by a randomized *pseudo-polynomial time* algorithm. We say a problem Π is strongly NP-hard, if Π_u is NP-hard. For example, 0/1-integer programming with fixed number of constraints is in pseudo-polynomial time, while general 0/1-integer programming is strongly NP-hard.

Note that these three quantities are closely related with the concept of the condition numbers studied in the smoothed analysis of the perceptron algorithm and the interior-point algorithms. Beier and Vöcking prove that the reciprocal of each of these quantities is polynomial with high probability. Consequently, if the binary optimization problem has k stochastic equations and n variables, then with high probability the winner is uniquely determined when revealing $O(\log(nk/\sigma))$ bits of each stochastic coefficient. So, if Π_u is in ZPP, then the ZPP algorithm would solve almost all perturbed instances.

Integer Programming

Röglin and Vöcking (2005) extend the result of Beier and Vöcking to integer linear programming. They consider programs of the form

$$\max \quad \mathbf{c}^T \mathbf{x} \quad \text{subject to} \quad \mathbf{Ax} \le \mathbf{b} \text{ and } \mathbf{x} \in \mathcal{D}^n, \tag{9.7}$$

where $\mathbf{A}$ is an $m \times n$ real matrix, $\mathbf{b} \in \mathbb{R}^m$, and $\mathcal{D} \subset \mathbb{Z}$.

Theorem 9.12 (Röglin and Vöcking) *For any constant c, let Π be a class of integer linear programs of form (9.7) with $|D| = O(n^c)$. Then, Π is in smoothed polynomial time in the probabilistic sense if and only if $\Pi_u \in$ ZPP.*

Smoothed Competitive Ratio of Online Scheduling

In online computing (Sleator & Tarjan (1985)) an input is given as a sequence of events $\mathbf{x} = x_1 \circ x_2 \circ \ldots \circ x_i \circ \ldots$. For all i, an online algorithm A must generate a response r_i based on $\{x_1, \ldots, x_i\}$ only. Let $\mathbf{r}_A(\mathbf{x}) = r_1 \circ r_2 \circ \ldots \circ r_i \circ \ldots$ be its response sequence. There is a positive-valued cost function $\mathbf{COST}(\mathbf{r})$ for each response $\mathbf{r}$.

Let $\mathbf{OPT}(\mathbf{x})$ be the cost of the best response, possibly generated by an optimal offline algorithm that has access of all events in $\mathbf{x}$. Sleator and Tarjan (1985) define the worst-case competitive ratio of an online algorithm to be

$$\sup_{\mathbf{x}} \left\{ \frac{\mathbf{COST}(\mathbf{r}_A(\mathbf{x}))}{\mathbf{OPT}(\mathbf{x})} \right\}.$$

Becchetti, Leonardi, Marchetti-Spaccamela, Schäfer, and Vredeveld (2003) apply smoothed analysis to evaluate the performance of online algorithms. For a perturbation model $\mathcal{D}$, they define the smoothed competitive-ratio as

$$\sup_{\bar{\mathbf{x}}} \left\{ E_{\mathbf{x} \in_D \mathcal{D}(\bar{\mathbf{x}})} \left[\frac{\mathbf{COST}\left(\mathbf{r}_A(\mathbf{x})\right)}{\mathbf{OPT}\left(\mathbf{x}\right)} \right] \right\}.$$

They then apply this measure to the following online scheduling problem. The input is a collection of jobs $\{j_1, \ldots, j_n\}$. Each job j_i has a release time R_i and processing time T_i. An online scheduler only knows the existence of j_i at its release time R_i. Its processing time T_i is not known until the job is completed. In the system, one is allowed to interrupt a running job and resume it later on the system. The system could have only one machine or m parallel machines. The scheduler decides when and which uncompleted job should be executed at an available machine, as well as when and which running job to interrupt. Each machine can only process at most one job at any time. Suppose the completion time of job j_i is C_i. The *flow time* of j_i is then $F_i = C_i - R_i$. The objective of the scheduler is to minimize the total flow time

$$\sum_i F_i = \sum_i \left(C_i - R_i\right).$$

Since the Multi-Level Feedback algorithm (MLF) is one of the commonly used processor-scheduling algorithms in operating systems such as Unix and Windows NT, Becchetti et al. (2003) choose to analyze its smoothed competitive ratio.

MLF maintains a set of queues $Q_0, \ldots, Q_{\ldots}$ and uses Q_i to book-keep jobs that have recently processed for 2^{i-1} time units. At each stage, the scheduler processes the job at the front of the lowest non-empty queue. This algorithm is *non-clairvoyant* as it makes decisions without full knowledge of the running time of each job.

Although MLF performs well in practice, its worst-case competitive ratio is rather poor. In fact no deterministic non-clairvoyant preemptive scheduling algorithm has good competitive ratio due to a lower bound of $\Omega(2^K)$ on the competitive ratio of any such scheduling algorithm when processing times of jobs are chosen from $[1, 2^K]$, as shown by Motwani, Phillips, and Torng (1993).

In their smoothed analysis, Becchetti et al. use the partial bit perturbation model introduced by Banderier, Beier, and Mehlhorn (2003) with magnitude parameter $k \leq K$. See Definition 9.4. Becchetti et al.'s results hold for any *well-conditioned distribution* over $[0, 2^{k-1}]$ whose density function f satisfies that f is symmetric around its mean $\mu(f) \in [0, 2^{k-1}]$ and f is non-decreasing in $[1, \mu(f)]$. Let σ denote the standard deviation of f.

Theorem 9.13 (Smoothed Performance of MLF) *For any K, k, and σ, under a partial k-bit perturbation model with a well-conditioned distribution of standard deviation σ, the smoothed competitive ratio of MLF is $O\left(2^{2k}\sigma^{-2}2^{K-k} + 2^{2k}\sigma^{-3}\right)$.*

For example, Theorem 9.13 implies that the smoothed competitive ratio of MLF is $O(2^{K-k})$ for the uniform partial k-bit randomization as the standard deviation of this distribution is $\Theta(2^k)$.

Metrical Task Systems

The *metrical task systems* introduced by Borodin, Linial, and Saks (1992) provide a general framework for modeling many online problems including various scheduling and paging problems. A metrical task system is defined by a weighted, connected and undirected graph $G = (V, E, \mathbf{d})$, where, for each $e \in E$, $\mathbf{d}(e) > 0$ specifies the length of edge e. For simplicity, one can assume $V = \{1, \ldots, n\}$. Naturally, via shortest paths, $\mathbf{d}$ also defines the distance $\mathbf{d}(u, v)$ between any two vertices u and v in the graph. A task can then be represented by an n-dimensional vector $\tau = (\tau(1), \ldots, \tau(n))$, where $\tau(i)$ specifies the service cost of performing τ at vertex i. In an online metrical task system, a server is initially positioned at a starting vertex v_0, and needs to service a sequence $(\tau_1, \ldots, \tau_i, \ldots)$ of tasks. Upon receiving τ_i, an online algorithm must decide which vertex v_i to service τ_i. So the cost to service τ_i depends on v_{i-1} and is equal $\mathbf{d}(v_i, v_{i-1}) + \tau_i(v_i)$. The objective of the optimization problem is to minimize the total service cost $\sum_i \left(\mathbf{d}(v_i, v_{i-1}) + \tau_i(v_i)\right)$.

The deterministic online algorithm with the best possible worst-case competitive ratio is the Work Function Algorithm (WFA) developed by Borodin, Linial, and Saks. The idea of this algorithm is very simple. Let $w_i(v)$ be the minimum offline cost to process $(\tau_1, \ldots, \tau_i)$ with starting position v_0 and ending position v. The vector $\mathbf{w}_i = (\ldots, w_i(v), \ldots)$ is called the work function. One can compute $\mathbf{w}_i$ incrementally by dynamic programming. The optimal off-line cost to process $(\tau_1, \ldots, \tau_i)$ is then $\min_v w_i(v)$. WFA simply chooses s_i to be the vertex that realizes $\min_v(w_i(v) + \mathbf{d}(s_{i-1}, v))$. Borodin et al. proved that the worst-case competitive ratio of WFA is $2n - 1$, and also proved that $2n - 1$ is the best possible competitive ratio for any deterministic online algorithm.

Schäfer and Sivadasan (2004) consider the smoothed competitive ratio assuming that the service cost of each task is subject to a random perturbation with mean 0 and standard deviation σ. If the perturbation makes a cost negative, then the cost would be reassigned to 0.

Theorem 9.14 (Schäfer and Sivadasan) *There exist constants c_1 and c_2 such that for all $(\dots, \bar{\tau}_i, \dots)$ with $\bar{\tau}_i \leq 1$, the smoothed competitive ratio of WFA is bounded above by the smaller of the following two quantities:*

$$c_1 \cdot \left(\frac{\mathbf{DIAMETER}\,(G)}{\lambda_{\min}} \left(\frac{\lambda_{\min}}{\sigma} + \log \Delta \right) \right), \quad and$$

$$\left[c_2 \cdot \left(\sqrt{n \cdot \frac{\lambda_{\max}}{\lambda_{\min}} \left(\frac{\lambda_{\min}}{\sigma} + \log \Delta \right)} \right), \right.$$

where $\lambda_{\min} = \min_{e \in E} \mathbf{d}(e)$, $\lambda_{\max} = \max_{e \in E} \mathbf{d}(e)$, $\mathbf{DIAMETER}\,(G)$ is the diameter, $\max_{u,v} \mathbf{d}(u,v)$, of G, and Δ is the maximum vertex degree of G.

Furthermore, if the service cost vector of each task contains at most k non-zero entries, then the smoothed competitive ratio of WFA is

$$O\left(k \cdot \frac{\lambda_{\max}}{\lambda_{\min}} \left(\frac{\lambda_{\min}}{\sigma} + \log \Delta \right) \right).$$

9.3.3 Scientific Computing

Scientific computing is another area where input parameters are often continuous. In addition, due to round-off errors in scientific computing, inputs to numerical algorithms are subject to random perturbations (Wilkinson (1963)). We now discuss results of smoothed analysis in solving linear systems and in parallel mesh generation.

Growth Factor and Bit-Complexity of Gaussian Elimination

Solving linear systems is the most fundamental problem in computational science and numerical linear algebra (Strang (1980), Golub & Van Loan (1989), Demmel (1997)). The most common method used to find a solution to a system $\mathbf{Ax} = \mathbf{b}$ is the classical Gaussian elimination. The method first uses elimination to reduce an n variables and n equations system to a smaller $n - 1$ by $n - 1$ system and then recursively solves the smaller system. In the elimination, it chooses one of the equations and one of the variables, and uses the chosen equation to eliminate the variable from other equations. The choice of the equation and the variable is determined by a pivoting rule.

The simplest pivoting rule is to use the ith equation to eliminate the ith variable. This process, often referred as *Gaussian elimination without pivoting*, factors the coefficient matrix $\mathbf{A}$ into

$$\mathbf{A} = \mathbf{LU},$$

where $\mathbf{L}$ is a lower triangular matrix and $\mathbf{U}$ is an upper triangular matrix.

The pivoting rule most used in practice is *partial pivoting*. In the ith step, it chooses the equation in which the ith variable has the largest coefficient in absolute value, and uses that equation to eliminate the ith variable. Gaussian elimination with partial pivoting defines a row-permutation matrix $\mathbf{P}$ and factors $\mathbf{PA}$ into

$$\mathbf{PA} = \mathbf{LU}.$$

Because of the partial pivoting, all entries in $\mathbf{L}$ have absolute value at most 1.

Another quite natural pivoting rule is the complete pivoting rule. In the ith step, it chooses the equation containing the the largest coefficient (in absolute value) from the entire system uses it to eliminate the variable that has that coefficient. Gaussian elimination with complete pivoting produces a row permutation matrix $\mathbf{P}$ (for the choices of equations) and a column permutation matrix $\mathbf{Q}$ (for the choices of variables) and factors $\mathbf{PAQ}$ into

$$\mathbf{PAQ} = \mathbf{LU}.$$

In his seminal work, Wilkinson (1961) considers the number of bits needed to obtain a solution of a given accuracy. He proves that it suffices to carry out Gaussian elimination with

$$b + \log_2(5n\kappa(\mathbf{A}) \left\|\mathbf{L}\right\|_\infty \left\|\mathbf{U}\right\|_\infty / \left\|\mathbf{A}\right\|_\infty + 3)$$

bits of accuracy to obtain a solution that is accurate to b bits. In the formula, $\kappa(\mathbf{A}) = \left\|\mathbf{A}\right\|_2 \left\|\mathbf{A}^{-1}\right\|_2$ is the condition number of $\mathbf{A}$ where $\left\|\mathbf{A}\right\|_2 = \max_{\mathbf{x}} \left\|\mathbf{A}\mathbf{x}\right\|_2 / \left\|\mathbf{x}\right\|_2$, and $\left\|\mathbf{A}\right\|_\infty$ is the maximum absolute row sum. The reciprocal of $\left\|\mathbf{A}^{-1}\right\|_2$ is also known as the smallest singular value of $\mathbf{A}$.

The quantity $\left\|\mathbf{L}\right\|_\infty \left\|\mathbf{U}\right\|_\infty / \left\|\mathbf{A}\right\|_\infty$ is called the growth factor of the elimination. It depends on the pivoting rule. We will use $\rho_{GEWP}(\mathbf{A})$, $\rho_{GEPP}(\mathbf{A})$, and $\rho_{GECP}(\mathbf{A})$ to respectively denote the growth factors of Gaussian elimination without pivoting, with partial pivoting, and with complete pivoting.

For some nonsingular matrix $\mathbf{A}$, $\rho_{GEWP}(\mathbf{A})$ could be unbounded as the pivoting coefficient could be zero or arbitrarily close to zero.

Wilkinson constructs a family of matrices, $\mathbf{W}_n = \mathbf{L}_n + \mathbf{C}_n$, where $\mathbf{L}_n$ is an $n \times n$ lower triangular matrix with diagonal entries equal to 1 and off-diagonal entries equal to -1, and $\mathbf{C}_n$ is a matrix with $\mathbf{C}_n[n, n] = 0$, $\mathbf{C}_n[i, j] = 0$ for $j < n$, and $\mathbf{C}_n[i, n] = 1$ for $i < n$. For $\mathbf{W}_n$, $\rho_{GEPP}(\mathbf{W}_n) = \Omega(2^n)$. On the positive side, Wilkinson also proves that for any non-singular matrix $\mathbf{A}$, $\rho_{GECP}(\mathbf{A}) = n^{O(\log n)}$.

Wilkinson's counterexample shows that in the worst-case one must use at least $\Omega(n)$ bits to accurately solve every linear system using the partial pivoting rule. However, in practice one usually obtains accurate answers using much less precision[1]. In fact, it is rare to find an implementation of Gaussian elimination that uses more than double precision, and high-precision solvers are rarely used or needed (Trefethen & Schreiber (1990) and Trefethen & Bau (1997)). For example, LAPACK uses 64 bits (Anderson et al. (1999)).

Sankar, Spielman, and Teng (2005) conduct smoothed analysis of growth factors for Gaussian eliminations without pivoting and with partial pivoting. They prove the following results.

Theorem 9.15 (Gaussian Elimination without Pivoting) *For $n > e^4$, let $\bar{\mathbf{A}}$ be an n-by-n matrix for which $\left\|\bar{\mathbf{A}}\right\|_2 \leq 1$, and let $\mathbf{A}$ be a σ-Gaussian perturbation of $\bar{\mathbf{A}}$, for $\sigma \leq 1/2$. Then,*

$$\mathrm{E}\left[\log \rho_{GEWP}(\mathbf{A})\right] \leq 3 \log_2 n + 2.5 \log_2 \left(\frac{1}{\sigma}\right) + \frac{1}{2} \log_2 \log_2 n + 1.81.$$

Theorem 9.16 (Gaussian Elimination with Partial Pivoting) *For any n-by-n matrix $\bar{\mathbf{A}}$ such that $\left\|\bar{\mathbf{A}}\right\|_2 \leq 1$, let $\mathbf{A}$ be a σ-Gaussian perturbation of $\bar{\mathbf{A}}$. Then, for $x > 1$*

$$\Pr\left[\rho_{GEPP}(\mathbf{A}) > x^{21} \left(\frac{n(1 + \sigma\sqrt{n})}{\sigma}\right)^{12 \log n}\right] \leq x^{-\log n}.$$

Condition Number of Matrices

A key step in the smoothed analysis of the growth factor is to obtain a smoothed bound on the condition number of a square matrix. The condition number $\kappa(\mathbf{A}) = \left\|\mathbf{A}\right\|_2 \left\|\mathbf{A}^{-1}\right\|_2$ measures how much the solution to a system $\mathbf{A}\mathbf{x} = \mathbf{b}$ changes as one makes slight changes to $\mathbf{A}$ and $\mathbf{b}$. A consequence is that if one solves the linear system using fewer than $\log(\kappa(\mathbf{A}))$ bits of precision, one is likely to obtain a result far from a solution (Golub & Van Loan (1989), Trefethen & Bau (1997), Demmel (1997)).

Sankar, Spielman, and Teng (2005) establish the following smoothed bound on the condition number:

Theorem 9.17 (Smoothed Analysis of Condition number) *Let $\bar{\mathbf{A}}$ be an $n \times n$ matrix satisfying $\left\|\bar{\mathbf{A}}\right\|_2 \leq \sqrt{n}$, and let $\mathbf{A}$ be a σ-Gaussian*

---

[1] Wright (1993) gives a collection of natural problems for which Gaussian elimination with partial pivoting is unstable.

perturbation of $\bar{\mathbf{A}}$, *with* $\sigma \leq 1$. *Then,*

$$\Pr\left[\kappa(\mathbf{A}) \geq x\right] \leq \frac{14.1n\left(1 + \sqrt{2\ln(x)/9n}\right)}{x\sigma}.$$

As bounds on the norm of a random matrix are standard, to prove Theorem 9.17, one only needs to focus on the norm of the inverse. Notice that $1/\left\|\mathbf{A}^{-1}\right\|_2 = \min_{\mathbf{x}} \left\|\mathbf{A}\mathbf{x}\right\|_2 / \left\|\mathbf{x}\right\|_2$ is the smallest singular value of $\mathbf{A}$. Sankar, Spielman, and Teng prove the following theorem:

Theorem 9.18 (Smallest singular value) *Let* $\bar{\mathbf{A}}$ *be an arbitrary square matrix in* $\mathbb{R}^{n \times n}$, *and let* $\mathbf{A}$ *be a* σ-*Gaussian perturbation of* $\bar{\mathbf{A}}$. *Then*

$$\Pr_{\mathbf{A}}\left[\left\|\mathbf{A}^{-1}\right\|_2 \geq x\right] \leq 2.35\frac{\sqrt{n}}{x\sigma}$$

Wschebor (2004) improves the smoothed bound on the condition number.

Theorem 9.19 (Wschebor) *Let* $\bar{\mathbf{A}}$ *be an* $n \times n$ *matrix and let* $\mathbf{A}$ *be a* σ-*Gaussian perturbation of* $\bar{\mathbf{A}}$ *for* $\sigma \leq 1$. *Then,*

$$\Pr\left[\kappa(\mathbf{A}) \geq x\right] \leq \frac{n}{x}\left(\frac{1}{4\sqrt{2\pi n}} + 7\left(5 + \frac{4\left\|\bar{\mathbf{A}}\right\|_2^2(1 + \log n)}{\sigma^2 n}\right)^{1/2}\right).$$

When $\left\|\bar{\mathbf{A}}\right\|_2 \leq \sqrt{n}$, his result implies

$$\Pr\left[\kappa(\mathbf{A}) \geq x\right] \leq O\left(\frac{n \log n}{x\sigma}\right).$$

Zero-Preserving Perturbations and Symmetric Linear Systems

Many matrices that occur in practice are both symmetric and sparse. Moreover, numerical algorithms take advantage of any assumed sparseness. Thus, it is natural to study the smoothed complexity of algorithms under perturbations that respect symmetry and non-zero structure. See Definition 9.8 for zero-preserving perturbations. One can express a symmetric matrix $\mathbf{A}$ as $\mathbf{T} + \mathbf{D} + \mathbf{T}^T$, where $\mathbf{T}$ is a lower-triangular matrix with zeros on the diagonal and $\mathbf{D}$ is a diagonal matrix. By making a zero-preserving perturbation to $\bar{\mathbf{T}}$, we preserve the symmetry and the zero-structure of the matrix.

Sankar, Spielman, Teng (2005) extend their results on condition number and growth factor to symmetric matrices with zero-preserving and symmetry-preserving perturbations.

Theorem 9.20 (Condition number and growth factor of symmetric matrices) *Let $\bar{\mathbf{A}} = \bar{\mathbf{T}} + \bar{\mathbf{D}} + \bar{\mathbf{T}}^T$ be an arbitrary n-by-n symmetric matrix satisfying $\left\| \bar{\mathbf{A}} \right\|_2 \leq \sqrt{n}$. Let $\sigma^2 \leq 1$, let $\mathbf{T}$ be a zero-preserving σ-Gaussian perturbation of $\bar{\mathbf{T}}$, let $\mathbf{G}_D$ be a diagonal matrix of Gaussian random variables of standard deviation σ and mean 0 that are independent of $\mathbf{T}$ and let $\mathbf{D} = \bar{\mathbf{D}} + \mathbf{G}_D$. Then, for $\mathbf{A} = \mathbf{T} + \mathbf{D} + \mathbf{T}^T$ and $x \geq 2$,*

$$\Pr\left[\kappa(\mathbf{A}) \geq x\right] \leq O\left(\frac{n^2 \sqrt{\log x}}{x\sigma}\right).$$

In addition,

$$\mathrm{E}\left[\log \rho_{GEWP}(\mathbf{A})\right] = O\left(\log\left(\frac{n}{\sigma}\right)\right).$$

Well-Shaped Mesh Generation: Parallel Delaunay Refinement

Mesh generation is a key problem in scientific computing (Edelsbrunner (2001) and Teng & Wong (2000)). In mesh generation, one is given a geometric domain, specified by its boundary. The goal is to produce a triangulation of the domain, wherein all triangles are well-shaped. One standard definition of well-shapedness is that the ratio of the circum-radius to the shortest side of each triangle is bounded by a prespecified constant, such as 2 (Miller et al. (1995)). One would also like to minimize the number of triangles in the mesh.

A special family of input domains is the periodic point set: If P is a finite set of points in the half-open unit box $[0,1)^d$ and $\mathbb{Z}^d$ is the d-dimensional integer grid, then $S = P + \mathbb{Z}^d$ is a periodic point set (Edelsbrunner (2001)). The periodic set S contains all points $p + v$, where $p \in P$ and v is an integer vector. Periodic point sets are often used to study some basic issues in well-shaped mesh generation and to simplify the presentation of several mesh generation algorithms (Cheng et al. (2000) and Edelsbrunner et al. (2000)).

The Delaunay triangulation of a periodic point set is also periodic. In general, this triangulation might not be well-shaped. A practical and popular technique to generate a well-shaped mesh is to iteratively apply Delaunay refinement[1]: Choose a triangle that is not well-shaped, add its circumcenter to the domain, and recompute the Delaunay triangulation. This process will be repeated until there are no more poorly-shaped elements.

---

[1] Chew (1989) and Ruppert (1993) pioneer this approach in two dimensions and Shewchuk (1998) extends it to three dimensions. Li and Teng (2001), Cheng and Dey (2002) resolve a major difficulty of Shewchuk's extension to ensure that all tetrahedra are well-shaped.

The standard Delaunay refinement algorithms are inherently sequential. Spielman, Teng, and Üngör (2002) define a parallel rule[1] for adding more points for refinement. They prove that if the minimum pair-wise distance of the input point set is s, then the use of the parallel rule takes $O(\log^2(1/s))$ parallel time. In the smoothed setting where input points are subject to small perturbations, their result implies:

Theorem 9.21 (Parallel Delaunay Refinements) *For any point set $\bar{P} \subseteq \mathbb{R}^2$, let P be a σ-Gaussian perturbation of $\bar{P}$. Then there is a parallel Delaunay refinement algorithm that, in $O(\log^2(1/\sigma))$ time, produces a well-shaped mesh for $P + \mathbb{Z}^2$. Moreover, the number of elements in this mesh is within a constant factor of the optimal solution. The number of processors needed is equal to the size of the resulting mesh.*

9.3.4 Discrete and Geometric Structures

The notion of perturbations and neighborhoods is far less straightforward for discrete structures than for continuous structures. Perhaps, the simplest model for perturbing a graph $\bar{G} = (V, \bar{E})$ is to insert every non-edge $(i, j) \notin \bar{E}$ into the graph with some probability σ and to delete every edge $(i, j) \in E$ from the graph with some probability σ. We denote this distribution on graphs by $\mathcal{P}(\bar{G}, \sigma)$ and call the resulting graph $G = (V, E)$ a σ-perturbation of $\bar{G}$.

Unfortunately, such perturbations can considerably change the properties of a graph. For example, the above perturbation model can not be used if we would like to study the performance of a bisection algorithm for planar graphs, because almost all perturbed graphs are non-planar.

We now discuss some modifications of this model, which have been proposed in an attempt to make the analysis more meaningful.

- **Property-Preserving Perturbations** (Spielman & Teng (2003a)): Given a property P and a basic model of perturbation, a P-preserving perturbation of an object $\bar{X}$ is a perturbation of $\bar{X}$ according to the basic perturbation model but subject to the condition $P(X) = P(\bar{X})$. In the case when $\bar{G}$ is a graph and the basic model is the σ-perturbation of $\bar{G}$, the probability of a graph G with $P(G) = P(\bar{G})$ is

$$\frac{\Pr_{G \leftarrow \mathcal{P}(\bar{G}, \sigma)}\left[G \text{ and } (P(G) = P(\bar{G}))\right]}{\Pr_{G \leftarrow (\bar{G}, \sigma)}\left[P(G) = P(\bar{G})\right]}$$

---

[1] By relaxing the refinement rule that the circumcenters of poorly shaped triangles must be added, then in Spielman, Teng, and Üngör (2004), they show another approach with $(\log(1/s))$ parallel time.

In the property preserving perturbation model for graphs, P can be any graph property such as planarity or can be a combination of several properties such as being planar and having a bisection of size at most B.

- **The Semi-Random Model** (Blum & Spencer (1995)): Blum and Spencer's semi-random graph model combines Santha and Vazirani's (1986) semi-random source with the random graph model that has a "planted solution" (Boppana (1987)). Since this is best described by an example, let us consider the k-Coloring Problem.

An adversary plants a solution by partitioning the set V of n vertices into k subsets $V_1, \ldots, V_k$. Let

$$F = \{(u, v) | u \text{ and } v \text{ are in different subsets}\}$$

be the set of potential inter-subset edges. A graph is then constructed by the following semi-random process that perturbs the decisions of the adversary: Initially let $H = F$. Then while H is not empty,

1. the adversary chooses an edge e from H, and decides whether it would like to include the edge in the graph.
2. The semi-random process then reverses the decision with probability σ.
3. The edge e is then removed from H.

We will refer such a graph as a *semi-random k-colorable graph*. In this model, one can also require that each V_i has size n/k or $\Theta(n/k)$. Such a graph is called a *balanced semi-random k-colorable graph*.

All semi-random k-colorable graphs have the same planted coloring: $\Pi(v) = i$ for all $v \in V_i$, because both the adversary and the semi-random process preserve this solution by only considering edges from F.

As with the smoothed model, one can work with the semi-random model, by varying σ from 0 to $1/2$, to interpolate between worst-case and average-case complexity for k-coloring. In fact, the semi-random model is related to the following perturbation model that *partially preserves* a particular solution:

Let $\bar{G} = (V, \bar{E})$ be a k-colorable graph. Let $\Pi : V \to \{1, \ldots, k\}$ be a k-coloring of $\bar{G}$ and let $V_i = \{v \mid \Pi(v) = i\}$. The model then returns a graph $G = (V, E)$ that is a σ-perturbation of $\bar{G}$ subject to Π also being a valid k-coloring of G. In other words, the perturbation is subject to

$$E \subseteq F = \{(u, v) | u \text{ and } v \text{ are in different subsets}\}.$$

This perturbation model is equivalent to the semi-random model in which the adversary is oblivious. An oblivious adversary simply chooses a set $\bar{E} \subseteq F$, and sends the decisions that only include edges in $\bar{E}$ (and hence exclude edges in $F - \bar{E}$) through the semi-random process. Thus, if a k-coloring algorithm can successfully color semi-random k-colorable graphs (with high probability), it must also be able to color graphs generated by the perturbation model that partially preserves a particular solution.

- **Solution-Preserving Perturbations**: The semi-random model only generates graphs that have the planted solution and hence only assigns non-zero probabilities to graphs that have the planted solution. In this model, the decision problem, such as whether a graph is k-colorable, is trivial. The search problem, in which one is asked to find the planted solution, is the subject of the study.

  An alternative model is to apply perturbations that preserve a planted solution. Continuing to use the graph coloring problem as our example, let $\Pi : V \to \{1, \ldots k\}$ be a k-coloring assignment. For a graph $\bar{G} = (V, \bar{E})$, if it is k-colored by Π, then G is a σ-perturbation of $\bar{G}$ subject to G having Π as a solution; otherwise G is a σ-perturbation of $\bar{G}$ subject to G not having Π as a solution.

- **Monotone Adversary Semi-Random Model** (Blum & Spencer (1995), Feige & Kilian (1998) and Feige & Krauthgamer (2002)): Blum and Spencer define another generation process of semi-random graphs of k-colorable graphs: First, partition the set of n vertices into k subsets $V_1, \ldots, V_k$ each having n/k vertices. Second, choose a set E_1 of edges by selecting each edge in

$$F = \{(u, v) | u \text{ and } v \text{ are in different subsets}\}$$

independently with probability σ. Then, the adversary chooses another set E_2 from F and returns $G = (V, E_1 \cup E_2)$.

The monotone adversary semi-random model can be applied to a graph problem P with a particular "planted" solution S. It is a two step model. In the first step, it generates a "random" graph $\tilde{G} = (V, \tilde{E})$ with S as a solution according to some distribution. In the second step, the adversary can only modify $\tilde{G}$ in a limited way – the modification has to respect the planted solution S.

For example, the following is the semi-random model for graph bisection analyzed by Feige and Kilian (1998): Let $0 \le q < p \le 1$, let V_1 be a subset of V of size $n/2$ and let $V_2 = V - V_1$. The random process builds a graph $\tilde{G} = (V, \tilde{E})$ by selecting each edge in $V_1 \times V_2$ with independent probability q and each edge in $(V_1 \times V_1) \cup (V_2 \times V_2)$ with probability p. The adversary is then given the chance to add a subset of edges from $(V_1 \times V_1) \cup (V_2 \times V_2)$ to the graph and remove a subset of edges of $V_1 \times V_2$ from the graph.

Results in the Semi-Random Model

We now summarize some results in the monotone adversary semi-random model.

Theorem 9.22 (Semi-Random Coloring: Blum-Spencer) *For any $\epsilon > 0$ and $p \geq n^{\frac{-2k}{(k+1)k-2}+\epsilon}$, there is a polynomial-time algorithm to k-color a balanced semi-random k-colorable graph with probability $1 - o(1)$. When $k = 3$, the condition on p is $p \geq n^{-1/3+\epsilon}$.*

Feige and Kilian (1998) improve the above result and show:

Theorem 9.23 (Semi-random Coloring: Feige and Kilian) *For any constant k, $\epsilon > 0$ and $p \geq ((1 + \epsilon)k \ln n)/n$, there is a polynomial-time algorithm to k-color a balanced semi-random k-colorable graph with high probability. But if $p < (1-\epsilon) \ln /n$, every random polynomial time algorithm will fail with high probability to k-color a balanced semi-random k-colorable graph, unless $NP \subseteq BPP$.*

Feige and Kilian (1998) also extend their semi-random analysis to the maximum independent set and Graph Bisection Problem. For the independent set problem, they develop a new two-phase algorithm. The algorithm first uses semidefinite programming as a tool to compute a constant number of nearly independent sets. It then uses a matching based optimization technique to purify the output of Phase I to extract the embedded maximum independent set.

Theorem 9.24 (Semi-Random Maximum Independent Set) *For any $\alpha > 0$, $\epsilon > 0$, and $p = (1 + \epsilon) \ln n/\alpha n$, there is a randomized polynomial time algorithm that finds the embedded independent set of size αn in a semirandom graph with high probability.*

For graph bisection, Feige and Kilian consider the monotone adversary semirandom model and analyze the performance of a semidefinite-programming-based bisection algorithm.

Theorem 9.25 (Semi-Random Bisection) *There exists a constant c such that for any $1 \geq p > q$ satisfying $p - q \geq c\sqrt{p \log n/n}$, with high probability, Feige-Kilian's algorithm finds the embedded bisection in a semirandom graph in polynomial time.*

Feige and Krauthgamer (2002) examine the Bandwidth Problem:. One is given a undirected graph $G = (V, E)$ of n vertices and is asked to find a linear ordering π from V onto $\{1, \ldots, n\}$ to minimize the *bandwidth*: $\max \{|\pi(u) - \pi(v)| : (u, v) \in E\}$.

In the semi-random model, using parameters B and p, a graph is generated by choosing a linear ordering π of V and selecting each pair (u, v) satisfying $|\pi(u) - \pi(v)| \leq B$ with probability p. Then the monotone adversary may add a set of pairs (w, z) satisfying $|\pi(w) - \pi(z)| \leq B$.

Theorem 9.26 (Bandwidth Problem) *There exists a constant c such that for any ϵ, B, and p satisfying $\ln^2 n \leq Bn/\ln n$ and $p \geq c\ln n/B$, with high probability, Feige and Krauthgamer's algorithm, in polynomial time, returns a linear ordering of vertices with bandwidth $(1+\epsilon)B$ for semirandom graphs.*

Results in the Property-Preserving Model

Spielman and Teng (2003a) prove that under property-preserving perturbations, several property-testing algorithms become sublinear-time decision algorithms with low smoothed error probability.

In a decision problem of a property P over an instance domain D_n, one is given an instance $x \in D_n$ and is asked to decide whether $P(x)$ is true or false. In graph theory, some graph properties such as Bipartite, the property of being bipartite, have a polynomial-time decision procedure. Other properties such as ρ-Clique, the property of having a clique of size ρn and ρ-Bisection, the property of having a bisection of at most ρn^2 edges, have no polynomial-time decision procedure unless $\mathsf{NP} = \mathsf{RP}$.

Rubinfeld and Sudan (1996) introduce *property testing*, a relaxation of the standard decision problem. The objective of property testing is to efficiently distinguish those instances that have a property from those that are "far" from having the property. An algorithm A is said to be a property tester for the property P with parameter $\epsilon \geq 0$ if

1. for all x with property P, then $\Pr[A(x, \epsilon) = 1] \geq 2/3$; and
2. for all x of distance at least ϵ from every instance that has property P, $\Pr[A(x, \epsilon) = 1] \leq 1/3$,

under some appropriate measure of distance on inputs. It follows from this definition that a property testing algorithm A satisfies

$$\Pr[A(X, \epsilon) \neq P(X)] < 1/3$$

for all instances that have property P and for all instances that are at least ϵ distance from those with property P. For graphs, one possible measure of the distance between two graphs $G_1 = (V, E_1)$ and $G_2 = (V, E_2)$ on n

vertices is the fraction of edges on which G_1 and G_2 differ: $|(E_1 \cup E_2) \setminus (E_1 \cap E_2)|/ \binom{n}{2}$.

A typical property-testing algorithm will use a randomized process to choose a small number of facets of x to examine, and then make its decision. For example, a property tester for a graph property may query whether or not certain edges exist in the graph. The quality of a property-testing algorithm is measured by its query complexity (the number of queries to the input) and its time complexity.

Under this relaxation, many properties can be tested by sub-linear algorithms that examine random portions of their input. For example, Goldreich, Goldwasser, and Ron (1998) prove the following theorem.

Theorem 9.27 (Goldreich-Goldwasser-Ron) *The properties ρ-Clique and ρ-Bisection have property-testing algorithms with query complexity polynomial in $1/\epsilon$ and time complexity $2^{\tilde{O}(1/\epsilon^3)}$, and the property Bipartite has a property testing algorithm with query and time complexities polynomial in $1/\epsilon$.*

Let P be a graph property. Let $\mathcal{P}_P(X, \sigma)$ be the distribution of P-preserving σ-perturbations of X. Spielman and Teng (2003a) use the following lemma to relate the smoothed error probability of using a testing algorithm for P as a decision procedure with the probability that a P-preserving perturbed instance is far from one having property P.

Lemma 9.2 (Testing for Decision: Smoothed Error Probability) *Let P be a property and $\mathcal{P}(\bar{X}, \sigma)$ be a family of distributions satisfying for all $\bar{X}$ without property P,*

$$\Pr_{X \leftarrow \mathcal{P}(\bar{X}, \sigma)} \left[X \text{ is } \epsilon\text{-close to } P | P(X) = P(\bar{X}) \right] \leq \lambda(\epsilon, \sigma, n).$$

Then for every P-testing algorithm A and every input $\bar{X}$,

$$\Pr_{X \leftarrow \mathcal{P}(\bar{X}, \sigma)} \left[A(X) \neq P(X) | P(X) = P(\bar{X}) \right] < 1/3 + \lambda(\epsilon, \sigma, n).$$

Spielman and Teng (2003a) show that for any graph $\bar{G}$ and for $P \in$ {Bipartite, ρ-Clique, ρ-Bisection}, then $\bar{G}$ not satisfying P implies that (for any ϵ for the first property, and $\epsilon < \sigma(1/4 - 2\rho)$ for the last two properties),

$$\Pr_{G \leftarrow \mathcal{P}_P(\bar{G}, \sigma)}[G \text{ is } \epsilon \text{ close to a graph with property } P] < 2^{-\Omega(n^2)}.$$

Clearly, if $\bar{G}$ satisfies P then G also satisfies P. Therefore,

Theorem 9.28 (Testing for Decision: Smoothed Perspective) *There exists an algorithm A that takes as input a graph G, examines $poly(1/\sigma)$ edges of G and runs in time $\tilde{O}(1/\epsilon^3)$ when P is Bipartite, and in $2^{\tilde{O}(1/\epsilon^2)}$ time when P is ρ-Clique or ρ-Bisection such that for every $\bar{G}$, if G is a P-property preserving σ-perturbation of $\bar{G}$, then*

$$\Pr\left[A(G) \neq P(G)\right] < 1/3 + o(1).$$

Applying the techniques developed by Feige and Kilian (1998) Spielman and Teng (2003a) demonstrate a testing algorithm with faster smoothed complexity for ρ-Clique.

Theorem 9.29 (Fast Clique Tester) *Let ρ and $\sigma < 1/2$ be constants. For any graph $\bar{G}$, let G be a ρ-Clique preserving σ-perturbation of $\bar{G}$. Then, there exists an algorithm A that examines the induced subgraph of G on a randomly chosen set of $\frac{8}{\rho\sigma} \log\left(\frac{4}{\rho\sigma}\right)$ vertices and runs in time polynomial in $\frac{1}{\rho\sigma}$ to achieve*

$$\Pr\left[A(G) \neq \rho\text{-}Clique(G)\right] < 1/4 + o(1).$$

Comparison of Perturbation Models

Suppose we have two models M_1 and M_2 for measuring a performance quality Q of an algorithm A. Then, M_1 is said to be *more adversarial*[1] than M_2 if $M_1(Q, A, \sigma) \geq M_2(Q, A, \sigma)$. For example, the worst-case measure is more adversarial than the smoothed measure and than an average measure. If M_1 is more adversarial than M_2, then any upper bound for model M_1 is also an upper bound for M_2. Conversely, any lower bound for M_2 is an lower bound for M_1.

As noted in Blum & Spencer (1995) and Feige & Kilian (1998), the monotone adversary semi-random model is at least as adversarial as the semi-random model. In turn, the semi-random model is more adversarial than the semi-random model with an oblivious adversary.

However, results for the semi-random model may not always extend to the property-preserving perturbation model, even though these two are seemingly related. This is partially due to the fact that the semi-random model only produces "positive" instances which share a common, planted solution. For example, for the k-coloring problem, the semi-random model

---

[1] One might extend our standard complexity notions such as O, o, Ω, and Θ for models of measures. For example, we could say $M_1 = \Theta(M_2)$ if there exist constants n_0, σ_0, and $C_1 \leq C_2$ such that for all $n \geq n_0$ and $\sigma \leq \sigma_0$, $C_1 M_2(Q, A, \sigma) \leq M_1(Q, A, \sigma) \leq C_2 M_2(Q, A, \sigma)$ when input size is n.

only generates graphs that are k-colorable and all graphs from this distribution share a common, planted k coloring. Thus, this model does not assign probabilities to graphs without this planted solution. As some of these "unassigned" graphs are still k-colorable, results in the semi-random model may not provide any performance information on them. If we want to extend a result from the semi-random model to the smoothed model with property-preserving perturbations, we need to understand the contribution of these instances. For some graph problems, such as the bisection problem, it remains open whether results in the semi-random model still hold in the property-preserving perturbation model. In fact, it is not even clear whether the results would extend to the more closely related solution-preserving perturbation model.

Finding a proper perturbation model for modeling practical discrete algorithms can be a challenging task. For certain graph problems such as coloring and bisection, the semi-random model defines a family of distributions using a random process that can be efficiently implemented. In contrast, the conditional density of an instance in the property-preserving distribution might be hard to compute in polynomial time. As argued in Spielman & Teng (2003a), the "practical" distributions may not have efficient descriptions. So the requirement that a perturbation model be efficiently implementable might not be reasonable or necessary. Of course, if one would like to conduct some experimental studies of these models, it would be helpful if one had an efficient procedure for the perturbation.

For certain graph properties such as planarity, it is relatively hard to define a semi-random model to study them. For more complex studies, such as those on the performance of a planar bisection algorithm, one might need to be more creative to define a suitable semi-random model. Being able to model these studies might be the strength of the property-preserving perturbation framework.

Number of Left-to-Right Maxima and Height of Binary Search Trees

For a sequence $\mathbf{a} = (a_1, \ldots, a_n)$, an element a_i is a *left-to-right maximum* of $\mathbf{a}$ if $a_i > a_j$, for all $j < i$. When $\mathbf{a} = (1, 2, \ldots, n)$, the number of left-to-right maxima is n. Manthey and Reischuk (2005) prove the following bounds that improve an early result of Banderier, Beier, and Mehlhorn (2003).

Theorem 9.30 (Manthey-Reischuk) *For any sequence $\bar{\mathbf{a}}$, let $\mathbf{a}$ be an σ-partial permutation of $\bar{\mathbf{a}}$. Let $\phi(\mathbf{a})$ be the number of left-to-right maxima* *of* **a**. *Then for all* $0 < \sigma < 1$,

$$0.4(1 - \sigma)\sqrt{n/\sigma} \leq \mathrm{E}\left[\phi(\mathbf{a})\right] \leq 3.6(1 - \sigma)\sqrt{n/\sigma}.$$

The most commonly used data structure for storing a set whose elements have a total ordering is the binary search tree. Perhaps the simplest way to form a binary search tree for a sequence $\mathbf{a} = (a_1, \ldots, a_n)$ is to insert elements of **a** one by one into an initially empty binary search tree. For $i = 2$ to n, we insert a_i into the binary search tree T_{i-1} of $\{a_1, \ldots, a_{i-1}\}$ to produce tree T_i. An important parameter of this data structure is the height of the tree. Let $T(\mathbf{a})$ denote the binary search tree so constructed. If **a** is sorted, either in increasing or decreasing order, then the height of $T(\mathbf{a})$ is n. Manthey and Reischuk (2005) prove the following result.

Theorem 9.31 (Manthey-Reischuk) *For any sequence* $\bar{\mathbf{a}}$, *let* **a** *be an* σ-*partial permutation of* $\bar{\mathbf{a}}$. *Let* $h(\mathbf{a})$ *be the height of the binary search tree* $T(\mathbf{a})$. *Then for all* $0 < \sigma < 1$,

$$0.8(1 - \sigma)\sqrt{n/\sigma} \leq \mathrm{E}\left[h(\mathbf{a})\right] \leq 6.7(1 - \sigma)\sqrt{n/\sigma}.$$

Number of Extreme Points in d Dimensions

Recall that the vertices of the convex hull of a set P of points in $\mathbb{R}^d$ are points in P that cannot be expressed as a convex combination of other points in P. These points are also called the *extreme points* of P. Let $v(P)$ be the number of extreme points of P. A well known result (Bentley et al. (1978)) in geometry states: If P is a set of n points chosen uniformly at random from the unit d-cube, then the expected value of $v(P)$ is $O(\log^{d-1} n)$.

Damerow and Sohler (2004) consider the following version of the smoothed value of $v(P)$ and prove the following theorem.

Theorem 9.32 (Damerow-Sohler) *For any set* $\bar{P} = \{\bar{\mathbf{p}}_1, \ldots, \bar{\mathbf{p}}_n\}$ *of* n *points from the unit* d-*cube, let* $\mathbf{r}_1, \ldots, \mathbf{r}_n$ *be* n *vectors chosen uniformly at random from the cube* $[-\epsilon, \epsilon]^d$. *Let* $P = \{\bar{\mathbf{p}}_1 + \mathbf{r}_1, \ldots, \bar{\mathbf{p}}_n + \mathbf{r}_n\}$ *be an perturbation of* $\bar{P}$. *Then*

$$\mathrm{E}\left[v(P)\right] = O\left(\left(\frac{n \log n}{\epsilon}\right)^{1 - 1/(d+1)}\right).$$

Motion Complexity

Kinetic data structures have become subjects of active research in computational geometry since their introduction by Basch, Guibas, and Hershberger (1997). The aim of these dynamic data structures is to efficiently maintain combinatorial and geometric information of continuously moving objects. The motion of objects is typically specified by some algebraic function. The complexity of the kinetic data structures depends on the initial positions of objects as well as the functions that govern their motion.

As a first step to study the smoothed complexity of kinetic data structures, Damerow, Meyer auf der Heide, Räcke, Scheideler, and Sohler (2003) consider the following problem: Given a set $P = \{\mathbf{p}_1, \ldots, \mathbf{p}_n\}$ of n points in $\mathbb{R}^d$ and a set $V = \{\mathbf{v}_1, \ldots, \mathbf{v}_n\}$ of initial velocity vectors, the position of the ith object at time t is $\mathbf{p}_i(t) = \mathbf{p}_i + t\mathbf{v}_i$. The *motion complexity* for maintaining the orthogonal bounding box is defined to be the number of combinatorial changes of the $2d$ sides of the bounding box.

In the worst case, this motion complexity is $\Omega(dn)$. When $\mathbf{v}_i$ and $\mathbf{p}_i$ are chosen uniformly at random from the unit cube $[-1, 1]^d$, the average complexity is $O(d \log n)$. Damerow et al. prove the following result.

Theorem 9.33 (Smoothed Motion Complexity) *For any positive integer d, let $\bar{P} = \{\bar{\mathbf{p}}_1, \ldots, \bar{\mathbf{p}}_n\} \subset [-1, 1]^d$ and $\bar{V} = \{\bar{\mathbf{v}}_1, \ldots, \bar{\mathbf{v}}_n\} \subset [-1, 1]^d$. Let $P = \{\mathbf{p}_1, \ldots, \mathbf{p}_n\}$ and $V = \{\mathbf{v}_1, \ldots, \mathbf{v}_n\}$ be σ-Gaussian perturbations of $\bar{P}$ and $\bar{V}$, respectively. Then the expected motion complexity for maintaining the orthogonal bounding box of (P, V) is $O(d(1 + 1/\sigma) \log^{1.5} n)$ and $\Omega(d\sqrt{\log n})$.*

Properties of Perturbed Graphs and Formula

Each perturbation model in the smoothed analysis of graph algorithms can be used to define a distribution of random graphs. For example, let $H = (V, E)$ be an undirected graph over vertices $V = \{1, \ldots, n\}$. For any $\sigma < 1$, the σ-perturbations of H can be viewed as a distribution of random graphs. Let us refer it as $G_{H,\sigma}$. Similarly, for any graph property P, the P-preserving σ-perturbations of a graph H is also a distribution of random graphs, denoted by $G_{P,H,\sigma}$.

For any m and $H = (V, E)$, Bohman, Frieze and Martin (2003) define a distribution $G_{H,m}$ of random graphs $(V, E \cup R)$ where R is a set of m edges chosen uniformly at random from the complement of E, i.e., chosen from $\bar{E} = \{(i, j) \notin E\}$.

Krivelevich, Sudakov, and Tetali (2005) prove a sharp threshold for the appearance of a fixed subgraph. For a fixed graph Γ with n_Γ vertices and e_Γ

edges, let $m(\Gamma) = \max\{e_{\Gamma'}/n_{\Gamma'} | \Gamma' \subseteq \Gamma, n_{\Gamma'} > 0\}$. For any positive integer r, let

$$m_r(\Gamma) = \min_{r\text{-way partition } (V_1, \ldots, V_r) \text{ of } V(\Gamma)} \max_{|V_i| > 0} m(\Gamma(V_i))$$

Theorem 9.34 (Emerging of a Subgraph) *Let r and α be constants such that $r \geq 2$ and $\alpha \in \left(\frac{r-2}{r-1}, \frac{r-1}{r}\right]$. Let Γ be a graph of no more than n vertices. Then, for every graph H over $\{1, \ldots, n\}$ of average degree αn, where $m = \omega(n^{2-1/m_r(\Gamma)})$, $G_{H,m}$ almost surely contains a copy of Γ. Moreover, there exists a graph H' of average degree αn such that if $m = o(n^{2-1/m_r(\Gamma)})$, then $G_{H',m}$ almost surely does not contain a copy of Γ. Here $f = \omega(g)$ is equivalent to $g = o(f)$.*

Krievlevich, Sudakov, and Tetalli also consider the problem of adding m randomly chosen disjunctions of k literals to a satisfiable instance, F, of a k-SAT formula on n variables. They obtain the following result.

Theorem 9.35 (Transition From Happy to Unhappy) *For some $c > 0$ and $\epsilon < 1/k$, let F be a satisfiable k-SAT formula on n variables with at least $cn^{k-\epsilon}$ clauses. Then almost surely the conjunction of F with a randomly chosen k-SAT formula of $m = \omega(n^{k\epsilon})$ clauses is not satisfiable. Moreover, there exists a satisfiable k-SAT formula F' of $\Omega(n^{k-\epsilon})$ clauses such that the conjunction of F with a randomly chosen k-SAT formula of $m = o(n^{k\epsilon})$ clauses is satisfiable with high probability.*

Several other discrete properties have been studied in the smoothed setting: for example, Flaxman and Frieze (2004) consider the diameter of perturbed digraphs and Sudakov and Vondrak (2005) examine the 2-colorability of dense hypergraphs.

Very recently, Arthur and Vassilvitskii (2005) analyze the smoothed iteration complexity of the k-means method (Lloyd (1982)) for clustering: Given a set P of n points in $\mathbb{R}^d$, the k-means method first chooses an arbitrary set of k centers and uses the Voronoi diagram of these centers to partition P into k clusters. Then the center-of-mass of each of these clusters becomes a new center, and the k-means method re-clusters the points and repeats this process until it stabilizes. They show that in the worst-case, the k-means method requires $2^{\Omega(\sqrt{n})}$ iterations to converge. In contrast, they prove that the k-means method has probably polynomial smoothed complexity.

9.4 Open Problems

In this section, we discuss some emerging open questions in smoothed analysis. For some problems and algorithms, we will present concrete conjectures, whereas for others, we will discuss possible directions for research.

P1: The Simplex Method and its Pivoting Rules

During a Phase II iteration of a simplex algorithm, a vertex $\mathbf{v}_{i-1}$ may have several neighboring vertices with better objective values. The algorithm needs to decide which of them should be chosen as the next vertex $\mathbf{v}_i$. Simplex algorithms differ by their pivoting rules which guide this decision when there are multiple choices. Several pivoting rules have been proposed in the literature (Dantzig (1991) and Chvatal (1983)). They include

- *Greedy*: Choose the vertex that maximizes the improvement of the objective value, $\mathbf{c}^T(\mathbf{v}_i - \mathbf{v}_{i-1})$.
- *Steepest-Edge*: Choose the vertex $\mathbf{v}_i$ whose edge $(\mathbf{v}_{i-1}, \mathbf{v}_i)$ forms the smallest angle with $\mathbf{c}$, that is, $\mathbf{v}_i$ maximizes

$$\frac{\mathbf{c}^T(\mathbf{v}_i - \mathbf{v}_{i-1})}{||\mathbf{v}_i - \mathbf{v}_{i-1}||_2}.$$

- *Random*: Choose randomly from the neighboring vertices with better objective values according to some distribution.

There are other rules such as the shadow-vertex rule (Gass & Saaty (1955)), Bland's rule (1977) and the self-dual simplex rule (Dantzig (1991) and Lemke (1965)).

The steepest-edge rule (Forrest & Goldfarb (1992)) is among the most commonly used in practice. But most existing pivoting rules have been shown to have exponential worst-case complexity (Klee & Minty (1972), Goldfarb & Sit (1979), Goldfarb (1983), Jeroslow (1973), Murty (1980), Amenta & Ziegler (1999)).

A natural open question is whether our results such as Theorem 9.1 on the smoothed complexity of the shadow-vertex rule can be extended to other pivoting rules.

Conjecture 9.36 (Smoothed Simplex Conjecture) *The smoothed complexity of the simplex algorithms with greedy, steepest-descent, and random pivoting rules are polynomial under Gaussian perturbations.*

The key step for proving Conjecture 9.36 is the Phase II complexity, since Phase I computation can often be reduced to a Phase II computation in which we already know that the constraints are feasible and have an initial vertex of the feasible region. Like the shadow-vertex simplex algorithm, these simplex algorithms iteratively produce a path of vertices of the feasible region that monotonically improves the objective value. We refer to such a path as a **c**-*monotonic path* of the feasible polyhedron, recalling that **c** is the objective direction.

There is a geometric reason, as pointed out in Spielman & Teng (2004), that the shadow-vertex simplex algorithm is the first simplex algorithm considered in both the average-case (Borgwardt (1980)) and smoothed analyses: The length of the **c**-monotonic path constructed by the shadow-vertex algorithm can be bounded above by the size of the shadow. Moreover, the shadow depends only on the initial vertex and the objective direction. So in the probabilistic analysis, we do not need to explicitly consider the iterative steps taken by the simplex algorithm.

The key to analyze the smoothed complexity of simplex algorithms with, say, the steepest-edge pivoting rule is to search for a simpler geometric parameter that upper bounds the length of its **c**-monotonic path. We could start with linear programs of form

$$\max \quad \mathbf{c}^T \mathbf{x} \qquad \text{subject to } \mathbf{A}\mathbf{x} \leq \mathbf{1}.$$

One approach is to relate the length of the steepest-edge **c**-monotonic path with the size of the shadow analyzed in Theorem 9.1.

Conjecture 9.37 (Shadow and Steepest-Edge Path: I) *For any $m \times n$ matrix $\bar{\mathbf{A}}$ and an n-vector $\bar{\mathbf{c}}$ such that $\left\| \bar{\mathbf{A}}, \bar{\mathbf{c}} \right\|_F \leq 1$, let $\mathbf{A}$ and $\mathbf{c}$ be σ-Gaussian perturbations of $\bar{\mathbf{A}}$ and $\bar{\mathbf{c}}$, respectively. For an m-vector $\mathbf{z}$, let $\mathbf{v}$ be the vertex of the polyhedron $\{\mathbf{x} : \mathbf{A}\mathbf{x} \leq \mathbf{1}\}$ that maximizes $\mathbf{z}^T \mathbf{v}$ and let P be the steepest-edge $\mathbf{c}$-monotonic path of the polyhedron $\{\mathbf{x} : \mathbf{A}\mathbf{x} \leq \mathbf{1}\}$ starting from $\mathbf{v}$. Then, there exists a constant $\alpha \geq 1$ and an $x_0(m, n, 1/\sigma)$ polynomial in m, n, and $1/\sigma$, such that for all $x > x_0(m, n, 1/\sigma)$*

$$\Pr\left[|P| > x\right] \leq \alpha \cdot \Pr\left[|\mathbf{Shadow}_{\mathbf{z},\mathbf{c}}(\mathbf{A})| > x\right].$$

Conjecture 9.38 (Shadow and Steepest-Edge Path: II) *Let $\mathbf{A}$, $\mathbf{c}$, $\mathbf{z}$, P be the same as defined in Conjecture 9.37. Then, there exists an $h(m, n, 1/\sigma)$ polynomial in m, n, and $1/\sigma$ and constants k_1 and k_2*

$$\Pr\left[|P| > h(m, n, 1/\sigma) \cdot |\mathbf{Shadow}_{\mathbf{z},\mathbf{c}}(\mathbf{A})|\right] = \frac{1}{m^{k_1} n^{k_2}}.$$

One can try to prove similar conjectures for the greedy **c**-monotonic path.

One might, of course, consider more direct approaches to solve the challenge posed by iterative methods. The main difficulty in conducting probabilistic analyses for iterative methods is not unique to the simplex method. In the study of the growth factor of Gaussian elimination with partial pivoting and interior-point methods for linear and convex programming, we have been facing a similar challenge: How do we resolve the dependency of the structures before and after an iterative step? In each iterative step, the algorithm explores some fraction of data which is initially either random or subject to some random perturbations. However, this exploration of the algorithm in determining an iterative step spoils the "randomness" in the same fraction of the data. We can be lucky with some iterative algorithms, such as the shadow-vertex simplex algorithm and Gaussian elimination without pivoting. For each of these, we have found a "flat" upper bound parameter to support an iteration-free analysis of the iterative algorithm. Developing a systematic framework for probabilistic analysis of general iterative methods is a challenging and rewarding research direction.

One simplex algorithm that comes with a flat geometric characterization is Lemke's self-dual parametric simplex algorithm (Dantzig (1991) and Lemke (1965)). Polynomial expected bound was established for a lexicographic variant of the self-dual model for solving a random linear program whose matrices are drawn from a spherically-symmetric distribution (Adler, Karp, & Shamir (1987), Adler & Megiddo (1985), and Todd (1986)).

Conjecture 9.39 (Lemke's Self-Dual Parametric Simplex Algorithm) *Lemke's self-dual parametric simplex algorithm has polynomial smoothed complexity under Gaussian perturbations.*

Another line of research in the smoothed analysis of the simplex algorithm is to extend the result from Gaussian perturbations to other perturbations. Recall the family of distributions introduced by Beier and Vöcking (2004): Let f be a piece-wise continuous univariate density function with finite mean $\int_{\mathbb{R}} |x| \, f(x)dx$ and $\sup_x f(x) = 1$. For $\sigma \leq 1$, let f_σ be a scaling of f such that $f_\sigma(x) = f(x/\sigma)/\sigma$. For any $\bar{x}$, an f-perturbation with magnitude $\sigma < 1$ is a random variable $x = \bar{x} + r$, where r is randomly chosen according to density f_σ.

Conjecture 9.40 (Perturbations and Simplex Methods) *For any piece-wise continuous univariate density function f with finite mean and $\sup_x f(x) = 1$, there exists a σ_0 such that for any $0 \leq \sigma \leq \sigma_0$ the simplex algorithm with the shadow-vertex pivoting rule has smoothed time complexity polynomial in m, n, and $1/\sigma$ under f-perturbations with magnitude σ.*

P2: Smoothed Hirsch Conjecture

The Hirsch conjecture states that for any convex polytope of m facets in n dimensions, the diameter of the graph induced by the 1-skeleton of the polytope is bounded above by $(m - n)$. In other words, any two vertices on the polytope are connected by a path of length at most $(m - n)$. The best bound on the diameter of the polyhedron known today is given by Kalai and Kleitman (1992). Their bound is $m^{\log_2 n + 1}$. In the smoothed setting, we would like to prove the following conjecture.

Conjecture 9.41 (Smoothed Hirsch Conjecture) *For any $m \times n$ matrix $\bar{\mathbf{A}}$ and any m-vector $\bar{\mathbf{b}} \in \mathbb{R}^m$ such that $\left\|\bar{\mathbf{A}}\right\|_F \leq 1$ and $\left\|\bar{\mathbf{b}}\right\|_2 \leq 1$, let $\mathbf{A}$ and $\mathbf{b}$ be σ-Gaussian perturbations of $\bar{\mathbf{A}}$ and $\bar{\mathbf{b}}$, respectively. If $\mathbf{A}\mathbf{x} \leq \mathbf{b}$ is feasible, then the expected diameter of the polyhedron defined by $\mathbf{A}\mathbf{x} \leq \mathbf{b}$ is polynomial in m, n, and $1/\sigma$.*

A special case of the Smoothed Hirsch Conjecture closely related with Theorem 9.2 is:

Conjecture 9.42 (Smoothed Hirsch Conjecture: Special Case) *For any $m \times n$ matrix $\bar{\mathbf{A}}$ such that $\left\|\bar{\mathbf{A}}\right\|_F \leq 1$, let $\mathbf{A}$ be a σ-Gaussian perturbation of $\bar{\mathbf{A}}$. Then the expected diameter of the polyhedron defined by $\mathbf{A}\mathbf{x} \leq \mathbf{1}$ is polynomial in m, n, and $1/\sigma$.*

Note that Conjecture 9.42 does not directly follow from Theorem 9.2, although the latter states that on the polyhedron $\{\mathbf{x} \mid \mathbf{A}\mathbf{x} \leq \mathbf{1}\}$, for any two n-vectors $\mathbf{c}$ and $\mathbf{z}$, the expected length of the shortest path connecting the vertex optimized by $\mathbf{c}$ and the vertex optimized by $\mathbf{z}$ is polynomial in m, n and $1/\sigma$. There could be an exponential number of pairs of vertices on the polyhedron. Without using additional geometric properties, the probability bound in Theorem 9.2 may not be strong enough.

P3: Number of Iterations of IPM and Parallel Programming Algorithms

Several interior-point methods take $O(\sqrt{n} \cdot L)$ number of iterations to solve a linear program. Some experiments suggest that the number of iterations could be as low as $O\left(\log^2(mn)\right)$ when long-step interior-point methods are used. Although progress has been made in recent years concerning the convergence of these methods, it remains open whether, in the worst-case, these long-step methods are better than the more prudent short-step methods (Wright (1997) and Ye (1997)). If practical observations are any indication (Todd (1991)), then the following conjecture could be true.

Conjecture 9.43 (IPM: Optimistic Convergence) *Under Gaussian perturbations, there is an interior-point algorithm with smoothed iteration complexity* $O\left(\log^2(mn/\sigma)\right)$ *for solving linear programming with input dimensions* $m \times n$.

The best worst-case iteration lower bound is $\Omega(n^{1/3})$ due to Todd (1994) and Todd and Ye (1996). However, the programs for which these lower bounds hold are very ill-conditioned. Dunagan, Spielman, and Teng (2002) observe that small perturbations could improve the condition numbers of the resulting linear program. Thus, the lower bound of Todd-Ye does not hold in the smoothed model.

There has not yet been a lower bound of $\Omega(n^\epsilon)$ for well-conditioned linear programs. We would like to know whether such a lower bound holds for well-conditioned programs, or whether there are interior-point algorithms that take fewer iterations when their input is well-conditioned. Perhaps one could start with the following weaker version of Conjecture 9.43:

Conjecture 9.44 (IPM: First step?) *There is an interior-point algorithm with smoothed iteration complexity* $O\left(n^\epsilon \log^2(mn/\sigma)\right)$ *for solving linear programs, for some* $\epsilon < 1/2$.

A closely related theoretical question is the smoothed parallel complexity of linear programming. In the worst-case, linear programming is complete for P under log-space reductions (Dobkin, Lipton & Reiss (1979)). In other words, it is unlikely that one can solve linear programs in poly-logarithmic time using a polynomial number of processors. Solving linear programming approximately is also complete for P (Serna (1991) and Megiddo (1992)).

If Conjecture 9.43 is true, then linear programming has a *smoothed NC* algorithm – one can solve a linear program in poly-logarithmic (in m, n, and $1/\sigma$)) time using a polynomial number of processors. One might relax the poly-logarithmic dependency on $1/\sigma$ and conjecture that:

Conjecture 9.45 (Parallel Linear Programming) *There exists a linear programming algorithm with smoothed parallel complexity, under* σ-*Gaussian perturbations,* $O\left(\sigma^{-k_1} \log^{k_2}(mn)\right)$ *for some constants* k_1 *and* k_2.

Luby and Nisan (1993) consider the parallel complexity of a special family of linear programs

$$\max \quad \mathbf{c}^T \mathbf{x} \quad \text{subject to } \mathbf{A}\mathbf{x} \le \mathbf{b},$$

where all input entries in $\mathbf{A}$, $\mathbf{b}$ and $\mathbf{c}$ are positive. They give a parallel algorithm that, for any $\epsilon > 0$, finds a $(1 + \epsilon)$-approximate solution in time polynomial in $1/\epsilon$ and $\log(mn)$ using $O(mn)$ processors. Although positive linear programming is also complete for P under log-space reductions (Trevisan & Xhafa (1998)) the following conjecture may still be true.

Conjecture 9.46 (Positive Linear Programming) *There exists a positive linear programming algorithm with smoothed parallel complexity* $O\big(\sigma^{-k_1}\log^{k_2}(mn)\big)$ *for some constant k_1 and k_2.*

The conjecture above may follow from the results of Spielman & Teng (2003b), Dunagan, Spielman, & Teng (2002) and Luby & Nisan (1993). It might be more challenging, however, to prove the following stronger conjecture:

Conjecture 9.47 (Positive Linear Programming: Stronger Version) *There is a positive linear programming algorithm with smoothed parallel complexity* $O\left(\log^c(mn/\sigma)\right)$ *for some constant c.*

We would like to conclude this subsection with what might be a simple question. Vaidya (1987) proves that any linear program $(\mathbf{A}, \mathbf{b}, \mathbf{c})$ of m constraints and n variables can be solved in $O((n+m)n^2+(n+m)^{1.5}n)L)$ time. Is it true that the L in Vaidya's bound can be replaced by $\log C(\mathbf{A}, \mathbf{b}, \mathbf{c})$? Recall that $C(\mathbf{A}, \mathbf{b}, \mathbf{c})$ is the condition number of the linear program defined by $(\mathbf{A}, \mathbf{b}, \mathbf{c})$. If true, we can apply the smoothed analysis of Dunagan, Spielman and Teng (2002) to show Vaidya's linear programming algorithm runs in $O((n + m)n^2 + (n + m)^{1.5}n)\log(mn/\sigma))$ smoothed time. Combining with Clarkson's reduction (1995), this would reduce the smoothed complexity for linear programming in low dimensions to

$$O\left(n^2 m + n^6 \log m \log(mn/\sigma)\right).$$

P4: Convex and Conic Programming

Can we extend Theorem 9.7 and Theorem 9.8 from linear programming to conic convex programming (Cucker & Peña (2001) and Freund & Vera (2000)) and to general convex programming (Nesterov & Nemirovskii (1994))? For conic convex programming, one can first consider the following form

$$\min \quad \mathbf{c}^T \mathbf{x} \qquad \text{subject to } \mathbf{A}\mathbf{x} - \mathbf{b} \in \mathbf{C}_1 \text{ and } \mathbf{x} \in \mathbf{C}_2$$

where $\mathbf{C}_1 \subset \mathbb{R}^m$ and $\mathbf{C}_2 \subset \mathbb{R}^n$ are closed convex cones.

The dual of this program is

$$\max \quad \mathbf{b}^T \mathbf{y} \qquad \text{subject to } \mathbf{c} - \mathbf{A}^T \mathbf{y} \in \mathbf{C}_2^* \text{ and } \mathbf{y} \in \mathbf{C}_1^*,$$

where for $i \in \{1, 2\}$, $\mathbf{C}_i^*$ is the dual cone of $\mathbf{C}_i$:

$$\mathbf{C}_i^* = \left\{\mathbf{u} : \mathbf{u}^T\mathbf{v} \geq 0 \text{ for any } \mathbf{v} \in \mathbf{C}_i.\right\}$$

Note that in this form the primal and dual programs have the same format. As the concept of the distance to ill-posedness can be easily extended from linear programs to conic programs, Renegar's condition number for linear programs naturally extends to conic programs. Similarly, it has been shown (Freund & Vera (200) and Nesterov & Nemirovskii (1994)) that the complexity of the interior-point method for conic programming depends logarithmically on the condition number of the input program.

For a convex body $\mathbf{K}$ in $\mathbb{R}^d$, let $\partial\mathbf{K}$ be its boundary. For any $\epsilon \geq 0$, let

$$\partial(\mathbf{K}, \epsilon) = \{\mathbf{x} : \exists \mathbf{x}' \in \partial\mathbf{K}, \|\mathbf{x} - \mathbf{x}'\|_2 \leq \epsilon\}$$

The proof of Theorem 9.7 uses the following key probabilistic bound of Ball (1993) in convex geometry.

Theorem 9.48 (Ball (1993)) *Let μ be the density function of a n-dimensional Gaussian random vector with center $\mathbf{0}$ and variance σ^2. Then for any convex body $\mathbf{K}$ in $\mathbb{R}^n$,*

$$\int_{\partial\mathbf{K}} \mu \leq 4n^{1/4}.$$

The smoothed analysis of Dunagan, Spielman and Teng (2002) applied the following corollary of Theorem 9.48 to estimate the probability that a perturbed linear program is poorly conditioned.

Corollary 9.1 (Darting the boundary of a convex set) *For a vector $\bar{\mathbf{x}} \in \mathbb{R}^n$, let $\mathbf{x}$ be a σ-Gaussian perturbation of $\bar{\mathbf{x}}$. Then for any convex body $\mathbf{K}$ in $\mathbb{R}^n$,*

$$\Pr_{\mathbf{x}}\left[\mathbf{x} \in \partial(\mathbf{K}, \epsilon) \setminus \mathbf{K}\right] \leq \frac{4n^{1/4}\epsilon}{\sigma} \text{ (outside boundary),}$$

$$\Pr_{\mathbf{x}}\left[\mathbf{x} \in \partial(\mathbf{K}, \epsilon) \cap \mathbf{K}\right] \leq \frac{4n^{1/4}\epsilon}{\sigma} \text{ (inside boundary).}$$

Proving the following conjecture would allow us to extend Theorem 9.7 and Theorem 9.8 to conic convex programming.

Conjecture 9.49 (Linear Transformation of Convex Cones) *For any convex cone $\mathbf{K}$ in $\mathbb{R}^n$, let $\mathbf{A}$ be a σ-Gaussian perturbation of an $m \times n$ matrix $\bar{\mathbf{A}}$ with $\|\bar{\mathbf{A}}\|_F \leq 1$. Then, there exist constants σ_0, m_0, n_0, c, k_1,*

k_2, and k_3 such that for any convex cone $\mathbf{C}$ with angle Θ, $n \geq n_0$, $m \geq m_0$, and $0 \leq \sigma \leq \sigma_0$

$$\Pr\left[(\mathbf{A} \cdot \mathbf{K}) \cap \mathbf{C} = \emptyset \ \& \ (\mathbf{A} \cdot \mathbf{K}) \cap \partial(\mathbf{C}, \epsilon) \neq \emptyset\right] \leq c \cdot m^{k_1} n^{k_2} \left(\frac{\epsilon}{\sigma}\right)^{k_3}, \text{ and}$$

$$\Pr\left[(\mathbf{A} \cdot \mathbf{K}) \cap \mathbf{C} \neq \emptyset \ \& \ ((\mathbf{A} \cdot \mathbf{K}) \cap \mathbf{C}) \subseteq \partial(\mathbf{C}, \epsilon \cdot \Theta)\right] \leq c \cdot m^{k_1} n^{k_2} \left(\frac{\epsilon}{\sigma}\right)^{k_3}.$$

Note that when $\mathbf{K}$ is a single vector, $(\mathbf{A} \cdot \mathbf{K})$ is a Gaussian perturbation of the vector $(\bar{\mathbf{A}} \cdot \mathbf{K})$. Thus, in this case, Conjecture 9.49 is a special case of Corollary 9.1 and hence is true.

Conjecture 9.50 (Smoothed condition number of conic programming) *For any* $(\bar{\mathbf{A}}, \bar{\mathbf{b}}, \bar{\mathbf{c}})$ *and* $\sigma \leq 1$, *let* $\mathbf{A}$, $\mathbf{b}$ *and* $\mathbf{c}$ *be* σ-*Gaussian perturbations of* $\bar{\mathbf{A}}$, $\bar{\mathbf{b}}$ *and* $\bar{\mathbf{c}}$. *Then, for any closed convex cones* $\mathbf{C}_1$ *and* $\mathbf{C}_2$, *the expectation of the the logarithm of the condition of the conic program defined by* $\mathbf{A}, \mathbf{b}, \mathbf{c}$ *together with* $\mathbf{C}_1$ *and* $\mathbf{C}_2$ *is* $O(\log(mn/\sigma))$.

An important family of conic convex programming problems is semi-definite programming (SDP). The standard primal form of a semi-definite program is (Todd (2001))

$$\max_{\mathbb{X}} \quad \mathbf{C} \bullet \mathbb{X} \qquad \text{subject to } \mathbf{A}_i \bullet \mathbb{X} = b_i, \ i = 1, \ldots, m \text{ and } \mathbb{X} \succeq \mathbf{0}$$

where $\mathbf{C}$, $\mathbf{A}_i$, and $\mathbb{X}$ are symmetric matrices and $\mathbb{X}$ is required to be positive semi-definite. Because the set of positive semi-definite matrices forms a convex cone, a semi-definite program is a conic convex program. One can define the condition number of a semi-definition program for $(\mathbf{C}, \{\mathbf{A}_i\}, \mathbf{b})$. We conjecture that the expectation of the the logarithm of this condition number is $O(\log(mn/\sigma))$ under Gaussian perturbations.

Even though every convex optimization problem can be transformed into an equivalent instance of conic programming, the transformation, like those among normal forms of linear programming, may not preserve the condition number of the programs. Freund and Ordóñez (2005) explicitly consider the condition number of convex programming in the following non-conic form:

$$\min \quad \mathbf{c}^T \mathbf{x} \qquad \text{subject to } \mathbf{A}\mathbf{x} - \mathbf{b} \in \mathbf{C} \text{ and } \mathbf{x} \in \mathbf{K}$$

where $\mathbf{C} \subset \mathbb{R}^m$ is a convex cone while $\mathbf{K}$ could be any closed convex set (including a cone).

It would be interesting to understand the smoothed behavior of the condition number of convex programs in this form.

P5: Two-Person Games and Multi-Person Games

A two-person game or bimatrix game (Nash (1951) and Lemke (1965)) is specified by two $m \times n$ matrices $\mathbf{A}$ and $\mathbf{B}$, where the m rows represent the pure strategies for the first player, and the n columns represent the pure strategies for the second player. In other words, if the first player chooses strategy i and the second player chooses strategy j, then the payoffs to the first and the second players are a_{ij} and b_{ij}, respectively.

Nash's theorem (1951) on non-cooperative games when specialized to two-person games states that there exists a profile of possibly mixed strategies so that neither player can gain by changing his/her (mixed) strategy, while the other player stays put. Such a profile of strategies is called a *Nash equilibrium*.

Mathematically, a mixed strategy for the first player can be expressed by a column probability vector $\mathbf{x} \in \mathbb{R}^m$, that is, a vector with non-negative entries that sum to 1, while a mixed strategy for the second player is a probability vector $\mathbf{y} \in \mathbb{R}^n$. A Nash equilibrium is then a pair of probability vectors $(\mathbf{x}, \mathbf{y})$ such that for all probabilities vectors $\mathbf{x}' \in \mathbb{R}^m$ and $\mathbf{y}' \in \mathbb{R}^n$,

$$\mathbf{x}^T \mathbf{A} \mathbf{y} \geq (\mathbf{x}')^T \mathbf{A} \mathbf{y} \quad \text{and} \quad \mathbf{x}^T \mathbf{B} \mathbf{y} \geq \mathbf{x}^T \mathbf{B} \mathbf{y}'.$$

The complexity of finding a Nash equilibrium of a two-person game remains open and is considered to be a major open question in theoretical computer science. Recently, Savani and von Stengel (2004) show that the classical Lemke-Howson algorithm (1964) needs an exponential number of steps in the worst case.

In smoothed analysis with Gaussian perturbations, we assume the payoff matrices $\mathbf{A}$ and $\mathbf{B}$ are subject to small Gaussian perturbations. The most optimistic conjecture is:

Conjecture 9.51 (Smoothed 2-Nash Conjecture) *The problem of finding a Nash equilibrium of a two-person game, 2-Nash, is in smoothed polynomial time under Gaussian perturbations.*

As the first step of this research, one could examine the worst-case instances of Savani-von Stengel in the smoothed model. We would like to understand whether such worst-case instances are stable when subject to perturbations. If one can build a stable instance with poor complexity for Lemke-Howson's algorithm, then its smoothed complexity would be poor.

Open Question 1 (Smoothed Complexity of Lemke-Howson) *Does Lemke-Howson's algorithm for 2-Nash have polynomial smoothed complexity under Gaussian perturbations?*

One could consider other algorithms in order to prove the Smoothed 2-Nash Conjecture. An encouraging recent development is the work of Barany, Vempala, and Vetta (2005) who show that when entries of $\mathbf{A}$ and $\mathbf{B}$ are Gaussian variables with mean 0, then 2-Nash has polynomial average-case complexity. So far, their technique does not quite apply to the smoothed model.

As a more general research direction, one can ask similar questions about other games, such as the combinatorially defined graphical games (Kearns, Littman & Singh (2001)), general multi-player finite games (Nash (1951)), or stochastic games (Jurdzinski (2005)).

The 4-person game, 4-Nash, was recently shown to be PPAD-complete by Daskalakis, Goldberg, and Papadimitriou (2005). This result implies that 4-Nash is as hard as the computation of Brouwer fixed points.

Open Question 2 (Game Theory and Algorithms) *What is the smoothed complexity of the computation of Nash equilibria? What is the impact of perturbations to mechanism design?*

In particular,

Open Question 3 (Smoothed Complexity of 4-Nash) *Is 4-Nash, or 3-Nash, in smoothed polynomial time under Gaussian perturbations?*

P6: Gaussian Elimination and Condition Numbers

Several very basic questions on the stability and growth factors of Gaussian elimination remain open. In the worst-case, there are matrices for which Gaussian elimination with partial pivoting has a larger growth factor than Gaussian elimination without pivoting. Similarly, there are matrices for which Gaussian elimination with complete pivoting has a larger growth factor than Gaussian elimination with partial pivoting. Experimentally, partial pivoting has been shown to be much more stable than no pivoting but less stable than complete pivoting.

The most important open problem in the smoothed analysis of Gaussian elimination is to improve the bound of Theorem 9.16. Experimental work seems to suggest that it is exponentially unlikely that Gaussian elimination with partial pivoting has a superpolynomial growth factor.

Conjecture 9.52 (Exponential Stability of GEPP) *For any $n \times n$ matrix $\bar{\mathbf{A}}$ such that $\|\bar{\mathbf{A}}\|_2 \le 1$, let $\mathbf{A}$ be a σ-Gaussian perturbation of $\bar{\mathbf{A}}$. Then, there exists constants c_1 and c_2 such that*

$$\Pr_{\mathbf{A}} \left[\rho_{GEPP}(\mathbf{A}) > x \left(\frac{n}{\sigma} \right)^{c_1} \right] \le 2^{-c_2 x}.$$

There are several other matrix factorization methods that also enjoy practical success. One example is the Bruhat's decomposition that factors $\mathbf{A}$ as $\mathbf{A} = \mathbf{V\Pi U}$ where $\mathbf{\Pi}$ is a permutation matrix, $\mathbf{V}$ and $\mathbf{U}$ are upper triangular matrices, and $\mathbf{\Pi}^T \mathbf{V\Pi}$ is a lower triangular matrix (van den Driessche,Odeh & Olesky (1998)). Another example is the superLU algorithm developed by Li and Demmel (Li (2005)). It first permutes a matrix $\mathbf{A}$ in order to move large elements to the diagonal. A maximum weighted matching algorithm is used for this step to produce a permutation matrix $\mathbf{P}$. The algorithm then symmetrically permutes $\mathbf{PA}$ into $\mathbf{Q(PA)Q}^T$ to improve the sparsity for elimination. Then $\mathbf{Q(PA)Q}^T$ is factored into $\mathbf{LU}$ using Gaussian elimination with no pivoting but with one modification: if during the elimination the current pivoting diagonal entry is smaller than $\epsilon \|\mathbf{A}\|_F$, for some ϵ, then it is replaced by $\sqrt{\epsilon} \|\mathbf{A}\|_2$ before the elimination step proceeds. To solve a linear system $\mathbf{Ax} = \mathbf{b}$, one can use this factorization to obtain an approximate solution by solving the two triangular systems, one defined by $\mathbf{U}$ and one defined by $\mathbf{L}$. Finally the algorithm may apply a few iterations to improve its solution.

Open Question 4 (Stability of Linear Solvers) *What is the smoothed performance of these practically-used factorization algorithms and linear solvers under Gaussian perturbations or under zero-preserving Gaussian perturbations?*

An alternative approach to improve the stability of LU factorization is to use randomization. For example, in the ith step of elimination, instead of choosing the equation with the largest i^{th} coefficient (in absolute value) as in partial pivoting, one can select the next equation from a random distribution that depends on the magnitudes the ith coefficients of the equation. Intuitively, the larger the magnitude of its ith coefficient, the higher is the chance that the equation is chosen. For example, suppose the ith coefficients are $a_{i,i}^{(i-1)}, \ldots, a_{n,i}^{(i-1)}$. For each $p > 0$, the *p-normal partial pivoting* chooses the equation with coefficient $a_{k,i}^{(i-1)}$ for $k \geq i$ with probability

$$\frac{\left(a_{k,i}^{(i-1)} \right)^p}{\left(a_{i,i}^{(i-1)} \right)^p + \ldots + \left(a_{n,i}^{(i-1)} \right)^p}.$$

Open Question 5 (Gaussian elimination with p-normal partial pivoting) *What is the expected growth factor of Gaussian elimination with p-normal partial pivoting?*

Is there a p such that the expected growth factor of Gaussian elimination with p-normal partial pivoting is polynomial in n?

Under Gaussian perturbations, what is the smoothed growth factor of Gaussian elimination with p-normal partial pivoting?

Does Gaussian elimination with p-normal partial pivoting have exponential stability as defined in Conjecture 9.52?

There are several questions still open dealing with the condition numbers and the smallest singular value of a square matrix.

Conjecture 9.53 (Condition Number) *Let* $\bar{\mathbf{A}}$ *be an* $n \times n$ *matrix satisfying* $\|\bar{\mathbf{A}}\|_2 \leq \sqrt{n}$, *and let* $\mathbf{A}$ *be a* σ-*Gaussian perturbation of* $\bar{\mathbf{A}}$ *for* $\sigma \leq 1$. *Then,*

$$\Pr\left[\kappa(\mathbf{A}) \geq x\right] \leq O\left(\frac{n}{x\sigma}\right).$$

Conjecture 9.54 (Smallest Singular Value) *Let* $\bar{\mathbf{A}}$ *be an arbitrary square matrix in* $\mathbb{R}^{n \times n}$, *and let* $\mathbf{A}$ *be a* σ-*Gaussian perturbation of* $\bar{\mathbf{A}}$. *Then*

$$\Pr_{\mathbf{A}}\left[\left\|\mathbf{A}^{-1}\right\|_2 \geq x\right] \leq \frac{\sqrt{n}}{x\sigma}.$$

In the average case where $\mathbf{G}$ is a Gaussian matrix with each of its entries an independent univariate Gaussian variable with mean 0 and standard deviation σ, Edelman (1988) proves

$$\Pr_{\mathbf{G}}\left[\left\|\mathbf{G}^{-1}\right\|_2 \geq x\right] \leq \frac{\sqrt{n}}{x\sigma}.$$

One possible way to prove Conjecture 9.54 would be to show that the Gaussian matrix considered by Edelman, is in fact, the worst-case distribution, as stated in the next conjecture.

Conjecture 9.55 (Gaussian Matrices and Gaussian Perturbations)
Let $\bar{\mathbf{A}}$ *be an arbitrary square matrix in* $\mathbb{R}^{n \times n}$, *and let* $\mathbf{A}$ *be a* σ-*Gaussian perturbation of* $\bar{\mathbf{A}}$. *Let* $\mathbf{G}$ *be a Gaussian matrix of variance* σ^2 *as above. Then for all* $x \geq 1$

$$\Pr_{\mathbf{A}}\left[\left\|\mathbf{A}^{-1}\right\|_2 \geq x\right] \leq \Pr_{\mathbf{G}}\left[\left\|\mathbf{G}^{-1}\right\|_2 \geq x\right].$$

Finally, we have a conjecture on the smallest singular value of a Boolean perturbation of binary matrices.

Conjecture 9.56 (Smallest Singular Value of Binary Matrices)
Let $\bar{\mathbf{A}}$ *be an arbitrary square matrix in* $\{-1, +1\}^{n \times n}$, *and let* $\mathbf{A}$ *be a* σ-*Boolean perturbation of* $\bar{\mathbf{A}}$. *Then there exists a constant* $\alpha < 1$ *such that*

$$\Pr_{\mathbf{A}}\left[\left\|\mathbf{A}^{-1}\right\|_2 \geq x\right] \leq \frac{\sqrt{n}}{x\sigma} + \alpha^n.$$

P7: Algebraic Eigenvalue Problems

Steve Vavasis[1] suggests studying the smoothed complexity of the classical QR iteration algorithm for solving algebraic eigenvalue problems. The eigenvalue problem is to find all eigenvalue-eigenvector pairs of a given $n \times n$ matrix $\mathbf{A}$, where the entries of $\mathbf{A}$ could be either complex or real. A scalar λ and an n-dimensional vector $\mathbf{x}$ form an eigenvalue-eigenvector pair of $\mathbf{A}$ if $\mathbf{A}\mathbf{x} = \lambda\mathbf{x}$. Note that the eigenvalue λ and the entries of its eigenvector $\mathbf{x}$ could be complex. The famous Schur decomposition theorem states:

Theorem 9.57 (Schur Decomposition) *If* $\mathbf{A}$ *is an* $n \times n$ *complex matrix, then there exists a unitary matrix* $\mathbf{Q}$ *such that*

$$\mathbf{Q}^H \mathbf{A} \mathbf{Q} = \mathbf{T},$$

where $\mathbf{T}$ *is an upper triangular matrix with all the eigenvalues of* $\mathbf{A}$ *appearing on its diagonal.*

In addition, if $\mathbf{A}$ *is a real matrix, then there exists an orthogonal matrix* $\mathbf{Q} \in \mathbb{R}^{n \times n}$ *such that*

$$\mathbf{Q}^T \mathbf{A} \mathbf{Q} = \begin{pmatrix} \mathbf{R}_{1,1} & \mathbf{R}_{1,2} & \cdots & \mathbf{R}_{1,k} \\ \mathbf{0} & \mathbf{R}_{2,2} & \cdots & \mathbf{R}_{2,k} \\ \vdots & \vdots & \vdots & \vdots \\ \mathbf{0} & \mathbf{0} & \cdots & \mathbf{R}_{k,k} \end{pmatrix},$$

where $\mathbf{R}_{i,i}$ *is either a scalar or a* 2×2 *matrix. When* $\mathbf{R}_{i,i}$ *is a scalar, it is an eigenvalue of* $\mathbf{A}$ *and when* $\mathbf{R}_{i,i}$ *is a* 2×2 *matrix, it has complex conjugate eigenvalues.*

The QR iteration algorithm was first developed by Francis (1961). Its basic form is very simple. Initially, let $\mathbf{A}_0 = \mathbf{A}$. Iteratively, in the kth step, the algorithm first computes an QR-decomposition of $\mathbf{A}_{i-1}$:

$$\mathbf{A}_{k-1} = \mathbf{Q}_{k-1} \mathbf{R}_{k-1},$$

where $\mathbf{Q}_{k-1}$ is a unitary matrix and $\mathbf{R}_{k-1}$ is an upper triangular matrix. It then defines

$$\mathbf{A}_k = \mathbf{R}_{k-1} \mathbf{Q}_{k-1}.$$

It is well known that in the complex case when $|\lambda_1| > |\lambda_2| > \cdots > |\lambda_n|$, the QR iteration algorithm converges and produces the Schur decomposition. In the real case, under some mild condition, the QR iteration algorithm converges to produce a real Schur decomposition (Wilkinson (1988)

---

[1] Personal Communication.

and Golub & Van Loan (1989)). Thus one can use the QR iteration algorithm to approximate all eigenvalues of $\mathbf{A}$ to an arbitrary precision.

Open Question 6 (Smoothed Complexity of QR iterations: Steve Vavasis) *What is the smoothed complexity of the QR iteration algorithm?*

The convergence of the QR iteration algorithm depends on the the minimum gaps among eigenvalues of the input matrix. For example, in the complex case when $|\lambda_1| > |\lambda_2| > \cdots > |\lambda_n|$, the lower off-diagonal (i,j)-entry of $\mathbf{A}_k$ $(i > j)$ is $O((\lambda_i/\lambda_j)^k)$ (Wilkinson (1988) and Golub & Van Loan (1989)).

Thus, understanding the eigenvalue gaps in the smoothed setting could hold the key to establishing the smoothed rate of convergence of the QR iteration algorithm.

Conjecture 9.58 (Minimum Complex Eigenvalue Gaps) *Let $\bar{\mathbf{A}}$ be an $n \times n$ complex matrix with $\left\| \bar{\mathbf{A}} \right\|_F \leq 1$. Let $\mathbf{A}$ be a σ-Gaussian perturbation of $\bar{\mathbf{A}}$. Let $\lambda_1, \ldots, \lambda_n$ be the eigenvalues of $\mathbf{A}$ and assume $|\lambda_1| \geq |\lambda_2| \geq \cdots \geq |\lambda_n|$. Then, there exist positive constants c, k_1, k_2, k_3 such that, for all $x > 1$,*

$$\Pr\left[\min_{i>1}\left|\frac{\lambda_{i-1}-\lambda_i}{\lambda_{i-1}}\right| \leq \frac{1}{x}\right] \leq c \cdot n^{k_1} \cdot x^{-k_2} \cdot \sigma^{-k_3}.$$

One can similarly make a conjecture for real matrices.

Conjecture 9.59 (Minimum Real Eigenvalue Gaps) *Let $\bar{\mathbf{A}} \in \mathbb{R}^{n \times n}$ with $\left\| \bar{\mathbf{A}} \right\|_F \leq 1$. Let $\mathbf{A}$ be a σ-Gaussian perturbation of $\bar{\mathbf{A}}$. Then, there exist positive constants c, k_1, k_2, k_3 such that, for all $x > 1$,*

$$\Pr\left[\min_{\text{non-conjugate eigenvalues } \lambda_i,\, \lambda_j}\left|\frac{\lambda_i-\lambda_j}{\lambda_i}\right| \leq \frac{1}{x}\right] \leq c \cdot n^{k_1} \cdot x^{-k_2} \cdot \sigma^{-k_3}.$$

For a symmetric matrix $\mathbf{A}$, all of its eigenvalues are real and QR iterations preserve the symmetry as

$$\mathbf{A}_i = \mathbf{R}_{i-1}\mathbf{Q}_{i-1} = \mathbf{Q}_{i-1}^T\mathbf{Q}_{i-1}\mathbf{R}_{i-1}\mathbf{Q}_{i-1} = \mathbf{Q}_{i-1}^T\mathbf{A}_{i-1}\mathbf{Q}_{i-1}.$$

If the eigenvalues of $\mathbf{A}$ are $\lambda_1, \ldots, \lambda_n$, then the QR iteration algorithm converges to $\mathbf{diag}\,(\lambda_1, \ldots, \lambda_n)$, the diagonal matrix whose diagonal entries are $\lambda_1, \ldots, \lambda_n$.

Proving the following conjecture could be useful to establish a smoothed rate of convergence of the QR iteration algorithm under zero-preserving Gaussian perturbations.

Conjecture 9.60 (Minimum Symmetric Eigenvalue Gaps) *Let* $\bar{\mathbf{A}}$ *be an* $n \times n$ *real and symmetric matrix with* $\|\bar{\mathbf{A}}\|_F \leq 1$. *Let* $\mathbf{A} = \bar{\mathbf{A}} + \sigma \mathbf{diag}(g_1, \ldots, g_n)$ *be a* σ-*Gaussian perturbation of the diagonal of* $\bar{\mathbf{A}}$. *Let* $\lambda_1, \ldots, \lambda_n$ *be the eigenvalues of* $\mathbf{A}$ *such that* $|\lambda_1| \geq |\lambda_2| \geq \cdots \geq |\lambda_n|$. *Then, there exist positive constants* c, k_1, k_2, k_3 *such that, for all* $x > 1$,

$$\Pr\left[\min_i \left|\frac{\lambda_{i-1} - \lambda_i}{\lambda_i}\right| \leq \frac{1}{x}\right] \leq c \cdot n^{k_1} \cdot x^{-k_2} \cdot \sigma^{-k_3}.$$

A closely related conjecture is about the singular value gaps.

Conjecture 9.61 (Minimum Singular Value Gaps) *Let* $\bar{\mathbf{A}}$ *be an* $m \times n$ *real matrix with* $\|\bar{\mathbf{A}}\|_F \leq 1$ *and* $m \leq n$. *Let* $\mathbf{A}$ *be a* σ-*Gaussian perturbation of* $\bar{\mathbf{A}}$. *Let* $s_1, \ldots, s_m$ *be the singular value of* $\mathbf{A}$ *such that* $s_1 \geq s_2 \geq \cdots \geq s_m$. *Then, there exist positive constants* c, k_1, k_2, k_3 *such that, for all* $x > 1$,

$$\Pr\left[\min_i \left(\frac{s_{i-1} - s_i}{s_i}\right) \leq \frac{1}{x}\right] \leq c \cdot n^{k_1} \cdot x^{-k_2} \cdot \sigma^{-k_3}.$$

In practice, one usually does not apply the QR iteration algorithm directly to an input matrix $\mathbf{A}$. The QR computation of each iteration could take $O(n^3)$ time, which might be too expensive. In fact, most practical implementations first use an orthogonal similarity transformation to reduce the matrix $\mathbf{A}$ to an upper-Hessenberg form (Wilkinson (1988) and Golub & Van Loan (1989)) $A_0 = \mathbf{Q}_0^H \mathbf{A} \mathbf{Q}_0$. A matrix $\mathcal{H} = (h_{i,j})$ is upper Hessenberg if $h_{i,j} = 0$ for $i > j + 1$. This step is important because the QR factorization of an upper Hessenberg matrix can be computed in $O(n^2)$ time, instead of $O(n^3)$. The standard approach uses Givens rotations at each step to perform QR factorization of an upper Hessenberg matrix. A nice property of the Givens process for QR factorization is that the resulting QR iteration algorithm preserves the upper Hessenberg form.

In general, the practical QR iteration algorithms go beyond just applying an initial Hessenberg reduction. What makes them more successful in practice is the collection of shifting strategies that are used to improve the rate of convergence (Wilkinson (1988) and Golub & Van Loan (1989)). Each shifted iteration consists of the following steps.

1. Determine a scalar μ_{i-1};
2. Compute $\mathbf{A}_{i-1} - \mu_{i-1}\mathbf{I} = \mathbf{Q}_{i-1}\mathbf{R}_{i-1}$;
3. Let $\mathbf{A}_i = \mathbf{R}_{i-1}\mathbf{Q}_{i-1} + \mu_{i-1}\mathbf{I}$.

In practice, QR iteration algorithms may perform double shifts during each iteration.

Open Question 7 (Smoothed Complexity of Practical QR Iteration Algorithms)

- *What is the smoothed complexity of these practical QR iteration algorithms?*

- *Is the smoothed rate of convergence of any practical QR iteration algorithm better than that of the classical QR iteration algorithm?*

- *What is the impact of Hessenberg reduction on the smoothed rate of convergence of QR iteration Algorithms?*

- *What are the smoothed rates of convergence of the classical or practical symmetric QR iteration algorithms under symmetry-preserving Gaussian perturbations and under symmetry-preserving and zero-preserving Gaussian perturbations?*

P8: Property-Preserving Perturbations

For some discrete problems, as we have discussed in Section 9.3.4, results from the semi-random model might not always extend to the corresponding property-preserving perturbations. Perhaps the most appealing problem is the Bisection Problem.

Open Question 8 (Bisection) *Is the ρ-Bisection Problem, under ρ-Bisection preserving perturbations, in smoothed polynomial time (in the probabilistic sense), for some constant $0 < \rho < 1$?*

A closely related problem is whether a ρ-Bisection property testing algorithm exists that runs in time polynomial in $1/\epsilon$ and $1/\sigma$ in the smoothed model under ρ-Bisection-preserving σ-perturbations. Another related problem is whether the ρ-Bisection Problem is in smoothed polynomial time (in the probabilistic sense) under the solution-preserving perturbations.

The property-preserving model is not limited to discrete settings. It can be applied to the continuous setting as well. For example, one can study the smoothed complexity of a linear programming algorithm under feasibility-preserving Gaussian perturbations.

Open Question 9 (Feasibility and Linear Programming) *Is the simplex method with the shadow-vertex pivoting rule still in smoothed polynomial time under feasibility-preserving Gaussian perturbations?*

Is the smoothed value of the logarithm of the condition number of linear programs still poly-logarithmic in m, n, and $1/\sigma$ as stated in Theorem 9.7, under feasibility-preserving Gaussian perturbations?

As the purpose of smoothed analysis is to shed light on the practical performance of an algorithm, it is more desirable to use a perturbation model that better fits the input instances. See the *Final Remarks* at the end of this paper for more discussion. Thus, if all or most practical instances to an algorithm share some common structures, such as being symmetric or being planar, then to have a meaningful theory, we may have to consider perturbations that preserve these structures. For example, the fact that many scientific computing algorithms use the sparsity of the input to achieve good performance encourages us to define the zero-preserving or magnitude-preserving perturbations such as relative Gaussian perturbations. So far, however, the smoothed complexities of various problems and algorithms under these perturbations remains wide open.

Open Question 10 (Structure-Preserving Perturbations) *What is the impact of structure-preserving perturbations, such as magnitude-preserving and zero-preserving perturbations, on the smoothed complexity of an algorithm?*

P9: Smoothed Complexity and Approximation Algorithms

Open Question 11 (Smoothed Complexity and Hardness of Approximation) *Is there any connection between the smoothed complexity of an optimization problem and the hardness of its approximation? Under what conditions does "hard to approximate" imply "high smoothed complexity" and vice versa?*

As smoothed time complexity measures the performance of an algorithm A on an input x by the expected performance of A over a "neighborhood" distribution of x, intuitively, if this complexity is low, then one could first perturb an instance and solve the optimization problem over the perturbed instance. The resulting algorithm then has low randomized complexity.

How good this randomized algorithm can be as an approximation algorithm may depend on the perturbation model, the property of the objective function and the structure of the solution space.

Suppose A is an algorithm for solving a minimization problem with an objective function f over an input domain $D = \cup_n D_n$. Suppose further, there is a family of neighborhoods $N_\sigma(\bar{x}) \subseteq \cup_{n'=\Theta(n)} D_{n'}$ for every $\bar{x} \in D_n$, such that for all $x \in N_\sigma(\bar{x})$, $|f(A(x)) - f(A(\bar{x}))| \leq h(\sigma)$ where $h : \mathbb{R} \to \mathbb{R}^+$ is a monotonically increasing function.

If there is a family of perturbations $\mathcal{R} = \cup_{n,\sigma} R_{n,\sigma}$, where, for each $\bar{x} \in D_n$, $R_{n,\sigma}$ defines a perturbation distribution over $N_\sigma(\bar{x})$ such that the smoothed complexity of A under this perturbation model is $T(n, \sigma)$, then A can be used as a family of randomized approximation algorithms of expected complexity $T(n, \sigma)$ that comes within $h(\sigma)$ of the optimal value for minimizing f in instance $\bar{x}$, provided $\mathcal{R}$ can be efficiently sampled.

For example, consider a two-person game given by two $m \times n$ matrices $\bar{\mathbf{A}} = (\bar{a}_{i,j})$ and $\bar{\mathbf{B}} = (\bar{b}_{i,j})$. Suppose $\mathbf{A} = (a_{i,j})$ and $\mathbf{B} = (b_{i,j})$ are σ-*uniform-cube perturbations* of $\bar{\mathbf{A}}$ and $\bar{\mathbf{B}}$, respectively, where $a_{i,j}$ (and $b_{i,j}$) is an independent random variable chosen uniformly from the interval $[\bar{a}_{i,j} - \sigma, \bar{a}_{i,j} + \sigma]$ (and $[\bar{b}_{i,j} - \sigma, \bar{b}_{i,j} + \sigma]$). Then, for every pair of mixed strategies $\mathbf{x}$ and $\mathbf{y}$, $\left| \mathbf{x}^T \mathbf{A} \mathbf{y} - \mathbf{x}^T \bar{\mathbf{A}} \mathbf{y} \right| \leq 2\sigma$. and $\left| \mathbf{x}^T \mathbf{B} \mathbf{y} - \mathbf{x}^T \bar{\mathbf{B}} \mathbf{y} \right| \leq 2\sigma$.

Now suppose $(\mathbf{x}, \mathbf{y})$ is a Nash equilibrium for $(\mathbf{A}, \mathbf{B})$. Then, for any $(\mathbf{x}', \mathbf{y}')$, we have

$$(\mathbf{x}')^T \bar{\mathbf{A}} \mathbf{y} - \mathbf{x}^T \bar{\mathbf{A}} \mathbf{y} \leq ((\mathbf{x}')^T \mathbf{A} \mathbf{y} - \mathbf{x}^T \mathbf{A} \mathbf{y}) + 4\sigma \leq 4\sigma,$$

as well as $\mathbf{x}^T \bar{\mathbf{B}} \mathbf{y}' - \mathbf{x}^T \bar{\mathbf{B}} \mathbf{y} \leq 4\sigma$. Thus, $(\mathbf{x}, \mathbf{y})$ is a (4σ)-*Nash equilibrium* for $(\bar{\mathbf{A}}, \bar{\mathbf{B}})$: a profile of mixed strategies such that no player can gain more than an amount 4σ by changing his/her strategy unilaterally. Similarly, if $(\mathbf{x}, \mathbf{y})$ is an ϵ-Nash equilibrium for $(\mathbf{A}, \mathbf{B})$, then $(\mathbf{x}, \mathbf{y})$ is an $(\epsilon + 4\sigma)$-Nash equilibrium for $(\bar{\mathbf{A}}, \bar{\mathbf{B}})$. Therefore,

Proposition 9.12 (Smoothed 2-Nash and Approximated 2-Nash) *If 2-Nash can be solved in smoothed time polynomial in m, n, and $g(1/\sigma)$ under σ-uniform-cube perturbations, then an ϵ-Nash equilibrium of two-person games can found in randomized time polynomial in m, n, and $g(1/\epsilon)$.*

However, for a constrained optimization problem, the optimal solution of a perturbed instance x may not be feasible for the original instance $\bar{x}$. Although running A on the perturbed instance x provides a good approximation of the optimal value for the original instance $\bar{x}$, one still needs an efficient procedure to "round" the solution for x to a feasible solution for $\bar{x}$ in order to approximately solve the optimization problem. For some problems, such as linear programming, the optimal solution for x might be quite "far" away from the optimal solution for $\bar{x}$, although their objective values are be close. This discrepancy might pose algorithmic challenges to approximations.

Another interesting direction of research is to examine the worst-case instances appearing in the literature to determine whether they are stable under perturbations. If all the known worst-case instances of a problem or an algorithm are not stable under some perturbations, then one could ask whether its smoothed complexity under these perturbations is low, or if there are other bad instances that are stable.

P10: Other Algorithms and Practical Algorithms

There are many other successful practical heuristics that we cannot discuss here in great detail. For example, Berthold Vöcking suggests, as interesting and relevant research directions, considering the smoothed complexity of heuristics like branch-and-bound or cutting-plane methods on structurally simple optimization problems like packing problems with a constant number of constraints.

Other very popular methods include the multilevel algorithms (Brandt (1988) and Teng (1998)), differential evolution (Price, Storn & Lampinen (2005)), and various local search and global optimization heuristics. We would like to understand the smoothed complexity of these methods. For example, the following conjecture is at the center of our research (in this area).

Conjecture 9.62 (Multilevel Bisection Conjecture) *There is a multilevel bisection algorithm with smoothed polynomial time complexity that finds a $(c \cdot \rho)$-bisection (for some constant c) under ρ-bisection-preserving perturbations as well as in the semi-random model.*

Conjecture 9.63 (Multilevel Sparsest-Cut Conjecture) *There is a multilevel partitioning algorithm with smoothed polynomial time complexity that finds a partition with sparsity $c \cdot \rho$ under ρ-sparsest-cut-preserving perturbations as well as in the semi-random model.*

Final Remarks

Developing rigorous mathematical theory that can model the observed performance of practical algorithms and heuristics has become an increasingly important task in Theoretical Computer Science. Unlike four decades ago, when theorists were introducing asymptotic complexity but practitioners could only solve linear systems with less than 500 variables, we now have computers that are capable of solving very large-scale problems. Moreover, as heuristic algorithms become ubiquitous in applications, we have increasing opportunities to obtain data, especially on large-scale problems, from these remarkable heuristics.

One of the main objectives of smoothed analysis is to encourage the development of theories for the practical behaviors of algorithms. We are especially interested in modeling those algorithms whose practical performance is much better than their worst-case complexity measures.

A key step is to build analyzable models that are able to capture some essential aspects of the algorithms and, of equal importance, the inherent

properties of practical instances. So necessarily, any such model should be
more instance-oriented than our traditional worst-case and average analyses, and should consider the formation process of input instances.

However, modeling observed data and practical instances is a challenging task. Practical inputs are often complex and there may be multiple
parameters that govern the process of their formation. Most of the current
work in smoothed analysis focuses on the randomness in the formation of
inputs and approximates the likelihood of any particular instance by its
similarity or distance to a "hidden blueprint" of a "targeted" instance of
the input-formation process. As the targeted instance might not be known
to the algorithm, in the same spirit of worst-case analysis, we used the maximum over all possible targeted instances in the definition of the smoothed
complexity.

This approach to characterize the randomness in the formation of input
instances promises to be a good first step to model or to approximate the
distribution of practical instances. One must understand the possible lim-
itations of any particular perturbation model, however, and not overstate
the practical implication of any particular analysis.

One way to improve the similarity-or-distance based perturbation models
is to develop an analysis framework that takes into account the formation
of input instances. For example, if the input instances to an algorithm A
come from the output of another algorithm B, then algorithm B, together
with a model of B's input instances, is the description of A's inputs. To be
concrete, consider finite-element calculations in scientific computing. The
input to its linear solver A are stiffness matrices which are produced by
a finite-element mesh generation algorithm B. The meshing algorithm B,
which could be a randomized algorithm, receives a geometric domain Ω
and a partial differential equation F as an input instances to construct a
stiffness matrix. So the distribution of the stiffness matrices to algorithm
A is determined by the distribution $\mathcal{D}$ of the geometric domains Ω and
the set F of partial differential equations, as well as the mesh generation
algorithm B. One can define the measure of the performance of A as

$$\mathrm{E}_{(\Omega,F)\leftarrow\mathcal{D}}\left[\mathrm{E}_{X\leftarrow B(\Omega,F)}\left[Q(A,X)\right]\right].$$

If, for example, $\bar{\Omega}$ is a design of an advanced rocket from a set $\mathcal{R}$ of
"blueprints" and F is from a set $\mathcal{F}$ of PDEs for physical parameters such
as pressure, speed, and temperature for the rocket, and Ω is generated by
a perturbation model $\mathcal{P}$ of the blueprints, then one may further measure
the performance of A by the smoothed value of the quantity above:

$$\max_{F\in\mathcal{F},\bar{\Omega}\in\mathcal{R}}\mathrm{E}_{\Omega\leftarrow\mathcal{P}(\bar{\Omega})}\left[\mathrm{E}_{X\leftarrow B(\Omega,F)}\left[Q(A,X)\right]\right].$$

There might be many different frameworks for modeling the formation process of the input instances. For example, one could use a Markov process, a branch tree with probabilistic nodes and binary branching nodes or some innovative diagrams or flowcharts. The better we can model our input data, the more accurately we can model the performance of an algorithm. But to be rigorous mathematically, we may have to come up with a conjecture that matches the practical observations, and find a way to prove the conjecture.

Another objective of smoothed analysis is to provide insights and motivations to design new algorithms, especially those with good smoothed complexity. For example, our analysis of the smoothed growth factor suggests a new and more stable solver for linear systems: Suppose we are given a linear system $\mathbf{A}\mathbf{x} = \mathbf{b}$. We first use the standard elimination-based algorithm – or software – to solve $\mathbf{A}\mathbf{x} = \mathbf{b}$. Suppose $\mathbf{x}^*$ is the solution computed. If $\|\mathbf{b} - \mathbf{A}\mathbf{x}^*\|$ is small enough, then we simply return $\mathbf{x}^*$. Otherwise, we can determine a parameter ϵ and generate a new linear system $(\mathbf{A} + \epsilon\mathbf{G})\mathbf{y} = \mathbf{b}$, where $\mathbf{G}$ is a Gaussian matrix with independent entries with mean 0 and variance 1. So instead of using the solution of $\mathbf{A}\mathbf{x} = \mathbf{b}$, we solve a perturbed linear system $(\mathbf{A} + \epsilon\mathbf{G})\mathbf{y} = \mathbf{b}$. It follows from the condition number analysis that if ϵ is (significantly) smaller than $\kappa(\mathbf{A})$, then the solution to the perturbed linear system is a good approximation to the original one. One can use practical experience or binary search to set ϵ.

The new algorithm has the property that its success depends only on the machine precision and the condition number of $\mathbf{A}$, while the original algorithm may fail due to large growth factors. For example, the following is a segment of matlab code that first solves a linear system whose matrix is the 70×70 Wilkinson matrix, using the Matlab linear solver, then solves it with our new algorithm.

```
>> % Using the Matlab Solver
>> n = 70; A = 2*eye(n)-tril(ones(n)); A(:,n)=1;
>> b = randn(70,1);
>> x = A\b;
>> norm(A*x-b)
>> 2.762797463910437e+004
>> % FAILED because of large growth factor
>> %Using the new solver
>> Ap = A + randn(n)/10^9;
>> y = Ap\b;
>> norm(Ap*y-b)
```

```
>> 6.343500222435404e-015
>> norm(A*y-b)
>> 4.434147778553908e-008
```

Because the Matlab linear solver uses Gaussian elimination with partial pivoting, it fails to solve the linear system because of the large growth factor. But our perturbation-based algorithm finds a good solution.

We conclude this paper with the open question that initially led us to smoothed analysis.

Open Question 12 (Linear Programming in Strongly Random Polynomial Time?) *Can the techniques from the smoothed analysis of the simplex and interior-point methods be used to develop a randomized strongly polynomial-time algorithm for linear programming?*

9.5 Acknowledgments

We thank Berthold Vöcking for his suggestions of research direction in smoothed analysis, Steve Vavasis for his suggestion of performing smoothed analysis on the QR iteration algorithm, Ben Hescott for mentioning the paper by Daskalakis, Goldberg, and Papadimitriou, and Jinyi Cai for bringing the Bruhat's decomposition to our attention. We thank Kyle Burke and Ben Hescott for their helpful comments and discussions on this writing. Finally, we would like to express our appreciation of all the valuable feedbacks that we received after presenting our work on smoothed analysis at various universities, institutions, and conferences.

Daniel Spielman is partially supported by the NSF grant CCR-0324914 and Shang-Hua Teng is partially supported by the NSF grants CCR-0311430 and ITR CCR-0325630.

References

L. Adleman and M.-D. Huang (1987). Recognizing primes in random polynomial time. In *Proceedings of the Nineteenth Annual ACM Symposium on Theory of Computing*, pages 462–469.

I. Adler, R. M. Karp, and R. Shamir (1987). A simplex variant solving an $m \times d$ linear program in $O(\min(m^2, d^2))$ expected number of steps. *Journal of Complexity*, 3:372–387.

I. Adler and N. Megiddo (1985). A simplex algorithm whose average number of steps is bounded between two quadratic functions of the smaller dimension. *J. ACM*, 32(4):871–895, October.

A. Aggarwal, B. Alpern, A. Chandra, and M. Snir (1987). A model for hierarchical memory. In *Proceedings of the Nineteenth Annual ACM Symposium on Theory of Computing*, pages 305–314.

S. Agmon (1954). The relaxation method for linear inequalities. *Canadian Journal of Mathematics*, 6:382–392.

M. Agrawal, N. Kayal and N. Saxena (2004). Primes is in P. In *Annals of Mathematics 160(2)*, pages 781–793.

A. V. Aho, J E. Hopcroft, and J. Ullman (1983). *Data Structures and Algorithms*. Addison-Wesley Longman.

N. Alon and J. H. Spencer (1992). *The Probabilistic Method*. John Wiley and Sons.

N. Amenta and G. Ziegler (1999). Deformed products and maximal shadows of polytopes. In B. Chazelle, J.E. Goodman, and R. Pollack, editors, *Advances in Discrete and Computational Geometry*, number 223 in Contemporary Mathematics, pages 57–90. Amer. Math. Soc.

E. Anderson, Z. Bai, C. Bischof, J. Demmel, J. Dongarra, J. Du Croz, A. Greenbaum, S. Hammarling, A. McKenney, S. Ostrouchov, and D. Sorensen (1999). *LAPACK Users' Guide, Third Edition*. SIAM, Philadelphia.

D. Arthur and S. Vassilvitskii (2005) On the Worst Case Complexity of the k-means Method. http://dbpubs.stanford.edu:8090/pub/2005-34.

K. Ball (1993). The reverse isoperimetric problem for gaussian measure. *Discrete and Computational Geometry*, 10(4):411–420.

C. Banderier, R. Beier, and K. Mehlhorn (2003). Smoothed analysis of three combinatorial problems. In *Proceedings of the Twenty-eighth International Symposium on Mathematical Foundations of Computer Science*, volume 2747, pages 198–207.

I. Barany, S. Vempala, and A. Vetta (2005). Nash equilibria in random games. In *Proceedings of Forty-sixth Annual IEEE Symposium on Foundations of Computer Science*.

J. Basch, L. J. Guibas, and J. Hershberger (1997). Data structures for mobile data. In *Proceedings of the Eighth Annual ACM-SIAM Symposium on Discrete Algorithms*, pages 747–756.

L. Becchetti, S. Leonardi, A. Marchetti-Spaccamela, G. Schäfer, and T. Vredeveld (2003). Average case and smoothed competitive analysis of the multi-level feedback algorithm. In *Proceedings of the Forty-fourth Annual IEEE Symposium on Foundations of Computer Science*, page 462.

R. Beier and B. Vöcking (2004). Typical properties of winners and losers in discrete optimization. In *Proceedings of the Thirty-sixth Annual ACM Symposium on Theory of Computing*, pages 343–352.

J. L. Bentley, H. T. Kung, M. Schkolnick, and C. D. Thompson (1978). On the average number of maxima in a set of vectors and applications. *J. ACM*, 25(4):536–543.

R. G. Bland (1977). New finite pivoting rules. *Mathematics of Operations Research*, 2:103 – 107.

H. D. Block (1962). The perceptron: A model for brain functioning. *Reviews of Modern Physics*, 34:123–135.

A. Blum and J. Dunagan (2002). Smoothed analysis of the perceptron algorithm for linear programming. In *Proceedings of the Thirteenth Annual ACM-SIAM Symposium on Discrete Algorithms*, pages 905–914.

A. Blum and J. Spencer (1995). Coloring random and semi-random k-colorable graphs. *J. Algorithms*, 19(2):204–234.

T. Bohman, A. Frieze, and R. Martin (2003). How many random edges make a dense graph hamiltonian? *Random Struct. Algorithms*, 22(1):33–42.

R. B. Boppana (1987). Eigenvalues and graph bisection: An average-case analysis. In *Proceedings of the Forty-seventh Annual IEEE Symposium on Foundations of Computer Science*, pages 280–285.

K. H. Borgwardt (1980). *The Simplex Method: a probabilistic analysis*. Springer-Verlag.

A. Borodin and R. El-Yaniv (1998). *Online computation and competitive analysis*. Cambridge University Press, New York, NY, USA.

A. Borodin, N. Linial, and M. E. Saks (1992). An optimal on-line algorithm for metrical task system. *J. ACM*, 39(4):745–763.

A. Brandt (1988). Multilevel computations: Review and recent developments. In *Multigrid Methods: Theory, Applications, and Supercomputing*, Marcel-Dekker, S. F. McCormick, editor, 541–555.

S.-W. Cheng and T. K. Dey (2002). Quality meshing with weighted delaunay refinement. In *Proceedings of the Thirteenth Annual ACM-SIAM Symposium on Discrete Algorithms*, pages 137–146.

S. W. Cheng, T. K. Dey, H. Edelsbrunner, M. A. Facello, and S.-H. Teng (2000). Sliver exudation. *J. ACM*, 47:883 – 904.

L.P. Chew (1989). Guaranteed-quality triangular meshes. Technical Report TR-89-983, Cornell University, Ithaca.

V. Chvatal (1983). *Linear Programming*. A Series of Books in the Mathematical Sciences. Freeman.

K. L. Clarkson (1995). Las Vegas algorithms for linear and integer programming when the dimension is small. *J. ACM*, 42(2):488–499.

A. Condon, H. Edelsbrunner, E. A. Emerson, L. Fortnow, S. Haber, R. Karp, D. Leivant, R. Lipton, , N. Lynch, I. Parberry, C. Papadimitriou, M. Rabin, A. Rosenberg, J. S. Royer, J. Savage, A. L. Selman, C. Smith, E. Tardos, and J. S. Vitter (1999). Challenges for theory of computing. In *Report of an NSF-Sponsored Workshop on Research in Theoretical Computer Science*.

T. H. Cormen, C. E. Leiserson, R. L. Rivest, and C. Stein (2001). *Introduction to Algorithms*. McGraw-Hill Higher Education.

F. Cucker and J. Peña (2001). A primal-dual algorithm for solving polyhedral conic systems with a finite-precision machine. *SIAM J. on Optimization*, 12(2):522–554.

V. Damerow, F. M. auf der Heide, H. Räcke, C. Scheideler, and C. Sohler (2003). Smoothed motion complexity. In *Proceedings of the Eleventh Annual European Symposium on Algorithms*, pages 161–171.

V. Damerow and C. Sohler (2004). Extreme points under random noise. In *European Symposium on Algorithms*, pages 264–274.

G. B. Dantzig (1951). Maximization of linear function of variables subject to linear inequalities. In T. C. Koopmans, editor, *Activity Analysis of Production and Allocation*, pages 339–347.

G. B. Dantzig (1991). *Linear Programming and Extensions*. Springer.

C. Daskalakis, C. H. Papadimitriou, P. W. Goldberg (2005). The complexity of computing a nash equilibrium. *Electronic Colloquium on Computational Complexity*, TR05-115.

J. Demmel (1997). *Applied Numerical Linear Algebra*. SIAM.

A. Deshpande and D. A. Spielman (2005). Improved smoothed analysis of the shadow vertex simplex method. In *Proceedings of Forty-sixth Annual IEEE Symposium on Foundations of Computer Science.*

D. Dobkin, R. J. Lipton and S. Reiss (1979). Linear programming is log–space hard for P. *Information Processing Letters*, 8:96–97.

J. Dunagan, D. A. Spielman, and S.-H. Teng (2002). Smoothed analysis of renegar's condition number for linear programming. available at `http://math.mit.edu/~spielman/SmoothedAnalysis`, submitted to *Mathematical Programming.*

J. Dunagan and S. Vempala (2004). A simple polynomial-time rescaling algorithm for solving linear programs. In *Proceedings of the Thirty-sixth Annual ACM Symposium on Theory of Computing*, pages 315–320.

A. Edelman (1988). Eigenvalues and condition numbers of random matrices. *SIAM J. Matrix Anal. Appl.*, 9(4):543–560.

H. Edelsbrunner (2001). *Geometry and topology for mesh generation.* Cambridge University Press.

H. Edelsbrunner, X.-Y. Li, G. L. Miller, A. Stathopoulos, D. Talmor, S.-H. Teng, A. Üngör, and N. Walkington (2000). Smoothing and cleaning up slivers. In *Proceedings of the Thirty-second Annual ACM Symposium on Theory of Computing*, pages 273–277.

U. Feige and J. Kilian (1998). Heuristics for finding large independent sets, with applications to coloring semi-random graphs. In *Proceedings of the Thirty-ninth Annual Symposium on Foundations of Computer Science*, page 674.

U. Feige and R. Krauthgamer (2002). A polylogarithmic approximation of the minimum bisection. *SIAM J. on Computing*, 31:1090–1118.

W. Feller (1968,1970). *An Introduction to Probability Theory and Its Applications*, volume 1,2. Wiley, New York.

A. Flaxman and A. M. Frieze (2004). The diameter of randomly perturbed digraphs and some applications.. In *APPROX-RANDOM*, pages 345–356.

J. J. Forrest, D. Goldfarb, D. (1992) Steepest-edge simplex algorithms for linear programming. *Mathematical Programming* 57, 341–374.

J. G. F. Francis (1961). The qr transformation a unitary analogue to the lr transformation: Part 1. *The Computer Journal*, 4(3):256 – 271.

R. M. Freund and F. Ordóñez (2005). On an extension of condition number theory to non-conic convex optimization. *Math of OR*, 30(1).

R. Freund and J. Vera (1999). On the complexity of computing estimates of condition measures of a conic linear system. Operations Research Center Working Paper, MIT, 1999, submitted to *Mathematics of Operations Research.*

R. Freund and J. Vera (2000). Condition-based complexity of convex optimization in conic linear form via the ellipsoid algorithm. *SIAM J. on Optimization*, 10(1):155–176.

M. Frigo, C. E. Leiserson, H. Prokop, and S. Ramachandran (1999). Cache-oblivious algorithms. In *Proceedings of the Fortieth Annual Symposium on Foundations of Computer Science*, page 285.

S. Gass and Th. Saaty (1955). The computational algorithm for the parametric objective function. *Naval Research Logistics Quarterly*, 2:39–45.

D. Goldfarb (1983). Worst case complexity of the shadow vertex simplex algorithm. Technical report, Columbia University.

D. Goldfarb and W. T. Sit (1979). Worst case behaviour of the steepest edge simplex method. *Discrete Applied Math*, 1:277–285.

O. Goldreich, S. Goldwasser, and D. Ron (1998). Property testing and its connection to learning and approximation. *J. ACM*, 45(4):653–750, July.

G. H. Golub and C. F. Van Loan (1989). *Matrix Computations*. second edition.

M. Hestenes and E. Stiefel (1952). Methods of conjugate gradients for solving linear systems. *the Journal of Research of the National Bureau of Standards*.

R. G. Jeroslow (1973). The simplex algorithm with the pivot rule of maximizing improvement criterion. *Discrete Math.*, 4:367–377.

M. Jurdzinski (2005). Stochastic games: A tutorial. http://www.games.rwth-aachen.de/Events/Bordeaux/t_mju.pdf.

G. Kalai (1992). A subexponential randomized simplex algorithm (extended abstract). In *Proceedings of the Twenty-fourth Annual ACM Symposium on Theory of Computing*, pages 475–482.

G. Kalai and D. J. Kleitman (1992). A quasi-polynomial bound for the diameter of graphs of polyhedra. *Bulletin Amer. Math. Soc.*, 26:315–316.

N. K. Karmarkar (1984). A new polynomial–time algorithm for linear programming. *Combinatorica*, 4:373–395.

M. J. Kearns, M. L. Littman, and S. P. Singh (2001). Graphical models for game theory. In *UAI '01: Proceedings of the Seventeenth Conference in Uncertainty in Artificial Intelligence*, pages 253–260.

L. G. Khachiyan (1979). A polynomial algorithm in linear programming. *Doklady Akademia Nauk SSSR*, pages 1093–1096.

V. Klee and G. J. Minty (1972). How good is the simplex algorithm ? In Shisha, O., editor, *Inequalities – III*, pages 159–175. Academic Press.

M. Krivelevich, B. Sudakov, and P. Tetali (2005). On smoothed analysis in dense graphs and formulas. http://www.math.princeton.edu/~bsudakov/smoothed-analysis.pdf.

C. E. Lemke (1965). Bimatrix equilibrium points and mathematical programming. *Management Science*, 11:681–689.

C. E. Lemke and JR. J. T. Howson (1964). Equilibrium points of bimatrix games. *J. Soc. Indust. Appl. Math.*, 12:413–423.

X.-Y. Li and S.-H. Teng (2001). Generate sliver free three dimensional meshes. In *Proceedings of the Twelfth ACM-SIAM Symp. on Discrete Algorithms*, pages 28–37.

X. S. Li (2005). An overview of superLU: Algorithms, implementation, and user interface. *ACM Trans. Math. Softw.*, 31(3):302–325.

S. Lloyd (1982) Least squares quantization in PCM *IEEE Transactions on Information Theory*, 28 (2) pages 129 –136.

M. Luby and N. Nisan (1993). A parallel approximation algorithm for positive linear programming. In *Proceedings of the Twenty-fifth Annual ACM Symposium on Theory of Computing*, pages 448–457.

B. Manthey and R. Reischuk (2005). Smoothed analysis of the height of binary search tress. *Electronic Colloquium on Computational Complexity*, TR05-063.

J. Matoušek, M. Sharir, and E. Welzl (1992). A subexponential bound for linear programming. In *Proceedings of the Eighth Annual Symposium on Computational Geometry*, pages 1–8.

N. Megiddo (1986). Improved asymptotic analysis of the average number of steps performed by the self-dual simplex algorithm. *Mathematical Programming*, 35(2):140–172.

N. Megiddo (1992). A note on approximate linear programming. *Inf. Process. Lett.*, 42(1):53, 1992.

Gary L. Miller (1975). Riemann's hypothesis and tests for primality. In *Proceedings of Seventh Annual ACM Symposium on Theory of Computing*, pages 234–239.

G. L. Miller, D. Talmor, S.-H. Teng, and N. Walkington (1995). A Delaunay based numerical method for three dimensions: generation, formulation, and partition. pages 683–692.

M. L. Minsky and S. A. Papert (1988). *Perceptrons: expanded edition*. MIT Press.

R. Motwani, S. Phillips, and E. Torng (1993). Non-clairvoyant scheduling. In *Proceedings of the Fourth Annual ACM-SIAM Symposium on Discrete Algorithms*, pages 422–431.

R. Motwani and P. Raghavan (1995). *Randomized algorithms*. Cambridge University Press.

K. G. Murty (1980). Computational complexity of parametric linear programming. *Math. Programming*, 19:213–219.

J. Nash (1951). Noncooperative games. *Annals of Mathematics*, 54:289–295.

Y. Nesterov and A. Nemirovskii (1994). *Interior Point Polynomial Methods in Convex Programming: Theory and Applications*. Society for Industrial and Applied Mathematics, Philadelphia.

A. B. Novikoff (1962). On convergence proofs on perceptrons. In *Symposium on the Mathematical Theory of Automata, 12*, pages 615–622.

P. van den Driessche and O. H. Odeh and D. D. Olesky (1998). Bruhat decomposition and numerical stability. *SIAM J. on Matrix Analysis and Applications*, 19(1):89–98.

C. H. Papadimitriou (1994). *Computational Complexity*. Addison-Wesley.

C. H. Papadimitriou and K. Steiglitz (1982). *Combinatorial optimization: algorithms and complexity*. Prentice-Hall.

K. Price, R. Storn, and J. Lampinen (2005). *Differential Evolution - A Practical Approach to Global Optimization*. Springer.

R. Ravi and M. X. Goemans (1996). The constrained minimum spanning tree problem (extended abstract). In *Proceedings of the Fifth Scandinavian Workshop on Algorithm Theory*, pages 66–75. Springer-Verlag.

J. Renegar (1994). Some perturbation theory for linear programming. *Math. Programming*, 65(1, Ser. A):73–91.

J. Renegar (1995a). Incorporating condition measures into the complexity theory of linear programming. *SIAM J. Optim.*, 5(3):506–524.

J. Renegar (1995b). Linear programming, complexity theory and elementary functional analysis. *Math. Programming*, 70(3, Ser. A):279–351.

H. Röglin, B. Vöcking (2005). Smoothed analysis of integer programming. In *Michael Junger and Volker Kaibel, editors, Proceedings of the of the*

Eleventh International Conference on Integer Programming and Combinatorial Optimization, volume 3509 of Lecture Notes in Computer Science, Springer, pages 276 – 290.

F. Rosenblatt (1962). *Principles of neurodynamics; perceptrons and the theory of brain mechanisms.* Spartan Books.

R. Rubinfeld and M. Sudan (1996). Robust characterizations of polynomials with applications to program testing. *SIAM J. on Computing,* 25(2):252–271, April.

J. Ruppert(1993). A new and simple algorithm for quality 2-dimensional mesh generation. In *Proceedings of the Fourth ACM-SIAM Symp. on Disc. Algorithms,* pages 83–92.

A. Sankar, D. A. Spielman, and S.-H Teng (2005). Smoothed analysis of the condition numbers and growth factors of matrices. *SIAM J. on Matrix Analysis and Applications,* to appear.

M. Santha and U. V Vazirani (1986). Generating quasi-random sequences from semi-random sources. *J. Comput. Syst. Sci.,* 33(1):75–87.

R. Savani and B. von Stengel (2004). Exponentially many steps for finding a nash equilibrium in a bimatrix game. In *Proceedings of the Forty-fifth Annual IEEE Symposium on Foundations of Computer Science,* pages 258–267.

F. Schäfer and N. Sivadasan (2004). Topology matters: Smoothed competitiveness of metrical task systems. In *Proceedings of the Twenty-first Annual Symposium on Theoretical Aspects of Computer Science, (Montpellier, France, March 25-27, 2004),* volume 2996 of *LNCS,* pages 489–500. Springer-Verlag.

A. Schrijver (1986). *Theory of Linear and Integer Programming.* Wiley, 1986.

S. Sen, S. Chatterjee, and N. Dumir (2002). Towards a theory of cache-efficient algorithms. *J. ACM,* 49(6):828–858.

M. Serna (1991). Approximating linear programming is log-space complete for p. *Inf. Process. Lett.,* 37(4):233–236.

J. R. Shewchuk (1998). Tetrahedral mesh generation by Delaunay refinement. In *Proceedings of the Fourteenth Annual ACM Symposium on Computational Geometry,* pages 86–95.

M. Sipser (1996). *Introduction to the Theory of Computation.* International Thomson Publishing.

D. D. Sleator and R. E. Tarjan (1985). Amortized efficiency of list update and paging rules. *Commun. ACM,* 28(2):202–208.

S. Smale (1982). The problem of the average speed of the simplex method. In *Proceedings of the Eleventh International Symposium on Mathematical Programming,* pages 530–539, August.

S. Smale (1983). On the average number of steps in the simplex method of linear programming. *Mathematical Programming,* 27:241–262.

R. Solovay and V. Strassen (1977). A fast Monte-Carlo test for primality. *SIAM J. Comput.* 6(1):84–85.

D. A. Spielman and S.-H. Teng (2003a). Smoothed analysis (motivation and discrete models. In *Algorithms and Data Structures, 8th International Workshop,* pages 256–270.

D. A. Spielman and S.-H. Teng (2003b). Smoothed analysis of termination of linear programming algorithms. *Mathematical Programming, Series B,* 97:375–404.

D. A. Spielman and S.-H. Teng (2004). Smoothed analysis of algorithms: Why the simplex algorithm usually takes polynomial time. *J. ACM*, 51(3):385–463.

D. A. Spielman, S.-H. Teng, and A. Üngör (2002). Parallel Delaunay refinement: Algorithms and analyses. In *Proceedings of the Eleventh International Meshing Roundtable, International Journal of Computational Geometry & Applications (to appear)*, pages 205–217.

D. A. Spielman, S.-H. Teng, and A. Üngör (2004). Time complexity of practical parallel steiner point insertion algorithms. In *Proceedings of the Sixteenth Annual ACM Symposium on Parallelism in Algorithms and Architectures*, pages 267–268.

G. Strang (1980). *Linear Algebra and its Application, 2nd. Ed.* Academic Press.

B. Sudakov and J. Vondrak (2005). How many random edges make a dense hypergraph non-2-colorable?. http://www.math.princeton.edu/~bsudakov/smoothed-analysis-hyper.pdf.

S.-H. Teng and C. W. Wong (2000). Unstructured mesh generation: Theory, practice, and perspectives. *Int. J. Computational Geometry & Applications*, 10(3):227.

S.-H. Teng (1998). Coarsening, sampling, and smoothing: Elements of the multilevel method. In R. S. Schreiber M. Heath, A Ranade, editor, *Algorithms for Parallel Processing, volume 105*, pages 247 – 276. volume 105 of IMA Volumes in Mathematics and its Applications, Springer.

M. J. Todd (1986). Polynomial expected behavior of a pivoting algorithm for linear complementarity and linear programming problems. *Mathematical Programming*, 35:173–192.

M. J. Todd (1991). Probabilistic models for linear programming. *Mathematics of Operations Research*, 16(4):671–693.

M. J. Todd and Y. Ye (1996). A lower bound on the number of iterations of long-step and polynomial interior-point methods for linear programming. *Annals of Operations Research*, 62:233–252.

M. Todd (2001). Semidefinite optimization. *Acta Numerica*, 10:515–560.

M. J. Todd (1994). A lower bound on the number of iterations of primal-dual interior-point methods for linear programming. In G. A. Watson and D. F. Griffiths, editors, *Numerical Analysis 1993*, pages 237 – 259. Longman Press, Harlow.

L. N. Trefethen and D. Bau (1997). *Numerical Linear Algebra*. SIAM, Philadelphia, PA.

L. N. Trefethen and R. S. Schreiber (1990). Average-case stability of Gaussian elimination. *SIAM J. on Matrix Analysis and Applications*, 11(3):335–360.

L. Trevisan and F. Xhafa (1998). The parallel complexity of positive linear programming. *Parallel Processing Letters*, 8(4):527–533.

P. M. Vaidya (1987). An algorithm for linear programming which requires $O(((m + n)n^2 + (m + n)^{1.5}n)L)$ arithmetic operations. In *Proceedings of the Nineteenth Annual ACM Symposium on Theory of Computing*, pages 29–38.

V. V. Vazirani (2001). *Approximation algorithms*. Springer-Verlag.

J. Vera (1996). Ill-posedness and the complexity of deciding existence of solutions to linear programs. *SIAM J. on Optimization*, 6(3).

J. H. Wilkinson (1961). Error analysis of direct methods of matrix inversion. *J. ACM.*, 8:261–330.

J. H. Wilkinson (1963). *Rounding Errors in Algebraic Processes.*

J. H. Wilkinson (1988). *The algebraic eigenvalue problem.* Oxford University Press.

S. J. Wright (1993). A collection of problems for which gaussian elimination with partial pivoting is unstable. *SIAM J. Sci. Comput.*, 14(1):231–238.

S. J. Wright (1997). *Primal-dual interior-point methods.* Society for Industrial and Applied Mathematics, Philadelphia.

M. Wschebor (2004). Smoothed analysis of $\kappa(\mathbf{a})$. *J. of Complexity*, 20(1):97–107, February.

Y. Ye (1997). *Interior point algorithms: theory and analysis.* John Wiley & Sons.

M. Ziegelmann (2001). *Constrained Shortest Paths and Related Problems.* PhD thesis, Universität des Saarlandes, July.

10

Finite Element Approximation of High-dimensional Transport-dominated Diffusion Problems

Endre Süli

Oxford University Computing Laboratory
University of Oxford
Oxford, United Kingdom
e-mail: Endre.Suli@comlab.ox.ac.uk

Abstract

High-dimensional partial differential equations with nonnegative characteristic form arise in numerous mathematical models in science. In problems of this kind, the computational challenge of beating the exponential growth of complexity as a function of dimension is exacerbated by the fact that the problem may be transport-dominated. We develop the analysis of stabilised sparse finite element methods for such high-dimensional, non-self-adjoint, possibly degenerate differential equations.

10.1 Introduction

Suppose that $\Omega := (0, 1)^d$, $d \geq 2$, and that $a = (a_{ij})_{i,j=1}^d$ is a symmetric positive semidefinite matrix with entries $a_{ij} \in \mathbb{R}$, $i, j = 1, \ldots, d$. In other words,

$$a^\top = a \qquad \text{and} \qquad \xi^\top a \xi \geq 0 \qquad \forall \xi \in \mathbb{R}^d.$$

Suppose further that $b \in \mathbb{R}^d$ and $c \in \mathbb{R}$, and let $f \in \mathrm{L}^2(\Omega)$. We shall consider the partial differential equation

$$-a : \nabla\nabla u + b \cdot \nabla u + cu = f(x), \qquad x \in \Omega, \tag{10.1}$$

subject to suitable boundary conditions on $\partial\Omega$ which will be stated below. Here $\nabla\nabla u$ is the $d \times d$ Hessian matrix of u whose (i, j) entry is $\partial^2 u/\partial x_i \, \partial x_j$, $i, j = 1, \ldots, d$. Given two $d \times d$ matrices A and B, we define their scalar product $A : B := \sum_{i,j=1}^d A_{ij} B_{ij}$. The associated matrix norm $|A| := (A : A)^{1/2}$ is called the Frobenius norm of A.

The real-valued polynomial $\alpha \in \mathcal{P}^2(\mathbb{R}^d; \mathbb{R})$ of degree ≤ 2 defined by

$$\xi \in \mathbb{R}^d \;\mapsto\; \alpha(\xi) = \xi^\top a\,\xi \in \mathbb{R}$$

is called the *characteristic polynomial* or *characteristic form* of the differential operator

$$u \;\mapsto\; \mathcal{L}u := -a : \nabla\nabla u + b \cdot \nabla u + cu$$

featuring in (10.1) and, under our hypotheses on the matrix a, the equation (10.1) is referred to as a *partial differential equation with nonnegative characteristic form* (cf. Oleĭnik & Radkevič (1973)).

For the sake of simplicity of presentation we shall confine ourselves to differential operators $\mathcal{L}$ with constant coefficients. In this case,

$$a : \nabla\nabla u = \nabla \cdot (a\nabla u) = \nabla\nabla : (au) \qquad \text{and} \qquad b \cdot \nabla u = \nabla \cdot (bu).$$

With additional technical difficulties most of our results can be extended to the case of variable coefficients, where $a = a(x)$, $b = b(x)$ and $c = c(x)$ for $x \in \Omega$.

Partial differential equations with nonnegative characteristic form frequently arise as mathematical models in physics and chemistry (particularly in the kinetic theory of polymers and coagulation-fragmentation problems), molecular biology and mathematical finance (cf. van Kampen (1992), Öttinger (1996), Laurençot & Mischler (2002), Elf, Lötstedt & Sjöberg (2003)). Important special cases of these equations include the following:

(a) when the diffusion matrix $a = a^\top$ is positive definite, (10.1) is an elliptic partial differential equation;

(b) when $a \equiv 0$ and the transport direction $b \neq 0$, the partial differential equation (10.1) is a first-order hyperbolic equation;

(c) when

$$a = \begin{pmatrix} \alpha & 0 \\ 0 & 0 \end{pmatrix},$$

where α is a $(d-1) \times (d-1)$ symmetric positive definite matrix and $b = (0, \ldots, 0, 1)^\top \in \mathbb{R}^d$, (10.1) is a parabolic partial differential equation, with time-like direction b.

In addition to these classical types, the family of partial differential equations with nonnegative characteristic form encompasses a range of other linear second-order partial differential equations, such as degenerate elliptic equations and ultra-parabolic equations. According to a result of Hörmander (2005) (cf. Theorem 11.1.10 on p. 67), second-order hypoelliptic operators have non-negative characteristic form, after possible multiplication by -1, so they too fall within this category.

For classical types of partial differential equations, such as those listed under (a), (b) and (c) above, rich families of reliable, stable and highly accurate numerical techniques have been developed. Yet, only isolated attempts have been made to explore computational aspects of the class of partial differential equations with nonnegative characteristic form as a whole (cf. Houston & Süli (2001) and Houston, Schwab & Süli (2002)). In particular, there has been no research to date on the numerical analysis of high-dimensional partial differential equations with nonnegative characteristic form.

The field of stochastic analysis is a particularly fertile source of equations of this kind (cf. Bass (1997)): the progressive Kolmogorov equation satisfied by the probability density function $\psi(x_1, \ldots, x_d, t)$ of a d-component vectorial stochastic process $X(t) = (X_1(t), \ldots, X_d(t))^\top$ which is the solution of a system of stochastic differential equations including Brownian noise is a partial differential equation with nonnegative characteristic form in the $d + 1$ variables $(x, t) = (x_1, \ldots, x_d, t)$. To be more precise, consider the stochastic differential equation:

$$\mathrm{d}X(t) = b(X(t))\,\mathrm{d}t + \sigma(X(t))\,\mathrm{d}W(t), \qquad X(0) = X,$$

where $W = (W_1, \ldots, W_p)^\top$ is a p-dimensional Wiener process adapted to a filtration $\{\mathcal{F}_t,\, t \geq 0\}$, $b \in \mathrm{C}_b^1(\mathbb{R}^d; \mathbb{R}^d)$ is the drift vector, and $\sigma \in \mathrm{C}_b^2(\mathbb{R}^d, \mathbb{R}^{d \times p})$ is the diffusion matrix. Here $\mathrm{C}_b^k(\mathbb{R}^n, \mathbb{R}^m)$ denotes the space of bounded and continuous mappings from $\mathbb{R}^n$ into $\mathbb{R}^m$, $m, n \geq 1$, all of whose partial derivatives of order k or less are bounded and continuous on $\mathbb{R}^n$. When the subscript b is absent, boundedness is not enforced.

Assuming that the random variable $X(t) = (X_1(t), \ldots, X_d(t))^\top$ has a probability density function $\psi \in \mathrm{C}^{2,1}(\mathbb{R}^d \times [0, \infty), \mathbb{R})$, then ψ is the solution of the initial-value problem

$$\frac{\partial \psi}{\partial t}(x, t) = (A\psi)(x, t), \qquad x \in \mathbb{R}^d,\, t > 0,$$
$$\psi(x, 0) = \psi_0(x), \qquad x \in \mathbb{R}^d,$$

where the differential operator $A : \mathrm{C}^2(\mathbb{R}^d; \mathbb{R}) \to \mathrm{C}^0(\mathbb{R}^d; \mathbb{R})$ is defined by

$$A\psi := -\sum_{j=1}^d \frac{\partial}{\partial x_j}\left(b_j(x)\psi\right) + \frac{1}{2}\sum_{i,j=1}^d \frac{\partial^2}{\partial x_i \partial x_j}\left(a_{ij}(x)\psi\right),$$

with $a(x) = \sigma(x)\sigma^\top(x) \geq 0$ (see Corollary 5.2.10 on p.135 in Lapeyre, Pardoux & Sentis (2003)). Thus, ψ is the solution of the initial-value problem

$$\frac{\partial \psi}{\partial t} + \sum_{j=1}^{d} \frac{\partial}{\partial x_j}(b_j(x)\psi) = \frac{1}{2}\sum_{i,j=1}^{d} \frac{\partial^2}{\partial x_i \partial x_j}(a_{ij}(x)\psi), \qquad x \in \mathbb{R}^d,\, t \geq 0,$$

$$\psi(x,0) = \psi_0(x), \qquad\qquad\qquad x \in \mathbb{R}^d,$$

where, for each $x \in \mathbb{R}^d$, $a(x)$ is a $d \times d$ symmetric positive semidefinite matrix. The progressive Kolmogorov equation $\frac{\partial \psi}{\partial t} = A\psi$ is a partial differential equation with nonnegative characteristic form, called a Fokker–Planck equation[1].

The operator A is generally nonsymmetric (since, typically, $b \neq 0$) and degenerate (since, in general, $a(x) = \sigma(x)\sigma^\top(x)$ has nontrivial kernel). In addition, since the (possibly large) number d of equations in the system of stochastic differential equations is equal to the number of components of the independent variable x of the probability density function ψ, the Fokker–Planck equation may be high-dimensional.

The focus of the present paper is the construction and the analysis of finite element approximations to *high-dimensional* partial differential equations with non-negative characteristic form. The paper is structured as follows. In order to provide a physical motivation for the mathematical questions considered here, we begin by presenting an example of a high-dimensional transport-dominated diffusion problem which arises from the kinetic theory of dilute polymers. We shall also explain briefly why such high-dimensional transport-dominated diffusion problems present a computational challenge. We shall then state in Section 11.3 the appropriate boundary conditions for the model equation (10.1), derive the weak formulation of the resulting boundary value problem and show the existence of a unique weak solution. Section 11.4 is devoted to the construction of a hierarchical finite element space for univariate functions. The tensorisation of this space and the subsequent sparsification of the resulting tensor-product space are described in Section 11.5; our chief objective is to reduce the computational complexity of the discretisation without adversely effecting the approximation properties of the finite element space. In Sections 10.6 and 10.7 we build a stabilised finite element method over the sparse tensor product space, and we explore its stability and convergence.

The origins of sparse tensor product constructions and hyperbolic cross spaces can be traced back to the works of Babenko (1960) and Smolyak (1963); we refer to the papers of Temlyakov (1989), DeVore, Konyagin & Temlyakov (1998) for the study of high-dimensional approximation problems, to the works of Wasilkowski & Woźniakowski (1995) and Novak &

---

[1] After the physicists Adriaan Daniël Fokker (1887–1972) and Max Planck (1858–1947).

Ritter (1998) for high-dimensional integration problems and associated complexity questions, to the paper of Zenger (1991) for an early contribution to the numerical solution of high-dimensional partial differential equations, to the articles by von Petersdorff & Schwab (2004) and Hoang & Schwab (2005) for the analysis of sparse-grid methods for high-dimensional elliptic multiscale problems and parabolic equations, respectively, and to the recent Acta Numerica article of Bungartz & Griebel (2004) for a detailed survey of the field of sparse-grid methods.

10.2 An example from the kinetic theory of polymers

We present an example of a high-dimensional partial differential equation with nonnegative characteristic form which originates from the kinetic theory of dilute polymeric fluids. The fluid is assumed to occupy a domain $O \subset \mathbb{R}^n$; for physical reasons, $n = 2$ or $n = 3$ here.

There is a hierarchy of mathematical models that describe the evolution of the flow of a dilute polymer, the complexity of the model being dependent on the level of model-reduction (coarse-graining) that has taken place. The simplest model of this kind to account for noninteracting polymer chains is the so-called dumbbell model where each polymer chain which is suspended in the viscous incompressible Newtonian solvent whose flow-velocity is $u(x, t)$, $x \in O$, $t \in [0, T]$, is modelled by a dumbbell; a dumbbell consists of two beads connected by an elastic spring. At time $t \in [0, T]$ the dumbbell is characterised by the position of its centre of mass $X(t) \in \mathbb{R}^d$ and its elongation vector $Q(t) \in \mathbb{R}^d$. When a dumbbell is placed into the given velocity field $u(x, t)$, three forces act on each bead: the first force is the drag force proportional to the difference between the bead velocity and the velocity of the surrounding fluid particles; the second force is the elastic force F due to the spring stiffness; the third force is due to thermal agitation and is modelled as Brownian noise. On rescaling the elongation vector, Newton's equations of motion for the beads give rise to the following system of stochastic differential equations:

$$\mathrm{d}X(t) = u(X(t), t)\,\mathrm{d}t, \tag{10.2}$$

$$\mathrm{d}Q(t) = \left(\nabla_X u(X(t), t)\, Q(t) - \frac{1}{2\lambda}\, F(Q(t)) \right) \mathrm{d}t + \frac{1}{\sqrt{\lambda}}\, \mathrm{d}W(t), \tag{10.3}$$

where $W = (W_1, \ldots, W_n)^\top$ is an n-dimensional Wiener process, $F(Q)$ denotes the elastic force acting on the chain due to elongation, and the positive parameter $\lambda = \xi/(4H)$ characterises the elastic property of the fluid, with $\xi \in \mathbb{R}_{>0}$ denoting the drag coefficient and $H \in \mathbb{R}_{>0}$ the spring stiffness.

Let $(x, q, t) \mapsto \psi(x, q, t)$ denote the probability density function of the vector-valued stochastic process $(X(t), Q(t))$; thus, $\psi(x, q, t)|\mathrm{d}x| \, |\mathrm{d}q|$ represents the probability, at time $t \in [0, T]$, of finding the centre of mass of a dumbbell in the volume element $x + \mathrm{d}x$ and having the endpoint of its elongation vector within the volume element $q + \mathrm{d}q$. Let us suppose that the elastic force $F : D \subseteq \mathbb{R}^d \to \mathbb{R}^d$, $d = 2, 3$, of the spring is defined on an open ball $D \subseteq \mathbb{R}^d$ through a (sufficiently smooth) potential $U : \mathbb{R}_{\geq 0} \to \mathbb{R}$ via $F(q) := U'(\frac{1}{2}|q|^2) \, q$. Then, the probability density function $\psi(x, q, t)$ of the stochastic process $(X(t), Q(t))$ defined by (10.2), (10.3) satisfies the Fokker–Planck equation

$$\frac{\partial \psi}{\partial t} + \nabla_x \cdot (u\psi) + \nabla_q \cdot \left((\nabla_x u) q \, \psi - \frac{1}{2\lambda} F(q) \psi \right) = \frac{1}{2\,\lambda} \Delta_q \psi, \qquad (10.4)$$

for $x \in O$, $q \in D$ and $t \in (0, T]$. The equation is supplemented by the initial condition $\psi(x, q, 0) = \psi_0(x, q) \geq 0$ and appropriate boundary conditions. Due to the fact that, unlike (10.3), the differential equation (10.2) does not involve random effects, the Fokker–Planck equation (10.4) for the associated probability density function is a degenerate parabolic equation for $\psi(x, q, t)$ with no diffusion in the x-direction.

In order to complete the definition of the dumbbell model, we note that the velocity field u appearing in (10.4) and the pressure p of the solvent are, in turn, found from the incompressible Navier–Stokes equations

$$\frac{\partial u}{\partial t} + (u \cdot \nabla_x) u - \nu \, \Delta_x u + \nabla_x p = \nabla_x \cdot \tau, \qquad \text{in } O \times (0, T],$$
$$\nabla_x \cdot u = 0, \qquad \text{in } O \times (0, T],$$
$$u = 0, \qquad \text{on } \partial O \times (0, T],$$
$$u(x, 0) = u_0(x), \qquad x \in O,$$

where the elastic extra-stress tensor $\tau = \tau(\psi)$ is defined in terms of the probability density function ψ as follows:

$$\tau(\psi) := k \, \mu \, (C(\psi) - \rho(\psi) \, I).$$

Here k, $\mu \in \mathbb{R}_{>0}$ are, respectively, the Boltzmann constant and the absolute temperature, I is the unit $n \times n$ tensor, and

$$C(\psi)(x, t) := \int_D \psi(x, q, t) \, U'(\tfrac{1}{2}|q|^2) \, q \, q^\top \, \mathrm{d}q,$$
$$\rho(\psi)(x, t) := \int_D \psi(x, q, t) \, \mathrm{d}q.$$

We refer to the recent paper of Barrett, Schwab & Süli (2005) for theoretical results concerning the existence of a global weak solution to this coupled

Fokker–Planck–Navier–Stokes problem; see also the work of Le Bris & Lions (2004) on related transport(-diffusion) problems with nonsmooth transport fields.

The Fokker–Planck equation (10.4) is a partial differential equation with nonnegative characteristic form in $2n + 1$ independent variables $x \in O \subset \mathbb{R}^n$, $q \in D \subset \mathbb{R}^n$ and $t \in (0, T] \subset \mathbb{R}_{>0}$. In order to provide a rough estimate of the computational complexity of a classical algorithm for the numerical solution of the equation (10.4) supplemented with an initial condition and suitable boundary conditions, let us suppose that the spatial domain is $O \times D = (-1/2, 1/2)^{2n}$ and[1] that a standard continuous piecewise linear Galerkin finite element method is used on each time level over a uniform axiparallel spatial mesh. Let us further suppose that the mesh has the relatively coarse spacing $h = 1/64$ in each of the $2n$ spatial co-ordinate directions and that a simple one-step method (such as the forward or backward Euler scheme, or the Crank–Nicolson scheme) is used to evolve the discrete solution in time. Ignoring degrees of freedom that lie on the boundary of $O \times D$, we see that the resulting system of linear equations involves around $63^4 = 15752962 \approx 1.5 \times 10^7$ unknowns on each time level when $n = 2$ (i.e., $2n = 4$) and around $63^6 = 62523502209 \approx 6.2 \times 10^{10}$ unknowns on each time level when $n = 3$ (i.e., $2n = 6$). Even on such coarse meshes the number of degrees of freedom in the numerical approximation to the analytical solution in 4 and 6 dimensions is very large, and grows very rapidly (exponentially fast, in fact,) as a function of $d = 2n + 1$, the number of independent variables. In general, on a uniform mesh of size $h = 1/N$ in each of the $2n$ spatial co-ordinate directions, the number of unknowns per time level (counting only those that are internal to $O \times D$) is $(N-1)^{2n}$. Over a unit time interval, and using the Crank–Nicolson scheme with time step $k = h$, this amounts to a total of approximately $N(N-1)^{2n} = \mathcal{O}(N^d)$ unknowns.

In addition to being high-dimensional, the equation (10.4) exhibits the features of a first-order hyperbolic equation with respect to $x \in \mathbb{R}^n$ (when variation with respect to q is suppressed), and those of a second-order parabolic transport-diffusion equation with respect to $q \in \mathbb{R}^n$ (when variation with respect to x is suppressed).

Our objective in this paper is to explore the algorithmic implications of this unpleasant combination of high-dimensionality and transport-dominated diffusion. In particular, our aim is to develop purely deterministic

---

[1] Here, for simplicity, we took $D = (-1/2, 1/2)^n$, — a ball in $\mathbb{R}^n$ of radius $1/2$ in the ℓ^∞-norm. In fact, D is a ball in $\mathbb{R}^n$ in the ℓ^2-norm of a certain fixed radius $q_{\max} \leq \infty$, the maximum admissible length of the elongation vector Q.

numerical algorithms based on the Galerkin method for high-dimensional transport-dominated diffusion problems of the form (10.1).

Alternative, stochastic, or mixed deterministic-stochastic computational approaches which have been pursued in the literature employ the intimate connection between the Fokker–Planck equation satisfied by the probability density function and the system of stochastic differential equations which govern the evolution of the underlying stochastic process (see, for example, the monograph of Öttinger (1996) and the survey paper by Jourdain, Le Bris & Lelièvre (2004)).

10.2.1 The curse of dominant transport

Classical Galerkin methods comprise a class of stable, reliable and accurate techniques for the numerical approximation of diffusion-dominated problems typified by symmetric elliptic equations (viz. equation (10.1) in the special case when a is a symmetric positive definite matrix and $b = 0$). In this case, a Galerkin method for the numerical solution of the equation (10.1), supplemented with a suitable boundary condition, coincides with the Ritz method based on energy minimisation over a finite-dimensional subspace of the infinite-dimensional Hilbert space $\mathcal{H}$ containing the weak solution u to the boundary value problem. The energy-norm is simply the norm induced by the symmetric and coercive bilinear form associated with the weak formulation of the problem, which acts as an inner product on $\mathcal{H}$. The Galerkin approximation to u is then the best approximation to u in the energy norm from the finite-dimensional subspace. If, on the other hand, $b \neq 0$, then a Galerkin method for the numerical solution of an elliptic equation of the form (10.1) cannot be rephrased in the language of energy minimisation over a finite-dimensional space; nevertheless, it will supply an accurate approximation to u, as long as a 'dominates' b in a certain sense.

In a Galerkin finite element method the finite-dimensional subspace from which the approximate solution u_h is sought consists of continuous piecewise polynomial functions of a fixed degree p which are defined over a partition of a certain fixed 'granularity' $h > 0$ of the computational domain $\Omega \subset \mathbb{R}^d$. Suppose, for example, that $d = 1$, $\Omega = (0, 1)$, $p = 1$, $a \in \mathbb{R}_{>0}$, $b \in \mathbb{R}$, $c = 0$, $f \in C[0,1]$, $f \geq 0$ and $h = 1/N$, where $N \in \mathbb{N}_{>1}$; let us also suppose for the sake of simplicity that homogeneous Dirichlet boundary conditions are imposed on $\partial\Omega = \{0, 1\}$. As long as $a \geq \frac{1}{2}h|b|$ (i.e., provided that the transport-diffusion problem is diffusion-dominated relative to the finite element partition), the qualitative behaviour of u_h will be correct, in the sense that u_h will obey a maximum principle analogous to the one satisfied by the analytical solution u.

This favourable behaviour of the approximate solution u_h is completely lost in the transport-dominated regime, when $a < \frac{1}{2}h|b|$; for such h, u_h exhibits maximum-principle-violating oscillations on the scale of the mesh. The oscillations will be particularly prominent in the boundary layer located in the vicinity of one of the endpoints of the interval $[0, 1]$, i.e., at $x = 0$ when $b < 0$ and $x = 1$ when $b > 0$.

An analogous situation is observed in the multidimensional case. Suppose, for example, that $\Omega = (0, 1)^d$ with $d > 1$, $p = 1$ (i.e., continuous piecewise linear polynomials in d variables are used on a simplicial partition of $\overline{\Omega}$), $a = a^\top \in \mathbb{R}^{d \times d}$ is a positive definite matrix, $b \in \mathbb{R}^d$, $c = 0$, $f \in C(\overline{\Omega})$ and $h \in \mathbb{R}_{>0}$ is a mesh-parameter measuring the granularity of the finite element mesh; again, we assume that a homogeneous Dirichlet boundary condition is imposed on $\partial\Omega$. When $|a| \ll h|b|$, maximum-principle-violating oscillations will be observed in the vicinity of boundary layers; the oscillations will extend into the interior of the computational domain along subcharacteristic curves (i.e., along the transport direction b). Of course, if the mesh parameter h is sufficiently reduced so that $h|b| \ll |a|$, then the numerical approximation u_h will recover its accuracy and will appear qualitatively correct. Unfortunately the reduction of the mesh-parameter h to this level may place unachievable demands on limited computational resources.

10.2.2 The curse of dimensionality

The computational complexity of a numerical algorithm for the approximate solution of a transport-dominated diffusion equation is particularly unfavourable when the problem is high-dimensional. If, for example, continuous piecewise polynomial finite element basis functions of degree p are used in d dimensions on a mesh of size h and u is sufficiently smooth, in the limit of $h \rightarrow 0$ and $p \rightarrow \infty$ the error $E = \|u - u_h\|_{L^2(\Omega)}$ will exhibit the optimal asymptotic convergence rate: $E \asymp C_p(u) \, (h/(p + 1))^{p+1}$, where $C_p(u) = \text{Const.} |u|_{H^{p+1}(\Omega)}$. Now, when $|b|/|a| \gg 1$, $C_p(u) \asymp \text{Const.} (|b|/|a|)^{p+1/2}$. Hence, for a preset tolerance TOL, the requirement that $E = \text{TOL}$ translates into requiring that

$$\frac{h}{p+1} \asymp \text{Const.} \left((|a|/|b|)^{1-1/(2(p+1))} \text{TOL}^{1/(p+1)} \right).$$

At the same time, the computational complexity of the numerical method will scale as $\text{Const.}((p + 1)/h)^d$. In terms of TOL this then gives

$$\text{Complexity} \asymp \text{Const.} \left((|b|/|a|)^{d(1-1/(2(p+1)))} \text{TOL}^{-d/(p+1)} \right). \tag{10.5}$$

Exponential growth of computational complexity as a function of the dimension of the problem is referred to as the *curse of dimensionality*. It is clear from (10.5) that for a transport-dominated diffusion problem, where $|b|/|a| \gg 1$, the curse of dimensionality may be particularly harmful. The focus of the paper is precisely this unfavourable situation, when the curse of dimensionality is exacerbated by dominant transport.

10.3 Boundary conditions and weak formulation

Before embarking on the construction of the numerical algorithm, we shall introduce the necessary boundary conditions and the weak formulation of the model boundary-value problem on $\Omega = (0,1)^d$ for the equation (10.1).

Let Γ denote the union of all $(d-1)$-dimensional open faces of the domain $\Omega = (0,1)^d$. On recalling that, by hypothesis, $a = a^\top$ and $\alpha(\xi) = \xi^\top a \xi \geq 0$ for all $\xi \in \mathbb{R}^d$, we define the subset Γ_0 of Γ by

$$\Gamma_0 := \{x \in \Gamma : \alpha(\nu(x)) > 0\} \, ;$$

here $\nu(x)$ denotes the unit normal vector to Γ at $x \in \Gamma$, pointing outward with respect to Ω. The set Γ_0 can be thought of as the *elliptic part* of Γ. The complement $\Gamma \backslash \Gamma_0$ of Γ_0 is referred to as the *hyperbolic part* of Γ. We note that, by definition, $\alpha = 0$ on $\Gamma \setminus \Gamma_0$.

On introducing the *Fichera function*

$$x \in \Gamma \mapsto \beta(x) := b \cdot \nu(x) \in \mathbb{R}$$

defined on Γ, we subdivide $\Gamma \backslash \Gamma_0$ as follows:

$$\Gamma_- := \{x \in \Gamma \backslash \Gamma_0 : \beta(x) < 0\}, \qquad \Gamma_+ := \{x \in \Gamma \backslash \Gamma_0 : \beta(x) \geq 0\} \, ;$$

the sets Γ_- and Γ_+ are referred to as the (hyperbolic) *inflow* and *outflow* boundary, respectively. Thereby, we obtain the following decomposition of Γ:

$$\Gamma = \Gamma_0 \cup \Gamma_- \cup \Gamma_+.$$

Lemma 10.1 *Each of the sets Γ_0, Γ_-, Γ_+ is a union of $(d-1)$-dimensional open faces of Ω. Moreover, each pair of mutually opposite $(d-1)$-dimensional open faces of Ω is contained either in the elliptic part Γ_0 of Γ or in its complement $\Gamma \setminus \Gamma_0 = \Gamma_- \cup \Gamma_+$, the hyperbolic part of Γ.*

Proof Since a is a constant matrix and ν is a face-wise constant vector, Γ_0 is a union of (disjoint) $(d-1)$-dimensional open faces of Γ. Indeed, if $x \in \Gamma_0$ and y is any point that lies on the same $(d-1)$-dimensional open face of Ω as x, then $\nu(y) = \nu(x)$ and therefore $\alpha(\nu(y)) = \alpha(\nu(x)) > 0$; hence $y \in \Gamma_0$ also.

A certain $(d-1)$-dimensional open face φ of Ω is contained in Γ_0 if, and only if, the opposite face $\hat{\varphi}$ is also contained in Γ_0. To prove this, let $\varphi \subset \Gamma_0$ and let $x = (x_1, \ldots, x_i, \ldots, x_d) \in \varphi$, with Ox_i signifying the (unique) co-ordinate direction such that $\nu(x) \| Ox_i$; here $O = (0, \ldots, 0)$. In other words, $x_i \in \{0, 1\}$, and the $(d-1)$-dimensional face φ to which x belongs is orthogonal to the co-ordinate direction Ox_i. Hence, the point $\hat{x} = (x_1, \ldots, |x_i - 1|, \ldots, x_d)$ lies on the $(d-1)$-dimensional open face $\hat{\varphi}$ of Ω that is opposite the face φ (i.e., $\hat{\varphi} \| \varphi$), and $\nu(\hat{x}) = -\nu(x)$. As α is a homogeneous function of degree 2 on Γ_0, it follows that

$$\alpha(\nu(\hat{x})) = \alpha(-\nu(x)) = (-1)^2 \alpha(\nu(x)) = \alpha(\nu(x)) > 0,$$

which implies that $\hat{x} \in \Gamma_0$. By what we have shown before, we deduce that the entire face $\hat{\varphi}$ is contained in Γ_0.

Similarly, since b is a constant vector, each of Γ_- and Γ_+ is a union of $(d-1)$-dimensional open faces of Γ. If a certain $(d-1)$-dimensional open face φ is contained in Γ_-, then the opposite face $\hat{\varphi}$ is contained in the set Γ_+.

We note in passing, however, that if $\varphi \subset \Gamma_+$ then the opposite face $\hat{\varphi}$ need not be contained in Γ_-; indeed, if $\varphi \subset \Gamma_+$ and $\beta = 0$ on φ then $\beta = 0$ on $\hat{\varphi}$ also, so then both φ and the opposite face $\hat{\varphi}$ are contained in Γ_+. Of course, if $\beta > 0$ on $\varphi \subset \Gamma_+$, then $\beta < 0$ on the opposite face $\hat{\varphi}$, and then $\hat{\varphi} \subset \Gamma_-$.   □

We consider the following boundary–value problem: find u such that

$$\mathcal{L}u \equiv -a : \nabla\nabla u + b \cdot \nabla u + cu = f \quad \text{in } \Omega, \tag{10.6}$$

$$u = 0 \quad \text{on } \Gamma_0 \cup \Gamma_-. \tag{10.7}$$

Before stating the variational formulation of (10.6), (10.7), we note the following simple result (see, e.g., Houston and Süli (2001) for a proof).

Lemma 10.2 *Suppose that $M \in \mathbb{R}^{d \times d}$ is a $d \times d$ symmetric positive semidefinite matrix. If $\xi \in \mathbb{R}^d$ satisfies $\xi^\top M \xi = 0$, then $M\xi = 0$.*

Since $a \in \mathbb{R}^{d \times d}$ is a symmetric positive semidefinite matrix and $\nu^\top a \nu = 0$ on $\Gamma \setminus \Gamma_0$, we deduce from Lemma 10.2 with $M = a$ and $\xi = \nu$ that

$$a\nu = 0 \quad \text{on } \Gamma \setminus \Gamma_0. \tag{10.8}$$

Let us suppose for a moment that (10.6), (10.7) has a solution u in $\mathrm{H}^2(\Omega)$. Thanks to our assumption that a is a constant matrix, we have that

$$a : \nabla\nabla u = \nabla \cdot (a\nabla u).$$

Furthermore, $a\nabla u \in [\mathrm{H}^1(\Omega)]^d$, which implies that the normal trace $\gamma_{\nu,\partial\Omega}(a\nabla u)$ of $a\nabla u$ on $\partial\Omega$ belongs to $\mathrm{H}^{1/2}(\partial\Omega)$. By virtue of (10.8),

$$\gamma_{\nu,\partial\Omega}(a\nabla u)|_{\Gamma\setminus\Gamma_0} = 0.$$

Note also that $\mathrm{meas}_{d-1}(\partial\Omega\setminus\Gamma) = 0$. Hence

$$\int_{\partial\Omega}\gamma_{\nu,\partial\Omega}(a\nabla u)\cdot\gamma_{0,\partial\Omega}(v)\mathrm{d}s = \int_\Gamma \gamma_{\nu,\partial\Omega}(a\nabla u)|_\Gamma\cdot\gamma_{0,\partial\Omega}(v)|_\Gamma\mathrm{d}s = 0 \quad (10.9)$$

for all $v \in \mathcal{V}$, where

$$\mathcal{V} = \left\{v \in \mathrm{H}^1(\Omega) : \gamma_{0,\partial\Omega}(v)|_{\Gamma_0} = 0\right\}.$$

This observation will be of key importance. On multiplying the partial differential equation (10.6) by $v \in \mathcal{V}$ and integrating by parts, we find that

$$(a\nabla u, \nabla v) - (u, \nabla\cdot(bv)) + (cu, v) + \langle u, v\rangle_+ = (f, v) \qquad \text{for all } v \in \mathcal{V},$$
$$(10.10)$$

where $(\cdot, \cdot)$ denotes the L^2 inner–product over Ω and

$$\langle w, v\rangle_\pm = \int_{\Gamma_\pm}|\beta|wv\,\mathrm{d}s,$$

with β signifying the Fichera function $b\cdot\nu$, as before. We note that in the transition to (10.10) the boundary integral term on Γ which arises in the course of partial integration from the $-\nabla\cdot(a\nabla u)$ term vanishes by virtue of (10.9), while the boundary integral term on $\Gamma\setminus\Gamma_+ = \Gamma_0\cup\Gamma_-$ resulting from the $b\cdot\nabla u$ term on partial integration disappears since $u = 0$ on this set by (10.7).

The form of (10.10) serves as motivation for the statement of the weak formulation of (10.6), (10.7) which is presented below. We consider the inner product $(\cdot, \cdot)_\mathcal{H}$ defined by

$$(w, v)_\mathcal{H} := (a\nabla w, \nabla v) + (w, v) + \langle w, v\rangle_{\Gamma_-\cup\Gamma_+}$$

and denote by $\mathcal{H}$ the closure of the space $\mathcal{V}$ in the norm $\|\cdot\|_\mathcal{H}$ defined by

$$\|w\|_\mathcal{H} := (w, w)_\mathcal{H}^{1/2}.$$

Clearly, $\mathcal{H}$ is a Hilbert space. For $w \in \mathcal{H}$ and $v \in \mathcal{V}$, we now consider the bilinear form $B(\cdot, \cdot) : \mathcal{H}\times\mathcal{V}\to\mathbb{R}$ defined by

$$B(w, v) := (a\nabla w, \nabla v) - (w, \nabla\cdot(bv)) + (cw, v) + \langle w, v\rangle_+,$$

and for $v \in \mathcal{V}$ we introduce the linear functional $L : \mathcal{V}\to\mathbb{R}$ by

$$L(v) := (f, v).$$

We shall say that $u \in \mathcal{H}$ is a *weak solution* to the boundary–value problem (10.6), (10.7) if

$$B(u, v) = L(v) \qquad \forall v \in \mathcal{V}. \tag{10.11}$$

The existence of a unique weak solution is guaranteed by the following theorem (cf. also Theorem 1.4.1 on p.29 of Oleĭnik & Radkevič (1973) and Houston & Süli (2001)).

Theorem 10.1 *Suppose that $c \in \mathbb{R}_{>0}$. For each $f \in \mathrm{L}^2(\Omega)$, there exists a unique u in a Hilbert subspace $\hat{\mathcal{H}}$ of $\mathcal{H}$ such that (10.11) holds.*

Proof For $v \in \mathcal{V}$ fixed, we deduce by means of the Cauchy-Schwarz inequality that

$$B(w, v) \le K_1 \|w\|_{\mathcal{H}} \|v\|_{\mathrm{H}^1(\Omega)} \qquad \forall w \in \mathcal{H},$$

where we have used the trace theorem for $\mathrm{H}^1(\Omega)$. Thus $B(\cdot, v)$ is a continuous linear functional on the Hilbert space $\mathcal{H}$. By the Riesz representation theorem, there exists a unique element $T(v)$ in $\mathcal{H}$ such that

$$B(w, v) = (w, T(v))_{\mathcal{H}} \qquad \forall w \in \mathcal{H}.$$

Since B is bilinear, it follows that $T : v \to T(v)$ is a linear operator from $\mathcal{V}$ into $\mathcal{H}$. Next we show that T is injective. Note that

$$B(v, v) = (a\nabla v, \nabla v) - (v, \nabla \cdot (bv)) + (cv, v) + \langle v, v \rangle_+ \qquad \forall v \in \mathcal{V}.$$

Upon integrating by parts in the second term on the right–hand side we deduce that

$$B(v, v) = (a\nabla v, \nabla v) + c\|v\|^2 + \tfrac{1}{2}\langle v, v \rangle_{\Gamma_- \cup \Gamma_+} \ge K_0 \|v\|_{\mathcal{H}}^2 \qquad \forall v \in \mathcal{V},$$

where $K_0 = \min(c, \tfrac{1}{2}) > 0$ and $\| \cdot \| = \| \cdot \|_{\mathrm{L}^2(\Omega)}$. Hence

$$(v, T(v))_{\mathcal{H}} \ge K_0 \|v\|_{\mathcal{H}}^2 \qquad \forall v \in \mathcal{V}. \tag{10.12}$$

Consequently, $T : v \mapsto T(v)$ is an injection from $\mathcal{V}$ onto the range $\mathcal{R}(T)$ of T contained in $\mathcal{H}$. Thus, $T : \mathcal{V} \to \mathcal{R}(T)$ is a bijection. Let $S = T^{-1} : \mathcal{R}(T) \to \mathcal{V}$, and let $\hat{\mathcal{H}}$ denote the closure of $\mathcal{R}(T)$ in $\mathcal{H}$. Since, by (10.12), $\|S(v)\|_{\mathcal{H}} \le (1/K_0)\|v\|_{\mathcal{H}}$ for all $v \in \mathcal{R}(T)$, it follows that $S : \mathcal{R}(T) \to \mathcal{V}$ is a continuous linear operator; therefore, it can be extended, from the dense subspace $\mathcal{R}(T)$ of $\hat{\mathcal{H}}$ to the whole of $\hat{\mathcal{H}}$, as a continuous linear operator $\hat{S} : \hat{\mathcal{H}} \to \mathcal{H}$. Furthermore, since

$$|L(v)| \le \|f\| \|v\|_{\mathcal{H}} \qquad \forall v \in \mathcal{H},$$

it follows that $L \circ \hat{S} : v \in \hat{\mathcal{H}} \mapsto L(\hat{S}(v)) \in \mathbb{R}$ is a continuous linear functional on $\hat{\mathcal{H}}$. Since $\hat{\mathcal{H}}$ is closed (by definition) in the norm of $\mathcal{H}$, it is a

Hilbert subspace of $\mathcal{H}$. Hence, by the Riesz representation theorem, there exists a unique $u \in \hat{\mathcal{H}}$ such that

$$L(\hat{S}(w)) = (u, w)_{\mathcal{H}} \qquad \forall w \in \hat{\mathcal{H}}.$$

Thus, by the definition of $\hat{S}$, $\hat{S}(w) = S(w)$ for all w in $\mathcal{R}(T)$; hence,

$$L(S(w)) = (u, w)_{\mathcal{H}} \qquad \forall w \in \mathcal{R}(T).$$

Equivalently, on writing $v = S(w)$,

$$(u, T(v))_{\mathcal{H}} = L(v) \qquad \forall v \in \mathcal{V}.$$

Thus we have shown the existence of a unique $u \in \hat{\mathcal{H}}(\subset \mathcal{H})$ such that

$$B(u, v) \equiv (u, Tv)_{\mathcal{H}} = L(v) \qquad \forall v \in \mathcal{V},$$

which completes the proof. $\qquad\qquad\qquad\qquad\qquad\qquad\qquad\qquad\qquad$ □

We note that the boundary condition $u|_{\Gamma_-} = 0$ on the inflow part Γ_- of the hyperbolic boundary $\Gamma \setminus \Gamma_0 = \Gamma_- \cup \Gamma_+$ is imposed weakly, through the definition of the bilinear form $B(\cdot, \cdot)$, while the boundary condition $u|_{\Gamma_0} = 0$ on the elliptic part Γ_0 of Γ is imposed strongly, through the choice of the function space $\mathcal{H}$. Indeed, all elements in $\mathcal{H}$ vanish on Γ_0. Hence, we deduce from Lemma 10.1 that

$$\bigotimes_{i=1}^{d} \mathrm{H}^1_{(0)}(0,1) \equiv \mathrm{H}^1_{(0)}(0,1) \otimes \cdots \otimes \mathrm{H}^1_{(0)}(0,1) \subset \mathcal{H}, \qquad (10.13)$$

where the ith component $\mathrm{H}^1_{(0)}(0,1)$ in the d-fold tensor product on the left-hand side of the inclusion is defined to be equal to $\mathrm{H}^1_0(0,1)$ if the co-ordinate direction Ox_i is orthogonal to a pair of $(d-1)$-dimensional open faces contained in the elliptic part Γ_0 of Γ; otherwise (i.e., when the direction Ox_i is orthogonal to a pair of $(d-1)$-dimensional open faces contained in the hyperbolic part $\Gamma \setminus \Gamma_0 = \Gamma_- \cup \Gamma_+$ of Γ), it is defined to be equal to $\mathrm{H}^1(0,1)$. Clearly, if φ and $\hat{\varphi}$ are a pair of $(d-1)$-dimensional open faces of Ω which are opposite each other (i.e., $\varphi \parallel \hat{\varphi}$), then there exists a unique $i \in \{1, \ldots, d\}$ such that the co-ordinate direction Ox_i is orthogonal to this pair of faces.

Next, we shall consider the discretisation of the problem (10.11). Motivated by the tensor product structure of the space on the left-hand side of the inclusion (10.13), we shall base our Galerkin discretisation on a finite-dimensional subspace of $\mathcal{H}$ which is the tensor product of univariate subspaces of $\mathrm{H}^1_{(0)}(0,1)$. Thus, we begin by setting up the necessary notation in the case of the univariate space $\mathrm{H}^1_{(0)}(0,1)$.

10.4 Univariate discretisation

Let $I = (0, 1)$ and consider the sequence of partitions $(\mathcal{T}^\ell)_{\ell \geq 0}$, where $\mathcal{T}^0 = \{I\}$ and where the partition $\mathcal{T}^{\ell+1}$ is obtained from the previous partition $\mathcal{T}^\ell = \{I_j^\ell : j = 0, \ldots, 2^\ell - 1\}$ by halving each of the intervals I_j^ℓ. We consider the finite-dimensional linear subspace $\mathcal{V}^\ell$ of $\mathrm{H}^1(0, 1)$ consisting of all continuous piecewise polynomials of degree $p = 1$ on the partition $\mathcal{T}^\ell$. We also consider its subspace $\mathcal{V}_0^\ell := \mathcal{V}^\ell \cap C_0[0, 1] \subset \mathrm{H}_0^1(0, 1)$ consisting of all continuous piecewise linear functions on $\mathcal{T}^\ell$ that vanish at both endpoints of the interval $[0, 1]$.

The mesh size in the partition $\mathcal{T}^\ell$ is $h_\ell := 2^{-\ell}$ and we define $N_0^\ell := \dim(\mathcal{V}_0^\ell)$. Clearly, $N_0^\ell = 2^\ell - 1$ for $\ell \geq 0$. We define $M_0^\ell := N_0^\ell - N_0^{\ell-1}$, $\ell \geq 1$, and let $M_0^0 := N_0^0 = 0$. Analogously, we define $N^\ell := \dim(\mathcal{V}^\ell)$ and $M^\ell := N^\ell - N^{\ell-1}$ for $\ell \geq 1$, with $M^0 = N^0 = 2$. Then, $N^\ell = N_0^\ell + 2 = 2^\ell + 1$ for all $\ell \geq 0$, and $M^\ell = M_0^\ell = 2^{\ell-1}$, $\ell \geq 1$. In what follows, we shall not distinguish between M_0^ℓ and M^ℓ for $\ell \geq 1$ and will simply write M^ℓ for both.

For $L \geq 1$ we consider the linearly independent set

$$\{\psi_j^\ell : j = 1, \ldots, M^\ell, \quad \ell = 1, \ldots, L\}$$

in $\mathcal{V}_0^L$, where, for $x \in [0, 1]$,

$$\psi_j^\ell(x) := \left(1 - 2^\ell \left| x - \frac{2j - 1}{2^\ell} \right| \right)_+, \quad j = 1, \ldots, 2^{\ell-1}, \quad \ell = 1, \ldots, L.$$

Clearly,

$$\mathcal{V}_0^L = \mathrm{span}\{\psi_j^\ell : j = 1, \ldots, M^\ell, \quad \ell = 1, \ldots, L\},$$

$$\mathrm{diam}(\mathrm{supp}\, \psi_j^\ell) \leq 2.2^{-\ell}, \qquad j = 1, \ldots, M^\ell, \quad \ell = 1, \ldots, L.$$

Any function $v \in \mathcal{V}_0^L$ has the representation

$$v(x) = \sum_{\ell=1}^{L} \sum_{j=1}^{M^\ell} v_j^\ell \psi_j^\ell(x),$$

with a uniquely defined set of coefficients $v_j^\ell \in \mathbb{R}$.

For $L \geq 1$, we consider the $\mathrm{L}^2(0, 1)$-orthogonal projector

$$P_0^L : \mathrm{L}^2(0, 1) \to \mathcal{V}_0^L.$$

This has the following approximation property (cf. Brenner & Scott (2002)):

$$\left\| v - P_0^L v \right\|_{\mathrm{H}^s(0,1)} \leq \mathrm{Const.} 2^{-(2-s)L} \|v\|_{\mathrm{H}^2(0,1)}, \tag{10.14}$$

where $L \geq 1$, $s \in \{0, 1\}$, and $v \in \mathrm{H}^2(0, 1) \cap \mathrm{H}_0^1(0, 1)$. In particular, $v =$

$\lim_{L\to\infty} P_0^L v$ for all $v \in \mathrm{H}^2(0,1) \cap \mathrm{H}_0^1(0,1)$, where the limit is considered in the $\mathrm{H}^s(0,1)$-norm, $s \in \{0,1\}$.

In order to extend the construction to the multidimensional case, it is helpful to define the *increment spaces* $\mathcal{W}_0^\ell$, $\ell \geq 0$, as follows:

$$\mathcal{W}_0^0 := \mathcal{V}_0^0 = \{0\},$$
$$\mathcal{W}^\ell := \mathrm{span}\{\psi_j^\ell : 1 \leq j \leq M^\ell\}, \qquad \ell \geq 1.$$

With this notation, we can write

$$\mathcal{V}_0^\ell = \mathcal{V}_0^{\ell-1} \oplus \mathcal{W}^\ell, \qquad \ell \geq 1.$$

Therefore,

$$\mathcal{V}_0^\ell = \mathcal{W}_0^0 \oplus \mathcal{W}^1 \oplus \cdots \oplus \mathcal{W}^\ell = \mathcal{W}^1 \oplus \cdots \oplus \mathcal{W}^\ell, \qquad \ell \geq 1. \tag{10.15}$$

We proceed similarly for functions v which do not vanish at the endpoints of the interval $[0,1]$. Any $v \in \mathcal{V}^L$, $L \geq 1$, has the representation

$$v(x) = (1-x)v(0) + xv(1) + \sum_{\ell=1}^{L}\sum_{j=1}^{M^\ell} v_j^\ell \psi_j^\ell(x),$$

with a uniquely defined set of coefficients $v_j^\ell \in \mathbb{R}$. For $L \geq 0$ we shall write this expansion in compact form as

$$v(x) = \sum_{\ell=0}^{L}\sum_{j=1}^{M^\ell} v_j^\ell \psi_j^\ell(x),$$

where $\psi_1^0(x) = 1 - x$, $\psi_2^0(x) = x$, $v_1^0 = v(0)$ and $v_2^0 = v(1)$. Thus,

$$\mathcal{V}^L = \mathrm{span}\{\psi_j^\ell : j = 1, \dots, M^\ell, \quad \ell = 0, \dots, L\}, \qquad L \geq 0.$$

For $L \geq 0$ we consider the $\mathrm{L}^2(0,1)$-orthogonal projector[1]

$$P^L : \mathrm{L}^2(0,1) \to \mathcal{V}^L.$$

This has the following approximation property (cf. Brenner & Scott (2002)):

$$\|v - P^L v\|_{\mathrm{H}^s(0,1)} \leq \mathrm{Const.} 2^{-(2-s)L} \|v\|_{\mathrm{H}^2(0,1)}, \tag{10.16}$$

where $L \geq 0$, $s \in \{0,1\}$ and $v \in \mathrm{H}^2(0,1)$. In particular, $v = \lim_{L\to\infty} P^L v$

---

[1] The choice of P^L is somewhat arbitrary; e.g., we could have defined $P^L : \mathrm{H}^1(0,1) \mapsto \mathcal{V}^L$ by $P^L v := I^0 v + P_0^L(v - I^0 v)$, where $(I^0 v)(x) = (1-x)v(0) + xv(1)$, and arrived at identical conclusions to those below. For example, (10.16) will follow from (10.14) on noting that $\|v - I^0 v\|_{\mathrm{H}^s(0,1)} \leq \mathrm{Const.}|v|_{\mathrm{H}^2(0,1)}$ for $s \in \{0,1\}$ and $v \in \mathrm{H}^2(0,1)$. In addition, this alternative projector has the appealing property:

$$P^L\big|_{\mathrm{H}_0^1(0,1)} = P_0^L \qquad \text{for all } L \geq 1.$$

for all $v \in \mathrm{H}^2(0,1)$, where the limit is considered in the $\mathrm{H}^s(0,1)$-norm for $s \in \{0,1\}$.

This time, we define the increment spaces $\mathcal{W}^\ell$, $\ell \geq 0$, as follows:

$$\mathcal{W}^0 := \mathcal{V}^0 = \mathrm{span}\{1-x, x\},$$
$$\mathcal{W}^\ell := \mathrm{span}\{\psi_j^\ell : 1 \leq j \leq M^\ell\}, \qquad \ell \geq 1.$$

Hence, we can write

$$\mathcal{V}^\ell = \mathcal{V}^{\ell-1} \oplus \mathcal{W}^\ell, \qquad \ell \geq 1.$$

Therefore,

$$\mathcal{V}^\ell = \mathcal{W}^0 \oplus \mathcal{W}^1 \oplus \cdots \oplus \mathcal{W}^\ell, \qquad \ell \geq 1. \tag{10.17}$$

10.5 Sparse tensor-product spaces

Now we return to the original multidimensional setting on $\Omega = (0,1)^d$ and consider the finite-dimensional subspace V_0^L of $\bigotimes_{i=1}^d \mathrm{H}_{(0)}^1(0,1)$ defined by

$$V_0^L := \bigotimes_{i=1}^d \mathcal{V}_{(0)}^L = \mathcal{V}_{(0)}^L \otimes \cdots \otimes \mathcal{V}_{(0)}^L, \tag{10.18}$$

where the ith component $\mathcal{V}_{(0)}^L$ in this tensor product is chosen to be $\mathcal{V}_0^L$ if the co-ordinate axis Ox_i is orthogonal to a pair of $(d-1)$-dimensional open faces of Ω which belong to Γ_0, and $\mathcal{V}_{(0)}^L$ is chosen as $\mathcal{V}^L$ otherwise. In particular, if $a = 0$ and therefore $\Gamma_0 = \emptyset$, then $\mathcal{V}_{(0)}^L = \mathcal{V}^L$ for each component in the tensor product. Conversely, if a is positive definite, then $\Gamma_0 = \Gamma$ and therefore $\mathcal{V}_{(0)}^L = \mathcal{V}_0^L$ for each component of the tensor product. In general, for $a \geq 0$ that is neither identically zero nor positive definite, $\mathcal{V}_{(0)}^L = \mathcal{V}_0^L$ for a certain number i of components in the tensor product, where $0 < i < d$, and $\mathcal{V}_{(0)}^L = \mathcal{V}^L$ for the rest.

Using the hierarchical decompositions (10.15) and (10.17), we have that

$$V_0^L = \bigoplus_{|\ell|_\infty \leq L} \mathcal{W}^{\ell_1} \otimes \cdots \otimes \mathcal{W}^{\ell_d}, \qquad \ell = (\ell_1, \ldots, \ell_d), \tag{10.19}$$

with the convention that $\mathcal{W}^{\ell_i=0} = \{0\}$ whenever Ox_i is a co-ordinate direction that is orthogonal to a pair of $(d-1)$-dimensional open faces contained in Γ_0; otherwise, $\mathcal{W}^{\ell_i=0} = \mathrm{span}\{1-x_i, x_i\}$.

The space V_0^L has $\mathcal{O}(2^{Ld})$ degrees of freedom, a number that grows exponentially as a function of d. In order to reduce the number of degrees of freedom, we shall replace V_0^L with a lower-dimensional subspace $\hat{V}_0^L$

defined as follows:

$$\hat{V}_0^L := \bigoplus_{|\ell|_1 \leq L} \mathcal{W}^{\ell_1} \otimes \cdots \otimes \mathcal{W}^{\ell_d}, \qquad \ell = (\ell_1, \ldots, \ell_d), \qquad (10.20)$$

again with the convention that $\mathcal{W}^{\ell_i=0} = \{0\}$ whenever Ox_i is a co-ordinate direction that is orthogonal to a pair of $(d-1)$-dimensional open faces contained in Γ_0; otherwise, $\mathcal{W}^{\ell_i=0} = \mathrm{span}\{1 - x_i, x_i\}$.

The space $\hat{V}_0^L$ is called a *sparse tensor product space*. It has

$$\dim(\hat{V}_0^L) = \mathcal{O}(2^L L^{d-1}) = \mathcal{O}(2^L (\log_2 2^L)^{d-1})$$

degrees of freedom, which is a considerably smaller number than

$$\dim(V_0^L) = \mathcal{O}(2^{Ld}) = \mathcal{O}(2^L (2^L)^{d-1}).$$

Let us consider the d-dimensional projector

$$P_{(0)}^L \cdots P_{(0)}^L : \bigotimes_{i=1}^d \mathrm{H}_{(0)}^1(0,1) \to \bigotimes_{i=1}^d \mathcal{V}_{(0)}^L = V_0^L,$$

where the ith component $P_{(0)}^L$ is equal to P_0^L if the co-ordinate direction Ox_i is orthogonal to a pair of $(d-1)$-dimensional open faces contained in Γ_0, and is equal to P^L otherwise.

Now, let

$$Q^\ell = \begin{cases} P^\ell - P^{\ell-1}, & \ell \geq 1, \\ P^0, & \ell = 0. \end{cases}$$

We also define

$$Q_0^\ell = \begin{cases} P_0^\ell - P_0^{\ell-1}, & \ell \geq 1, \\ P_0^0, & \ell = 0, \end{cases}$$

with the convention that $P_0^0 = 0$. Thus,

$$P_{(0)}^L = \sum_{\ell=0}^L Q_{(0)}^\ell,$$

where $Q_{(0)}^\ell = Q_0^\ell$ when $P_{(0)}^\ell = P_0^\ell$ and $Q_{(0)}^\ell = Q^\ell$ when $P_{(0)}^\ell = P^\ell$.

Hence,

$$P_{(0)}^L \cdots P_{(0)}^L = \sum_{|\ell|_\infty \leq L} Q_{(0)}^{\ell_1} \cdots Q_{(0)}^{\ell_d}, \qquad \ell = (\ell_1, \ldots, \ell_d),$$

where $Q_{(0)}^{\ell_i}$ is equal to Q_0^ℓ when the co-ordinate direction Ox_i is orthogonal to a pair of $(d-1)$-dimensional open faces in Γ_0, and equal to Q^ℓ otherwise.

The sparse counterpart $\hat{P}_0^L$ of the tensor-product projector $P_{(0)}^L \cdots P_{(0)}^L$ is then defined by truncating the index set $\{\ell : |\ell|_\infty \leq L\}$ of the sum to $\{\ell : |\ell|_1 \leq L\}$:

$$\hat{P}_0^L := \sum_{|\ell|_1 \leq L} Q_{(0)}^{\ell_1} \cdots Q_{(0)}^{\ell_d} : \bigotimes_{i=1}^{d} \mathrm{H}_{(0)}^1(0,1) \to \hat{V}_0^L, \qquad \ell = (\ell_1, \ldots, \ell_d),$$

where $Q_{(0)}^{\ell_i}$ is equal to Q_0^ℓ when the co-ordinate direction Ox_i is orthogonal to a pair of $(d-1)$-dimensional open faces contained in Γ_0, and equal to Q^ℓ otherwise. In order to formulate the approximation properties of the projector $\hat{P}_0^L$, for $k \in \mathbb{N}_{\geq 1}$ we define the space $\mathcal{H}^k(\Omega)$ of functions with *square-integrable mixed k^{th} derivatives*

$$\mathcal{H}^k(\Omega) = \{v \in \mathrm{L}^2(\Omega) : \mathrm{D}^\alpha v \in \mathrm{L}^2(\Omega),\ |\alpha|_\infty \leq k\}$$

equipped with the norm

$$\|v\|_{\mathcal{H}^k(\Omega)} = \left(\sum_{|\alpha|_\infty \leq k} \|\mathrm{D}^\alpha v\|_{\mathrm{L}^2(\Omega)}^2 \right)^{1/2}.$$

Now we are ready to state our main approximation result.

Proposition 10.1 *Suppose that $u \in \mathcal{H}^2(\Omega) \cap \bigotimes_{i=1}^{d} \mathrm{H}_{(0)}^1(0,1)$. Then, for $s \in \{0,1\}$,*

$$\|u - \hat{P}_0^L u\|_{\mathrm{H}^s(\Omega)} \leq \begin{cases} \mathrm{Const.} h_L^2 |\log_2 h_L|^{d-1} \|u\|_{\mathcal{H}^2(\Omega)}, & \text{if } s = 0, \\ \mathrm{Const.} h_L^{2-s} \|u\|_{\mathcal{H}^2(\Omega)}, & \text{if } s = 1, \end{cases} \tag{10.21}$$

where $h_L = 2^{-L}$.

Proof We follow the line of argument in the proof of Proposition 3.2 in the paper by von Petersdorff & Schwab (2004), suitably modified to accommodate our nonstandard function space $\mathcal{H}^2(\Omega) \cap \bigotimes_{i=1}^{d} \mathrm{H}_{(0)}^1(0,1)$, as well as the fact that the norm-equivalence properties in the $\mathrm{L}^2(0,1)$ and $\mathrm{H}^1(0,1)$ norms, employed for the wavelet basis therein, do not apply here.

In the one-dimensional case, on writing

$$Q_{(0)}^\ell u = \left(P_{(0)}^\ell u - u \right) + \left(u - P_{(0)}^{\ell-1} u \right), \qquad \ell = 1, 2, \ldots,$$

we deduce from the approximation properties of P_0^ℓ and P^ℓ that, for $u \in \mathrm{H}^2(0,1) \cap \mathrm{H}_{(0)}^1(0,1)$,

$$\left\| Q_{(0)}^\ell u \right\|_{\mathrm{H}^s(0,1)} \leq \mathrm{Const.} 2^{(s-2)\ell} \|u\|_{\mathrm{H}^2(0,1)}, \qquad \ell = 0, 1, \ldots, \tag{10.22}$$

where $s \in \{0, 1\}$. We recall that

$$u = \lim_{L \to \infty} P^L_{(0)} u = \lim_{L \to \infty} \sum_{\ell=0}^{L} Q^\ell_{(0)} u = \sum_{\ell=0}^{\infty} Q^\ell_{(0)} u$$

and hence

$$u - P^L_{(0)} u = \sum_{\ell > L} Q^\ell_{(0)} u$$

for all $u \in \mathrm{H}^2(0,1) \cap \mathrm{H}^1_{(0)}(0,1)$, where the limits of the infinite series are considered in the $\mathrm{H}^s(0,1)$-norm, $s \in \{0,1\}$.

In the multidimensional case, we deduce from (10.22) that

$$\left\| Q^{\ell_1}_{(0)} \otimes \cdots \otimes Q^{\ell_d}_{(0)} u \right\|_{\mathrm{H}^s(\Omega)} \le \mathrm{Const.} 2^{s|\ell|_\infty - 2|\ell|_1} \|u\|_{\mathcal{H}^2(\Omega)};$$

also,

$$u - \hat{P}^L_0 u = \sum_{|\ell|_1 > L} Q^{\ell_1}_{(0)} \otimes \cdots \otimes Q^{\ell_d}_{(0)} u$$

for all $u \in \mathcal{H}^2(\Omega) \cap \bigotimes_{i=1}^{d} \mathrm{H}^1_{(0)}(0,1)$, where the limit of the infinite sum is considered in the $\bigotimes_{i=1}^{d} \mathrm{H}^s(0,1)$-norm, $s \in \{0,1\}$. Noting that for $\ell = (\ell_1, \ldots, \ell_d)$, such that $|\ell|_1 = m$,

$$2^{s|\ell|_\infty - 2|\ell|_1} = 2^{(s-2)L + (s-2)(m-L) + s(|\ell|_\infty - m)},$$

we have that

$$\left\| u - \hat{P}^L_0 u \right\|_{\mathrm{H}^s(\Omega)} \le \mathrm{Const.} \left(\sum_{|\ell|_1 > L} 2^{s|\ell|_\infty - 2|\ell|_1} \right) \|u\|_{\mathcal{H}^2(\Omega)}$$

$$= \mathrm{Const.} \left(\sum_{m=L+1}^{\infty} \sum_{|\ell|_1 = m} 2^{s|\ell|_\infty - 2|\ell|_1} \right) \|u\|_{\mathcal{H}^2(\Omega)}$$

$$= \mathrm{Const.} 2^{(s-2)L} \left(\sum_{m=L+1}^{\infty} 2^{(s-2)(m-L)} \sigma_m \right) \|u\|_{\mathcal{H}^2(\Omega)},$$

where $\sigma_m = \sum_{|\ell|_1 = m} 2^{s(|\ell|_\infty - m)}$.

For $s = 0$ we have $\sigma_m \le \mathrm{Const.} m^{d-1}$, while for $s > 0$ the bound $\sigma_m \le \mathrm{Const.}$ holds, independent of m. The final forms of the inequalities (10.21) follow, with $2^{(s-2)L} = h_L^{2-s}$ and $L = |\log_2 h_L|$, on observing that $\sum_{m=L+1}^{\infty} 2^{(s-2)(m-L)} \sigma_m$ is bounded by $\mathrm{Const.} L^{d-1}$ when $s = 0$ and by a constant independent of L when $s > 0$. $\qquad \square$

Since the space $\mathcal{H}^k(\Omega)$ of functions of square-integrable mixed k^{th} derivatives is a proper subspace of the classical Sobolev space $\mathrm{H}^k(\Omega) = \{v \in$

$L^2(\Omega) : D^\alpha v \in L^2(\Omega), |\alpha|_1 \leq k\}$, Proposition 10.1 indicates that preserving the optimal approximation order $\mathcal{O}(h^{2-s})$ of the full tensor-product space V_0^L in the $H^s(\Omega)$-norm, $s = 0, 1$, upon sparsification (with a mild polylogarithmic loss of $|\log_2 h_L|^{d-1}$ in the case of $s = 0$) comes at the expense of increased smoothness requirements on the function u which is approximated from the sparse tensor-product space $\hat{V}_0^L$.

10.6 Sparse stabilised finite element method

Having defined the finite-dimensional space $\hat{V}_0^L$ from which the approximate solution will be sought, we now introduce the remaining ingredients of our Galerkin method: a bilinear form $b_\delta(\cdot, \cdot)$ which approximates the bilinear form $B(\cdot, \cdot)$ from the weak formulation (10.11) of the boundary value problem (10.6), (10.7) and a linear functional $l_\delta(\cdot)$ which approximates the linear functional $L(\cdot)$ from (10.11).

Let us consider the bilinear form

$$b_\delta(w, v) := B(w, v) + \delta_L \sum_{\kappa \in \mathcal{T}^L} (\mathcal{L}w, b \cdot \nabla v)_\kappa.$$

Here $\delta_L \in [0, 1/c]$ is a ('streamline-diffusion') parameter to be chosen below, and $\kappa \in \mathcal{T}^L$ are d-dimensional axiparallel cubic elements of edge-length h_L in the partition of the computational domain $\Omega = (0, 1)^d$; there are 2^{Ld} such elements κ in $\mathcal{T}^L$, a number that grows exponentially with d. The second term can be thought of as least-square stabilisation in the direction of subcharacteristics ('streamlines').

We also define the linear functional

$$l_\delta(v) := L(v) + \delta_L \sum_{\kappa \in \mathcal{T}^L} (f, b \cdot \nabla v)_\kappa \quad (= L(v) + \delta_L(f, b \cdot \nabla v)),$$

and consider the finite-dimensional problem: find $u_h \in \hat{V}_0^L$ such that

$$b_\delta(u_h, v_h) = l_\delta(v_h) \qquad \forall v_h \in \hat{V}_0^L. \tag{10.23}$$

The idea behind the method (10.23) is to introduce mesh-dependent numerical diffusion into the standard Galerkin finite element method along subcharacteristic directions, with the aim to suppress maximum-principle-violating oscillations on the scale of the mesh, and let $\delta_L \to 0$ with $h_L \to 0$. For an analysis of the method in the case of standard finite element spaces and (low-dimensional) elliptic transport-dominated diffusion equations we refer to the monograph of Roos, Stynes & Tobiska (1996).

It would have been more accurate to write u_{h_L} and v_{h_L} instead of u_h and v_h in (10.23). However, to avoid notational clutter, we shall refrain from doing so. Instead, we adopt the convention that the dependence of $h = h_L$ on the index L will be implied, even when not explicitly noted.

We begin with the stability-analysis of the method. Since $u_h|_\kappa$ is multilinear in each $\kappa \in \mathcal{T}^L$ and a is a constant matrix, it follows that

$$\nabla \cdot (a\nabla u_h)|_\kappa = a : \nabla\nabla u_h|_\kappa = 0.$$

Therefore,

$$b_\delta(u_h, v_h) = B(u_h, v_h) + \delta_L(b \cdot \nabla u_h + cu_h, b \cdot \nabla v_h)$$

for all $u_h, v_h \in \hat{V}_0^L$. We note in passing that this simplification of $b_\delta(\cdot, \cdot)$ over $\hat{V}_0^L \times \hat{V}_0^L$, in comparison with its original definition, has useful computational consequences: it helps to avoid summation over the 2^{Ld} elements κ comprising the mesh $\mathcal{T}^L$ in the implementation of the method. Now,

$$\begin{aligned}
b_\delta(v_h, v_h) &= (a\nabla v_h, \nabla v_h) + (cv_h, v_h) + \delta_L\|b \cdot \nabla v_h\|^2 \\
&\quad + \frac{1}{2}\int_{\Gamma_+ \cup \Gamma_-} |\beta||v_h|^2 \, ds + \frac{1}{2}c\delta_L \int_\Gamma \beta|v_h|^2 \, ds \\
&\geq (a\nabla v_h, \nabla v_h) + c\|v_h\|^2 + \delta_L\|b \cdot \nabla v_h\|^2 \qquad (10.24) \\
&\quad + \frac{1}{2}(1 + c\delta_L)\int_{\Gamma_+} |\beta||v_h|^2 \, ds + \frac{1}{2}(1 - c\delta_L)\int_{\Gamma_-} |\beta||v_h|^2 \, ds,
\end{aligned}$$

where we have made use of the facts that $\beta \leq |\beta|$ on Γ_- and $v_h|_{\Gamma_0} = 0$. Since (10.23) is a linear problem in a finite-dimensional linear space, (10.24) implies the existence and uniqueness of a solution u_h to (10.23) in $\hat{V}_0^L$.

Let us also note that

$$|l_\delta(v_h)| \leq \left(\frac{1}{c} + \delta_L\right)^{1/2} \|f\| \left(c\|v_h\|^2 + \delta_L\|b \cdot \nabla v_h\|^2\right)^{1/2} \qquad (10.25)$$

for all $v_h \in \hat{V}_0^L$. On noting that, by hypothesis, $1 - c\delta_L \geq 0$ and combining (10.24) and (10.25) we deduce that

$$\begin{aligned}
|||u_h|||_{\text{SD}}^2 &:= (a\nabla u_h, \nabla u_h) + c\|u_h\|^2 + \delta_L\|b \cdot \nabla u_h\|^2 \\
&\quad + \frac{1}{2}(1 + c\delta_L)\int_{\Gamma_+} |\beta||u_h|^2 \, ds + \frac{1}{2}(1 - c\delta_L)\int_{\Gamma_-} |\beta||u_h|^2 \, ds \\
&\leq \left(\frac{1}{c} + \delta_L\right)\|f\|^2.
\end{aligned}$$

Hence,

$$|||u_h|||_{\text{SD}} \leq (2/c)^{1/2}\|f\|, \qquad (10.26)$$

which establishes the stability of the method (10.23), for all $\delta_L \in [0, 1/c]$.

The next section is devoted to the error analysis of the method. We shall require the following multiplicative trace inequality.

Lemma 10.3 (Multiplicative trace inequality) *Let* $\Omega = (0,1)^d$ *where* $d \geq 2$ *and suppose that* Γ_+ *is the hyperbolic outflow part of* Γ. *Then,*

$$\int_{\Gamma_+} |v|^2 \, \mathrm{d}s \leq 4d\|v\| \, \|v\|_{\mathrm{H}^1(\Omega)} \qquad \forall v \in \mathrm{H}^1(\Omega).$$

Proof We shall prove the inequality for $v \in \mathrm{C}^1(\overline{\Omega})$. For $v \in \mathrm{H}^1(\Omega)$ the result follows by density of $\mathrm{C}^1(\overline{\Omega})$ in $\mathrm{H}^1(\Omega)$. As we have noted before, Γ_+ is a union of $(d-1)$-dimensional open faces of Ω. Let us suppose without loss of generality that the face $x_1 = 0$ of Ω belongs to Γ_+. Then,

$$v^2(0, x') = v^2(x_1, x') + \int_{x_1}^0 \frac{\partial}{\partial x_1} v^2(\xi, x') \mathrm{d}\xi, \qquad x' = (x_2, \ldots, x_n).$$

Hence, on integrating this over $x = (x_1, x') \in (0,1) \times (0,1)^{d-1} = \Omega$,

$$\int_{x' \in (0,1)^{d-1}} v^2(0, x') \, \mathrm{d}x' = \int_0^1 \int_{x' \in (0,1)^{d-1}} v^2(x_1, x') \, \mathrm{d}x' \, \mathrm{d}x_1$$

$$+ 2 \int_0^1 \int_{x' \in (0,1)^{d-1}} \int_{x_1}^0 v(\xi, x') \frac{\partial}{\partial x_1} v(\xi, x') \mathrm{d}\xi \, \mathrm{d}x' \, \mathrm{d}x_1$$

$$\leq \|v\|^2 + 2\|v\| \, \|v_{x_1}\|.$$

In the generic case when $\beta > 0$ on the whole of Γ_+, the set Γ_+ will contain at most d of the $2d$ faces of Ω, — at most one complete face of Ω orthogonal to the ith co-ordinate direction, $i = 1, \ldots, d$. Otherwise, if $\beta = 0$ on certain faces that belong to Γ_+, the set Γ_+ may contain as many as $2d - 1$ of the $2d$ faces of Ω. Thus, in the worst case,

$$\int_{\Gamma_+} |v|^2 \, \mathrm{d}s \leq (2d-1)\|v\|^2 + 4\|v\| \sum_{i=1}^d \|u_{x_i}\|.$$

Therefore,

$$\int_{\Gamma_+} |v|^2 \, \mathrm{d}s \leq 2d\sqrt{2} \max\left\{1, \frac{2}{d^{1/2}}\right\} \|v\| \, \|v\|_{\mathrm{H}^1(\Omega)} \leq 4d\|v\| \, \|v\|_{\mathrm{H}^1(\Omega)}.$$

Hence the required result. $\qquad\qquad\qquad\qquad\qquad\qquad\qquad\qquad\qquad\qquad$ □

10.7 Error analysis

Our goal in this section is to estimate the size of the error between the analytical solution $u \in \mathcal{H}$ and its approximation $u_h \in \hat{V}_0^L$. We shall assume throughout that $f \in \mathrm{L}^2(\Omega)$ and the corresponding solution $u \in \mathcal{H}^2(\Omega) \cap$

$\bigotimes_{i=1}^{d} H_{(0)}^1(0,1) \subset \mathcal{H}$. Clearly,

$$b_\delta(u - u_h, v_h) = B(u, v_h) - L(v_h) + \delta_L \sum_{\kappa \in \mathcal{T}^L} (\mathcal{L}u - f, b \cdot \nabla v_h)_\kappa$$

for all $v_h \in \hat{V}_0^L \subset \mathcal{V}$. Hence we deduce from (10.11) the following *Galerkin orthogonality* property:

$$b_\delta(u - u_h, v_h) = 0 \qquad \forall v_h \in \hat{V}_0^L. \tag{10.27}$$

Let us decompose the error $u - u_h$ as follows:

$$u - u_h = (u - \hat{P}^L u) + (\hat{P}^L u - u_h) = \eta + \xi,$$

where $\eta := u - \hat{P}^L u$ and $\xi := \hat{P}^L u - u_h$. By the triangle inequality,

$$|||u - u_h|||_{\mathrm{SD}} \le |||\eta|||_{\mathrm{SD}} + |||\xi|||_{\mathrm{SD}}. \tag{10.28}$$

We begin by bounding $|||\xi|||_{\mathrm{SD}}$. By (10.24) and (10.27), we have that

$$|||\xi|||_{\mathrm{SD}}^2 \le b_\delta(\xi, \xi) = b_\delta(u - u_h, \xi) - b_\delta(\eta, \xi) = -b_\delta(\eta, \xi).$$

Therefore,

$$|||\xi|||_{\mathrm{SD}}^2 \le |b_\delta(\eta, \xi)|. \tag{10.29}$$

Now since $\nabla\nabla(P_L u)|_\kappa = 0$ for each $\kappa \in \mathcal{T}^L$, we have that $\nabla\nabla\eta|_\kappa = \nabla\nabla u|_\kappa$ on each $\kappa \in \mathcal{T}^L$, and therefore

$$b_\delta(\eta, \xi) = (a\nabla\eta, \nabla\xi) - (\eta, b \cdot \nabla\xi) + (c\eta, \xi) + \int_{\Gamma_+} |\beta|\eta\xi\, \mathrm{d}s$$
$$+ \delta_L(-a : \nabla\nabla u + b \cdot \nabla\eta + c\eta, b \cdot \nabla\xi)$$
$$= \mathrm{I} + \mathrm{II} + \mathrm{III} + \mathrm{IV} + \mathrm{V} + \mathrm{VI} + \mathrm{VII}.$$

We shall estimate each of the terms I to VII in turn:

$$\mathrm{I} \le \left(|a|^{1/2}\|\nabla\eta\|\right) |||\xi|||_{\mathrm{SD}},$$

$$\mathrm{II} \le \left(\delta_L^{-1/2}\|\eta\|\right) |||\xi|||_{\mathrm{SD}},$$

$$\mathrm{III} \le \left(c^{1/2}\|\eta\|\right) |||\xi|||_{\mathrm{SD}},$$

$$\mathrm{V} \le \left(\delta_L^{1/2}|a|\, |u|_{\mathrm{H}^2(\Omega)}\right) |||\xi|||_{\mathrm{SD}},$$

$$\mathrm{VI} \le \left(\delta_L^{1/2}|b|\, \|\nabla\eta\|\right) |||\xi|||_{\mathrm{SD}},$$

$$\mathrm{VII} \le \left(c\delta_L^{1/2}\|\eta\|\right) |||\xi|||_{\mathrm{SD}}.$$

Here $|a|$ is the Frobenius norm of the matrix a and $|b|$ is the Euclidean norm of the vector b. It remains to estimate IV:

$$\text{IV} \leq \left(\frac{2|b|}{1+c\delta_L}\right)^{1/2} \left(\int_{\Gamma_+} |\eta|^2 \, \mathrm{d}s\right)^{1/2} |||\xi|||_{\mathrm{SD}}$$

$$\leq (2|b|)^{1/2} (4d)^{1/2} \|\eta\|^{1/2} \|\eta\|_{\mathrm{H}^1(\Omega)}^{1/2} |||\xi|||_{\mathrm{SD}},$$

where in the transition to the last line we used the multiplicative trace inequality from Lemma 10.3. Hence, by (10.29),

$$|||\xi|||_{\mathrm{SD}} \leq |a|^{1/2} \|\nabla\eta\| + \delta_L^{-1/2} \|\eta\| + c^{1/2} \|\eta\| + (8d)^{1/2} |b|^{1/2} \|\eta\|^{1/2} \|\eta\|_{\mathrm{H}^1(\Omega)}^{1/2}$$

$$+ \delta_L^{1/2} |a| |u|_{\mathrm{H}^2(\Omega)} + \delta_L^{1/2} |b| \|\nabla\eta\| + c\delta_L^{1/2} \|\eta\|. \tag{10.30}$$

On selecting

$$\delta_L := K_\delta \min\left(\frac{h_L^2}{|a|}, \frac{h_L |\log_2 h_L|^{d-1}}{d|b|}, \frac{1}{c}\right), \tag{10.31}$$

with $K_\delta \in \mathbb{R}_{>0}$ a constant, independent of h_L and d, we get that

$$|||\xi|||_{\mathrm{SD}}^2 \leq C(u)\left(|a|h_L^2 + \frac{h_L^4 |\log_2 h_L|^{2(d-1)}}{\delta_L}\right),$$

where $C(u) := \text{Const.}\|u\|_{\mathcal{H}^2(\Omega)}^2$, and Const. is a positive constant independent of h_L. An identical bound holds for $|||\eta|||_{\mathrm{SD}}$. Thus we have proved the following theorem.

Theorem 10.2 *Suppose that $f \in \mathrm{L}^2(\Omega)$, $c > 0$ and $u \in \mathcal{H}^2(\Omega) \cap \mathcal{H}$. Then, the following bound holds for the error $u - u_h$ between the analytical solution u of (10.11) and its sparse finite element approximation $u_h \in \hat{V}_0^L$ defined by (10.23), with $L \geq 1$ and $h = h_L = 2^{-L}$:*

$$|||u - u_h|||_{\mathrm{SD}}^2 \leq C(u)\left(|a|h_L^2 + h_L^4 |\log_2 h_L|^{2(d-1)}\right.$$

$$\left. \times \max\left(\frac{|a|}{h_L^2}, \frac{d|b|}{h_L |\log_2 h_L|^{d-1}}, c\right)\right),$$

with the streamline-diffusion parameter δ_L defined by the formula (10.31) and $C(u) = \text{Const.}\|u\|_{\mathcal{H}^2(\Omega)}^2$ where Const. is a positive constant independent of the discretisation parameter h_L.

10.8 Final remarks

We close with some remarks on Theorem 10.2 and on possible extensions of the results presented here. We begin by noting that, save for the polylogarithmic factors, the definition of δ_L and the structure of the error bound in the $||| \cdot |||_{\mathrm{SD}}$ norm are exactly the same as if we used the full tensor-product finite element space V_0^L instead of the sparse tensor product space $\hat{V}_0^L$ (cf. Houston & Süli (2001)). On the other hand, as we have commented earlier, through the use of the sparse space $\hat{V}_0^L$, computational complexity has been reduced from $\mathcal{O}(2^{Ld})$ to $\mathcal{O}(2^L(\log_2 2^L)^{d-1})$. Hence, in comparison with a streamline-diffusion method based on the full tensor-product space, a substantial computational saving has been achieved at the cost of only a marginal loss in accuracy.

a) In the diffusion-dominated case, that is when $|a| \approx 1$ and $|b| \approx 0$, we see from Theorem 10.2 that the error, in the streamline-diffusion norm $||| \cdot |||_{\mathrm{SD}}$, is $\mathcal{O}(h_L |\log_2 h_L|^{d-1})$ as $h_L \to 0$, provided that the streamline-diffusion parameter is chosen as

$$\delta_L = K_\delta h_L^2 / |a|.$$

This asymptotic convergence rate, as $h_L \to 0$, is slower, by the polylogarithmic factor $|\log_2 h_L|^{d-1}$, than the $\mathcal{O}(h_L)$ bound on the $\| \cdot \|_{\mathrm{H}^1(\Omega)}$ norm of the error in a standard sparse Galerkin finite element approximation of Poisson's equation on $\Omega = (0,1)^d$.

b) In the transport-dominated case, that is when $|a| \approx 0$ and $|b| \approx 1$,

$$\delta_L = K_\delta h_L |\log_2 h_L|^{d-1} / (d|b|),$$

so the error of the method, measured in the streamline-diffusion norm, is $\mathcal{O}(h_L^{3/2} |\log_2 h_L|^{d-1})$ when the diffusivity matrix a degenerates.

c) We have confined ourselves to finite element approximations based on tensor-product piecewise polynomials of degree $p = 1$ in each of the d co-ordinate directions. The analysis in the case of $p \geq 1$ is more technical and will be presented elsewhere.

d) For simplicity, we have restricted ourselves to *uniform* tensor-product partitions of $[0,1]^d$. Numerical experiments indicate that, in the presence of boundary-layers, the accuracy of the proposed sparse streamline-diffusion method can be improved by using high-dimensional versions of Shishkin-type boundary-layer-fitted tensor-product nonuniform grids. For technical details concerning the efficient implementation of sparse-grid finite element methods, we refer to the articles of Zumbusch (2000) and Bungartz & Griebel (2004).

Acknowledgements

I am grateful to Andrea Cangiani (Univerity of Pavia), Max Jensen (Humboldt University, Berlin), Christoph Ortner (University of Oxford) and Christoph Schwab (ETH Zürich) for helpful and constructive comments.

References

K. Babenko (1960), 'Approximation by trigonometric polynomials is a certain class of periodic functions of several variables', *Soviet Math, Dokl.* **1**, 672–675. Russian original in *Dokl. Akad. Nauk SSSR* **132**, 982–985.

J.W. Barrett, C. Schwab and E. Süli (2005), 'Existence of global weak solutions for some polymeric flow models', M3AS: Mathematical Models and Methods in Applied Sciences, 6(15).

R.F. Bass (1997), *Diffusion and Elliptic Operators*, Springer–Verlag, New York.

S.C. Brenner and L.R. Scott (2002), *The Mathematical Theory of Finite Element Methods*, 2nd Edition, Volume 15 of *Texts in Applied Mathematics*, Springer–Verlag, New York.

H.-J. Bungartz and M. Griebel (2004), 'Sparse grids', *Acta Numerica*, 1–123.

R. DeVore, S. Konyagin and V. Temlyakov (1998), 'Hyperbolic wavelet approximation', *Constr. Approx.* **14**, 1–26.

J. Elf, P. Lötstedt and P. Sjöberg (2003), 'Problems of high dimension in molecular biology', *Proceedings of the* 19th *GAMM-Seminar Leipzig* (W. Hackbusch, ed.), pp. 21–30.

V.H. Hoang and C. Schwab (2005), 'High dimensional finite elements for elliptic problems with multiple scales', *Multiscale Modeling and Simulation: A SIAM Interdisciplinary Journal* **3**(1), 168–194.

L. Hörmander (2005), *The Analysis of Linear Partial Differential Operators II: Differential Operators with Constant Coefficients*, Reprint of the 1983 edition, Springer–Verlag, Berlin.

P. Houston, C. Schwab and E. Süli (2002), 'Discontinuous *hp*-finite element methods for advection-diffusion-reaction problems', *SIAM Journal of Numerical Analysis* **39**(6), 2133–2163.

P. Houston and E. Süli (2001), 'Stabilized *hp*-finite element approximation of partial differential equations with non-negative characteristic form', *Computing.* **66**(2), 99–119.

B. Jourdain, C. Le Bris and T Lelièvre (2004), 'Coupling PDEs and SDEs: the illustrative example of the multiscale simulation of viscoelastic flows. Preprint.

N.G. van Kampen (1992), *Stochastic Processes in Physics and Chemistry*, Elsevier, Amsterdam.

B. Lapeyre, É. Pardoux and R. Sentis (2003), *Introduction to Monte-Carlo Methods for Transport and Diffusion Equations*, Oxford Texts in Applied and Engineering Mathematics, Oxford University Press, Oxford.

P. Laurençot and S. Mischler (2002), 'The continuous coagulation fragmentation equations with diffusion', *Arch. Rational Mech. Anal.* **162**, 45–99.

C. Le Bris and P.-L. Lions (2004), 'Renormalized solutions of some transport equations with $W^{1,1}$ velocities and applications', *Annali di Matematica* **183**, 97–130.

E. Novak and K. Ritter (1998), The curse of dimension and a universal method for numerical integration, in *Multivariate Approximation and Splines* (G. Nürnberger, J. Schmidt and G. Walz, eds), International Series in Numerical Mathematics, Birkhäuser, Basel, pp. 177–188.

O.A. Oleĭnik and E.V. Radkevič (1973), *Second Order Equations with Nonnegative Characteristic Form.* American Mathematical Society, Providence, RI.

H.-C. Öttinger (1996), *Stochastic Processes in Polymeric Fluids*, Springer-Verlag, New York.

T. von Petersdorff and C. Schwab (2004), 'Numerical solution of parabolic equations in high dimensions', M2AN Mathematical Modelling and Numerical Analysis **38**, 93–128.

H.-G. Roos, M. Stynes, and L. Tobiska (1996), *Numerical Methods for Singularly Perturbed Differential Equations. Convection–Diffusion and Flow Problems*, Volume 24 of *Springer Series in Computational Mathematics*. Springer–Verlag, New York.

S. Smolyak (1963), 'Quadrature and interpolation formulas for tensor products of certain classes of functions', *Soviet Math. Dokl.* **4**, 240-243. Russian original in *Dokl. Akad. Nauk SSSR* **148**, 1042–1045.

V. Temlyakov (1989), 'Approximation of functions with bounded mixed derivative', Volume 178 of *Proc. Steklov Inst. of Math.*, AMS, Providence, RI.

G. Wasilkowski and H. Woźniakowski (1995), 'Explicit cost bounds of algorithms for multivariate tensor product problems', *J. Complexity* **11**, 1–56.

C. Zenger (1991), Sparse grids, in *Parallel Algorithms for Partial Differential Equations* (W. Hackbusch, ed.), Vol. 31 of *Notes on Numerical Fluid Mechanics*, Vieweg, Braunschweig/Wiesbaden.

G. W. Zumbusch (2000), A sparse grid PDE solver, in *Advances in Software Tools for Scientific Computing* (H. P. Langtangen, A. M. Bruaset, and E. Quak, eds.), Vol. 10 of *Lecture Notes in Computational Science and Engineering*, Ch. 4, pp. 133–177. Springer, Berlin. (Proceedings SciTools '98).

11

Greedy Approximations

V. N. Temlyakov

Department of Mathematics
University of South Carolina
Columbia, SC, USA
e-mail: temlyak@math.sc.edu

Abstract

In nonlinear approximation we seek ways to approximate complicated functions by simpler functions using methods that depend nonlinearly on the function being approximated. Recently, a particular kind of nonlinear approximation, namely greedy approximation has attracted a lot of attention in both theoretical and applied settings. Greedy type algorithms have proven to be very useful in various applications such as image compression, signal processing, design of neural networks, and the numerical solution of nonlinear partial differential equations. A theory of greedy approximation is now emerging. Some fundamental convergence results have already been established and many fundamental problems remain unsolved. In this survey we place emphasis on the study of the efficiency of greedy algorithms with regards to redundant systems (dictionaries). Redundancy, on the one hand, offers much promise for greater efficiency in terms of the rate of approximation. On the other hand, it gives rise to highly nontrivial theoretical and practical problems. We note that there is solid justification for the importance of redundant systems in both theoretical questions and practical applications. This survey is a continuation of the survey Temlyakov (2003a) on nonlinear approximations. Here we concentrate on more recent results on greedy approximation.

11.1 Introduction

In the last decade we have seen successes in the study of nonlinear approximation (see the surveys DeVore (1998) and Temlyakov (2003a)). This study was motivated by numerous applications. Nonlinear approximation is important in applications because of its efficiency. Two types of nonlinear approximation are frequently employed in applications. Adaptive methods are used in PDE solvers. The m-term approximation considered in this paper is used in image and signal processing as well as in the design of neural networks. The fundamental question of nonlinear approximation is how to construct good methods (algorithms) of nonlinear approximation. This problem has two levels of nonlinearity. The first level of nonlinearity is m-term approximation with respect to bases. In this problem one can use the function expansion (unique) with regards to a given basis to build an approximant. In this case nonlinearity is introduced by looking for an m-term approximant with terms (basis elements in the approximant) allowed to depend upon a given function. This idea is utilized in the Thresholding Greedy Algorithm (see the survey Konyagin & Temlyakov (2002) for a detailed discussion). At a second level, nonlinearity is introduced when we replace a basis by a more general system which is not necessarily minimal (redundant system, dictionary). This case is much more complicated than the first (bases case). However, there is solid justification for the importance of redundant systems in both theoretical questions and in practical applications (see for instance Schmidt (1906-1907), Huber (1985), Donoho (2001)).

Recent results have established (see Temlyakov (2003a)) that greedy type algorithms are suitable methods of nonlinear approximation in both m-term approximation with regard to bases and m-term approximation with regard to redundant systems. The survey Temlyakov (2003a) contains a discussion of results on greedy approximation published before 2001. In this short survey we try to complement the paper Temlyakov (2003a) with a discussion of new results on greedy approximation with regard to a redundant system (dictionary). We refer the reader to the survey Konyagin & Temlyakov (2002) for a discussion of recent results on greedy approximation with regard to a basis.

In order to address the contemporary needs of approximation theory and computational mathematics, a very general model of approximation with regards to a redundant system (dictionary) has been considered in many recent papers. For a survey of some of these results we refer the reader to DeVore (1998) and Temlyakov (2003a). As such a model we choose a Banach space X with elements as target functions and an arbitrary system $\mathcal{D}$ of elements of this space, such that $\overline{\mathrm{span}}\,\mathcal{D} = X$, as an approximating system. We would like to have an algorithm for constructing m-term approximants that adds, at each step, only one new element from $\mathcal{D}$ and keeps the elements of $\mathcal{D}$ obtained at the previous steps. This requirement is an analog of an *on-line* computation property that is very desirable in practical algorithms. Clearly, we are looking for good algorithms which satisfy the minimum requirement that they converge for each target function. It is not obvious that such an algorithm exists in a setting at the level of generality above $(X, \mathcal{D}$ arbitrary$)$. It turns out that there is a fundamental principle that allows us to construct good algorithms both for arbitrary redundant systems and for very simple well structured bases like the Haar basis. This principle is the use of a greedy step in searching for a new element to be added to a given m-term approximant. The common feature of all algorithms of m-term approximation discussed in this paper is the presence of a greedy step. By a greedy step in choosing an mth element $g_m(f) \in \mathcal{D}$ to be used in an m-term approximant, we mean one which maximizes a certain functional determined by information from the previous steps of the algorithm. We obtain different types of greedy algorithms by varying the functional mentioned above and also by using different ways of constructing (choosing coefficients of the linear combination) the m-term approximant from the, already found, m elements of the dictionary.

Let X be a Banach space with norm $\|\cdot\|$. We say that a set of elements (functions) $\mathcal{D}$ from X is a dictionary if each $g \in \mathcal{D}$ has norm one ($\|g\| = 1$), and $\overline{\operatorname{span}}\,\mathcal{D} = X$. In addition we assume, for convenience (when we work in a Banach space), that a dictionary is symmetric:

$$g \in \mathcal{D} \quad \text{implies} \quad -g \in \mathcal{D}.$$

Thus from the definition of a dictionary it follows that any element $f \in X$ can be approximated arbitrarily well by finite linear combinations of the dictionary elements. Our primary goal in Sections 11.2 and 11.3 is to study representations of an element $f \in X$ by a series

$$f \sim \sum_{j=1}^{\infty} c_j(f) g_j(f), \quad g_j(f) \in \mathcal{D}, \quad c_j(f) > 0, \quad j = 1, 2, \ldots. \quad (11.1)$$

In building a representation (11.1) we should construct two sequences: $\{g_j(f)\}_{j=1}^{\infty}$ and $\{c_j(f)\}_{j=1}^{\infty}$. In this paper the construction of $\{g_j(f)\}_{j=1}^{\infty}$ will be based upon ideas used in greedy-type nonlinear approximation (greedy-type algorithms). This justifies the use of the term *greedy expansion* for (11.1) considered in the paper. The construction of $\{g_j(f)\}_{j=1}^{\infty}$ is, clearly, the most important and difficult part in building a representation (11.1). On the basis of the contemporary theory of nonlinear approximation with regards to redundant dictionaries we may conclude that the method of using a norming functional in the greedy steps of an algorithm is the most productive in approximation in Banach spaces. This method was used, for the first time, in Donahue, Gurvits, Darken & Sontag (1997).

We will study in Sections 11.2 and 11.3 of this paper greedy algorithms with regard to $\mathcal{D}$ that provide greedy expansions. For a nonzero element

$f \in X$ we denote by F_f a norming (peak) functional for f:

$$\|F_f\| = 1, \qquad F_f(f) = \|f\|.$$

The existence of such a functional is guaranteed by the Hahn-Banach Theorem. Let

$$r_{\mathcal{D}}(f) := \sup_{F_f} \sup_{g \in \mathcal{D}} F_f(g).$$

We note that in general a norming functional F_f is not unique. This is why we take $\sup_{F_f}$ over all norming functionals of f in the definition of $r_{\mathcal{D}}(f)$. It is known that in the case of uniformly smooth Banach spaces (our primary object here) the norming functional F_f is unique. In such a case we do not need $\sup_{F_f}$ in the definition of $r_{\mathcal{D}}(f)$. We consider here approximation in uniformly smooth Banach spaces. For a Banach space X we define the modulus of smoothness

$$\rho(u) := \sup_{\|x\|=\|y\|=1} \left(\frac{1}{2}(\|x + uy\| + \|x - uy\|) - 1 \right).$$

A uniformly smooth Banach space is one with the property

$$\lim_{u \to 0} \rho(u)/u = 0.$$

In Section 11.4 we discuss modifications of greedy algorithms that are motivated by practical applications of these algorithms. First of all, we assume that our evaluations are not exact. We use a functional that is an approximant of a norming functional and we use a near-best approximant instead of the best approximant. Secondly, we address the very important issue of the evaluation of $\sup_{g \in \mathcal{D}} F_f(g)$. To make this evaluation feasible we restrict our search to a finite subset $\mathcal{D}(N)$ of $\mathcal{D}$. In other words, we evaluate $\sup_{g \in \mathcal{D}(N)} F_f(g)$. We present the corresponding convergence and rate of convergence results in Section 11.4.

Section 11.5 is devoted to a discussion of specific dictionaries that are close (in a certain sense) to orthogonal bases in the case of Hilbert spaces. We also include new results on M-coherent dictionaries in Banach spaces.

We stress that in this short survey we discuss only a fraction of the recent results on greedy approximation. For instance, we do not cover the following very interesting topics that deserve separate surveys: greedy approximation with regard to a randomly chosen dictionary, simultaneous greedy approximations.

11.2 Greedy expansions in Hilbert spaces

11.2.1 Pure Greedy Algorithm

We define first the Pure Greedy Algorithm (PGA) in a Hilbert space H. We describe this algorithm for a general dictionary $\mathcal{D}$. If $f \in H$, we let $g(f) \in \mathcal{D}$ be an element from $\mathcal{D}$ which maximizes $|\langle f, g \rangle|$. We will assume, for simplicity, that such a maximizer exists; if not, suitable modifications are necessary (see Weak Greedy Algorithm in Temlyakov (2003a), p. 54) in the algorithm that follows. We define

$$G(f, \mathcal{D}) := \langle f, g(f) \rangle g(f)$$

and

$$R(f, \mathcal{D}) := f - G(f, \mathcal{D}).$$

Pure Greedy Algorithm (PGA) *We define* $R_0(f, \mathcal{D}) := f$ *and* $G_0(f, \mathcal{D}) := 0$. *Then, for each* $m \geq 1$, *we inductively define*

$$G_m(f, \mathcal{D}) := G_{m-1}(f, \mathcal{D}) + G(R_{m-1}(f, \mathcal{D}), \mathcal{D})$$

$$R_m(f, \mathcal{D}) := f - G_m(f, \mathcal{D}) = R(R_{m-1}(f, \mathcal{D}), \mathcal{D}).$$

For a general dictionary $\mathcal{D}$ we define the class of functions

$$\mathcal{A}_1^o(\mathcal{D}, M) := \{ f \in H : f = \sum_{k \in \Lambda} c_k w_k, \ w_k \in \mathcal{D},$$

$$\#\Lambda < \infty \text{ and } \sum_{k \in \Lambda} |c_k| \leq M \}$$

and we define $\mathcal{A}_1(\mathcal{D}, M)$ as the closure (in H) of $\mathcal{A}_1^o(\mathcal{D}, M)$. Furthermore, we define $\mathcal{A}_1(\mathcal{D})$ as the union of the classes $\mathcal{A}_1(\mathcal{D}, M)$ over all $M > 0$. For $f \in \mathcal{A}_1(\mathcal{D})$, we define the norm

$$|f|_{\mathcal{A}_1(\mathcal{D})}$$

as the smallest M such that $f \in \mathcal{A}_1(\mathcal{D}, M)$. It will be convenient to use the notation $A_1(\mathcal{D}) := \mathcal{A}_1(\mathcal{D}, 1)$.

It was proved in DeVore & Temlyakov (1996) that for a general dictionary $\mathcal{D}$ the Pure Greedy Algorithm provides the estimate

$$\|f - G_m(f, \mathcal{D})\| \leq |f|_{\mathcal{A}_1(\mathcal{D})} m^{-1/6}.$$

(In this and similar estimates we consider that the inequality holds for all possible choices of $\{G_m\}$.)

The above estimate was improved a little in Konyagin & Temlyakov (1999) to

$$\|f - G_m(f, \mathcal{D})\| \leq 4|f|_{\mathcal{A}_1(\mathcal{D})} m^{-11/62}.$$

In this subsection we discuss recent progress on the following open problem (see Temlyakov (2003a), p. 65, Open Problem 3.1). This problem is a central theoretical problem in greedy approximation in Hilbert spaces.

Open problem. *Find the order of decay of the sequence*

$$\gamma(m) := \sup_{f,\mathcal{D},\{G_m\}} (\|f - G_m(f,\mathcal{D})\| \|f\|_{\mathcal{A}_1(\mathcal{D})}^{-1}),$$

where sup is taken over all dictionaries $\mathcal{D}$, all elements $f \in \mathcal{A}_1(\mathcal{D}) \setminus \{0\}$ and all possible choices of $\{G_m\}$.

Recently, the known upper bounds in approximation by the Pure Greedy Algorithm have been improved upon by Sil'nichenko (2004). Sil'nichenko proved the estimate

$$\gamma(m) \leq Cm^{-\frac{s}{2(2+s)}}$$

where s is a solution from the interval $[1, 1.5]$ of the equation

$$(1+x)^{\frac{1}{2+x}} \left(\frac{2+x}{1+x}\right) - \frac{1+x}{x} = 0.$$

Numerical calculations of s (see Sil'nichenko (2004)) give

$$\frac{s}{2(2+s)} = 0.182\cdots > 11/62 = 0.177\ldots .$$

The technique used in Sil'nichenko (2004) is a further development of a method from Konyagin & Temlyakov (1999).

There is also some progress in the lower estimates. The estimate

$$\gamma(m) \geq Cm^{-0.27},$$

with a positive constant C, has been proved in Livshitz & Temlyakov (2003). For previous lower estimates see Temlyakov (2003a), p. 59.

11.2.2 Greedy Expansions

The PGA and its generalization the Weak Greedy Algorithm (WGA) (see Temlyakov (2003a), p. 54 or the definition of the WGA(b) below with $b = 1$) give, for every element $f \in H$, a convergent expansion in a series with respect to a dictionary $\mathcal{D}$. In this subsection we discuss a further generalization of the WGA that also provides a convergent expansion. We consider here a generalization of the WGA obtained by introducing to it a tuning parameter $b \in (0,1]$ (see Temlyakov (2003b)). Let a sequence $\tau = \{t_k\}_{k=1}^{\infty}$, $0 \leq t_k \leq 1$, and a parameter $b \in (0,1]$ be given. We define the Weak Greedy Algorithm with parameter b as follows.

Weak Greedy Algorithm with Parameter b **(WGA(b))** *We define* $f_0^{\tau,b} := f.$ *Then, for each* $m \geq 1,$ *we inductively define:*

1) $\varphi_m^{\tau,b} \in \mathcal{D}$ *is any element satisfying*

$$|\langle f_{m-1}^{\tau,b}, \varphi_m^{\tau,b} \rangle| \geq t_m \sup_{g \in \mathcal{D}} |\langle f_{m-1}^{\tau,b}, g \rangle|;$$

2) Set

$$f_m^{\tau,b} := f_{m-1}^{\tau,b} - b\langle f_{m-1}^{\tau,b}, \varphi_m^{\tau,b} \rangle \varphi_m^{\tau,b};$$

3) Define

$$G_m^{\tau,b}(f, \mathcal{D}) := b \sum_{j=1}^{m} \langle f_{j-1}^{\tau,b}, \varphi_j^{\tau,b} \rangle \varphi_j^{\tau,b}.$$

We note that the WGA(b) can be seen as a realization of the Approximate Greedy Algorithm studied in Gribonval & Nielsen (2001) and Galatenko & Livshitz (2003).

We point out that the WGA(b) contains, in addition to the first (greedy) step, the second step (see 2), 3) in the above definition) where we update an approximant by adding an orthogonal projection of the residual $f_{m-1}^{\tau,b}$ onto $\varphi_m^{\tau,b}$ multiplied by b. The WGA(b) therefore provides, for each $f \in H$, an expansion into a series (greedy expansion)

$$f \sim \sum_{j=1}^{\infty} c_j(f) \varphi_j^{\tau,b}, \quad c_j(f) := b\langle f_{j-1}^{\tau,b}, \varphi_j^{\tau,b} \rangle.$$

We begin with a convergence result from Temlyakov (2003b). We define by $\mathcal{V}$ the class of sequences $x = \{x_k\}_{k=1}^{\infty},$ $x_k \geq 0,$ $k = 1, 2, \ldots,$ with the following property: there exists a sequence $0 = q_0 < q_1 < \ldots$ that may depend on x such that

$$\sum_{s=1}^{\infty} \frac{2^s}{\Delta q_s} < \infty;$$

and

$$\sum_{s=1}^{\infty} 2^{-s} \sum_{k=1}^{q_s} x_k^2 < \infty,$$

where $\Delta q_s := q_s - q_{s-1}.$

Theorem 11.1 *Let* $\tau \notin \mathcal{V}.$ *Then, the WGA(b) with* $b \in (0, 1]$ *converges for each* f *and all Hilbert spaces* H *and dictionaries* $\mathcal{D}.$

Theorem 11.1 is an extension of the corresponding result for the WGA (see Temlyakov (2003a), p. 58).

We proved in Temlyakov (2003b) the following convergence rate of the WGA(b).

Theorem 11.2 *Let $\mathcal{D}$ be an arbitrary dictionary in H. Assume $\tau :=$ $\{t_k\}_{k=1}^{\infty}$ is a nonincreasing sequence and $b \in (0, 1]$. Then, for $f \in A_1(\mathcal{D})$, we have*

$$\|f - G_m^{\tau, b}(f, \mathcal{D})\| \leq \left(1 + b(2 - b) \sum_{k=1}^{m} t_k^2 \right)^{-(2-b)t_m / 2(2 + (2-b)t_m)}.$$

This theorem is an extension of the corresponding result for the WGA (see Temlyakov (2003a), p. 59). In the particular case $t_k = 1$, $k = 1, 2, \ldots$, we get the following rate of convergence

$$\|f - G_m^{1, b}(f, \mathcal{D})\| \leq Cm^{-r(b)}, \quad r(b) := \frac{2 - b}{2(4 - b)}.$$

We note that $r(1) = 1/6$ and $r(b) \to 1/4$ as $b \to 0$. Thus we can offer the following observation. At each step of the Pure Greedy Algorithm we can choose a fixed fraction of the optimal coefficient for that step instead of the optimal coefficient itself. Surprisingly, this leads to better upper estimates than those known for the Pure Greedy Algorithm.

11.2.3 Other Expansions

Important features of approximation schemes discussed in the previous subsection are the following. An approximation method (algorithm) provides an expansion and in addition each dictionary element participating in this expansion is chosen at a greedy step (it is a maximizer or a weak maximizer of $|\langle f_{m-1}, g \rangle|$ over $g \in \mathcal{D}$). The best upper estimates for such schemes (for $f \in A_1(\mathcal{D})$) are not better than $m^{-1/4}$. It is known that some greedy type algorithms, for instance, the Orthogonal Greedy Algorithm, provide a better rate of convergence $m^{-1/2}$ (see Temlyakov (2003a), p. 60). However, these algorithms do not provide an expansion. In this subsection we discuss very recent results of E. D. Livshitz on a construction of expansions (not greedy expansions) that provide a rate of convergence close to $m^{-1/2}$. This construction contains, in addition to the greedy step, another step that is not a greedy step. The following algorithm has been studied in Livshitz (2004).

Weak Reversing Greedy Algorithm (WRsGA) *Let $t \in (0, 1]$. For a given $\mathcal{D}$ and $f \in H$ we set $f_0 := f$. Assume that for $n \geq 0$ a sequence of residuals $f_0, \ldots, f_n$, a sequence of dictionary elements $g_1, \ldots, g_n$, and a* *sequence of coefficients* $c_1, \ldots, c_n$ *have already been defined in such a way that*

$$f_k = f_0 - \sum_{i=1}^{k} c_i g_i, \quad 1 \leq k \leq n.$$

Denote

$$D_n := \{g_i\}_{i=1}^{n},$$

$$v_n(g) := \sum_{i \leq n, g_i = g} c_i,$$

$$b_n := \sum_{g \in D_n} |v_n(g)|,$$

$$a_n := \langle f_n, f_n \rangle.$$

Let $n \geq 1$. *Consider the two inequalities*

$$\min(|v_n(g)|, |\langle f_n, g \rangle|) \geq \frac{a_n t^2}{9 b_n}, \tag{11.2}$$

$$\mathrm{sign}(v_n(g)) = -\mathrm{sign}(\langle f_n, g \rangle). \tag{11.3}$$

If there exists a $g \in D_n$ *satisfying both (11.2) and (11.3) then we set*

$$g_{n+1} := g, \quad c_{n+1} := \mathrm{sign}(\langle f_n, g \rangle) \min(|v_n(g)|, |\langle f_n, g \rangle|)$$

and

$$f_{n+1} := f_n - c_{n+1} g_{n+1}.$$

If there is no $g \in D_n$ *satisfying both (11.2) and (11.3) then we apply a greedy step and choose as* g_{n+1} *any element of the dictionary satisfying*

$$|\langle f_n, g_{n+1} \rangle| \geq t \sup_{g \in \mathcal{D}} |\langle f_n, g \rangle|.$$

Define

$$c_{n+1} := \langle f_n, g_{n+1} \rangle, \quad f_{n+1} := f_n - c_{n+1} g_{n+1}.$$

This algorithm has two different ways of choosing a new element g_{n+1} from the dictionary. One of them is the greedy step. A new feature of the above algorithm is the other step that we call a *reversing step*. The following result has been obtained in Livshitz (2004).

Theorem 11.3 *Let* $t \in (0, 1]$ *and* $f \in A_1(\mathcal{D})$. *Then,*

$$\|f_m\| \leq C(t) m^{-1/2} \ln m.$$

11.3 Greedy expansions in Banach spaces

Greedy expansions in Hilbert spaces have been well studied. The theory of greedy expansions in Hilbert spaces began in 1987 with the celebrated paper Jones (1987) on the convergence of the PGA. The WGA has also been studied extensively (see survey Temlyakov (2003a)). The corresponding convergence theorems and estimates for the rate of convergence are known. Recent results on greedy expansions in Hilbert spaces have been discussed in Section 11.2. Much less is known about greedy expansions in Banach spaces. There were only a few scattered results (see Dilworth, Kutzarova & Temlyakov (2002)) on greedy expansions in Banach spaces until a recent breakthrough result in Ganichev & Kalton (2003). We proceed to the formulation of this result. We begin with a definition of the Weak Dual Greedy Algorithm (see Temlyakov (2003a), p. 66). We assume here that $\mathcal{D}$ is a symmetric dictionary.

Weak Dual Greedy Algorithm (WDGA) *We define $f_0^D := f_0^{D,\tau} := f$. Then, for each $m \geq 1$, we inductively define:*

1) $\varphi_m^D := \varphi_m^{D,\tau} \in \mathcal{D}$ is any element satisfying

$$F_{f_{m-1}^D}(\varphi_m^D) \geq t_m \sup_{g \in \mathcal{D}} F_{f_{m-1}^D}(g);$$

2) a_m is such that

$$\left\| f_{m-1}^D - a_m \varphi_m^D \right\| = \min_{a \in \mathbb{R}} \left\| f_{m-1}^D - a \varphi_m^D \right\|;$$

3) Denote

$$f_m^D := f_m^{D,\tau} := f_{m-1}^D - a_m \varphi_m^D.$$

In the case $\tau = \{1\}$ the WDGA coincides with the Dual Greedy Algorithm (DGA). The following conjecture was formulated in Temlyakov (2003a), p. 73, Open problem 4.3: the Dual Greedy Algorithm converges for all dictionaries $\mathcal{D}$ and each element $f \in X$ in uniformly smooth Banach spaces X with modulus of smoothness of fixed power type q, $1 < q \leq 2$, $(\rho(u) \leq \gamma u^q)$.

Recently, the following very interesting result has been proved in Ganichev & Kalton (2003).

Theorem 11.4 *Let $\tau = \{t\}$, $t \in (0,1]$, and $X = L_p$, $1 < p < \infty$. Then, the WDGA converges for any dictionary $\mathcal{D}$ for all functions $f \in L_p$.*

This result is a first step in studying greedy expansions in Banach spaces. We also mention that E. D. Livshitz (see Livshitz (2003)) constructed expansions in Banach spaces based on different greedy-type algorithms. Theorem 11.4 gives a partial answer to the above conjecture. However, the general case of uniformly smooth Banach spaces X is still open. Theorem 11.4 also leaves open the following two important problems.

1. Is there a greedy type algorithm that provides convergent greedy expansions for each $\mathcal{D}$ and all f in a uniformly smooth Banach space?
2. What is a rate of convergence of greedy expansions for $f \in A_1(\mathcal{D})$?

In the paper Temlyakov (2003b) we addressed these two problems. We considered there a modification of the WDGA. The new feature of that modification is a method of selecting the mth coefficient $c_m(f)$ of the expansion (11.1). An approach developed in the paper Temlyakov (2003b) works in any uniformly smooth Banach space.

We proceed to the definition of the algorithm that we studied in Temlyakov (2003b).

Dual Greedy Algorithm with Parameters (t, b, μ) **(DGA**(t, b, μ)**)** *Let X be a uniformly smooth Banach space with modulus of smoothness $\rho(u)$ and let $\mu(u)$ be a continuous majorant of $\rho(u)$: $\rho(u) \leq \mu(u)$, $u \in [0, \infty)$. For parameters $t \in (0, 1]$, $b \in (0, 1]$ we define sequences $\{f_m\}_{m=0}^{\infty}$, $\{\varphi_m\}_{m=1}^{\infty}$, $\{c_m\}_{m=1}^{\infty}$ inductively. Let $f_0 := f$. If for $m \geq 1$ $f_{m-1} = 0$ then we set $f_j = 0$ for $j \geq m$ and stop. If $f_{m-1} \neq 0$ then we perform the following three steps.*

1) Take any $\varphi_m \in \mathcal{D}$ such that

$$F_{f_{m-1}}(\varphi_m) \geq t r_{\mathcal{D}}(f_{m-1}). \tag{11.4}$$

2) Choose $c_m > 0$ from the equation

$$\|f_{m-1}\| \mu(c_m / \|f_{m-1}\|) = \frac{tb}{2} c_m r_{\mathcal{D}}(f_{m-1}).$$

3) Define

$$f_m := f_{m-1} - c_m \varphi_m.$$

In Temlyakov (2003b) we proved the following convergence result.

Theorem 11.5 *Let X be a uniformly smooth Banach space with modulus of smoothness $\rho(u)$ and let $\mu(u)$ be a continuous majorant of $\rho(u)$ with the property $\mu(u)/u \downarrow 0$ as $u \to +0$. Then, for any $t \in (0, 1]$ and $b \in (0, 1)$, the DGA(t, b, μ) converges for each dictionary $\mathcal{D}$ and all $f \in X$.*

The following result from Temlyakov (2003b) gives a rate of convergence.

Theorem 11.6 *Assume X has modulus of smoothness $\rho(u) \leq \gamma u^q$, $q \in (1, 2]$. Denote $\mu(u) = \gamma u^q$. Then, for any dictionary $\mathcal{D}$ and any $f \in A_1(\mathcal{D})$,* *the rate of convergence of the DGA(t, b, μ) is bounded by*

$$\|f_m\| \leq C(t, b, \gamma, q) m^{-\frac{t(1-b)}{p(1+t(1-b))}}, \quad p := \frac{q}{q-1}.$$

Let us discuss an application of Theorem 11.6 in the case of a Hilbert space. It is well known and easy to check that for a Hilbert space H one has

$$\rho(u) \leq (1 + u^2)^{1/2} - 1 \leq u^2/2.$$

Therefore, by Theorem 11.6 with $\mu(u) = u^2/2$, the DGA(t, b, μ) provides the error estimate

$$\|f_m\| \leq C(t, b) m^{-\frac{t(1-b)}{2(1+t(1-b))}} \quad \text{for} \quad f \in A_1(\mathcal{D}). \tag{11.5}$$

The estimate (11.5) with $t = 1$ gives

$$\|f_m\| \leq C(b) m^{-\frac{1-b}{2(2-b)}} \quad \text{for} \quad f \in A_1(\mathcal{D}). \tag{11.6}$$

The exponent $\frac{1-b}{2(2-b)}$ in this estimate approaches $1/4$ as b approaches 0. Comparing (11.6) with the upper estimates from Section 11.2 for the PGA we observe that the DGA$(1, b, u^2/2)$ with small b provides a better upper estimate for the rate of convergence than the known estimates for the PGA. Thus, for the general method DGA$(1, b, u^2/2)$ designed in a Banach space we have a phenomenon similar to that for the WGA(b).

Let us determine how the DGA$(1, b, u^2/2)$ works in a Hilbert space. Consider its mth step. Let $\varphi_m \in \mathcal{D}$ be as in (11.4). It is then clear that φ_m maximizes the $\langle f_{m-1}, g \rangle$ over the dictionary $\mathcal{D}$ and

$$\langle f_{m-1}, \varphi_m \rangle = \|f_{m-1}\| r_{\mathcal{D}}(f_{m-1}).$$

The PGA would use φ_m with the coefficient $\langle f_{m-1}, \varphi_m \rangle$ at this step. The DGA$(1, b, u^2/2)$ uses the same φ_m and only a fraction of $\langle f_{m-1}, \varphi_m \rangle$:

$$c_m = b\|f_{m-1}\| r_{\mathcal{D}}(f_{m-1}). \tag{11.7}$$

Thus the choice $b = 1$ in (11.7) corresponds to the PGA. However, our technique from Temlyakov (2003b), designed for general Banach spaces, does not work in the case $b = 1$. From the above discussion we have the following surprising observation. The use of a small fraction ($c_m = b\langle f_{m-1}, \varphi_m \rangle$) of an optimal coefficient results in the improvement of the upper estimate for the rate of convergence.

11.4 Greedy algorithms with approximate evaluations and restricted search

Let $\tau := \{t_k\}_{k=1}^{\infty}$ be a given sequence of nonnegative numbers $t_k \leq 1$, $k = 1, \ldots$. We first define (see Temlyakov (2003a), p. 66) the Weak Chebyshev Greedy Algorithm (WCGA) that is a generalization for Banach spaces of the Weak Orthogonal Greedy Algorithm.

Weak Chebyshev Greedy Algorithm (WCGA) *We define* $f_0^c := f_0^{c,\tau} := f$. *Then, for each* $m \geq 1$ *we inductively define:*
1) $\varphi_m^c := \varphi_m^{c,\tau} \in \mathcal{D}$ *is any element satisfying*

$$F_{f_{m-1}^c}(\varphi_m^c) \geq t_m \sup_{g \in \mathcal{D}} F_{f_{m-1}^c}(g).$$

2) Define

$$\Phi_m := \Phi_m^{\tau} := \operatorname{span}\{\varphi_j^c\}_{j=1}^m,$$

and define $G_m^c := G_m^{c,\tau}$ *to be the best approximant to* f *from* Φ_m.
3) Denote

$$f_m^c := f_m^{c,\tau} := f - G_m^c.$$

In the case $t_k = 1$, $k = 1, 2, \ldots$, we call the WCGA the Chebyshev Greedy Algorithm (CGA). Both the WCGA and the CGA are theoretical greedy approximation methods. The term *weak* in the above definition means that at step 1) we do not shoot for the optimal element of the dictionary which realizes the corresponding supremum but are satisfied with a weaker property than being optimal. The obvious reason for this is that we do not know in general that the optimal element exists. Another practical reason is that the weaker the assumption the easier it is to satisfy it and, therefore, easier to realize in practice. However, it is clear that in the case of an infinite dictionary $\mathcal{D}$ there is no direct computationally feasible way to evaluate $\sup_{g \in \mathcal{D}} F_{f_{m-1}^c}(g)$.

At the second step we are looking for the best approximant of f from Φ_m. We know that such an approximant exists. However, in practice we cannot find it exactly. We can only find it approximately with error.

The above observations motivated us to consider a variant of the WCGA with an eye towards practically implementable algorithms.

We studied in Temlyakov (2002) the following modification of the WCGA. Let three sequences $\tau = \{t_k\}_{k=1}^{\infty}$, $\delta = \{\delta_k\}_{k=0}^{\infty}$, $\eta = \{\eta_k\}_{k=1}^{\infty}$ of numbers from $[0, 1]$ be given.

Approximate Weak Chebyshev Greedy Algorithm (AWCGA) *We define* $f_0 := f_0^{\tau, \delta, \eta} := f$. *Then, for each* $m \geq 1$ *we inductively define:*

1) F_{m-1} is a functional with properties

$$\|F_{m-1}\| \leq 1, \qquad F_{m-1}(f_{m-1}) \geq \|f_{m-1}\|(1 - \delta_{m-1});$$

and $\varphi_m := \varphi_m^{\tau,\delta,\eta} \in \mathcal{D}$ is any element satisfying

$$F_{m-1}(\varphi_m) \geq t_m \sup_{g \in \mathcal{D}} F_{m-1}(g).$$

2) Define

$$\Phi_m := \operatorname{span}\{\varphi_j\}_{j=1}^m,$$

and denote

$$E_m(f) := \inf_{\varphi \in \Phi_m} \|f - \varphi\|.$$

Let $G_m \in \Phi_m$ be such that

$$\|f - G_m\| \leq E_m(f)(1 + \eta_m).$$

3) Denote

$$f_m := f_m^{\tau,\delta,\eta} := f - G_m.$$

The term *approximate* in this definition means that we use a functional F_{m-1} that is an approximation to the norming (peak) functional $F_{f_{m-1}}$ and also that we use an approximant $G_m \in \Phi_m$ which satisfies a weaker assumption than being a best approximant to f from Φ_m. Thus, in the *approximate* version of the WCGA, we have addressed the issue of inexact evaluation of the norming functional and the best approximant. We did not address the issue of finding the $\sup_{g \in \mathcal{D}} F_{f_{m-1}^c}(g)$. In the paper Temlyakov (2004) we addressed this issue. We did it in two steps. First we considered the corresponding modification of the WCGA, and then the modification of the AWCGA. These modifications are done in the style of the concept of *depth search* from Donoho (2001).

We now consider a countable dictionary $\mathcal{D} = \{\pm\psi_j\}_{j=1}^\infty$. We denote $\mathcal{D}(N) := \{\pm\psi_j\}_{j=1}^N$. Let $\mathcal{N} := \{N_j\}_{j=1}^\infty$ be a sequence of natural numbers.

Restricted Weak Chebyshev Greedy Algorithm (RWCGA) *We define $f_0 := f_0^{c,\tau,\mathcal{N}} := f$. Then, for each $m \geq 1$ we inductively define:*
1) $\varphi_m := \varphi_m^{c,\tau,\mathcal{N}} \in \mathcal{D}(N_m)$ is any element satisfying

$$F_{f_{m-1}}(\varphi_m) \geq t_m \sup_{g \in \mathcal{D}(N_m)} F_{f_{m-1}}(g).$$

2) Define

$$\Phi_m := \Phi_m^{\tau,\mathcal{N}} := \operatorname{span}\{\varphi_j\}_{j=1}^m,$$

and define $G_m := G_m^{c,\tau,\mathcal{N}}$ to be the best approximant to f from Φ_m.

3) Denote

$$f_m := f_m^{c,\tau,\mathcal{N}} := f - G_m.$$

We formulate some results from Temlyakov (2002) and Temlyakov (2004) in the particular case of a uniformly smooth Banach space with modulus of smoothness of power type (see Temlyakov (2002), Temlyakov (2004) for the general case). The following theorem was proved in Temlyakov (2002).

Theorem 11.7 *Let a Banach space X have modulus of smoothness $\rho(u)$ of power type $1 < q \le 2$; $(\rho(u) \le \gamma u^q)$. Assume that*

$$\sum_{m=1}^{\infty} t_m^p = \infty, \quad p = \frac{q}{q-1};$$

and

$$\delta_m = o(t_m^p), \qquad \eta_m = o(t_m^p).$$

Then, the AWCGA converges for any $f \in X$.

We now give two theorems from Temlyakov (2004) on greedy algorithms with restricted search.

Theorem 11.8 *Let a Banach space X have modulus of smoothness $\rho(u)$ of power type $1 < q \le 2$; $(\rho(u) \le \gamma u^q)$. Assume that $\lim_{m \to \infty} N_m = \infty$ and*

$$\sum_{m=1}^{\infty} t_m^p = \infty, \quad p = \frac{q}{q-1}.$$

Then, the RWCGA converges for any $f \in X$.

For $b > 0$, $K > 0$, we define the class

$$\mathcal{A}_1^b(K, \mathcal{D}) := \{f : d(f, \mathcal{A}_1(\mathcal{D}(n))) \le K n^{-b}, \quad n = 1, 2, \dots\}.$$

Here, $\mathcal{A}_1(\mathcal{D}(n))$ is a convex hull of $\{\pm \psi_j\}_{j=1}^n$ and for a compact set F

$$d(f, F) := \inf_{\phi \in F} \|f - \phi\|.$$

Theorem 11.9 *Let X be a uniformly smooth Banach space with modulus of smoothness $\rho(u) \le \gamma u^q$, $1 < q \le 2$. Then, for $t \in (0,1]$ there exist $C_1(t, \gamma, q, K)$, $C_2(t, \gamma, q, K)$ such that for $\mathcal{N}$ with $N_m \ge C_1(t, \gamma, q, K) m^{r/b}$, $m = 1, 2, \dots$ we have, for any $f \in \mathcal{A}_1^b(K, \mathcal{D})$,*

$$\|f_m^{c,\tau,\mathcal{N}}\| \le C_2(t, \gamma, q, K) m^{-r}, \quad \tau = \{t\}, \quad r := 1 - 1/q.$$

We note that we can choose an algorithm from Theorem 11.9 that satisfies the *polynomial depth search* condition $N_m \leq Cm^a$ from Donoho (2001).

We proceed to an algorithm that combines approximate evaluations with restricted search. Let three sequences $\tau = \{t_k\}_{k=1}^{\infty}$, $\delta = \{\delta_k\}_{k=0}^{\infty}$, $\eta = \{\eta_k\}_{k=1}^{\infty}$ of numbers from $[0, 1]$ be given. Let $\mathcal{N} := \{N_j\}_{j=1}^{\infty}$ be a sequence of natural numbers.

Restricted Approximate Weak Chebyshev Greedy Algorithm (RAWCGA) *We define* $f_0 := f_0^{\tau, \delta, \eta, \mathcal{N}} := f$. *Then, for each* $m \geq 1$ *we inductively define:*

1) F_{m-1} *is a functional with properties*

$$\|F_{m-1}\| \leq 1, \qquad F_{m-1}(f_{m-1}) \geq \|f_{m-1}\|(1 - \delta_{m-1});$$

and $\varphi_m := \varphi_m^{\tau, \delta, \eta, \mathcal{N}} \in \mathcal{D}(N_m)$ *is any element satisfying*

$$F_{m-1}(\varphi_m) \geq t_m \sup_{g \in \mathcal{D}(N_m)} F_{m-1}(g).$$

2) Define

$$\Phi_m := \operatorname{span}\{\varphi_j\}_{j=1}^{m},$$

and denote

$$E_m(f) := \inf_{\varphi \in \Phi_m} \|f - \varphi\|.$$

Let $G_m \in \Phi_m$ *be such that*

$$\|f - G_m\| \leq E_m(f)(1 + \eta_m).$$

3) Denote

$$f_m := f_m^{\tau, \delta, \eta, \mathcal{N}} := f - G_m.$$

Theorem 11.10 *Let a Banach space* X *have modulus of smoothness* $\rho(u)$ *of power type* $1 < q \leq 2$; ($\rho(u) \leq \gamma u^q$). *Assume that* $\lim_{m \to \infty} N_m = \infty$,

$$\sum_{m=1}^{\infty} t_m^p = \infty, \qquad p = \frac{q}{q-1},$$

and

$$\delta_m = o(t_m^p), \qquad \eta_m = o(t_m^p).$$

Then, the RAWCGA converges for any $f \in X$.

We now make some general remarks on m-term approximation with the depth search constraint. The depth search constraint means that for a given m we restrict ourselves to systems of elements (subdictionaries) containing at most $N := N(m)$ elements. Let X be a linear metric space and for a set

$\mathcal{D} \subset X$, let $\mathcal{L}_m(\mathcal{D})$ denote the collection of all linear spaces spanned by m elements of $\mathcal{D}$. For a linear space $L \subset X$, the ϵ-neighborhood $U_\epsilon(L)$ of L is the set of all $x \in X$ which are at a distance not exceeding ϵ from L (i.e., those $x \in X$ which can be approximated to an error not exceeding ϵ by the elements of L). For any compact set $F \subset X$ and any integers $N, m \geq 1$, we define the (N, m)-entropy numbers (see Temlyakov (2003a), p. 94)

$$\epsilon_{N,m}(F, X) := \inf_{\#\mathcal{D}=N} \inf \left\{ \epsilon : F \subset \cup_{L \in \mathcal{L}_m(\mathcal{D})} U_\epsilon(L) \right\}.$$

We let $\Sigma_m(\mathcal{D})$ denote the collection of all functions (elements) in X which can be expressed as a linear combination of at most m elements of $\mathcal{D}$. Thus each function $s \in \Sigma_m(\mathcal{D})$ can be written in the form

$$s = \sum_{g \in \Lambda} c_g g, \quad \Lambda \subset \mathcal{D}, \quad \#\Lambda \leq m,$$

where the c_g are real or complex numbers. In some cases, it may be possible to write an element from $\Sigma_m(\mathcal{D})$ in this form in more than one way. The space $\Sigma_m(\mathcal{D})$ is not linear: the sum of two functions from $\Sigma_m(\mathcal{D})$ is generally not in $\Sigma_m(\mathcal{D})$.

For a function $f \in X$ we define its best m-term approximation error

$$\sigma_m(f) := \sigma_m(f, \mathcal{D}) := \inf_{s \in \Sigma_m(\mathcal{D})} \|f - s\|.$$

For a function class $F \subset X$ we define

$$\sigma_m(F) := \sigma_m(F, \mathcal{D}) := \sup_{f \in F} \sigma_m(f, \mathcal{D}).$$

We can express $\sigma_m(F, \mathcal{D})$ as

$$\sigma_m(F, \mathcal{D}) = \inf \left\{ \epsilon : F \subset \cup_{L \in \mathcal{L}_m(\mathcal{D})} U_\epsilon(L) \right\}.$$

It follows therefore that

$$\inf_{\#\mathcal{D}=N} \sigma_m(F, \mathcal{D}) = \epsilon_{N,m}(F, X).$$

In other words, finding best dictionaries consisting of N elements for m-term approximation of F is the same as finding sets $\mathcal{D}$ which attain the (N, m)-entropy numbers $\epsilon_{N,m}(F, X)$. It is easy to see that $\epsilon_{m,m}(F, X) = d_m(F, X)$ where $d_m(F, X)$ is the Kolmogorov width of F in X. This establishes a connection between (N, m)-entropy numbers and the Kolmogorov widths. One can find a further discussion on the nonlinear Kolmogorov (N, m)-widths and the entropy numbers in Temlyakov (2003a).

11.5 A specific class of dictionaries

In Sections 11.2–11.4 we considered a setting as general as possible: an arbitrary Hilbert or Banach space and any dictionary. We pointed out in Temlyakov (2003a) that such a generality is justified by the contemporary needs of approximation theory and computational mathematics. The generality of the setting allows us to apply the corresponding results for any dictionary. In this section we discuss the question of how we can use the specifics of a given dictionary $\mathcal{D}$ in order to improve our estimates for the approximation rate of some of our favorite algorithms. One can find a discussion of this question in the case of λ-quasi-orthogonal dictionary $\mathcal{D}$ in Temlyakov (2003a), p. 62–64.

In this section we consider dictionaries that have become popular in signal processing. Denote by

$$M(\mathcal{D}) := \sup_{g \neq h; g, h \in \mathcal{D}} |\langle g, h \rangle|$$

the coherence parameter of a dictionary $\mathcal{D}$. For an orthonormal basis $\mathcal{B}$ we have $M(\mathcal{B}) = 0$. It is clear that the smaller the $M(\mathcal{D})$ the more the dictionary $\mathcal{D}$ resembles an orthonormal basis. However, we should note that, in the case $M(\mathcal{D}) > 0$, $\mathcal{D}$ can be a redundant dictionary.

We now proceed to a discussion of the Orthogonal Greedy Algorithm (OGA). If H_0 is a finite-dimensional subspace of H, we let P_{H_0} be the orthogonal projector from H onto H_0. That is, $P_{H_0}(f)$ is the best approximation to f from H_0. As above, we let $g(f) \in \mathcal{D}$ be an element from $\mathcal{D}$ which maximizes $|\langle f, g \rangle|$. We shall assume, for simplicity, that such a maximizer exists; if not, suitable modifications are necessary (see Weak Orthogonal Greedy Algorithm in Temlyakov (2003a), p. 55) in the algorithm that follows.

Orthogonal Greedy Algorithm (OGA) *We define* $f_0^o := R_0^o(f) := R_0^o(f, \mathcal{D}) := f$ *and* $G_0^o(f) := G_0^o(f, \mathcal{D}) := 0$. *Then, for each* $m \geq 1$, *we inductively define:*

$$H_m := H_m(f) := \operatorname{span}\left\{ g\big(R_0^o(f)\big), \ldots, g\big(R_{m-1}^o(f)\big) \right\},$$

$$G_m^o(f) := G_m^o(f, \mathcal{D}) := P_{H_m}(f),$$

$$f_m^o := R_m^o(f) := R_m^o(f, \mathcal{D}) := f - G_m^o(f).$$

We discuss the performance of the OGA with respect to a dictionary with coherence parameter M. Let $\{f_m^o\}_{m=1}^{\infty}$ be a sequence of residuals of the OGA applied to an element f. It is natural to compare $\|f_m\|$ with the corresponding $\sigma_m(f, \mathcal{D})$. The idea of such a comparison for evaluating the quality of an approximation method goes back to Lebesgue. He proved the following inequality. For any 2π-periodic continuous function f one has

$$\|f - S_n(f)\|_\infty \le \left(4 + \frac{4}{\pi^2}\ln n\right) E_n(f)_\infty, \tag{11.8}$$

where $S_n(f)$ is the nth partial sum of the Fourier series of f and $E_n(f)_\infty$ is the error of the best approximation of f by trigonometric polynomials of order n in the uniform norm $\|\cdot\|_\infty$. The inequality (11.8) relates the error of a particular method (S_n) of approximation by trigonometric polynomials of order n to the best possible error $E_n(f)_\infty$ of approximation by trigonometric polynomials of order n. By a Lebesgue type inequality we mean an inequality that provides an upper bound for the error of a particular method of approximation of f by elements of a special form, say, form $\mathcal{A}$, by the best possible approximation of f by elements of the form $\mathcal{A}$. In the case of approximation with regards to bases (or minimal systems) Lebesgue type inequalities are known both in linear and in nonlinear settings (see the surveys Konyagin & Temlyakov (2002), Temlyakov (2003a)). It would be very interesting to prove Lebesgue type inequalities for redundant systems (dictionaries). However, there are substantial difficulties on this way. To illustrate these difficulties we give an example from DeVore & Temlyakov (1996).

Let $\mathcal{B} := \{h_k\}_{k=1}^\infty$ be an orthonormal basis in a Hilbert space H. Consider the following element

$$g := Ah_1 + Ah_2 + aA \sum_{k \ge 3} (k(k+1))^{-1/2} h_k$$

with

$$A := (33/89)^{1/2} \quad \text{and} \quad a := (23/11)^{1/2}.$$

Then, $\|g\| = 1$. We define the dictionary $\mathcal{D} = \mathcal{B} \cup \{g\}$. It has been proved in DeVore & Temlyakov (1996) that for the function

$$f = h_1 + h_2$$

we have

$$\|f - G_m(f, \mathcal{D})\| \ge m^{-1/2}, \quad m \ge 4.$$

It is clear that $\sigma_2(f, \mathcal{D}) = 0$.

We, therefore, look for conditions on a dictionary $\mathcal{D}$ that allow us to prove Lebesgue type inequalities. The condition that $\mathcal{D} = \mathcal{B}$ is an orthonormal basis for H guarantees that

$$\|R_m(f, \mathcal{B})\| = \sigma_m(f, \mathcal{B}).$$

This is an ideal situation. The results that we will discuss here concern the case when we replace an orthonormal basis $\mathcal{B}$ by a dictionary that is, in a certain sense, not far from an orthonormal basis.

The first general Lebesgue type inequality for the OGA for an M-coherent dictionary was obtained in Gilbert, Muthukrishnan & Strauss (2003). They proved that

$$\|f_m^o\| \leq 8m^{1/2}\sigma_m(f) \quad \text{for} \quad m < 1/(32M). \tag{11.9}$$

The constants in this inequality were improved upon in Tropp (2004) (see also Donoho, Elad & Temlyakov (2004)):

$$\|f_m^o\| \leq (1 + 6m)^{1/2}\sigma_m(f) \quad \text{for} \quad m < 1/(3M). \tag{11.10}$$

We proved in Donoho, Elad & Temlyakov (2005) the following Lebesgue type inequality for the OGA.

Theorem 11.11 *Let a dictionary $\mathcal{D}$ have mutual coherence $M = M(\mathcal{D})$. Then, for any $S \leq 1/(2M)$ we have the following inequalities*

$$\left\|f_S^o\right\|^2 \leq 2\|f_k^o\|(\sigma_{S-k}(f_k^o) + 3MS\|f_k^o\|), \quad 0 \leq k \leq S. \tag{11.11}$$

The inequalities (11.11) can be used for improving (11.10) for small m. We proved in Donoho, Elad & Temlyakov (2005) the following inequalities.

Theorem 11.12 *Let a dictionary $\mathcal{D}$ have mutual coherence $M = M(\mathcal{D})$. Assume $m \leq 0.05M^{-2/3}$. Then, for $l \geq 1$ satisfying $2^l \leq \log m$ we have*

$$\left\|f_{m(2^l-1)}^o\right\| \leq 6m^{2^{-l}}\sigma_m(f).$$

Corollary 11.1 *Let a dictionary $\mathcal{D}$ have mutual coherence $M = M(\mathcal{D})$. Assume $m \leq 0.05M^{-2/3}$. Then, we have*

$$\left\|f_{[m \log m]}^o\right\| \leq 24\sigma_m(f).$$

We note that, in particular, the inequality (11.10) implies the following nice property of the OGA.

Theorem 11.13 (E-property (Exact recovery of S-term polynomial)) *Let a dictionary $\mathcal{D}$ have mutual coherence $M = M(\mathcal{D})$. Assume that*

$$f = \sum_{j=1}^{S} a_j g_j, \quad g_j \in \mathcal{D}, \quad j = 1, \dots, S,$$

and $S < 1/(3M)$. Then, the OGA with respect to $\mathcal{D}$ recovers f exactly after S iterations ($f_S = 0$).

It is known (see Tropp (2004), Donoho, Elad & Temlyakov (2004)) that the OGA has the E-property with $S < (1+1/M)/2$ and that the condition $S < (1 + 1/M)/2$ is sharp. The proof of the E-property of the OGA is simpler than the proof of (11.10). We present here a generalization of the concept of M-coherent dictionary and the E-property to the case of Banach spaces.

Let $\mathcal{D}$ be a dictionary in a Banach space X. We define the coherence parameter of this dictionary in the following way

$$M(\mathcal{D}) := \sup_{g \neq h; g, h \in \mathcal{D}} \sup_{F_g} |F_g(h)|.$$

We note that, in general, a norming functional F_g is not unique. This is why we take $\sup_{F_g}$ over all norming functionals of g in the definition of $M(\mathcal{D})$. We do not need $\sup_{F_g}$ in the definition of $M(\mathcal{D})$ if for each $g \in \mathcal{D}$ there is a unique norming functional $F_g \in X'$. Then, we define $\mathcal{D}' := \{F_g, g \in \mathcal{D}\}$ and call $\mathcal{D}'$ a *dual dictionary* to a dictionary $\mathcal{D}$. It is known that the uniqueness of the norming functional F_g is equivalent to the property that g is a point of Gateaux smoothness:

$$\lim_{u \to 0} (\|g + uy\| + \|g - uy\| - 2\|g\|)/u = 0$$

for any $y \in X$. In particular, if X is uniformly smooth then F_f is unique for any $f \neq 0$. We consider the following greedy algorithm.

Weak Quasi-Orthogonal Greedy Algorithm (WQGA) *Let $t \in (0, 1]$. Denote $f_0 := f_0^{q,t} := f$ and find $\varphi_1 := \varphi_1^{q,t} \in \mathcal{D}$ such that*

$$|F_{\varphi_1}(f_0)| \geq t \sup_{g \in \mathcal{D}} |F_g(f_0)|.$$

Next, we find c_1 satisfying

$$F_{\varphi_1}(f - c_1\varphi_1) = 0.$$

Denote $f_1 := f_1^{q,t} := f - c_1\varphi_1$.

We continue this construction in an inductive way. Assume that we have already constructed residuals $f_0, f_1, \ldots, f_{m-1}$ and dictionary elements $\varphi_1, \ldots, \varphi_{m-1}$. Now, we pick an element $\varphi_m := \varphi_m^{q,t} \in \mathcal{D}$ such that

$$|F_{\varphi_m}(f_{m-1})| \geq t \sup_{g \in \mathcal{D}} |F_g(f_{m-1})|.$$

Next, we look for $c_1^m, \ldots, c_m^m$ satisfying

$$F_{\varphi_j}\left(f - \sum_{i=1}^{m} c_i^m \varphi_i\right) = 0, \quad j = 1, \ldots, m. \tag{11.12}$$

If there is no solution to (11.12) then we stop, otherwise we denote $f_m :=$ $f_m^{q,t} := f - \sum_{i=1}^{m} c_i^m \varphi_i$ *with* $c_1^m, \ldots, c_m^m$ *satisfying (11.12).*

We note that (11.12) has a unique solution if $\det \|F_{\varphi_j}(\varphi_i)\|_{i,j=1}^{m} \neq 0$. We apply the WQGA in the case of a dictionary with the coherence parameter $M := M(\mathcal{D})$. Then, by a simple well known argument on the linear independence of the rows of the matrix $\|F_{\varphi_j}(\varphi_i)\|_{i,j=1}^{m}$, we conclude that (11.12) has a unique solution for any $m < 1 + 1/M$. Thus, in the case of an M-coherent dictionary $\mathcal{D}$, we can run the WQGA for at least $[1/M]$ iterations.

Lemma 11.1 *Let* $t \in (0,1]$. *Assume that* $\mathcal{D}$ *has coherence parameter* M. *Let* $S < \frac{t}{1+t}(1 + 1/M)$. *Then, for any* f *of the form*

$$f = \sum_{i=1}^{S} a_i \psi_i,$$

where ψ_i *are distinct elements of* $\mathcal{D}$, *we have that* $\varphi_1^{q,t} = \psi_j$ *with some* $j \in [1, S]$.

Proof Let $A := \max_i |a_i| = |a_p| > 0$. Then,

$$|F_{\psi_p}(f)| \geq |F_{\psi_p}(a_p \psi_p)| - \sum_{j \neq p} |a_j F_{\psi_p}(\psi_j)| \geq A(1 - M(S-1)).$$

Therefore,

$$\max_i |F_{\psi_i}(f)| \geq A(1 - M(S-1)). \tag{11.13}$$

On the other hand, for any $g \in \mathcal{D}$ different from ψ_i, $i = 1, \ldots, S$, we get from our assumptions

$$|F_g(f)| \leq \sum_{i=1}^{S} |a_i F_g(\psi_i)| \leq AMS < tA(1 - M(S-1)). \tag{11.14}$$

Comparing (11.13) with (11.14) we conclude the proof. □

Theorem 11.14 *Let* $t \in (0,1]$. *Assume that* $\mathcal{D}$ *has coherence parameter* M. *Let* $S < \frac{t}{1+t}(1 + 1/M)$. *Then, for any* f *of the form*

$$f = \sum_{i=1}^{S} a_i \psi_i,$$

where ψ_i *are distinct elements of* $\mathcal{D}$, *we have that* $f_S^{q,t} = 0$.

Proof By Lemma 11.1 we obtain that each $f_m^{q,t}$, $m \leq S$, has a form

$$f_m^{q,t} = \sum_{i=1}^{S} a_i^m \psi_i.$$

We note that, by (11.13), $|F_{\varphi_{m+1}^{q,t}}(f_m^{q,t})| > 0$ for $m < S$ provided $f_m^{q,t} \neq 0$. Therefore, $\varphi_{m+1}^{q,t}$ is different from all the previous $\varphi_1^{q,t}, \ldots, \varphi_m^{q,t}$. Thus, assuming without loss of generality that $f_{S-1}^{q,t} \neq 0$, we conclude that the set of $\varphi_1^{q,t}, \ldots, \varphi_S^{q,t}$ coincides with the set $\psi_1, \ldots, \psi_S$.

The condition

$$F_{\psi_j}(\sum_{i=1}^{S} (a_i - a_i^S)\psi_i) = 0, \quad j = 1, \ldots, S,$$

implies $a_i = a_i^S$, $i = 1, \ldots, S$. In the above we used the fact that $\det \|F_{\varphi_j}(\varphi_i)\|_{i,j=1}^{S} \neq 0$. $\square$

References

R. A. DeVore (1998), 'Nonlinear approximation', *Acta Numerica* **7**, 51–150.

R. A. DeVore and V. N. Temlyakov (1996), 'Some remarks on greedy algorithms', *Advances in Computational Mathematics* **5**, 173–187.

S. J. Dilworth, D. Kutzarova, and V. N. Temlyakov (2002), 'Convergence of some greedy algorithms in Banach spaces', *J. Fourier Analysis and Applications* **8**, 489–505.

M. Donahue, L. Gurvits, C. Darken, and E. Sontag (1997), 'Rate of convex approximation in non-Hilbert spaces', *Constr. Approx.* **13**, 187–220.

D. L. Donoho (2001), 'Sparse components of images and optimal atomic decompositions', *Constr. Approx.* **17**, 353–382.

D. L. Donoho, M. Elad, and V. N. Temlyakov (2004), 'Stable recovery of sparse overcomplete representations in the presence of noise', IMI-Preprints Series, **6**, 1–42.

D. L. Donoho, M. Elad, and V. N. Temlyakov (2005), 'On the Lebesgue type inequalities for greedy approximation', manuscript, 1–11.

V. V. Galatenko and E. D. Livshitz (2003), 'On convergence of approximate weak greedy algorithms', *East J. Approx.* **9**, 43–49.

M. Ganichev and N. J. Kalton (2003), 'Convergence of the weak dual greedy algorithm in L_p-spaces', *J. Approx. Theory* **124**, 89–95.

A. C. Gilbert, S. Muthukrishnan, and M. J. Strauss (2003), 'Approximation of functions over redundant dictionaries using coherence', *The 14th Annual ACM-SIAM Symposium on Discrete Algorithms.*, 243–252.

R. Gribonval and M. Nielsen (2001), 'Approximate weak greedy algorithms', *Advances in Comput. Math.* **14**, 361–368.

P. J. Huber (1985), 'Projection pursuit', *Annals of Stat.* **13**, 435–475.

L. Jones (1987), 'On a conjecture of Huber concerning the convergence of projection pursuit regression', *Annals of Stat.* 15, 880–882.

S. V. Konyagin and V. N. Temlyakov (1999), 'Rate of convergence of pure greedy algorithm', *East J. Approx.* **5**, 493–499.

S. V. Konyagin and V. N. Temlyakov (2002), 'Greedy approximation with regard to bases and general minimal systems', *Serdica Math. J.* **28**, 305–328.

E. D. Livshitz (2003), 'On convergence of greedy algorithms in Banach spaces', *Matem. Zametki* **73**, 371-389.

E. D. Livshitz (2004), 'On reversing greedy algorithm' manuscript, 1–27.

E. D. Livshitz and V. N. Temlyakov (2003), 'Two lower estimates in greedy approximation', *Constr. Approx.* **19**, 509–523.

E. Schmidt (1906-1907), 'Zur Theorie der linearen und nichtlinearen Integralgleichungen. I', *Math. Annalen* **63**, 433–476.

A. V. Sil'nichenko (2004), 'On the rate of convergence of greedy algorithms', manuscript, 1–5.

V. N. Temlyakov (2002), 'Greedy type algorithms in Banach spaces and applications', IMI Preprints Series, **18**, 1–36.

V. N. Temlyakov (2003a), 'Nonlinear methods of approximation', *Found. Comput. Math.* **3**, 33–107.

V. N. Temlyakov (2003b), 'Greedy expansions in Banach spaces', IMI Preprints Series, **6**, 1–21.

V. N. Temlyakov (2004), 'On greedy algorithms with restricted depth search', IMI Preprints Series, **27**, 1–16.

J. A. Tropp (2004), 'Greed is good: Algorithmic results for sparse approximation', *IEEE Trans. Inform. Theory* **50**(10), 2231–2242.